Lecture Notes in Physics

Edited by J. Ehlers, München, K. Hepp, Zürich and
H. A. Weidenmüller, Heidelberg
Managing Editor: W. Beiglböck, Heidelberg

...l Models and
...ce

... Symposium held at the
...fornia, San Diego (La Jolla)

Edited by M. Rosenblatt and C. Van Atta

Springer-Verlag
Berlin · Heidelberg · New York 1972

Organizing Committee of the Symposium:

J. Lumley

M. Rosenblatt, Chairman

C. Van Atta

A. Yaglom

ISBN 3-540-05716-1 Springer-Verlag Berlin · Heidelberg · New York
ISBN 0-387-05716-1 Springer-Verlag New York · Heidelberg · Berlin

Offsetdruck: Julius Beltz, Hemsbach/Bergstr.

Preface

The symposium on Statistical Models and Turbulence was partially sponsored by the Institute for the Application of Statistics in the Physical Sciences. The organization of such a symposium was motivated in part by J. Neyman, then president of the IASPS, as one of a set of meetings accompanying the meeting of the International Statistical Institute in Washington, D.C. in 1971. The symposium was supported by the National Science Foundation through grant GP 27337 and the Office of Naval Research through contract N00014-71-C-0104 Task No. NR 042-268 and Task Order N00014-69-A-0200-6016 in the Mathematics Department of the University of California, San Diego, with M. Rosenblatt as principal investigator. The meeting could not have been held without support from these grants. The organizing committee consisted of

J. Lumley

M. Rosenblatt, Chairman

C. Van Atta

A. Yaglom

The object of the symposium was to provide a milieu in which people with a background in statistics, probability theory or fluid mechanics and an interest in the problem of turbulence would present and discuss results and ideas of current relevance. The meeting was held at UCSD in La Jolla from July 15th through Wednesday July 21st, 1971. The list of lectures presented is given below.

Thursday July 15

J. Lumley, Pennsylvania State University,
"Application of central limit theorems to turbulence problems"

M. Rosenblatt, University of California, San Diego
"Probability limit theorems and some questions in fluid mechanics"

Friday July 16

J. Dutton, Pennsylvania State University
"Some observed characteristics of atmospheric turbulence"

B. Mandelbrot, Watson Research Center, IBM
"Intermittent and sporadic turbulence: generative stochastic models"

C. Van Atta, University of California, San Diego
"Statistical selfsimilarity and inertial subrange turbulence"

E. J. Hannan, Australian National University
"Spectra changing over narrow bands"

E. Parzen, State University of New York, Buffalo
"Some recent advances in time series analysis"

Saturday July 17

D. Ruelle, Institut des Hautes Etudes Scientifiques
"Strange attractors as a mathematical explanation of turbulence"

K. Case, Rockefeller University
"Burger's equation: generalizations and solutions"

W. Munk, University of California, San Diego
"Turbulence in a stratified ocean"

E. Lorenz, Massachusetts Institute of Technology
"Investigating the predictability of turbulent motion"

W. Meecham, University of California, Los Angeles
"Use of Cameron-Martin-Wiener representations for nonlinear random process applications"

Sunday July 18

S. Corrsin, Johns Hopkins University
"Some random geometry problems suggested by turbulent flows"

W. B. Thompson, University of California, San Diego
"The statistical mechanics of the guiding center plasma"

Monday July 19

R. H. Kraichnan
"Comparison of some approximations for isotropic turbulence"

S. Orszag, Massachusetts Institute of Technology
"Numerical simulation of turbulence"

G. S. Deem, Bell Telephone Labs, IBM
"Numerical simulation of 2D incompressible Navier-Stokes and magnetohydrodynamic turbulence (comparisons and insights via movies)"

C. H. Gibson, University of California, San Diego
"Observations of the variability of the dissipation rate of turbulent fields"

F. Frenkiel and P. Klebanoff, Naval Ship Research and Development Center, and National Bureau of Standards
"Probability distribution of turbulent fields"

J. Wyngaard, Air Force Cambridge Research Laboratories
"Some measurements of the fine structure of large Reynolds number turbulence"

Tuesday July 20

J. M. Burgers, University of Maryland
"Statistical problems connected with asymptotic solutions of the one-dimensional nonlinear diffusion equation"

M. Kac, Rockefeller University
"Singular perturbation in some problems of statistical mechanics"

L. S. G. Kovasznay, Johns Hopkins University
"A simple statistical model for turbulence shear flows"

F. Busse, University of California, Los Angeles
"The bounding theory of turbulence and its physical significance in the case of turbulent couette flow"

W. J. Cocke, University of Arizona
"Non-analytic character of the shear-tensor distribution function in incompressible turbulence"

H. K. Moffatt, University of Cambridge
"Dynamo instability and feedback in a stochastically driven system"

Wednesday July 21

G. Kallianpur, University of Minnesota
"Homogeneous chaos expansions"

T. S. Lundgren, University of Minnesota
"A closure hypothesis for the hierarchy of equations for turbulent probability distribution functions"

Thanks are due to D. Coles, S. Corrsin, C. Eckart, R. Getoor, J. Herring, C. Leith, P. Lewis, J. Miles, P. Saffman, T. C. Sun and W. Thompson who chaired Sessions of the meeting. Roy H. Pearce and Paul Saltman spoke for the University of California and William H. Pell for the National Science Foundation in welcoming the meeting. We are especially indebted to Lillian C. Johnson and Elaine Morici who helped solve the many small problems that arose during the meeting.

TABLE OF CONTENTS

APPLICATION OF CENTRAL LIMIT THEOREMS TO TURBULENCE PROBLEMS*

by

J. L. Lumley**

The Pennsylvania State University
University Park

Abstract:

It is shown that (to the extent that the moments involved exist) the existence ($\neq 0$) of all (generalized) integral scales is necessary (and sufficient if all moments exist) for integrals over adjacent segments of a stationary process to become asymptotically independent, and sufficient to ensure that existing moments of integrals will become Gaussian. The conditions under which several recent central limit and related theorems for dependent variables have been proven, are shown to be closely related to this requirement. As a consequence of this examination, a slight weakening is suggested of the common condition that the spectrum be non-zero. Several physical problems are described, which may be resolved by the application of such a central limit theorem: longitudinal dispersion in a channel flow (previously treated semi-empirically); the spreading of hot spots, or the expansion of macromolecules; the weak interaction hypothesis (of Kraichnan) for Fourier components. Finally, it is shown that dispersion in homogeneous turbulence is unlikely to be explicable on the basis of a central limit theorem.

---

*This work supported by the U.S. National Science Foundation, under Grant No. GA18109.

**Professor of Aerospace Engineering.

EXAMINATION OF THE CONDITIONS

Introduction:

A number of central limit and related theorems have now been proven for dependent variables (for a partial list of references, see for example, Lumley, 1970). It is often difficult for a physicist to see how the conditions imposed in the various theorems are related to each other, or to see the physical implications of the conditions, so that he can judge whether a particular physical situation might be expected to satisfy them. In this section, I am going to consider the conditions imposed in several of these theorems, and place them all in the same form, so far as possible, showing how they are all related to a single condition, which may be obtained in a physically appealing manner.

Let us consider first the central limit theorem proposed by Rosenblatt (1956), in particular in the form given by him for a stationary process in continuous time (1961b): that, subject to appropriate conditions, weighted integrals of stationary functions (i.e. - filtered stationary functions) become asymptotically normal. The conditions are:

A) Strong mixing. $u(t)$ is strongly mixing if there is a function $g(t)$, non-negative, monotone decreasing, $g(t) \to 0$, $t \to \infty$, such that, if A is any event dependent only on values of $u(t)$ for $t \leq t_1$, and B any event dependent only on values of $u(t)$ for $t \geq t_2$, $t_2 > t_1$, then

$$|P(AB) - P(A)P(B)| < g(t_2 - t_1)$$

B) The correlation and the fourth order cumulant are absolutely integrable.

C) The spectral density is non-zero at any frequency at which one is filtering.

D) Certain conditions on the shape of the filter, or weighting function which need not conern us in detail; they amount to saying that the weighting function must accumulate progressively more and more of the function, or that the filter must become progressively narrower.

To a physicist B), C) and D) are relatively familiar assumptions; clearly A) is the assumption of interest. This is a general way of saying that sufficiently separated events become asymptotically independent as the separation increases, in a uniform manner. So stated, most physicists would probably be willing to assume that a stationary turbulent velocity was strongly mixing. The assumption is somewhat abstract, however, and it would be more satisfactory if one could point to some implications of the assumption closer to experience.

Strong mixing is used to show that integrals over adjacent segments of $u(t)$ become asymptotically statistically independent, although, as we shall see, an additional assumption is needed for this. Although it is intuitively appealing that this type of asymptotic independence is necessary for integrals to become asymptotically normal, this is something that is evidently not true, since it is easy to construct a Gaussian process such that integrals over adjacent segments do not become independent, though the integrals are, of course, normal. We will, in any event, proceed on the (intuitive) assumption that this independence is necessary for processes that are not already normal, and attempt to derive in a direct manner an equivalent statement of the condition.

Correlation of Adjacent Segments:

Suppose the interval $[0,T]$ is divided into n segments of length Δ, and we arrange matters so that both n and $\Delta \to \infty$ as $T \to \infty$, so that we have more and more segments, but each of them is longer and longer. We can manage this if we set $\Delta = a\,T^{\alpha}$, $0 < \alpha < 1$. If $u(t)$ is strongly mixing, then, roughly speaking, only a short piece of each segment of $u(t)$ at contiguous ends will be well correlated, and we may expect the correlation between adjacent segments to go to zero as $\Delta \to \infty$. Let us see what this implies. Normalize the integral

$$X = \int_0^T u(t)dt/g(T), \quad g^2(T) = 2\int_0^T (T-t)R(t)dt$$

$$R(t) = \overline{u(\tau)u(\tau+t)} \tag{1}$$

so that $\overline{X^2} = 1$. If we now define

$$X_\kappa = \int_{(\kappa-1)\Delta}^{\kappa\Delta} u(t)dt/g(T), \quad X = \Sigma_{\kappa=1}^{n} X_\kappa \tag{2}$$

we have

$$\overline{X_\kappa^2} = g^2(\Delta)/g^2(T) \tag{3}$$

Now, if the sum of all covariances between X_κ for unequal κ is to vanish, so that the variance of X is the sum of the variances of the X_κ, we must have

$$ng^2(\Delta)/g^2(T) \to 1 \qquad \text{or}$$

$$T^{1-\alpha} g^2(aT^\alpha)/g^2(T) \to a \tag{4}$$

The only solution of this is $g^2(T)/T \to \text{const.} \neq 0$. That is, (4) states that $g^2(T)/T \to g^2(\Delta)/\Delta$. The variables Δ and T go to infinity at different rates, which bear an arbitrary relation to each other, to the extent that α is only constrained by $0 < \alpha < 1$. The only way in which the sides can become equal is for each to become constant.

The requirement $g^2(T)/T \to \text{const.} \neq 0$ can be stated as

$$\lim_{T\to\infty} 2\int_0^T (1-t/T)R(t)dt \quad \text{exists}, \neq 0 \tag{5}$$

or

$$\int_{-\infty \atop (C,1)}^{+\infty} R(t)dt \quad \text{exists}, \neq 0 \tag{5a}$$

where the symbol indicates the Cesàro-1 value of the integral, defined by (5) (c.f. Hobson, 1926, §264, 265); by the permanence theorem, if the integral exists in the ordinary sense, it exists (C,1) and has the same value; there are cases, however, where (5) may exist, and the ordinary integral will not (for example, if u(t) has a periodic component). The point will arise again later.

If R(0), the variance, is finite, then (5) serves to define an integral scale: writing $R(t) = R(0)\,\rho(t)$, and

$$\mathcal{T} = \lim_{T\to\infty} \int_0^T (1-t/T)\rho(t)dt = \int_{0 \atop (C,1)}^{\infty} \rho(t)dt \quad \text{exists}, \neq 0 \tag{6}$$

Thus, a necessary and sufficient condition that integrals over adjacent segments become uncorrelated is that the covariance be integrable (C,1) to a non-zero value; if the variance is finite, that the integral scale exist (C,1), $\neq 0$. To a physicist, this is a familiar and unrestrictive assumption.

Asymptotic lack of correlation, of course, is not asymptotic independence . However, we may carry the above ideas to higher order moments, or, what is equivalent and algebraically less complicated, to cumulants. If the segments are to become asymptotically independent, we will have

$$F = \ln \overline{e^{ikX}} \rightarrow n \ln \overline{e^{ikX_\kappa}} = nF_\kappa \tag{7}$$

for the moment generating functions. The cumulant of order p is the pth derivative of F with respect to ik at the origin. For any cumulant that exists, then, we have as a necessary consequence of asymptotic independence of integrals over adjacent segments,

$$C_p(T)/g^p(T) \to nC_p(\Delta)/g^p(T) = (T/\Delta)C_p(\Delta)/g^p(T) \tag{8}$$

where $C_p(T)$ is the pth cumulant of $_o\int^T u(t)dt$. That is, the two expressions in (8) must have the same asymptotic behavior, if cross-product terms are to be of higher order. Equation (8) is of exactly the same form as (4) and the solution is the same: $C_p(T)/T \to$ const. $\neq 0$ (again using $\Delta = a\, T^\alpha$, $0 < \alpha < 1$).

We can immediately draw some interesting conclusions from this. The pth cumulant of X is $C_p(T)/g^p$, and we have*

$$C_p(T)/g^p = 0(T^{1-p/2}) \tag{9}$$

That is, all cumulants of order greater than second vanish asymptotically. This is a way of saying that the moments of the distribution of X become those of a Gaussian distribution. Hence, asymptotic independence of integrals over adjacent segments is a sufficient condition for the moments (that exist) to become Gaussian. If all the moments exist, $C_p(T)/T \to$ const. $\neq 0$ is a necessary and sufficient condition for asymptotic independence of integrals over adjacent segments. The existence of the limit $C_p(T)/T \neq 0$ for any cumulant that exists is necessary for asymptotic independence of integrals over adjacent segments, and is sufficient for the corresponding moment to become Gaussian.

Let us examine the quantity $C_p(T)/T$. If C_p is the pth cumulant of u(t) at p different times, then it is a function only of p-1 differences, and is symmetric under interchange of any two times τ_α and t_β, $\alpha \neq \beta$. In this way we can write, by successive interchange,

---

*Note that the theorem given in Lumley (1970), section 3.16 is incorrect; the conditions stated are not strong enough, and the predicted convergence is too rapid. The error lies in the matching.

$$C_p(T)/T = T^{-1} \underbrace{{}_o\int^T \cdots {}_o\int^T}_{p} C_p dt_1 \ldots dt_p = 2T^{-1} \underbrace{{}_o\int^T \cdots {}_o\int^T}_{p-1} {}_o\int^{t_1} C_p dt_1 \ldots dt_p$$

$$= 2^{p-1} T^{-1} {}_o\int^T dt_1 \underbrace{{}_o\int^{t_1} \cdots {}_o\int^{t_1}}_{p-1} C_p dt_2 \ldots dt_p \tag{10}$$

Introducing new variables, $\tau_j = t_1 - t_j$, $j = 2, \ldots, p$, we can write

$$C_p(T)/T = 2^{p-1} T^{-1} {}_o\int^T dt \underbrace{{}_o\int^t \cdots {}_o\int^t}_{p-1} C_p(\tau_2, \ldots, \tau_p) d\tau_2 \ldots d\tau_p \tag{11}$$

The limit of this is also a Cesaro-1 limit, and we can write

$$\lim_{T\to\infty} C_p(T)/T = 2^{p-1} \underbrace{{}_o\int^\infty \cdots {}_o\int^\infty}_{\substack{p-1 \\ (C,1)}} C_p(\tau_2, \ldots, \tau_p) d\tau_2 \ldots d\tau_p \tag{12}$$

Again, if the first p moments of u(t) exist at the origin, we can write $C_p(\underline{\tau}) = C_p(\underline{0})\, \rho_p(\underline{\tau})$, and set

$$\mathcal{T}_p = \Big[\underbrace{{}_o\int^\infty \cdots {}_o\int^\infty}_{\substack{p-1 \\ (C,1)}} \rho_p(\underline{\tau}) d\tau_2 \ldots d\tau_p \Big]^{1/(p-1)} \tag{13}$$

This also defines an integral scale, an integral volume, or hypervolume. For a u(t) having p finite moments, the existence of these scales is necessary for asymptotic independence of integrals over adjacent segments. If all moments exist, the existence of all scales is also sufficient for independence. For those moments that exist, the existence of the corresponding scale is sufficient for the vanishing of the corresponding cumulant, so that the corresponding moment will become Gaussian.

For complete consistency, we should write $R = C_2$, $\mathcal{T} = \mathcal{T}_2$, $g^2 = C_2$, so that $C_2 \to 2\,\mathcal{T}_2 C_2(0)T$, and in general

$$C_p(T) \to 2^{p-1} \mathcal{T}_p^{\,p-1} C_p(0) T \tag{14}$$

so that the cumulant of X can be written as

$$C_p(T)(C_2(T))^{-p/2} \to C_p(0)(C_2(0))^{-p/2}(2\mathcal{T}_2/T)^{p/2-1}(\mathcal{T}_p/\mathcal{T}_2)^{p-1} \tag{15}$$

This is a convenient expression in physical application, since turbulence is ordinarily generated by a mechanism of known scale, and from experience one expects that the largest scales of the turbulent motion will be of the order of that geometrically determined scale. Hence, one would anticipate that all the scales $\mathcal{T}_p$ would be roughly equal, so that $(\mathcal{T}_p/\mathcal{T}_2)$ will be a constant of order unity*.

For future reference, we can write the expression (15) for p = 4

$$K_X - 3 \to (K_u - 3)(2\mathcal{T}_2/T)(\mathcal{T}_4/\mathcal{T}_2)^3 \tag{16}$$

K_X and K_u are clearly the kurtosis of X and u respectively. Later we will present experimental evidence in support of this relation.

Gaussian Processes

It is interesting to examine Gaussian processes, about which many more definite statements can be made. In the case of a Gaussian process, for example, asymptotic lack of correlation of integrals over adjacent segments implies asymptotic independence of the integrals, so that the conditions on $C_p(T)/T$, $p > 2$, are not required.

Ordinary mixing (not strong) simply requires that the absolute value in A) vanish asymptotically, without the uniformity assumption. Maruyama (1949) has

---

*In turbulence work, this phrase is taken to mean "somewhere between 1/5 and 5."

shown that a necessary and sufficient condition for a stationary Gaussian process to be mixing is that the covariance $R(t) \to 0$, $|t| \to \infty$. This does not imply the existence of (5), which could diverge even though $R(t) \to 0$. On the other hand, (5) can exist and be $\neq 0$, without having $R(t) \to 0$. Thus, integrals over adjacent segments can become independent in the sense we have discussed without the process being mixing. The closest we can come to Maruyama's condition is this: if $R(t)$ is integrable (C,1), then $\lim_{(C,2)} R(t) = 0$*. That is, asymptotic independence of integrals over adjacent segments entails only $\lim_{(C,2)} R(t) = 0$, not $\lim R(t) = 0$.

Kolmogorov and Rozanov (1960) have shown that a stationary Gaussian process with zero mean (necessarily having finite variance) is strongly mixing if there is a function $\phi(z)$, analytic in the lower half-plane, so that, on the real axis $|S(\omega)/\phi(\omega)| \geq \varepsilon > 0$ and S/ϕ is uniformly continuous. We may translate at least part of this into more familiar terms if we take $S(\omega)$ as uniformly continuous (whether it is possible to have $\varphi(\omega)$ that are not uniformly continuous appears to be a moot point), and note that $S(\omega)$ is absolutely integrable, since $S(\omega) \geq 0$, and the variance is finite; hence R is uniformly continuous and absolutely integrable also**. Thus, all powers (≥ 1) of R are also integrable, so that all integral scales (6) based on powers of R (≥ 1) exist and are finite ($S(\omega)$ is bounded, by uniform continuity and absolute integrability).

By a construction of Sun (1965) (see theorem 1, pg. 72), if $G(u(t)) = Y(t)$ is a real, instantaneous function, $\overline{Y} = 0$, $\overline{Y^2} < \infty$, we may write

$$2 \int_0^T (1-t/T) R_G \, dt = \sum_{n=0}^{\infty} a_n^2 \, 2 \int_0^T (1-t/T) R^n \, dt, \quad G(u) = \sum_{n=0}^{\infty} a_n h_n(u) \tag{17}$$

where h_n are the Hermite polynomials, and $\overline{Y^2} = \sum_{n=1}^{\infty} a_n^2 R^n(0) < \infty$. R_G is the

---

*See Hobson (1926), §§ 264, 265.

**Since it is integrable (C,1) to $S(\omega)$ (see Hobson, 1926, §§ 480, 481); hence a uniformly continuous function integrable (C,1) (that is, R) produces an absolutely integrable uniformly continuous function (that is, S). Turning this around, we have the desired result.

covariance of G. Thus, the uniform continuity of the spectrum assures the existence of integral scales of all processes obtained by instantaneous functions of the process. Without becoming involved in the algebra, it is clear that moments of order greater than 2 may also be expressed as products of powers of R by the same technique that produced (17), so that uniform continuity also assures the existence of integral scales based on the higher order covariants of the processes obtained by taking instantaneous functions. Note that this is probably a little stronger than necessary; it should be enough to require that R be square integrable (cf. the theorem of Sun, 1965, below). We still do not know whether integral scales may be zero.

Evidently, if $S(\omega)$ has a zero, it must be the kind of zero which a function analytic in the lower half-plane may possess on the real line. That is, the zeros must be denumerable, and must be no worse than powers of $(\omega-\omega_o)^{2\kappa}$*. Physically, this has a fairly simple interpretation. If $S(\omega)$ has a zero of order 2κ at ω_o, then $ue^{i\omega_o t}$ is the κth derivative of another function, which has a non-vanishing integral scale, even though the integral scale of $ue^{i\omega_o t}$ vanishes (since this is proportional to the value of the spectrum). Hence, the condition that $|S(\omega)/\phi(\omega)/\geq \varepsilon > 0$ can be roughly rephrased, that if the integral scales vanish, for some value ω_o, u(t) can still be strongly mixing if it is obtainable by a finite order linear differential operator from a process all of whose integral scales exist and are $\neq 0$ (and which evidently has, therefore, asymptotically independent integrals over adjacent segments).

On this basis it is tempting to speculate, for stationary non-Gaussian processes, that for strong mixing all cumulants must be integrable (C,1) (all the integral scales must exist, if the moments are finite); they may be zero, but the process must be obtainable by a finite order linear differential operator from a process having non-zero integral scales.

Kolmogorov and Rozanov remark that the presence of discontinuities in $S(\omega)$ precludes strong mixing. A discontinuity means that R is integrable (C,1) but not absolutely integrable, so that the integral scale of R^2 will not exist.

---

*This appears to be somewhat simplistic; according to Sarason (c.f. H. Helson and D. Sarason "Past and Future". Math. Scand. 21 (1967), 5-16, and Sarason's "Addendum to 'Past and Future'" (unpublished)) other, rather subtle, sorts of behavior are possible.

Sun (1965) has proven the following theorem: if u(t) is a real stationary Gaussian process, continuous in the mean, and if $\overline{u} = 0$, and $S(\omega)$ exists, $\varepsilon L^1 \cap L^2$, such that

$$R(t) = \int_{-\infty}^{+\infty} e^{i\omega t} S(\omega) d\omega$$

and

$$\lim_{T\to\infty} T^{-1} \int_{-\infty}^{+\infty} \sin^2(T\omega/2)(\omega/2)^{-2} S(\omega) d\omega \tag{18}$$

exists, = a, $0 < a < \infty$, then if $Y(t) = G(u(t))$, $\overline{Y} = 0$, $\overline{Y^2} < \infty$, $T^{-1/2} \int_0^T Y(t)dt$ is asymptotically Gaussian.

Phrased this way, these conditions are a little obscure, but they may be easily related to ours by using some results of classical analysis*. Sun's conditions turn out to be exactly equivalent to requiring that R(t) and $R^2(t)$ be integrable (C,1) to non-zero values. Since R^2 is integrable, R^n is also for $n > 2$, so that by Sun's construction (17), cumulants of all orders for the process G(u(t)) are integrable (C,1) to non-zero values; in fact, Sun's conditions are exactly equivalent to the latter statement. Thus, Sun's conditions are necessary and sufficient that integrals over adjacent segments of the processes G(u(t)) become asymptotically independent.

---

*See, e.g. - Hobson, 1926, §§ 480, 481. First note that the Gaussian process necessarily has finite variance. Then, if S^2 is integrable, then S is integrable (C,1), and the (C,1) Fourier transform converges to R, R^2 integrable. But $S \geq 0$ if and only if R is a covariance. Since R(0) is finite, the integral (C,1) of S exists; since $S \geq 0$, $S \varepsilon L^1$. Thus $R(0) < \infty$, $S \varepsilon L^1 \cap L^2$, $R(t) = \int e^{i\omega t} S(\omega) d\omega$ implies R^2 integrable, continuous (since $S \varepsilon L^1$). (Note that integrable and integrable (C,1) are the same for a non-negative function). On the other hand, if R^2 is integrable, then the (C,1) transform of R converges to S, where S^2 is integrable, and the (C,1) transform of S converges to R. But since R(0) is finite, the (C,1) integral of S converges; $S \geq 0$, since R is a covariance, hence $S \varepsilon L^1$. Thus, R^2 integrable (and $R(0) < \infty$) implies $S \varepsilon L^1 \cap L^2$, $R = \int e^{i\omega t} S(\omega) d\omega$, R continuous. Note that the assumption of continuity of u(t) in the mean (equivalent to continuity of R at the origin, and hence everywhere: Loeve, 1955, p. 470) appears to be redundant. The additional assumption, (18), is simply the statement (5).

Rosenblatt (1961a) has presented an example which is mixing, but not strongly mixing, for which integrals over adjacent segments do not become independent, and which has a limiting non-Gaussian distribution. The example violates (5), but, if forced to satisfy it, would become Gaussian.

The General Case:

Let us return to Rosenblatt's theorem. It now seems likely that, for a process with finite moments, strong mixing is equivalent to assuming that the process could be obtained by a finite order differential operator from one having finite, non-zero integral scales. In the general case, strong mixing is evidently a way of assuring that the process becomes independent of itself at an equivalent rate, without assuming that moments exist.

The role of the moment assumption (B) is now clear: it assures that the first two integral scales will exist, though they may vanish. From the discussions above, it seems possible that this condition (absolute integrability) is a little stronger than necessary.

The third assumption (C) evidently assures that the integral scale will not vanish (since the value of the spectrum at ω_o is proportional to the integral scale of the process $u(t)\, e^{i\omega_o t}$). It is designed to exclude the following and similar cases: suppose that $u(t) = v'(t)$, where $v(t)$ is strongly mixing, and has finite moments, and finite, non-zero integral scales, but is not Gaussian. Then $u(t)$ satisfies A) and B), but not C); it has zero integral scale. Clearly its integral is $v(t)$, which is non-Gaussian, by assumption. Evidently integrals over adjacent segments do not become asymptotically independent.

The condition that integral scales exist and be non-zero is in a certain sense too strong. If $v(t)$ is strongly mixing and has a non-zero integral scale (other things being equal) it can be averaged, and will become Gaussian. But an average of $v(t)$ is an average of $u(t) = v'(t)$; instead of a top-hat weighting function, however, the weighting function is triangular. Evidently, by use of the right weighting function, it is possible for weighted integrals to become Gaussian even for processes with vanishing integral scales. The spectrum of u(t) vanished as ω^2 at the origin; the filter corresponding to the top-hat weighting function has tails $\sim \omega^{-2}$. Thus, the product of this filter and the spectrum is not a function which becomes progressively narrower as $T \to \infty$. On the other hand, the filter corresponding to the triangular weighting function has tails $\sim \omega^{-4}$; the product of this filter and the spectrum does become progressively narrower as $T \to \infty$.

Evidently, the following is true: let $S(\omega)$ be the spectrum of a process and $F(\omega-\omega_o)$ be a filter function about ω_o. Suppose $S(\omega) \propto (\omega-\omega_o)^{2n}$ near ω_o. If $(\omega-\omega_o)^{2n} F(\omega-\omega_o)$ is still a filter, i.e. - still integrable, then the averaging of S by F can be regarded as the averaging of $S/(\omega-\omega_o)^{2n}$ by $(\omega-\omega_o)^{2n}$ F. $S/(\omega-\omega_o)^{2n}$ has a non-zero integral scale, and integrals over adjacent segments will become asymptotically uncorrelated. If the process corresponding to $S/(\omega-\omega_o)^{2n}$ satisfies the other conditions, the filtration of the original process will result in a normal distribution.

From the similarity of this construction to that of the theorem of Kolmogorov and Rozanov (1960), it appears likely that one could say "a process is strongly mixing if and only if integrals (C,κ) over adjacent segments become asymptotically independent for some finite κ"*, or that a strongly mixing process may be averaged (C,κ) for some κ to a normal distribution.

---

*The limit (C,κ) of f(x) is $\lim (\kappa!/x^\kappa) \int_0^x dx' \underset{\kappa}{\ldots} \int_0^{x'} dx'' f(x'')$.

SOME APPLICATIONS

*Longitudinal Dispersion in a Channel Flow**:

Let us consider dispersion in a turbulent channel flow. $\underline{U} = \{U(x_2),0,0\}$ and the height of the channel is 2h. The position of a moving point is given by

$$X_1(\underline{a},t) - a_1 = \int_0^t \{U(X_2(\underline{a},t')) + u_1(\underline{X}(\underline{a},t'),t')\}dt' \quad (19)$$

$$X_2(\underline{a},t) - a_2 = \int_0^t v_2(\underline{a},t')dt' \quad (20)$$

$$X_3(\underline{a},t) - a_3 = \int_0^t v_3(\underline{a},t')dt' \quad (21)$$

Here, $u_i(\underline{x},t)$ is the Eulerian velocity fluctuation (i.e. - at a point fixed in laboratory coordinates); $u_i(\underline{X}(\underline{a},t),t)$ is the Eulerian fluctuation at the position of the moving point; $v_i(\underline{a},t)$, the Lagrangian velocity fluctuation. The velocities v_i may be taken to be stationary for sufficiently large t, since the flow is homogeneous in the streamwise direction, and the moving point may be expected to wander back and forth throughout the interior of the channel, forgetting its initial location. There is no reason to suppose that any integral scales of v_3 do not exist, or vanish, so that we may expect $X_3 - a_3$ to be asymptotically Gaussian. Although $X_2 - a_2$ is also the integral of a stationary function, it cannot have a Gaussian distribution, since the integral scale of v_2 must vanish. To see this, note that $X_2 - a_2$ must be stationary, since it is confined by the channel; thus v_2

---

* A more discursive presentation of this material will appear in Tennekes, Lumley (1971).

must be the derivative of a stationary process. Since X_2 is stationary, however, $U(X_2)$ is stationary also; thus the integrand of (19) is stationary. We expect $X_1 - a_1$ to disperse, so that the integrand is not the derivative of a stationary process; there is no other reason to assume that any of its integral scales vanishes, or that they fail to exist. Hence, $X_1 - \overline{X}_1$ will be asymptotically Gaussian. The mean axial position can be shown to be

$$X_1 = a_1 + U_b t \tag{22}$$

where U_b is the bulk velocity (see Lumley, 1962); the variance is asymptotically

$$\overline{(X_1 - \overline{X}_1)^2} \sim 2\overline{v_1^2} tT \tag{23}$$

where T is the integral scale, and it is straightforward to show (Tennekes, Lumley, 1971)

$$\overline{v_1^2} = (2h)^{-1} \int_{-h}^{h} \{\overline{u_1^2} + (U - U_b)^2\} dx_2 \tag{24}$$

(note that $v_1(\underline{a},t)$, the Lagrangian velocity fluctuation, is not $u_1(\underline{X}(\underline{a},t),t)$, but $U(X_2(\underline{a},t)) - U_b + u_1(\underline{X}(\underline{a},t),t) = v_1(\underline{a},t)$). If we make a standard large-Reynolds number approximation to the dependence of the functions in (24), we can write

$$\overline{v_1^2} = A u_*^2 \tag{25}$$

where $A \sim 5$, and u_* is the friction velocity. This presumes that the viscous sublayer is a small fraction of the total. The integral scale may be estimated as $T \sim h/u_*$, since these are the appropriate scales in the core region of the flow; thus, we should expect

$$\overline{(X_1 - \overline{X}_1)^2} \sim 10 u_* ht \tag{26}$$

which is in excellent agreement with experiment (Monin and Yaglom, 1971). In Figure 1, results from Taylor (1954) are presented showing the approach to a Gaussian distribution.

*The Spreading of Hot Spots, or the Expansion of Macromolecules**:

In discussing the spread of hot spots in turbulence, the following equation arises for the moment of inertia tensor of the heat distribution:

$$\dot{I}_{pq} - I_{pj}u_{q,j} - I_{qj}u_{p,j} = 2\gamma\delta_{pq} \tag{27}$$

where $u_{p,j}$ is the local velocity gradient, and γ the thermometric diffusivity.

In discussing the extension of macromolecules in turbulence, an equation of a form very similar to (27) is obtained for the moment of inertia tensor of the probability density of molecule end-to-end location - i.e. - the covariance tensor of the molecule distribution:

$$\dot{I}_{pq} - I_{pj}u_{q,j} - I_{qj}u_{p,j} + I_{pq}/T = \delta_{pq}r^2/3T \tag{28}$$

$u_{p,j}$ is taken at the spot or molecule location, so that it is an Eulerian quantity at a Lagrangian location.

It is desired to solve (27) and (28) for $u_{p,j}$ a stationary random (tensor) function of time, without assumptions regarding the symmetry or lack of it. This is in general very difficult, but by assuming high Reynolds number, the change in $u_{p,j}$ may be taken to be slow; an asymptotic expansion may be constructed, and the expectation of the solution given in terms of the general form

---

* A discussion of this section in full detail may be found in Lumley (1972).

$$\overline{f(t)f(t')\exp\left(\int_{t'}^{t} f(t'')dt''\right)} \tag{29}$$

where f(t) is stationary, and is a function only of the six real invariants of the tensor $u_{p,j}$ at time t. This type of solution is typical of equations in which the forcing function appears as a coefficient.

Now, the form (29) may be obtained as

$$-(\partial^2/\partial t\partial t')\overline{\exp\left(\int_{t'}^{t} f(t'')dt''\right)} \tag{30}$$

and if the invariants of $u_{p,j}$, and hence f, may be taken as having all finite, non-zero integral scales (more exactly, the variance of f may not exist; all we really need is the existence of non-zero values for the covariance integrals), then $\int_{t'}^{t} f(t'')dt''$ will be asymptotically Gaussian, as $t-t' \to \infty$. Thus

$$\overline{\exp\left(\int_{t'}^{t} f(t'')dt''\right)} \sim \exp\{(\overline{f} + \overline{f'^2}\mathcal{T}_f)(t-t')\} \tag{31}$$

where $\overline{f}$ is the mean, $\overline{f'^2}$ the mean square fluctuation, and $\mathcal{T}_f$ the integral scale, of f. Differentiation should only improve the convergence; hence

$$\overline{f(t)f(t')\exp\left(\int_{t'}^{t} f(t'')dt''\right)} \sim (\overline{f} + \overline{f'^2}\mathcal{T}_f)^2 \exp\{(\overline{f} + \overline{f'^2}\mathcal{T}_f)(t - t')\} \tag{32}$$

The rationale for this result is easy to see: as $t-t' \to \infty$, we expect f(t) and $f(t')$ to become independent, and the correlation of either with $\int_{t'}^{t} f(t'')dt''$ to be constant, independent of t and t', since each is only correlated with a (relatively) small region of $[t',t]$ at the appropriate end.

It is straightforward to extend the result to combinations of functions, as

$$\overline{p(t)q(t')\exp(\int_{t'}^{t} f(t'')dt'')} \qquad (33)$$

so long as p(t), q(t), f(t) are stationary, and are all determined by the values of the invariants of $u_{p,j}$ at time t. In this way one obtains finally that

$$\overline{I_{pp}} \sim C\{\exp(2\overline{|\lambda|^2}Tt) - 1\} \qquad (34)$$

as the solution to equation (27), and

$$\overline{I_{pp}} \sim C'[\exp\{(2\overline{|\lambda|^2}T-1/T)t\}-1] \qquad (35)$$

as the solution of equation (28), where C and C' are constants, and λ is the (complex) eigenvalue of the tensor $u_{p,j}$, and T the integral scale of its variation. When (35) is applied to a pipe flow (to assure stationarity) and the various quantities estimated on dimensional grounds, the prediction of the ratio of parameters at which $\overline{I_{pp}}$ will increase is in reasonably good agreement with observed phenomena (as they are currently interpreted).

Weak Interaction

In 1959, Kraichnan suggested as part of his model of turbulence dynamics the concept of "weak interaction;" that is, narrow-band-averaged Fourier coefficients, which must, of course, be uncorrelated, actually become statistically independent of each other asymptotically as the band width goes to zero. Let us examine this.

Consider the function

$$v(t) = \mathcal{Re}(au(t)e^{i\omega_1 t} + bu(t)e^{i\omega_2 t}) \qquad (36)$$

If v(t) has finite, non-zero values for all its integral scales (which entails u(t) having a non-zero spectrum at ω_1 and ω_2) the normalized integral of v(t) will become asymptotically Gaussian. But the integral over [0,T] of v(t) is just the linear combination

$$\int_0^T v(t)dt \sim aS_T(\omega_1) + bS_T(\omega_2) \tag{37}$$

where $S_T(\omega)$ is the narrow band averaged Fourier coefficient; i.e. - the integral of the Fourier coefficients over a band of width $\Delta\omega \sim T^{-1}$ about ω:

$$S_T(\omega) = \int_{-\infty}^{+\infty} H_T(\omega-\omega')\tilde{u}(\omega')d\omega', H_T(\omega) = (2/\omega T)\sin(\omega T/2) \tag{38}$$

and $\tilde{u}(\omega)$ is the generalized Fourier transform of u(t). But, if the linear combination (37) becomes Gaussian as $T \to \infty$, then $S_T(\omega_1)$ and $S_T(\omega_2)$ become jointly Gaussian, and since they are in addition uncorrelated, they become independent. This is exactly in the sense of weak interaction: there is just enough dependence between the Fourier coefficients $\tilde{u}(\omega)$ at various frequencies for $S_T(\omega)$ for any finite T to be non-Gaussian (if the $\tilde{u}(\omega)$ were truly independent, then, presuming the moments to be well behaved, $S_T(\omega)$ should be Gaussian for finite T), but to become Gaussian as $T \to \infty$. Evidently, the weak interaction hypothesis is roughly equivalent to asymptotic independence of adjacent segments.

We may simultaneously check the validity of the weak interaction hypothesis and the ideas expressed in the first section of this paper, since they are evidently the same, by measuring the kurtosis of narrow band filtered signals; since $T\Delta\omega \sim 1$, we expect, from (16),

$$K(\Delta\omega) - 3 = (K(\infty) - 3)(2T_2\Delta\omega)\beta \tag{39}$$

where β is a constant = $(T_4/T_2)^3$, and T_2 is the integral scale defined from the spectrum, i.e. -

$$T_2 = \pi S(\omega)/\overline{u^2}, \quad \text{where } \overline{u^2} = \int_{-\infty}^{+\infty} S(\omega)d\omega \tag{40}$$

We have made measurements of this quantity for the derivative of a turbulent velocity in a laboratory flow, at a value of $R_\lambda \sim 500$ (based on the Taylor microscale) at a frequency corresponding to $\kappa\eta \sim 0.6$; the large bandwidth value $K(\infty)$ was approximately 6. The results are shown in Figure 2. The slope is clearly tending toward one, as predicted. The constant β is evidently close to 10^2, so that $T_4/T_2 \sim 4.65$. The integral scale T of the velocity itself, corresponding roughly to the flow geometry, is much larger: $T/T_4 \sim 61.8$.

Dispersion in Homogeneous Turbulence:

Dispersion in decaying homogeneous turbulence is often cited as a suitable place for the application of a central limit theorem since distributions of wandering points are observed to be Gaussian. Let us consider, for example, the cross-stream position of a point

$$X_2(\underline{a},t) = a_2 + \int_{t_o}^{t} v_2(\underline{a},t')dt' \tag{41}$$

Decaying homogeneous turbulence in an Eulerian framework obeys an approximate similarity law

$$u'^2 = a^2 t^{-n}, \quad \ell = (2a/3n)t^{1-n/2}$$
$$u'/\ell = 3n/2t \tag{42}$$

where u'^2 is the r.m.s. fluctuating velocity, and ℓ is an integral length scale of the energy containing eddies. n does not differ from unity by more than 30% (Comte-Bellot and Corrsin, 1966). It is reasonable to expect that the Lagrangian velocities will partake of the same similarity. Batchelor and Townsend (1956) have suggested that v_2, referred to local scales of velocity and time, will be a stationary quantity. The time scale may be defined by

$$u' dt/\ell = d\eta, \text{ or } (3n/2)\ell n(t/t_o) = \eta, \; t = t_o e^{2\eta/3n} \tag{43}$$

where we have picked the zero of η at the release time (c.f. (41)). Then the quantity

$$v_2(\underline{a},t)/u'(t) = \omega(\eta) \tag{44}$$

may be expected to be a stationary function of η. In terms of this we may rewrite (41) as

$$\xi(\eta) = (X_2(\underline{a},t)-a_2)/\ell(t) = \int_o^\eta e^{-(2-n)\eta/3n} \omega(\eta-\eta')d\eta' \tag{45}$$

Now, as the wandering point begins, $\xi(\eta)$ grows from zero. The mean square value of ξ is given by

$$\overline{\xi^2(\eta)} = 3\int_o^\eta dx\rho(x)(e^{-(2-n)x/3n} - e^{(x-2\eta)(2-n)/3n}) \sim 3\int_o^\infty dx\rho(x)e^{-(2-n)x/3n} \tag{46}$$

where $\rho(\eta') = \overline{\omega(\eta)\omega(\eta + \eta')}$.

Initially, the growth rate is relatively rapid; by the time η has reached approximately 3, however, it has begun to slow, and by the time it has reached about 9 (where $e^{-(2-n)\eta/3n} \sim 0.05$) it has essentially stopped growing. $\xi(\eta)$ in fact becomes stationary; referred to the local length scale, it has stopped growing and the wandering point is really no longer being dispersed in the sense in which this word is usually used. This is because, although the limit on the integral in (45) continues to grow, in fact the true range of integration does not after η

reaches ~ 9. That is, the present value of ξ is determined by $\omega(\eta)$ over only the most recent 9η. Since ℓ/u' is approximately the Lagrangian time integral scale (Corrsin, 1963), $\eta = 9$ corresponds to roughly 9 Lagrangian time integral scales. Thus, there is no real reason to expect a central limit theorem to be applicable.

In Rosenblatt (1961) conditions are given on permissable weighting functions which are designed to avoid this situation. The exact nature of the conditions need not concern us; they amount to a requirement that an expression like (31) must grow unboundedly. The weighting function in (30) specifically violates the condition.

Batchelor (1956) has pointed out that this discussion is academic in any event, since the integral in (41) depends progressively on lower and lower frequencies, which are known not to obey the similarity law (42). The effect may be seen in another manner by noting that homogeneous turbulence undergoes a change to a viscous-dominated mode of decay, obeying a different similarity law, sufficiently long after its formation. If grid produced turbulence is taken as satisfying the similarity law (42) roughly from 20 mesh lengths to 400 mesh lengths behind the grid, about the most that can be achieved experimentally, then we have

$$\eta = (3n/2)\ell n\ 20 \sim 4.5 \tag{40}$$

so that even the asymptotic state is not practically achievable.

Acknowledgments:

Figure 2 was drawn from data taken and analyzed by Mr. Kunihiko Takeuchi and will form part of his doctoral dissertation. Computer programming for this analysis was done by Miss L. Katherine Jones.

I am grateful to Murray Rosenblatt, who read the manuscript and brought several errors of commission and omission, and several references, to my attention.

BIBLIOGRAPHY

Batchelor, G. K. and A. A. Townsend (1956), "Turbulent Diffusion," *Surveys in Mechanics*, Cambridge, The University Press, 352-399.

Comte-Bellot, G. and S. Corrsin (1966), "The Use of a Contraction to Improve the Isotropy of Grid Generated Turbulence," *J. Fluid Mech.*, 25, 657-682.

Corrsin, S. (1963), "Estimates of the Relations between Eulerian and Lagrangian Scales in Large Reynolds Number Turbulence," *J. Atmos. Sci.*, 20, 115-119.

Hobson, E. W. (1926), *The Theory of Functions of a Real Variable and the Theory of Fourier's Series*, The University Press, Cambridge.

Kolmogorov, A. N. and Yu. A. Rozanov (1960), "On Strong Mixing Conditions for Stationary Gaussian Processes," *Theory of Probability and Its Applications*, V, 204-208.

Kraichnan, R. H. (1959), "The Structure of Isotropic Turbulence at Very High Reynolds Numbers," *J. Fluid Mech.*, 5, 497-543.

Loeve, M. (1955), *Probability Theory*, Van Nostrand, New York.

Lumley, J. L. (1962), "The Mathematical Nature of the Problem of Relating Eulerian and Lagrangian Statistical Functions in Turbulence," in *Mécanique de la Turbulence*, Edition du CNRS, Paris, 17-26.

Lumley, J. L. (1970), *Stochastic Tools in Turbulence*, Academic Press, New York.

Lumley, J. L. (1972), "On the Solution of Equations Describing Small-Scale Deformation," in *Symposia Mathematica: Proceedings of Convegno sulla Teoria della Turbolenza al Istituto Nazionale di Alta Matematica, Roma, April 1971.* New York, Academic Press.

Maruyama, G. (1949), "The Harmonic Analysis of Stationary Stochastic Processes," *Mem. Fac. Sic. Kyushu Univ. Ser A.*, 4, 45-106.

Monin, A. S. and A. M. Yaglom (1971), *Statistical Fluid Mechanics*, Cambridge, M.I.T. Press.

Rosenblatt, M. (1956), "A Central Limit Theorem and a Strong Mixing Condition," *Proc. Natl. Acad. Sci. U.S.A.*, 42, 43-47.

Rosenblatt, M. (1961a), "Independence and Dependence," in *Proceedings of the Fourth Berkeley Symposium on Mathematical Statistics and Probability* (Neyman, ed.), II, Pt. III, U. Cal. Press, Berkeley, 431-443.

Rosenblatt, M. (1961b), "Some Comments on Narrow Band Pass Filters," *Q. Appl. Math.*, XVIII, 387-393.

Sun, T. C. (1965), "Some Further Results on Central Limit Theorems for Non-linear Functions of a Normal Stationary Process," *J. Math. Mech.*, 14, 71-85.

Taylor, G. I. (1954), "The Dispersion of Matter in Turbulent Flow through a Pipe," *Proc. Roy. Soc. A*, 223, 446-468.

Tennekes, H. and J. L. Lumley (1971), *A First Course in Turbulence*, M.I.T. Press, Cambridge.

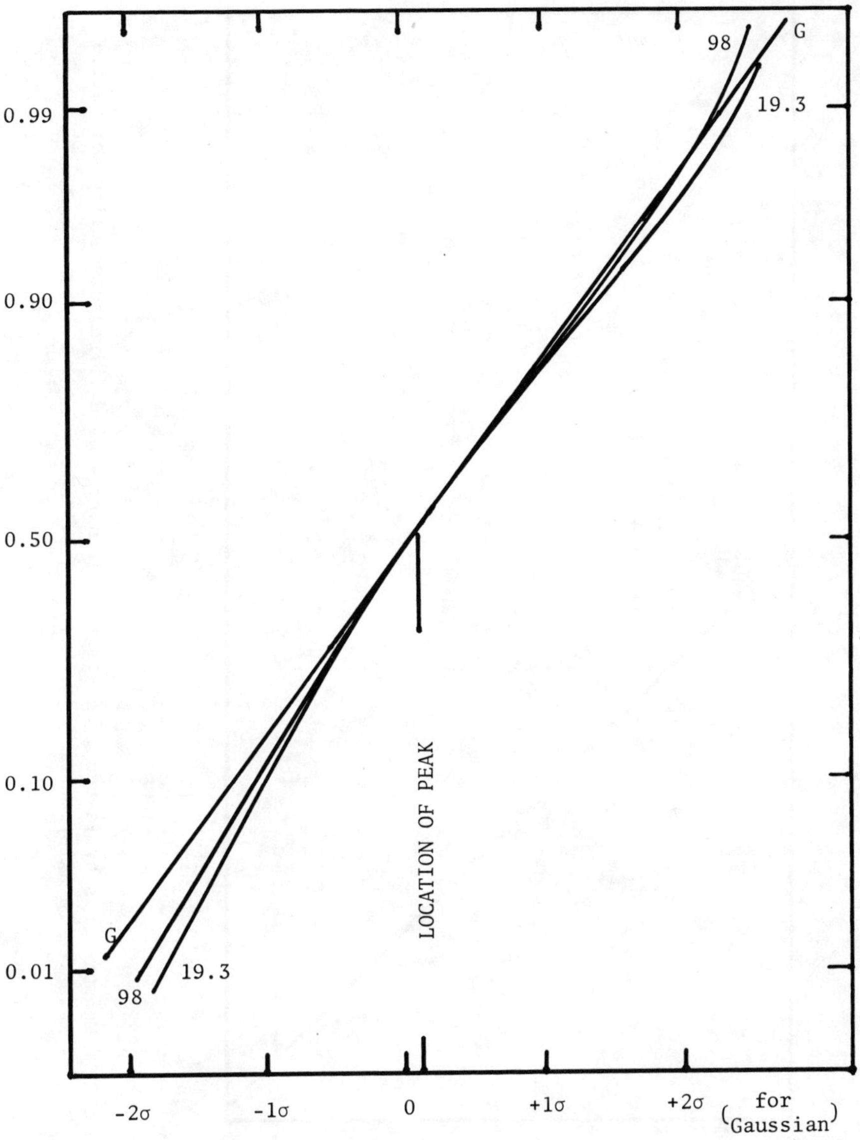

Figure 1. Concentration distributions measured by Taylor (1954) at two downstream locations in a pipe compared with Gaussian (G) of same maximum height and area. Tails on arriving side are shorter, and those on departing side longer, than Gaussian. Figures indicate values of $T/2\mathcal{T}$.

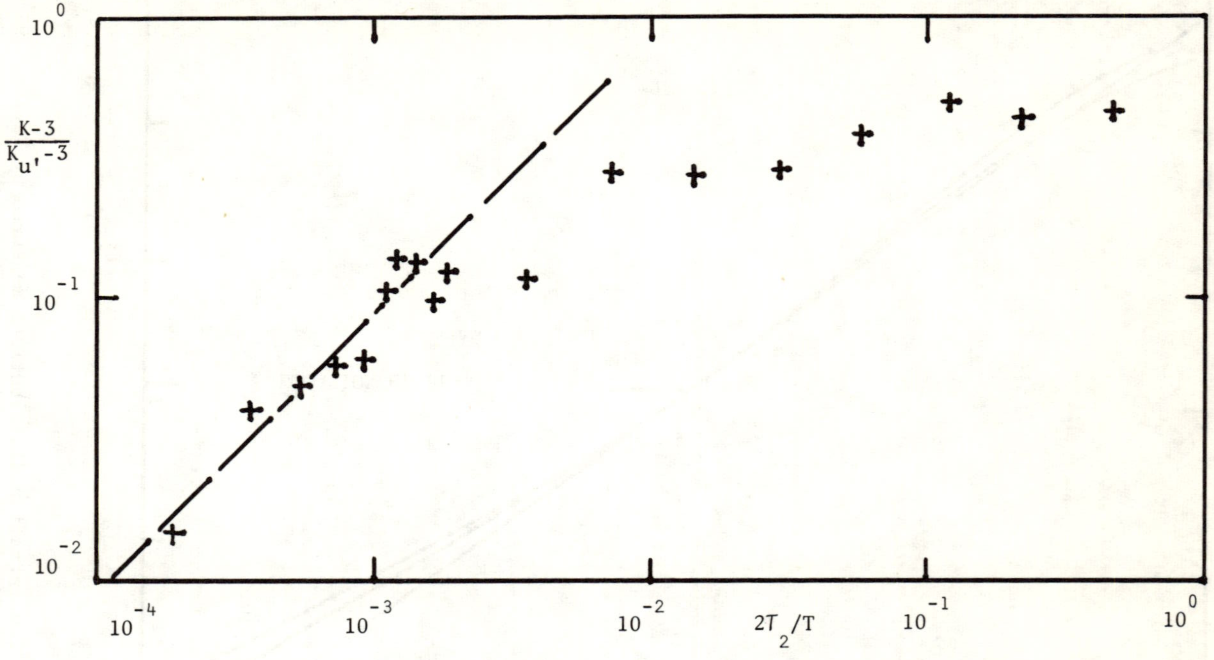

Figure 2. Measurements by Takeuchi of kurtosis of narrow band filtered velocity derivative measured in axisymmetric jet.

PROBABILITY LIMIT THEOREMS AND SOME QUESTIONS IN FLUID MECHANICS[1)]

by

M. Rosenblatt
University of California, San Diego (La Jolla)

0. Abstract

A number of problems in fluid mechanics which have been dealt with by making use of a central limit theorem (and asymptotic normality) are mentioned. A discussion of central limit theorems for stationary processes and the need for some form of asymptotic independence is given. The concepts of uniform ergodicity and strong mixing are introduced. An example of asymptotic nonnormality is given when the form of asymptotic independence is not sufficiently strong. The derivation of a new result indicating that uniform ergodicity is strong mixing when there is trivial tail field is briefly sketched.

1. Problems in Fluid Mechanics and Asymptotic Normality

There are a number of problems in fluid mechanics in which an appeal is made to a limit theorem like the central limit theorem for dependent processes as a folk theorem without too explicitly considering the conditions required for its validity. We mention a few of these problems. Batchelor considers the velocity field in the final period of decay of turbulence in [1] pp. 97-98. The nonlinear terms in the Navier-Stokes equation are neglected so that the velocity field $u(\bar{x},t)$ satisfies the heat equation and has the representation

$$(1) \qquad u(\bar{x},t) = [4\pi\nu(t-t_o)]^{-3/2}\int u(\bar{x}',t_o)\exp\left[-\frac{(\bar{x}-\bar{x}')^2}{4\nu(t-t_o)}\right]d\bar{x}'$$

in terms of the velocity field at time t_o for $t > t_o$. Of course, (1) amounts to a type of narrow band filtering as $t \to \infty$ (see Rozanov and Volkonski [12]). At this point an appeal is made to asymptotic normality as a folk theorem assuming that the velocity field can be regarded as a stochastic process stationary in x.

Still another application of such a central limit theorem is made by Cocke [2] when discussing the stretching of line and surface elements in a fluid in turbulent

1) Research supported by Office of Naval Research Contract N00014-69-A-0200-6016, Task NR-042-217.

motion. There is also the refinement of the Kolmogorov 5/3 law for the velocity spectrum in turbulence (see Kolmogorov [5]) obtained by considering the energy-dissipation fluctuations. A paper of Yaglom [14] discusses this refinement and the proposed log normal character of the energy-dissipation fluctuations by invoking some version of a central limit theorem.

2. The Central Limit Theorem for Independent and Dependent Sequences

There is an extensive and almost complete study of limit laws for partial sums of independent random variables that suggests the possibility of corresponding results for random processes under certain conditions. There are many open and unresolved questions of this type concerning random processes. However, to motivate our discussion on limit laws for random processes, we shall mention a few of the results that hold for independent random variables. Let $X_{1,n}, X_{2,n}, \cdots, X_{n,n}$, be independent and identically distributed with distribution function

$$F_n(x) = P[X_{k,n} \leq x], \qquad k = 1,\cdots,n \, . \tag{2}$$

The class of probability distributions that one can obtain as limit laws of

$$A_n^{-1} \sum_{k=1}^{n} X_{k,n} - B_n \tag{3}$$

when centering and normalizing by proper constants A_n, B_n as $n \to \infty$ is called the set of infinitely divisible laws. They can be characterized in terms of their Fourier-Stieltjes transforms $\varphi(t)$ by

$$\log \varphi(t) = \exp\left\{ it\gamma + \int_{-\infty}^{\infty} \left(e^{itx} - 1 - \frac{itx}{1+x^2} \right) \frac{1+x^2}{x^2} \, dG(x) \right\} \tag{4}$$

where γ is a real constant and G a bounded nondecreasing function. The most well known of these are the normal and Poisson distributions and it is the normal limit law that will be of greatest interest to us. It should be noted that generally there needn't be a nontrivial limiting distribution for a sequence of centered and normed sums (3) as $n \to \infty$ generally. The distribution of $\sum_{j=1}^{n} X_{j,n}$ is $F_n^{(n)}(x)$ ($F_n^{(n)}(x)$ is the nth convolution of F_n with itself). Even though there needn't be a nontrivial limiting distribution, there is a remarkable theorem of

Kolmogorov (see [11]) stating that there is an infinitely divisible distribution function $H_n(x)$ such that

(5)
$$\sup_x |F_n^{(n)}(x)-H_n(x)| \le C n^{-1/3}$$

where C is an absolute constant independent of F_n. The classical elementary version of a central limit theorem for independent random variables runs as follows: if $X_1, X_2, \cdots$ are independent, identically distributed random variables with common mean m and variance σ^2, $0 < \sigma^2 < \infty$, then

(6)
$$\lim_{n\to\infty} P\left\{\frac{1}{\sqrt{n}\,\sigma}\sum_{k=1}^{n}(X_k-m)\le x\right\} = \frac{1}{\sqrt{2\pi}}\int_{-\infty}^{x} e^{-u^2/2}du \ .$$

If the second moment of the distribution function $F(x)$ is not finite, there is no assurance of a limiting distribution for the appropriately normed and centered partial sums of identically distributed and independent random X_j with common distribution F

$$A_n^{-1}\sum_{k=1}^{n} X_k - B_n \ .$$

However, if there is a nontrivial limiting distribution G, it must belong to a class of distributions referred to as stable. A distribution G is stable if the convolution $G(c_1x) * G(c_2x)$ equals $G(cx+\gamma)$ for any $c_1,c_2 > 0$ with c,γ constants depending on c_1, c_2. The normal family of distributions contains the only nontrivial stable laws with finite second (and higher order) moments.

Most of our discussion of central limit theorems for random processes will be carried out for sequences. However, in most situations the obvious analogue for a continuous time parameter process will be valid. Limit laws for dependent processes that correspond to classical results for independent random variables are obtained under some assumption of asymptotic independence. We shall introduce some strong but rather natural conditions of asymptotic independence here. Let $\{X_k;\ k = \cdots,-1,0,1,\cdots\}$ be a stationary sequence of random variables, that is for any integer n, the joint distribution of the n-tuples

$$X_{k_1+m}, \ \cdots, \ X_{k_n+m}$$

is independent of m, $m = \cdots,-1,0,1,\cdots$. Let $\mathcal{B}_n$ and $\mathcal{F}_n$ be the Borel fields of events $\mathcal{B}_n = \mathcal{B}(X_j, j\leq n)$ and $\mathcal{F}_n = \mathcal{B}(X_j, j\geq n)$ generated by X_j, $j\leq n$, and X_j, $j\geq n$, respectively. They represent the information carried by the past and present relative to n, and future and present relative to n, respectively. Let τ be the shift transformation corresponding to a one step shift forward in time. The process $\{X_k; k = \cdots,-1,0,1,\cdots\}$ is said to be uniformly ergodic if

$$\sup_{\substack{A\in\mathcal{B}_0 \\ B\in\mathcal{F}_0}} \left| \frac{1}{n} \sum_{j=1}^{n} P(A\cap\tau^j B)-P(A)P(B) \right| = a(n) \to 0 \tag{7}$$

as $n\to\infty$. This condition was introduced by Cogburn (see [3] and [11]) in a discussion of a generalized central limit problem for stationary processes. He showed that if $\{X_j\}$ is a stationary Markov sequence and f_n a sequence of functions with $f_n(X_j)\to 0$ in probability as $n\to\infty$, then the only possible limit laws (if any) of $\sum_{j=1}^{n} f_n(X_j)$ will be infinitely divisible if and only if $\{X_j\}$ is uniformly ergodic. The process $\{X_k\}$ is said to be strongly mixing if

$$\sup_{\substack{A\in\mathcal{B}_0 \\ B\in\mathcal{F}_n}} |P(A\cap B)-P(A)P(B)| = b(n) \to 0 \tag{8}$$

as $n\to\infty$. Strong mixing was introduced by M. Rosenblatt [9] and has been used to obtain a number of versions of the central limit theorem for random processes (see [4], [9] and [12]). One will get asymptotic normality for partial sums of a stationary sequence if, for example, 4th moments exist and there is strong mixing with the coefficient $b(n) = O(n^{-3})$ as $n\to\infty$ and the variances of the partial sums $\sum_{j=1}^{n} X_j$ diverge as $n\to\infty$.

We give a brief sketch of an argument of Nagaev [8] under slightly altered conditions generalizing a result already mentioned on limit laws for sums of independent, identically distributed random variables.

Theorem 1. Let $\{X_k; k = \cdots,-1,0,1,\cdots\}$ be a strictly stationary random process that is uniformly ergodic. The only possible limit laws for normed and centered partial sums

(9) $$A_n^{-1} \sum_{k=1}^{n} X_k - B_n$$

with $A_n \to \infty$ are stable distributions.

Assume that there are sequences A_n, B_n with $A_n \to \infty$ such that one has a nontrivial limiting distribution

(10) $$\lim_{n\to\infty} P\{A_n^{-1}S_n - B_n \le x\} = F(x)$$

with $S_n = \sum_{j=1}^{n} X_j$. We symmetrize the random variables X_j (consider $\overline{X}_j = X_j - X_j'$ where X_j''s have the same joint distribution as the X_j's but are independent of them). Then

(11) $$\lim_{n\to\infty} P\{A_n^{-1}\overline{S}_n \le x\} = F(x) * F(-x)$$

where $*$ denotes the convolution operation and $\overline{S}_n = \sum_{j=1}^{n} \overline{X}_j$. However (11) cannot be valid unless $A_n/A_{n+1} \to 1$ as $n \to \infty$. Given any fixed $\alpha > 0$, we can then find a sequence $m(n) \to \infty$ as $n \to \infty$ such that

$$A_{m(n)}/A_n \to \alpha$$

as $n \to \infty$. Consider now

(12) $$A_n^{-1}S_{n+\ell(n)+m(n)} - C_n$$

where the sequences $\ell(n)$ and C_n are yet to be determined. Set

$$C_n = B_n + B_{\ell(n)} \frac{A_{\ell(n)}}{A_n} + B_{m(n)} \frac{A_{m(n)}}{A_n} .$$

Choose $\ell(n) \to \infty$ so slowly as $n \to \infty$ that $A_{\ell(n)}/A_n \to 0$. Then the distribution of (12) is asymptotically the same as that of

(13) $$(A_n^{-1}S_n - B_n) + \frac{A_{m(n)}}{A_n}\left(A_{m(n)}^{-1} \sum_{n+\ell(n)+1}^{n+\ell(n)+m(n)} X_j - B_{m(n)}\right).$$

A standard argument (see [Cogburn 3]) using uniform ergodicity implies that the asymptotic distribution of the two terms of (13) are asymptotically independent as

$n \to \infty$. But this implies that the asymptotic distribution of (13) is

$$F(x) * F(\frac{x}{\alpha}) .$$

On the other hand, by (10), the asymptotic distribution of (13) must be

$$F(\beta x+\gamma)$$

for some $\beta > 0$ and γ depending on α. However this means that F must be stable.

The concepts of uniform ergodicity and strong mixing for a continuous time parameter strictly stationary process $\{X(t), -\infty<t<\infty\}$ are analogous to the definitions in the discrete parameter case. The backward fields $\mathcal{B}_t = \mathcal{B}(X_\tau, \tau\leq t)$ and forward fields $\mathcal{F}_t = \mathcal{B}(X_\tau, \tau\geq t)$, $-\infty<t<\infty$. A continuous time parameter stationary process $\{X(t), -\infty<t<\infty\}$ is <u>strongly mixing</u> if

(14) $$\sup_{\substack{A\in\mathcal{B}_0 \\ B\in\mathcal{F}_t}} |P(A\cap B)-P(A)P(B)| = b(t) \to 0$$

as $t \to \infty$. There is a discretized version of strong mixing that we now discuss. Let $\mathcal{B}_k(\Delta) = \mathcal{B}\{X_{j\Delta};\ j\leq k\}$, $\mathcal{F}_k(\Delta) = \mathcal{B}\{X_{j\Delta};\ j\geq k\}$. Consider the condition

(15) $$\sup_{0<k\Delta\leq t}\ \sup_{\substack{A\in\mathcal{B}_0(\Delta) \\ B\in\mathcal{F}_k(\Delta)}} |P(A\cap B)-P(A)P(B)| = b(t) \to 0$$

as $t \to \infty$. If the process $X(t)$ is separable (see Doob's <u>Stochastic Processes</u>) in the sense that events can be approximated in probability by events determined by conditions at a countable set of time points, conditions (14) and (15) are equivalent. The continuous time version of uniform ergodicity is unfortunately not as natural. Uniform ergodicity in the continuous time case amounts to

(16) $$\sup_{\substack{A\in\mathcal{B}_0 \\ B\in\mathcal{F}_0}} \left|\frac{1}{N}\int_0^N \{P(A\cap\tau^t B)-P(A)P(B)\}dt\right| = a(N) \to 0$$

as $N \to \infty$ where τ^t is the shift transformation corresponding to a time shift of t units forward in time. A discretized version of this condition is

$$(17) \qquad \sup_{0<\Delta\le 1}\ \sup_{\substack{A\in\mathcal{B}_0(\Delta)\\ B\in\mathcal{F}_0(\Delta)}} \left| \frac{1}{T}\sum_{k=0}^{[T/\Delta]} \{P(A\cap\tau^k B) - P(A)P(B)\} \right| = a(T) \to 0$$

as $T\to\infty$ where $[x]$ is the greatest integer less than or equal to x. Condition (17) certainly implies (16) if the process $\{X(t), -\infty<t<\infty\}$ is separable. However, a remark of Cogburn (see [3]) indicates that the converse is not true and this indicates that the continuous time version of uniform ergodicity is not too natural and should perhaps be replaced by (17). Let $\{X(t)\}$ be the stationary process with instantaneous distribution uniformly distributed on the circumference of a circle (of length one) with the transition mechanism a deterministic rotation of t units in t time units. This process is uniformly ergodic in the sense of (16). However, a discrete time parameter process $\{X_k\}$ with instantaneous distribution uniformly distributed on the circumference and one-step transition mechanism a deterministic rotation of α units with α irrational is not uniformly ergodic using (7). This implies that the continuous time parameter process $\{X(t)\}$ does not satisfy (17). It should be noted that uniform ergodicity and strong mixing are in general much stronger than the ordinary notions of ergodicity and mixing as used in ergodic theory (Rosenblatt [11]). A continuous time parameter version of Theorem 1 is valid if the process $X(t)$ is <u>continuous in probability</u>, that is, for each $\epsilon > 0$

$$\lim_{\tau\to t} P[|X(t)-X(\tau)|>\epsilon] \to 0$$

for each t. Continuity in probability does not imply that the sample functions or trajectories of the process are continuous as can be seen by looking at the Poisson process.

We now mention an interesting central limit theorem for a rather special class of stationary processes due to Sun [13]. Let $X(t), -\infty<t<\infty$, be a real stationary normal process continuous in the mean that satisfies the following conditions:

(i) $EX(t) \equiv 0$

(ii) $r(h) = EX(t)X(t+h) = \int_{-\infty}^{\infty} e^{ih\lambda} f(\lambda)\,d\lambda$ with the spectral density

$f \in L^1(-\infty,\infty) \cap L^2(-\infty,\infty)$

(iii) $\lim_{T\to\infty} \frac{1}{T} \int_{-\infty}^{\infty} \frac{\sin^2 \frac{T}{2}\lambda}{(\frac{1}{2}\lambda)^2} f(\lambda)d\lambda = f(0)$ exists, is positive and finite.

Theorem 2. Let $X(t)$, $-\infty<t<\infty$, be a normal stationary process satisfying conditions (i), (ii), and (iii). Let $Y(t) = G(X(t))$ be such that $EY(t) \equiv 0$, $EY^2(t) < \infty$, with G a real instantaneous function. Then

$$T^{-1/2} \int_0^T Y(t)dt$$

is asymptotically normally distributed with mean zero and the finite variance

$$\rho = 2\pi \left[a_1^2 f(0) + \sum_{n=2}^{\infty} a_n^2 \int_{-\infty}^{\infty}\cdots\int \prod_{j=1}^{n-1} f(\lambda_j)\cdot f\left(-\sum_{j=1}^{n-1} \lambda_j\right) d\lambda_1 \cdots d\lambda_{n-1} \right]$$

with a_j the Fourier-Hermite coefficients of the function $G(x)$.

3. An Example of Asymptotic Nonnormality

As indicated in section 1, there are a number of problems in which it is natural to try to apply a probabilistic limit theorem. Given a sufficiently strong form of asymptotic independence, for example uniform ergodicity, Theorem 1 implies that the limit law (if it exists) for partial sums must be stable. If second moments exist and a sufficiently strong rate of decrease of $a(k)$ to zero as $k \to \infty$ is assumed, the limit law must be Gaussian. However, it should be strongly kept in mind that it is not true if a sufficiently strong form of asymptotic independence is not assumed. For example, in some works (see Lumley [7]) it is assumed that ergodicity itself might be enough. Specifically, Lumley in section 3.16 of his book considers $X(t)$ a stationary ergodic process with zero mean that has a covariance function with a continuous second derivative. The behavior of the covariance function at ∞ is assumed to be no worse that algebraic. It is then conjectured that if k is infinitely differentiable with bounded support, then

$$\int_{-\infty}^{\infty} X(t)k(t/T)dt$$

will be asymptotically Gaussian as $T \to \infty$ when properly normalized if its variance diverges as $T \to \infty$.

However, the following example (a modification of one given in [10]) of a continuous time parameter process indicates that this is not so. Let $\{X(t), -\infty<t<\infty\}$ be a continuous time parameter process Gaussian process (all finite dimensional distributions jointly Gaussian) with mean zero $EX(t) \equiv 0$ and covariance function

$$r(t) = EX(s+t)X(s) = (1+|t|)^{-2\alpha}, \qquad \alpha > 0. \tag{18}$$

The function $r(t)$ can be seen to be a covariance function by noting that it is real, convex for $t > 0$ with $\lim_{t\to 0} r(t) = 0$, and applying a result of Polya's (see Loéve [6] pp. 217-218). The spectral density of the process

$$f(x) = 2\int_0^\infty \frac{\cos tx}{(1+t)^{2\alpha}}\,dt$$

$$= 4\alpha\, x^{2\alpha-1}\int_0^\infty \frac{\sin v}{(x+v)^{2\alpha+1}}\,dv, \qquad x > 0,$$

so that $f(x)$ behaves like $|x|^{2\alpha-1}$ as $|x| \to 0$. Also since

$$\int_0^\infty \frac{\sin v}{(x+v)^{2\alpha+1}}\,dv = \int_0^\pi \sin v \sum_{k=0}^\infty (x+v+k\pi)^{-2\alpha-1}(-1)^k dv$$

and for large x

$$(x+v)^{-2\alpha-1} - (x+v+\pi)^{-2\alpha-1} = (x+v)^{-2\alpha-1}\left[1 - \left(1 + \frac{\pi}{x+v}\right)^{-2\alpha-1}\right]$$

$$\cong (x+v)^{-2\alpha-1}(2\alpha+1)\,\frac{\pi}{x+v} = \pi(2\alpha+1)(x+v)^{-2\alpha-2}$$

it follows that $f(x)$ is of the order $|x|^{-3}$ precisely as $|x| \to \infty$. The process $\{X(t)\}$ is ergodic, mixing and even purely nondeterministic. Consider the process $Y(t) = X(t)^2$. It is, of course, also ergodic, mixing and purely nondeterministic. The process $Y(t)$ has covariance $2r(t)^2$. Let $0 < \alpha < \frac{1}{4}$. Notice that then the process $X(t)$ does not satisfy the assumptions of Theorem 2 since its spectral density $f \notin L^2$. Let us now consider

$$\int_0^T Y(t)dt - T = \int_0^T X(t)^2 dt - T.$$

The characteristic function of this centered integral is

$$\exp\{\tfrac{1}{2}\sum_{k=2}^{\infty}(2it)^k a_k/k\}$$

where

$$a_k = a_k(T) = \underbrace{\int_0^T \cdots \int_0^T}_{k} r(u_1-u_2)\cdots r(u_{k-1}-u_k)\, r(u_k-u_1)\,du_1\cdots du_k \ .$$

If we now consider the normalized random variable

$$T^{-1+2\alpha}\left\{\int_0^T Y(t)dt - T\right\} \tag{19}$$

as $T \to \infty$, it has the limiting distribution with characteristic function

$$\exp\left\{\frac{1}{2}\sum_{k=2}^{\infty}(2it)^k b_k/k\right\} \tag{20}$$

where

$$b_k = \underbrace{\int_0^1 \cdots \int_0^1}_{k} |x_1-x_2|^{-2\alpha}|x_2-x_3|^{-2\alpha}\cdots|x_k-x_{k-1}|^{-2\alpha}\, dx_1\cdots dx_k \ .$$

Notice that the limiting distribution has all its moments finite but is not Gaussian and hence not stable. We are outside the range of Theorem 1 and so the process $X(t)$ is not uniformly ergodic. Of course, the process $X(t)$ does not have a covariance function $r(t)$ (see formula (18)) which has a continuous second derivative. However, this is easy to arrange. Consider the covariance function

$$\rho(t) = r(t) * (1+t^2)^{-m}$$

with m a positive integer and $*$ denoting the convolution operation. This covariance function is continuously differentiable $2m-2$ times and has the same asymptotic behavior at ∞ as $r(t)$. If $X(t)$ is now a Gaussian stationary process with mean zero and covariance $\rho(t)$, we will have the same type of limiting characteristic function (20) for (19). Suppose that $k(t)$ is an infinitely differentiable function with $0 \le k(t) \le 1$ for all t and

$$k(t) = \begin{cases} 1 & \text{for} \quad 0 \leq t \leq 1 \\ 0 & \text{for} \quad t \leq -\varepsilon \quad \text{or} \quad t \geq 1+\varepsilon \end{cases}$$

with ε a small fixed positive number. Then

$$T^{-1+2\alpha}\left\{\int_{-\infty}^{\infty} Y(k)k(t/T)dt - \int_{-\infty}^{\infty} k(t/T)dt\right\}$$

will have a limiting nonGaussian distribution as $T \to \infty$ with properties similar to that given by (20).

4. Strong Mixing and Uniform Ergodicity

The conditions of strong mixing and uniform ergodicity are attractive and meaningful conditions for stationary sequences. An exception has to be made for uniform ergodicity in the case of a continuous time parameter process in view of the remarks already made in section 2. It is obvious that strong mixing is a stronger condition than uniform ergodicity. However, it is of great interest to get a clearer picture of the relationship between these two conditions. A remark of Cogburn [3] already indicates that a stationary sequence that is uniformly ergodic but not mixing can have at worst a finite number of cyclically moving sets. We shall give here a sketch of the proof of the following theorem for stationary Markov sequences.

Theorem 3. A uniformly ergodic stationary sequence that is mixing must be strongly mixing.

The details of the proof of this result will appear in a later publication. However, we state a number of Lemmas with a few interpolatory remarks which it is hoped will indicate the basic idea of the derivation for stationary Markov sequences. Let T be the operator on functions (bounded or essentially bounded) induced by a transition probability function $P(\cdot,\cdot)$ with invariant probability measure μ, that is,

$$(Tf)(x) = \int P(x,dy)f(y)$$

where

$$\|f\|^2 = \int |f(x)|^2 \mu(dx).$$

Lemma 1. Consider $1 > \epsilon > 0$, $M > 0$. Let

$$\|f\|^2 - \|Tf\|^2,\ \|g\|^2 - \|Tg\|^2 < \epsilon^2.$$

Then if $|f(x)|$, $|g(x)| \leq M < \infty$ for all (almost all) x and α, β are two real numbers with $|\alpha|$, $|\beta| \leq M$ there is a polynomial function $a(M)$ of M such that

$$\|\alpha f+\beta g\|^2 - \|T(\alpha f+\beta g)\|^2 \leq a(M)\ \epsilon$$

and

$$\|fg\|^2 - \|T(fg)\|^2 \leq a(M)\ \epsilon\ .$$

Lemma 1 can be used to derive the following result for stationary Markov sequences.

Lemma 2. If a stationary Markov sequence is not strongly mixing there are numbers α, $0 < \alpha < 1$, such that for each $\epsilon > 0$ there are sequences of sets $A_k(\epsilon)$, $B_k(\epsilon)$, $k = 1,2,\cdots$, such that

$$E|P(\tau^k A_k(\epsilon)|\mathcal{B}_o) - I_{B_k(\epsilon)}| \leq \epsilon$$

and

$$|P(B_k(\epsilon))-\alpha| \leq \epsilon\ .$$

In Lemma 2 τ stands for the shift transformation (a shift of one time unit forward in time) and I_A for the indicator function of the set A. The conditions stated for a number α as given in Lemma 2 are given a more convenient formulation in Lemma 3.

Lemma 3. Let α, $0 < \alpha < 1$, be a number satisfying the conditions given in Lemma 2. An equivalent formulation is that for each sequence $\epsilon_k \to 0$ as $k \to \infty$ there are sequences of sets A_k, B_k with

(21) $$E|P(\tau^k A_k|\mathcal{B}_o)-I_{B_k}| \leq \epsilon_k$$

and

(22) $$|P(B_k)-\alpha| \leq \epsilon_k.$$

Call the set of values α, $0 < \alpha < 1$, for which (21), (22) hold right tail values. It is clear that all limit points β, $0 < \beta < 1$, of right tail values are also right tail values.

Lemma 4. Zero cannot be a limit point of the right tail values of a stationary Markov sequence if it is uniformly ergodic.

Lemma 5. The set of all right tail values α, $0 < \alpha < 1$, of a uniformly ergodic stationary Markov sequence must be of the form

$$\alpha = \frac{k}{m} , \quad k = 1,2,\cdots,m-1 , \tag{23}$$

for some finite integer $m > 0$. The right tail values (23) must correspond to m cyclically moving sets with a cycle of m units in time.

Corollary. If a stationary uniformly ergodic Markov sequence has no cyclically moving sets (no right tail values), it must be strongly mixing.

REFERENCES

1. Batchelor, G. K. The Theory of Homogeneous Turbulence, Cambridge (1953).

2. Cocke, W. J. "Turbulent hydrodynamic line-stretching: the random-walk limit", Physics of Fluids, to be published.

3. Cogburn, R. "Conditional probability operators", Ann. Math. Statist. 33, 634-658 (1962).

4. Ibragimov, I. A. "Some limit theorems for stationary processes", Theor. Probability Appl. 7, 349-382 (1962).

5. Kolmogorov, A. N. "Mecanique de la turbulence" (Colloque Intern. du CNRS a Marseille), J. Fluid Mech. 13, (1962).

6. Loeve, M. Probability Theory, 3rd Edition, Van Nostrand.

7. Lumley, J. L. Stochastic Tools in Turbulence, Academic (1970).

8. Nagaev, S. V. "Some limit theorems for stationary Markov chains", Theory Probability Appl. 2, 378-406 (1957).

9. Rosenblatt, M. "A central limit theorem and a strong mixing condition", Proc. Nat. Acad. Sci. U.S.A. 42, 43-47 (1956).

10. Rosenblatt, M. "Independence and dependence", Proc. 4th Berkeley Symposium on Mathematical Statistics and Probability, 431-443 (1960).

11. Rosenblatt, M. Markov Processes: Structure and Asymptotic Behavior, Springer (1971).

12. Rozanov, Yu. A. and Volkonski, V. A. "Some limit theorems for random functions I", Theor. Probability Appl. 4, 178-197 (1959).

13. Sun, T. C. "Some further results on central limit theorems for non-linear functions of a normal stationary process", J. Math. Mech. 14, 71-85 (1965).

14. Yaglom, A. M. "The influence of fluctuations in energy dissipation on the shape of turbulence characteristics in the inertial interval", Soviet Physics-Doklady 11, 26-29 (1966).

STATISTICAL PROBLEMS CONNECTED WITH ASYMPTOTIC SOLUTIONS OF THE ONE-DIMENSIONAL NONLINEAR DIFFUSION EQUATION.

By J.M. Burgers

Institute for Fluid Dynamics and Applied Mathematics

University of Maryland
College Park, Maryland
20742

Introduction - Geometric solution of the differential equation. - The statistical problems considered in the following pages refer to features of sets of solutions of the one-dimensional nonlinear diffusion equation

(1) $$u_t + u\,u_x = \nu\,u_{xx} \quad ,$$

whose solutions are supposed to be deduced from sets of oscillating initial data $u_o(x)$, for the case where ν goes to zero (always remaining a positive quantity). As is well known, the solutions of (1) under these circumstances obtain the form of a "sawtooth profile", with segments of upward slope $1/t$ alternating with downward jumps or "shocks". The strengths of these shocks and the distances between them depend upon the initial data. When sets of initial data have been provided, strengths and positions of the shocks will vary from case to case, and it can be asked whether we can predict some statistical aspects of the results. The basic problem presenting itself in this connection is whether we can find a distribution function for the strengths of the shocks. This will give a means to calculate average values of arbitrary powers of the shock strength. Other statistical problems can be taken in view once this basic problem has been solved.

We make use of a geometrical method for solving eq. (1) in the case $\nu \to +0$, which is based upon the Hopf-Cole solution. The substitution

(2) $$u = -\,2\,\nu\,(w_x/w)$$

transforms eq. (1) into a third order equation for w , which can be integrated once and be reduced to the form

(3) $$w_t = \nu\,w_{xx} \quad .$$

The initial data for u can be transformed into initial data for w. Making use of a standard solution of the ordinary diffusion equation, we obtain the following expression for $u(x,t)$:

$$(4)\qquad u = \frac{\displaystyle\int_{-\infty}^{+\infty} d\xi \, \frac{x-\xi}{t} \exp\left[-\frac{1}{2\nu}\left\{\frac{(x-\xi)^2}{2t} - \int_0^{\xi} a(\xi_1)\, d\xi_1\right\}\right]}{\displaystyle\int_{-\infty}^{+\infty} d\xi \, \exp\left[-\frac{1}{2\nu}\left\{\frac{(x-\xi)^2}{2t} - \int_0^{\xi} a(\xi_1)\, d\xi_1\right\}\right]} ,$$

where $a(\xi_1)$ stands for $-u_o(\xi_1)$. In the case $\nu \to +0$ the most important contribution to the integrals comes from the immediate neighborhood of the point ξ_m, which, for the chosen t and x, gives the absolute minimum of the expression

$$(5)\qquad Z(\xi) = \frac{(x-\xi)^2}{2t} - \int_0^{\xi} a(\xi_1)\, d\xi_1 .$$

We then obtain:

$$(6)\qquad u = u_o(\xi_m) = \frac{x-\xi_m}{t} .$$

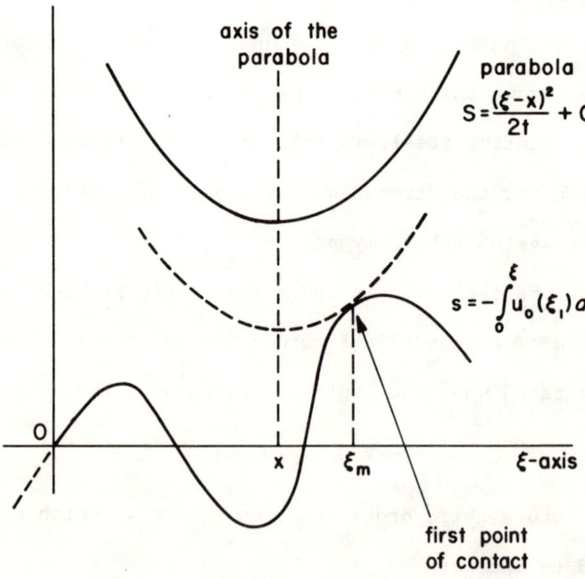

Fig. 1. - Initial data curve ("s-curve") and parabola S .

To find ξ_m we consider a parabola

$$S(\xi) = (\xi - x)^2/2t + \text{constant} \ ; \tag{7}$$

and a curve

$$s(\xi) = \int_0^\xi a(\xi_1)\, d\xi_1 = -\int_0^\xi u_o(\xi_1)\, d\xi_1 \quad , \tag{8}$$

taking the constant in (7) so high that the parabola everywhere is above the s-curve. We then decrease the constant until the parabola in coming down touches the s-curve from above for the first time. The contact point gives ξ_m . This is illustrated in Fig. 1.

When we are interested in the values of u as a function of x , for constant t , we let the parabola glide over the s-curve, always touching it from above, without ever crossing it.

If t exceeds a certain minimum, so that the parabola is not too narrow (rather broad), values of x will be found for which the parabola touches the s-curve in <u>two</u> points simultaneously (see Fig. 2). This indicates a local

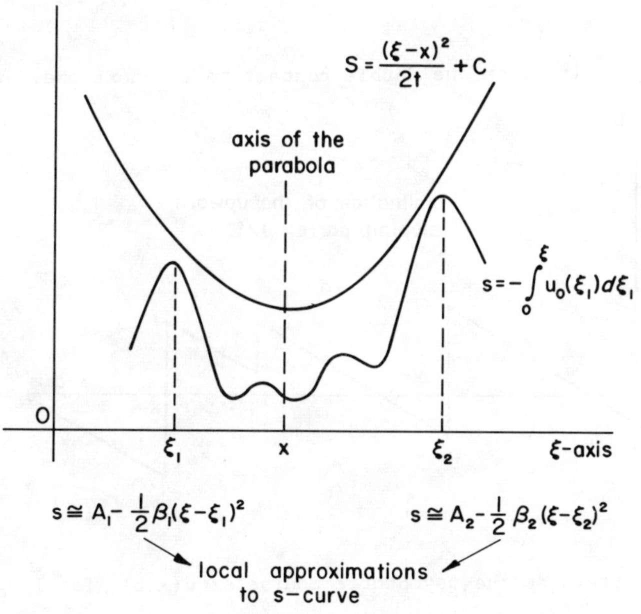

Fig. 2. - Parabola approaching two tops of the s-curve simultaneously.

jump downward of $u(x,t)$ at such an x. When t is large, a supposition which we make in all following considerations, the parts of the s-curve between two double contacts give a very nearly linear increase of u with slope $1/t$ and the resulting curve for u approaches the sawtooth profile mentioned above (see Figs. 3 and 4).

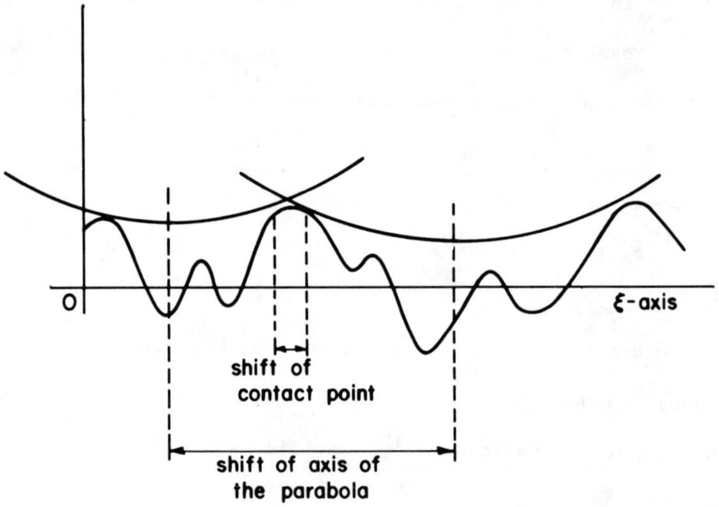

Fig. 3. - Shift from one double contact to the next one.

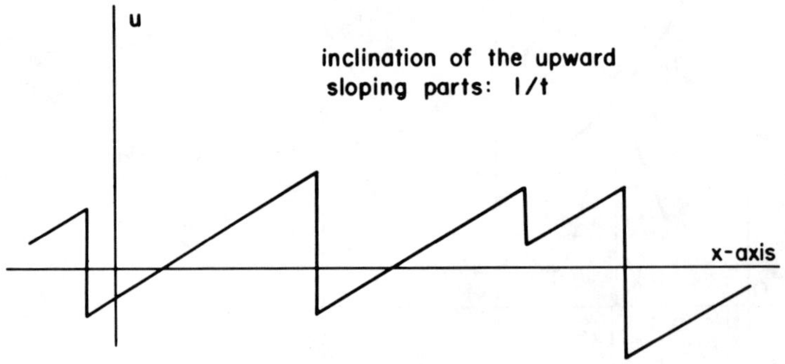

Fig. 4. - Sawtooth pattern as the asymptotic course of $u(x,t)$ for large t.

<u>The main statistical problem.</u> - We characterize each shock by two parameters, one being the horizontal distance ξ (≥ 0) between the two contact points defining the shock; the other being their vertical distance η (which can be positive or negative). The shock strength is equal to ξ/t. When sets of initial data have been given, the first statistical problem which presents itself is to find a distribution function $p(\xi,\eta)\, d\xi\, d\eta$ for the number of shocks of type ξ,η, per unit length of the x-axis. The total number of shocks per unit length is

$$N_s = \int_0^\infty d\xi \int_{-\infty}^{+\infty} d\eta\ p(\xi,\eta) \quad ; \tag{9a}$$

and it will be seen that the function $p(\xi,\eta)$ is normalized by the condition

$$\int_0^\infty d\xi \int_{-\infty}^{+\infty} d\eta\ \xi\ p(\xi,\eta) = 1 \quad . \tag{9b}$$

To obtain an expression for $p(\xi,\eta)$ we consider an "inversion" of this problem and ask: "What is the probability that a curve $s(\xi)$ with random oscillations in the vertical direction, after having made contact with a parabola $S(\xi) = \xi^2/2t$, makes another contact at a specified horizontal distance, without crossing the parabola"? If there were no parabola in the field, the problem could be reduced to the following one: "What is the probability that a curve with random oscillations in the vertical direction, starting from a given point, reaches a point at the height η above the starting point, at a horizontal distance ξ ?" This probability is given by a solution $\psi(\xi_2, \eta_2\ ;\ \xi_1, \eta_1)$ of the ordinary diffusion equation

$$\frac{\partial\psi}{\partial\xi} = J\,\frac{\partial^2\psi}{\partial\eta^2} \quad , \tag{10}$$

where ξ_1, η_1 are the coordinates of the starting point, at which a unit source is supposed to be located; and ξ_2, η_2 are those of the point to be reached, with $\xi_2 - \xi_1 = \xi$; $\eta_2 - \eta_1 = \eta$. The constant J in this equation is the integral of the correlation function for the oscillations of the s-curve; in the case under consideration with large t this is the only feature derived from the initial data which enters into the result.

To take account of the condition that the parabolic curve shall not be crossed, we introduce this curve as an absorbing boundary of the field and require that the solution $\psi(\xi_2\,,\ \eta_2\,;\ \xi_1\,,\ \eta_1)$ of the differential equation (10) shall be zero on the parabola. The domain of ψ is supposed to be situated below the parabola. To take account of the requirement that the s-curve must "contact" the parabola at both points, we do not use the function $\psi(\xi_2\,,\eta_2\,;\ \xi_1\,,\eta_1)$ itself (which is zero at the parabola), but take instead the double derivative

$$(11) \qquad \Psi(\xi_2;\xi_1) = \frac{\partial^2\psi}{\partial\eta_1\ \partial\eta_2} \qquad \left(\begin{array}{c}\text{with } \eta_1 \text{ and } \eta_2 \text{ on the} \\ \text{parabola.}\end{array}\right)$$

Since the s-curve can be extended both to the right beyond the second contact point, and to the left from the first contact point, we also must introduce two factors to exclude crossing of the parabola in these exterior regions. The first factor is defined by

$$(12) \qquad E(\xi_2) = -\frac{\partial}{\partial\eta_2}\left[\lim_{\xi=\infty}\int_{-\infty}^{S(\xi)} d\eta\ \psi(\xi,\eta;\ \xi_2,\eta_2)\right]\left(\begin{array}{c}\text{with } \eta_2 \text{ on the} \\ \text{parabola}\end{array}\right) .$$

The second factor $F(\xi_1)$ is defined by an analogous formula, based upon a diffusion function operating toward the left. We now proclaim that the function $p(\xi,\eta)$, apart from a constant factor, can be represented by the product $E\ \Psi\ F$ and write

$$(13) \qquad p(\xi,\eta) = m\ J^2\ E(\xi_2)\ \Psi(\xi_2\ ;\ \xi_1)\ F(\xi_1) \quad .$$

with $\xi = \xi_2 - \xi_1$; $\eta = S(\xi_2) - S(\xi_1)$. - The numerical factor m must be determined by means of the normalizing condition (9b). The introduction of the factor J^2 is necessary in connection with the dimensions (in terms of length and time) of the quantities occurring on both sides of the equation.

The question of dimensions appears to be of interest in the treatment of the statistical problem. However, instead of using dimensional expressions based upon units of length and time, it is more convenient to base them upon the quantity J and the time, since J is a constant of the problem. It is found that distances in the horizontal direction (measured on the ξ-scale) have dimensions $J^{1/3}\, t^{2/3}$, while distances in the vertical direction (measured on the η-scale) have dimensions $J^{2/3}\, t^{1/3}$. A result of these formulas is that the average value ℓ of the "wavelength" or distance between successive shocks is proportional to $t^{2/3}$. The same holds for the average value of ξ (which is equal to ℓ) . It follows that the average shock strength is proportional to $t^{-1/3}$. If the saw-tooth profile is considered as representing velocities, the average "kinetic energy" per unit length of the x-axis is proportional to $t^{-2/3}$; and the average dissipation per unit length of the x-axis is proportional to $t^{-5/3}$. Once the dimensional relations have been found, most of the further work can be carried out in non-dimensional formulas, obtained by setting $J = 1$; $t = 1$.

Construction of the distribution function. - A number of properties of the functions ψ , Ψ , E , F can be derived from an investigation of the behavior of solutions of eq. (10) with a boundary curve $S(\xi)$ of more general form than the parabola considered above, in particular with reference to the behavior of these functions when small changes are introduced into the boundary. As an example we mention that the following expression is obtained for Ψ (in non-dimensional form)

(14) $$\Psi = \frac{1}{2\sqrt{\pi}\,\xi^{3/2}} \exp\left(-\frac{\eta^2}{4\xi}\right) \theta(\xi) ,$$

where $\theta(\xi)$ is an absolutely convergent series in powers of $\xi^{3/2}$, the coefficients of which can be derived by means of a recurrence scheme. However, the series is not convenient for further work, and a solution of different form has been constructed.

For this purpose we introduce an auxiliary variable $v = \xi^2/2 - \eta$ so that the parabola is represented by $v = 0$, and write

(15) $$\psi = q(v) \exp\left(\frac{\xi v}{2} - \sigma\xi - \frac{\xi^3}{12}\right) .$$

The function $q(v)$ must then satisfy the equation

$$(16) \qquad d^2q/dv^2 = (v/2 - \sigma)\, q \quad ,$$

and must vanish both at $v = 0$ and $v = \infty$. This can be obtained by relating it to the Airy function, which is defined by the integral

$$(17) \qquad q_o(v) = \sqrt{3} \int_0^\infty dz \cos[z^3 + (3/2)^{1/3} zv] \quad ,$$

and satisfies the equation

$$(18) \qquad d^2q_o/dv^2 = (v/2)\, q_o \quad .$$

This function can also be expressed in terms of the Bessel functions $J_{1/3}$ and $J_{-1/3}$, or in terms of $K_{1/3}$. It vanishes exponentially for $v = +\infty$, algebraically for $v = -\infty$, and it has an infinite set of zero's on the negative real axis: $v = -2\sigma_1, -2\sigma_2, -2\sigma_3, \ldots$.

We define functions

$$(19) \qquad q_n(v) = q_o(v - 2\sigma_n) \quad ,$$

which vanish at $v = 0$ and $v = \infty$; and a set of numbers

$$(20) \qquad q'_n = (dq_n/dv)_{v=0} = (dq_o/dv)_{v=-2\sigma_n} \quad .$$

The following expressions are now obtained for the functions ψ, Ψ, E, F, written with x and y instead of ξ,η (so that now $v = x^2/2 - y$):

$$(21) \qquad \begin{cases} \psi(x_2,y_2;\, x_1,y_1) = \\ = \frac{1}{2} \sum_\ell \frac{q_\ell(v_2)\, q_\ell(v_1)}{(q'_\ell)^2} \exp\left\{ \frac{-x_2^3 + x_1^3}{12} - \frac{x_2v_2 - x_1v_1}{2} - \sigma_\ell(x_2 - x_1) \right\} ; \end{cases}$$

$$(22) \qquad \Psi(x_2;x_1) = \frac{1}{2} \sum_\ell \exp\left\{ \frac{-x_2^3 + x_1^3}{12} - \sigma_\ell(x_2 - x_1) \right\} \quad ;$$

$$E(x_2) = \frac{1}{2} \sum_m \frac{I_m}{q'_m} \exp\left(\frac{x_2^3}{12} + \sigma_m x_2\right) \quad ; \tag{23}$$

$$F(x_1) = \frac{1}{2} \sum_n \frac{I_n}{q'_n} \exp\left(-\frac{x_1^3}{12} - \sigma_n x_1\right) \quad , \tag{24}$$

where

$$I_m = \int_0^V dv\, q_m(v) \exp\left(\frac{xv}{2} - \sigma_m x - \frac{x^3}{12}\right) \quad . \tag{25}$$

The meaning of the quantities x and V in eqs. (23) - (25) will be explained below.

Expressions (22) - (25) form the foundation upon which the integrations must be built for the evaluation of various mean values. To link them with the function $p(\xi,\eta)$ we have the relations

$$\begin{cases} \xi = x_2 - x_1 \ ; \quad \eta = (x_2^2 - x_1^2)2 \quad ; \\ x_2 = \xi/2 + \eta/\xi \ ; \quad x_1 = -\xi/2 + \eta/\xi \quad . \end{cases} \tag{26}$$

When we are interested in the mean value, taken over the set of shocks, of a quantity $Q_o(\xi,\eta)$ depending upon the shock parameters ξ and η, we must evaluate an integral [see eq. (43)]:

$$P = \int_0^\infty d\xi \int_{-\infty}^{+\infty} d\eta\, Q_o(\xi,\eta)\, p(\xi,\eta) \quad . \tag{27}$$

This can be expressed in terms of x_2 , x_1 as follows:

$$P = \int_{-\infty}^{+\infty} dx_1 \int_{x_1}^{+\infty} dx_2 (x_2 - x_1)\, Q_o(x_2;x_1)\, E(x_2)\, \Psi(x_2;x_1)\, F(x_1) \quad . \tag{28}$$

The integration limits make $x_2 - x_1 = \xi \geq 0$. The factor $(x_2 - x_1)$ arises from the Jacobian of the transformation (26). Q_o is supposed now to be expressed in terms of x_2 , x_1 . As is seen from Eq. (14), and can be deduced also from the series occurring in (22), the function ψ diverges when ξ goes to zero; however, the factor $x_2 - x_1 - \xi$ makes integration possible.

Evaluation of the integral (28). - The formal integration scheme presupposes that the summations involved in the definitions of Ψ , E and F are carried out first. However, no methods are available for a direct summation, and in certain cases it leads to divergent expressions. Another way therefore must be followed. We first carry out the integrations with respect to x_1 and x_2 , using a domain of large finite extent, from $-x$ to $+x$. This leads to expressions depending upon x , in which various summations can be carried out. At the end x is made infinite, while at the same time V must become infinite, in such a way that $V - x^2/2$ is a large positive quantity. All terms with positive powers of x and V (and also terms of the form V/x^2) are found to disappear from the final result through exact cancellation. Terms with negative powers of x automatically vanish when $x \to \infty$, and the only terms remaining are finite quantities, from which the final result is obtained. The calculations are laborious. To give some idea of the steps to be taken, a few details will be mentioned for the case where the function Q_o in (27) is equal to unity. The corresponding integral will be denoted by P_1.

It is of importance to note that with integrals of the form indicated in (28), terms with third powers of x_2 and x_1 in the exponential functions drop out when the product $E\,\Psi\,F$ is formed, which is a great advantage. The expression for P_1 takes the form:

$$(29)\qquad P_1 = \lim_{x=\infty} \sum_{\ell,m,n} \int_{-x}^{+x} dx_1 \int_{x_1}^{+x} dx_2\ (x_2-x_1)\ \frac{I_m\ I_n}{8q'_m\ q'_n} \exp\left\{(\sigma_m-\sigma_\ell)x_2 + (\sigma_\ell-\sigma_n)x_1\right\} .$$

The integrations are elementary. It is necessary to distinguish between cases where $m = \ell$ and $m \neq \ell$, etc. The result of the x_2- integration is

$$(30)\qquad \sum_{\ell,n} \int_{-x}^{+x} dx_1\ \frac{I_n}{8q'_\ell q'_n} \left\{\frac{(x-x_1)^2}{2}\ I_\ell + (x-x_1)(S_1)_\ell - (S_2)_\ell + (s_2)_\ell I_\ell\right\} \exp\{(\sigma_\ell-\sigma_n)x_1\} ,$$

where the quantities $(s_p)_\ell$ and $(S_p)_\ell$, respectively, stand for the sums

$$\left.\begin{array}{ll}(31a) & (s_p)_\ell = \sum_{m\neq\ell} \dfrac{1}{(\sigma_m - \sigma_\ell)^p} \\ (31b) & (S_p)_\ell = q'_\ell \sum_{m\neq\ell} \dfrac{I_m \exp\{(\sigma_m-\sigma_\ell)x\}}{q'_m \cdot (\sigma_m - \sigma_\ell)^p}\end{array}\right\}$$

here ℓ is fixed; the summation is over all m, excepting $m = \ell$.

Next carrying out the x_1-integration, we arrive at

$$\sum_\ell \frac{I_\ell}{8(q'_\ell)^2}\Big\{\frac{4}{3} x^3 I_\ell + 2x(s_2)_\ell\, I_\ell +$$

$$+ \; 4x^2(S_1)_\ell - 4x(S_2)_\ell + 2(S_3)_\ell + (s_2)_\ell\,(S_1)_\ell\Big\} \; +$$

$$+ \sum_\ell \frac{(S_1)_\ell}{8(q'_\ell)^2}\Big\{2x(S_1)_\ell - 2(S_2)_\ell\Big\} \; +$$

$$+ \sum_{\ell\neq n} \frac{I_n}{8q'_\ell q'_n}\Big\{\frac{(S_1)_\ell}{(\sigma_\ell-\sigma_n)^2} - \frac{(S_2)_\ell - (s_2)_\ell I}{\sigma_\ell - \sigma_n}\Big\}\exp\{(\sigma_\ell-\sigma_n)x\} \;.$$

By extensive rearrangement the double sum in the last line can be reduced to

$$- \sum_\ell \frac{I_\ell}{8(q'_\ell)^2}\Big\{2(s_3)_\ell\, I_\ell - (s_2)_\ell\,(S_1)_\ell\Big\} \;.$$

Collecting terms we obtain

$$(32)\quad \left\{\begin{array}{l} P_1 = \sum_\ell \dfrac{1}{(q'_\ell)^2}\Big[\dfrac{1}{4}(xS_1^2 - S_1S_2) \; + \\ \quad + \dfrac{1}{2}(x^2S_1 + \dfrac{1}{2}s_2S_1 - xS_2 + \dfrac{1}{2}S_3)I + (\dfrac{1}{6}x^3 + \dfrac{1}{4}xs_2 - \dfrac{1}{4}s_3)\,I^2\Big]\end{array}\right.$$

(to simplify the writing the summation subscript ℓ has been omitted in the terms within the square brackets).

While eq. (32) looks as if only a single summation is left, the quantities s and S themselves are sums, as indicated in eqs. (31a), (31b). Explicit expressions can be derived for these quantities. The sums $(s_p)_\ell$, with $p \geq 2$ [$(s_1)_\ell$ is divergent, but does never occur], can be obtained from a series development according to powers of z of the equation

$$\sum_{m\neq\ell} \frac{z}{2(\sigma_m-\sigma_\ell)(2\sigma_m-2\sigma_\ell+z)} = \frac{1}{z} - \frac{q_\ell'(z)}{q_\ell(z)} \quad .$$

A few examples are

(33) $$(s_2)_\ell = \frac{4}{3}\sigma_\ell \; ; \quad (s_3)_\ell = 1 \; ; \quad (s_4)_\ell = \frac{16}{45}\sigma_\ell^2 \quad .$$

The sums $(S_p)_\ell$, with $p \geq 1$, have been derived from series developments according to powers of h , connected with integrals of the equation

$$\sum_{m\neq\ell} \frac{q_m(z)}{q_m'\cdot(h-2\sigma_\ell+2\sigma_m)} = \frac{q_\ell(z+h)}{q_\ell(h)} - \frac{q_\ell(z)}{q_\ell'\cdot h} \quad .$$

Some examples are

(34) $$(S_1)_\ell = 2R_\ell - xI_\ell \; ; \quad (S_2)_\ell = -2R_\ell^* + xR_\ell - \left(\frac{x^2}{2} + \frac{2\sigma_\ell}{3}\right) I_\ell \; ; \text{ etc.} \quad ,$$

where

(35a) $$R_\ell = q_\ell(V) \cdot \exp\left(\frac{xV}{2} - \sigma_\ell x - \frac{x^3}{12}\right) \quad ;$$

(35b) $$R_\ell^* = q_\ell'(V) \cdot \exp\left(\frac{xV}{2} - \sigma_\ell x - \frac{x^3}{12}\right) \quad .$$

Calculations have been carried out for $p = 1$ through 8 [the expression for $(S_8)_\ell$ has 41 terms]. When the results for the quantities $(s_p)_\ell$ and $(S_p)_\ell$ are substituted into eq. (32), we arrive at an equation

(36) $$P_1 = \sum_\ell \frac{1}{(q_\ell')^2} [AR^{*2} + BR^*R + CR^2 + DR^*I + ERI + FI^2] \quad .$$

in which the coefficients A,...F depend upon x , V and σ_ℓ . With the substitution

$$\sigma_\ell = \frac{1}{2} V - \frac{1}{4} x^2 + x^{1/2}\theta \quad ,$$

it appears that asymptotic series can be constructed for the quantities R^* , R and I , in terms of θ ; and a development can be obtained for R^*/R [the θ occurring here is not related to that used in eq. (14)]. With the aid of the latter result, (36) can be transformed into

$$P_1 = \sum_{\ell} \frac{1}{(q'_\ell)^2} \quad [MR^2 + NRI + FI^2] \quad . \tag{37}$$

After extensive calculations it is found that the quantities M , N , and F have the following values for P_1 :

$$M = 0 \; ; \quad N = \frac{2}{3}\,(V - \frac{1}{2}\,x^2) + \frac{4}{3}\,x^{\frac{1}{2}}\theta \; ; \quad F = -\frac{1}{3} \quad . \tag{38}$$

The remaining summation with respect to the subscript ℓ can be replaced by integrations with respect to θ . In the case of P_1 these integrations give:

$$\left\{ \begin{aligned} &\sum \frac{NRI}{(q')^2} = \frac{2}{3}\left(V - \frac{1}{2}\,x^2\right) - \frac{4}{3}\sqrt{\frac{x}{2\pi}} \quad ; \\ &\sum \frac{FI^2}{(q')^2} = -\frac{2}{3}\left(V - \frac{1}{2}\,x^2\right) + \frac{4}{3}\sqrt{\frac{x}{2\pi}} + m_1 \quad , \end{aligned} \right. \tag{39}$$

with the result:

$$P_1 = m_1 \quad , \tag{40}$$

where m_1 is a transcendental number defined by

$$m_1 = \left[2^{2/3}\, 3^{-1/3}\, \pi^{2/3} (L + \tfrac{1}{4})^{2/3} - 2^{2/3}\, 3^{-2/3}\, \pi^2 \sum_{\ell=1}^{L} \frac{1}{(q'_\ell)^2} \right] \quad , \tag{41}$$

for L increasing indefinitely. Approximately:

$$m_1 \cong 1.053 \quad .$$

A number of other integrals have been worked out in a similar way (with increasing numbers of terms). A selection is given in Table I.

The result $\tilde{P}_2 = 0$ is due to symmetry relations. - The result $P_2 = 1$ and the ratio $P^*/P_1 = 8/3$ could be deduced from general properties of the functions Ψ , E , F , as alluded to in connection with eq. (14); the integrations have confirmed these results. - The integrals P_3 , P_3^{o} and P_3^{oo} are related by linear equations; also P_5 , P_5^{o} , P_5^{oo} , P_5^{ooo} and P^{**} are linearly related.

Table I - Examples of integrals of type (28).

$$
\begin{array}{rcc}
 & \text{factor } Q_o \text{ in } (27) & \text{integral} \\
(42) & \left\{ \begin{array}{c} 1 \\ x_2 - x_1 \\ (x_2 - x_1)^2 \\ (x_2 - x_1)^3 \\ (x_2 - x_1)^4 \\ (x_2 - x_1)^5 \\ x_2 + x_1 \\ x_2^2 + x_2x_1 + x_1^2 \\ x_2x_1 \\ (x_2 - x_1)(x_2 + x_1)^2 \\ x_2^4 + x_2^3x_1 + x_2^2x_1^2 + x_2x_1^3 + x_1^4 \\ x_2x_1(x_2^2 + x_2x_1 + x_1^2) \\ x_2^2x_1^2 \\ (x_2^2 - x_1^2)^2 \end{array} \right. & \begin{array}{l} P_1 = m_1 \\ P_2 = 1 \\ P_3 = m_3 \\ P_4 = 4m_1 \\ P_5 = \frac{296}{21} + m_5 \\ P_6 = 20m_3 \\ \tilde{P}_2 = 0 \\ P_3^o = \frac{7}{4} m_3 \\ P_3^{oo} = \frac{1}{4} m_3 \\ P^* = \frac{8}{3} m_1 \\ P_5^o = \frac{416}{21} + \frac{31}{16} m_5 \\ P_5^{oo} = \frac{152}{21} + \frac{53}{80} m_5 \\ P_5^{ooo} = \frac{64}{21} + \frac{19}{80} m_5 \\ P^{**} = \frac{136}{21} + \frac{4}{5} m_5 \quad . \end{array}
\end{array}
$$

m_3 and m_5 are transcendental numbers, defined by equations analogous to (41), $(m_3 \simeq 1.8)$.

The result $P_2 = 1$ leads to $m = 1$ in eq. (13).

Mean values of certain quanities. - With these results we have carried out a major part of the program. The outcome of the calculations has shown that the expressions for Ψ , E and F , given in eqs. (22) - (24), are applicable and lead to results which can be considered as correct. It should be noted that no arbitrary adjustable factors have come in.

We now can evaluate various mean values connected with the shocks. The general formula for the average value of any quantity $Q_o(\xi,\eta)$ connected with a shock is given by ratio

$$\langle Q_o(\xi,\eta)\rangle = P/P_1 , \tag{43}$$

where P is defined by (27) or (28). Actually, this gives only the numerical coefficient of a dimensionless expression; to obtain the actual value we must introduce the dimensional factors with powers of J and t . In this way we find:

$$\langle \xi \rangle \text{ (obtained from } P_2/P_1) = \frac{1}{m_1} J^{1/3} t^{2/3} . \tag{44}$$

This quantity is equal to the mean wavelength ℓ between successive shocks. It follows that the average number of shocks per unit length of the x-axis is given by

$$N_s = m_1 J^{-1/3} t^{-2/3} . \tag{44a}$$

Similarly we obtain

$$\left\{\begin{aligned} &\langle \xi^2 \rangle \text{ (obtained from } P_3/P_1) = \frac{m_3}{m_1} J^{2/3} t^{4/3} ; \\ &\langle \xi^3 \rangle \text{ (obtained from } P_4/P_1) = 4 J t^2 ; \\ &\langle \xi^4 \rangle \text{ (obtained from } P_5/P_1 = \left(\frac{296}{21m_1} + \frac{m_5}{m_1}\right) J^{4/3} t^{8/3} . \end{aligned}\right. \tag{45}$$

When we write $\zeta = -\tfrac{1}{2}(x_2 + x_1)$ (the minus sign is connected with other relations); $\eta = (x_2^2 - x_1^2)/2t$, we further find

$$\left\{\begin{aligned} &\langle \zeta \rangle \text{ (from symmetry)} = 0 \\ &\langle \zeta^2 \rangle \text{ [obtained from } (P_3^o + P_3^{oo})/4P_1] = \frac{m_3}{2m_1} J^{2/3} t^{4/3} ; \\ &\langle \xi\zeta^2 \rangle \text{ [obtained from } P^*/4P_1] = \frac{2}{3} J t^2 ; \\ &\langle \eta^2 \rangle = \langle \xi^2\zeta^2 \rangle \text{ [obtained from } P^{**}/4P_1] = \left(\frac{34}{21m_1} + \frac{m_5}{5m_1}\right) J^{4/3} t^{8/3} . \end{aligned}\right. \tag{46}$$

From an integration of u^2 with respect to x , it is found that the average value of the integral over a wavelength can be expressed as:

$$\left\langle \int u^2 dx \right\rangle = \left\{ \langle \xi\zeta^2 \rangle + \frac{1}{12} \langle \xi^3 \rangle \right\} t^{-2} = J \quad .$$

Division by ℓ gives the mean value u^2 over x-axis. Further division by 2 gives the "mean kinetic energy" per unit length of the x-axis:

(47) $$E = \frac{1}{2\ell} \left\langle \int u^2 dx \right\rangle = \frac{1}{2} m_1 J^{2/3} t^{-2/3} \quad .$$

When the calculation for the "sawtooth" profile is carried out in such a way that the diffusion coefficient ν (the "viscosity") is retained to some extent, it appears that the "dissipation integral" for a single shock in the limit has a finite value $\xi^3/12t^3$. Hence the average dissipation per unit length of the x-axis becomes

(48) $$\varepsilon = \frac{1}{12\ell} \langle \xi^3 \rangle t^{-3} = \frac{1}{3} m_1 J^{2/3} t^{-5/3} \quad .$$

The results for E and ε evidently are in agreement.

We have not arrived at a closed expression for the distribution function $p(\xi,\eta)$. A formal expression, containing a double summation, can be obtained for the distribution function of the quantities ξ (irrespective of the values of η) , but no closed explicit formula is available.

Other statistical problems. - The statistical problems arising in connection with the solutions of eq. (1) are not exhausted by those that can be treated on the basis of eq. (27) or (28). This equation referred to problems relating to a single shock, or, expressed in connection with the geometric method for solving eq. (1), relating to a single parabolic arc. Many problems require the consideration of sets of two or more successive arcs. This is already the case when we are interested in statistical problems concerning wavelengths (the distances between successive shocks).

Statistical weighting functions referring to sets of two or more adjacent arcs can be constructed by making use of still another one of the general properties of the functions Ψ , E and F which can be deduced without introducing the series in terms of Airy functions. In non-dimensional form we have

$$(49)\quad \begin{cases} dE/dx_2 = \int_{x_2}^{x} dx'' \, (x'' - x_2)\, E(x'')\, \Psi(x''; x_2) & ; \\ - dF/dx_1 = \int_{-x}^{x_1} dx' \, (x_1 - x')\, \Psi(x_1; x')\, F(x') & , \end{cases}$$

(as before , the quantity x in the limits of the integral must go to infinity at the end of the calculations).

Starting from an identity, viz. the integral P_1 , divided by m_1 :

$$(50)\qquad 1 = \frac{1}{m_1} \int_{-x}^{+x} dx_1 \int_{x_1}^{+x} dx_2 (x_2 - x_1)\, E(x_2)\, \Psi(x_2; x_1)\, F(x_1) \quad ,$$

we can use the first equation of (49) - or, after a change of the order of integration, we can use the second equation - to prove the identity:

$$(51)\quad 1 = \frac{1}{m_1} \int_{-x}^{+x} dx_1' \int_{x_1'}^{+x} dx_1'' \int_{-x}^{x_1''} dx_2' \int_{x_2'}^{+x} dx_2'' (x_2''-x_2')(x_1''-x_1') E(x_2'') \Psi(x_2'';x_2') \Psi(x_1'';x_1') F(x_1') \quad .$$

We consider the multiple integral on the right hand side of this equation as an integral operator, which can be applied to an arbitrary integrand depending upon the quantities x_2'' , x_2' , x_1'' , x_1' (or any function of them), to serve as a weighting function or distribution function connected with two adjacent parabolic arcs. For instance, we can consider the wavelength or distance between the two successive shocks associated with these arcs,

$$\lambda = x_1'' - x_2' \quad .$$

If we apply some changes in the order of the integrations and introduce

$$(52)\qquad \Phi(\lambda) = \frac{1}{m_1} \int_{\lambda - x}^{x} dx_1'' \int_{-x}^{x_1''} dx_1' \int_{x_2'}^{x} dx_2'' \; (x_2''-x_2')(x_1''-x_1') E_2\, \Psi_2\, \Psi_1\, F_1 \quad ,$$

we can write eq. (51) in the form

$$(53) \qquad 1 = \int_0^{2\lambda} d\lambda\ \Phi(\lambda) \quad .$$

Hence we can take $\Phi(\lambda)$ as a distribution function for the wavelengths. The limits used in the integrals ensure that

$$x_2'' - x_2' = \xi_2 \geq 0 \ ; \quad x_1'' - x_1' = \xi_1 \geq 0 \ ; \quad x_1'' - x_2' = \lambda \geq 0 \ .$$

The same transformation can be used repeatedly. For instance, the next step gives

$$(54) \quad 1 = \frac{1}{m_1} \int_{-x}^{+x} dx_1' \int_{x_1'}^{+x} dx_1'' \int_{-x}^{x_1''} dx_2' \int_{x_2'}^{+x} dx_2'' \int_{-x}^{x_2''} dx_3' \int_{x_3'}^{+x} dx_3'' \cdot (x_3''-x_3')(x_2''-x_2')(x_1''-x_1')\ E_3\Psi_3\Psi_2\Psi_1F_1 \quad ,$$

which leads to a weighting funcion for a set of three successive parabolic arcs, with three shocks and two wavelengths between them. In this way we can continue. Weighting functions of these types are needed for the calculation of correlation functions.

Unfortunately, when we look at the expressions for the functions Ψ, E, F in (22) - (24), it is seen that not all third powers of variables in the exponential factors disappear. In the case of the integral (51) the third powers of x_2'' and x_1' cancel, but there remains a term $[(x_2')^3 - (x_1'')^3]/12$; while in (54) there remain both $[(x_3')^3 - (x_2'')^3]/12$ and $[(x_2')^3 - (x_1'')^3]/12$. Such terms, the origin of which is found in the application of eq. (15), present serious obstacles against evaluation of the integrals.

The only cases which have remained tractable, are certain relations coming up where we have to do with a single wave of <u>vanishing length.</u> These relations are connected with the vanishing of the shocks, occurring when two adjacent parabolic arcs merge into a single arc. An important parameter in these cases is the value of $-(d\lambda/dt)$ for $\lambda \to 0$, which leads to an expression for $-dN_s/dt$ (N_s being the average number of shocks per unit length of the x-axis). In eqs. (51) and (52) we then have $x_1'' = x_2'$, so that the troublesome term disappears from the exponential function. These calculations could be used to confirm certain relations referring to the disappearance of shocks and to the increase of $\langle \xi^2 \rangle$ with time, which had been found from dimensional formulas.

To gain some insight into the general case with finite wavelengths, an attempt has been made to introduce approximations for the functions Ψ , E , F , in order to get rid of the third power terms in the exponential functions and make the integrals tractable. In a sense this is substituting a "model" for the mathematical analysis. Notwithstanding the circumstance that the numerical results differ from what could be expected for the exact calculations, information is obtained concerning the structure of the system. The calculations, however, are elaborate even with the approximate formulas and it was not considered worthwhile to work them out in every detail. Here I will leave aside these calculations.

Annotated References.- An extensive report on the calculations summarized in this paper will probably appear as a Technical Note of the Institute for Fluid Dynamics and Applied Mathematics of the University of Maryland (approximately 290 pages and 25 figures). Scientists interested in procuring a copy are asked to write to the Institute (College Park, Maryland 20742).

The nonlinear equation was mentioned by the author in a paper "Application of a Model System to Illustrate some Points of the Statistical Theory of Free Turbulence", Proc. Roy. Netherl. Academy of Sciences (Amsterdam) 43, p. 8, 1940. It is often quoted from a contribution "A Mathematical Model Illustrating the Theory of Turbulence", in "Advances in Applied Mechanics", ed. by R. von Mises and Th. von Karman, vol. I, pp. 171-199, 1948.

References to various publications which appeared in the Netherlands (1939-1955) are given in the Technical Note, in connection with specific equations. The latest publications by the author prior to this Technical Note bearing upon the subject, are: "Statistical Problems Connected with the Solution of a Nonlinear Partial Differential Equation", in "Nonlinear Problems in Engineering", ed. by W.F. Ames (Acedemic Press, Inc., New York, 1964), which presents the basic thoughts connected with eqs. (9a,b) - (14) of the text above; and "Functions and Integrals Connected with Solutions of the Diffusion or Heat Flow Equation", Tech. Note BN-398 of the Institute for Fluid Dynamics and Applied Mathematics, University of Maryland, May 1965, containing the background mathematics referring to eqs. (10) - (41) and part of Table I of the present text. This Technical Note is no longer available, but its contents are taken up in Chapter 4, 5 and 6 of the Technical Note to appear.

The nonlinear equation has received much attention in present literature, in particular for finite values of the coefficient ν (the present work is concerned with the asymptotic case $\nu \to 0$). A survey of published solutions has been given E.R. Benton and C.W. Platzman, in "A Table of Solutions of the One-dimensional Burgers Equation", Quart. Appl. Math. (to appear). According to these authors the equation first occurred in a paper by H. Bateman, "Some Recent Researches on the Motion of Fluids", Mon. Weather Rev. 43, pp. 163-170, 1915.

For the Hopf-Cole transformation see: E. Hopf, Comm. Pure and Applied Mathematics 3, pp. 201-230, 1950 (in particular p. 203); J.D. Cole, Quart. Applied Mathematics, 9, pp. 225-236, 1951 (in particular p. 230).

For the Airy function see G.N. Watson, Theory of Bessel Functions (Cambridge University Press, 1944), pp. 188-190 and Table III.

ACKNOWLEDGEMENTS. - Various papers in which the basic ideas of this work found their first expressions, were written at the Technical University of Delft in the Netherlands before 1956.

A six months stay at the Hydrodynamics Laboratories of the California Institute of Technology, Pasadena, California (1950-1951) was supported by a Contract with the Department of the Air Force, No. AF33(038)-17207.

Since the end of 1955 until 1970 the work was carried on at the Institute for Fluid Dynamics and Applied Mathematics of the University of Maryland, and was partly supported by the Air Force Office of Scientific Research, as a subject coming under Contracts AF-18(600)993, AF-49(638)401, F-44620-67-C-0023, and Grants AFOSR-141-63, AFOSR-141-64, AFOSR-141-65 and AFOSR-141-66. For all support and for the generous interest of many colleagues, deep gratitude is expressed.

BURGER'S EQUATION: GENERALIZATIONS AND SOLUTIONS

Kenneth M. Case
The Rockefeller University
New York, N. Y. 10021

and

Institute for Defense Analyses
Arlington, Virginia 22202

I. INTRODUCTION

To understand turbulence and to evaluate approximate procedures for dealing with the Navier-Stokes equations it has been helpful to have had a simple model equation due to Burgers[1,3]. Among other properties it is so simple that analytical solutions are possible. Here we look at some simple generalizations of this equation and their solutions in special cases.

It should be noted that the essential point in these models is that by what is essentially the Euler trick[2] of dealing with the Riccati equation one can reduce the problem to a linear one.

II. CONSTRUCTION OF MODELS

There are two approaches one can take in generalizing the Burgers equation:

1. One can take the original equation and ask what (hopefully physical) modifications can be made and still get a soluble equation.

2. One can start with essentially the most general equation one thinks might be soluble and ask what generalization of Burgers equation this yields.

Here we will follow the second course since the treatment is much simpler and the two approaches yield almost exactly the same equations.

Consider the linear second order equation

$$\frac{\partial \Psi}{\partial t} = \nu \nabla^2 \Psi + \underset{\sim}{B} \cdot \nabla \Psi + C \Psi \tag{1}$$

Here, for generality, we consider ν, $\underset{\sim}{B}$, and C as given complex functions of $\underset{\sim}{r}$ and t

Let us make the Euler substitution

$$\Psi = e^{\lambda \int \underset{\sim}{U} \cdot d\underset{\sim}{r}} \tag{2}$$

Then $\lambda \underset{\sim}{U} = \nabla \ln \Psi$ (3)

and thus $\nabla \times \underset{\sim}{U} = 0$. (4)

(Here λ is a constant which we will choose as convenient later.)

Putting the expression for Ψ in Eq. 1 yields

$$\frac{\partial \underset{\sim}{U}}{\partial t} - \lambda \, \nabla(\nu \, \underset{\sim}{U}^2) = \nabla(\nu \, \nabla \cdot \underset{\sim}{U}) + \frac{1}{\lambda} \nabla C \, . \tag{5}$$

This is particularly simple in the case that ν is constant--which we will hereafter assume unless otherwise stated. Then

$$\nabla \nu \nabla \cdot \underset{\sim}{U} = \nu \nabla \nabla \cdot \underset{\sim}{U} = \nu \nabla^2 \underset{\sim}{U} \tag{6}$$

in virtue of Eq. 4. We choose $\lambda = -\frac{1}{2\nu}$ and then Eq. 5 becomes

$$\frac{\partial \underset{\sim}{U}}{\partial t} + \frac{1}{2} \nabla \underset{\sim}{U}^2 = \nu \nabla^2 \underset{\sim}{U} + \nabla(\underset{\sim}{B} \cdot \underset{\sim}{U}) - 2\nu \nabla C \tag{7}$$

If the real and imaginary parts of all quantities be denoted by subscripts 1 and 2, respectively, we find that Eq.7 is equivalent to the two real equations

$$\begin{aligned} \frac{\partial \underset{\sim}{U}_1}{\partial t} + \frac{1}{2} \nabla \{\underset{\sim}{U}_1{}^2 - \underset{\sim}{U}_2{}^2\} &= \nabla^2 \{\nu_1 \underset{\sim}{U}_1 - \nu_2 \underset{\sim}{U}_2\} \\ &+ \nabla \{\underset{\sim}{B}_1 \cdot \underset{\sim}{U}_1 - \underset{\sim}{B}_2 \cdot \underset{\sim}{U}_2\} \\ &- 2\nabla \{\nu_1 C_1 - \nu_2 C_2\} \end{aligned} \tag{8}$$

and

$$\frac{\partial \underset{\sim}{U}_2}{\partial t} + \nabla(\underset{\sim}{U}_1 \cdot \underset{\sim}{U}_2) = \nabla^2\{\nu_1 \underset{\sim}{U}_2 + \nu_2 \underset{\sim}{U}_1\}$$

$$+ \nabla\{\underset{\sim}{B}_1 \cdot \underset{\sim}{U}_2 + \underset{\sim}{B}_2 \cdot \underset{\sim}{U}_1\}$$

$$- 2\nabla\{\nu_1 C_2 + \nu_2 C_1\} \quad . \tag{9}$$

Some special cases of interest:

a. All quantities real. Then we reduce to the single Eq. 7. With $\overline{B} = \overline{0} = \overline{C}$ we have just Burgers' original equation. Equation 7 with all quantities real is then a generalization of the Burgers' equation including forcing terms described by $\underset{\sim}{B}$ and C.

b. $V_1 = C_1 = 0 = \underset{\sim}{B}_2$,

$$\nu_2 = \frac{\hbar}{2m} \; , \; \underset{\sim}{B}_1 = \frac{e}{mc}\underset{\sim}{A} \; , \; C_2 = -\frac{1}{\hbar}\left(V + \frac{e^2}{2mc^2}\underset{\sim}{A}^2\right) \tag{10}$$

In this case Eq. 1 is just the Schrödinger equation

$$i\hbar \frac{\partial \Psi}{\partial t} = \frac{\left(\frac{\hbar}{i}\nabla - \frac{e}{c}\underset{\sim}{A}\right)^2}{2m}\Psi + V\Psi \tag{11}$$

corresponding to a single particle moving in force field described by the vector potential $\underset{\sim}{A}$ and potential energy V. The corresponding forms that Eqs. 8 and 9 take are

$$\frac{\partial \underset{\sim}{U}_1}{\partial t} + \frac{1}{2}\nabla\{(\underset{\sim}{U}_1 - \frac{e}{mc}\underset{\sim}{A})^2 - \underset{\sim}{U}_2{}^2\} = -\frac{\hbar}{\partial m}\nabla^2\underset{\sim}{U}_2 - \frac{1}{m}\nabla V \quad , \tag{12}$$

$$\frac{\partial \underset{\sim}{U}_2}{\partial t} + \nabla\underset{\sim}{U}_2 \cdot (\underset{\sim}{U}_1 - \frac{e}{mc}\underset{\sim}{A}) = \frac{\hbar}{\partial m}\nabla^2\underset{\sim}{U}_1 \quad . \tag{13}$$

These equations give a sort of hydrodynamic formulation of Quantum Mechanics. However, for our present purpose the important point is that we now have considerable insight into when the general Eqs. 8 and 9 will have closed form solutions. This will be when a closely related Quantum Mechanical problem admits of such.

III. SOME SIMPLE SOLUTIONS

Let us consider the real form of Eq. 7. To solve the initial value problem it suffices by virtue of Eq. 3 to find the solution of Eq. 1 subject to the condition

$$\lim_{t \to t_0} \Psi(\underset{\sim}{r},t) = \delta(\underset{\sim}{r} - \underset{\sim}{r}_0) \ . \tag{14}$$

The fact that the interesting cases are for V small and that $\underset{\sim}{U}$ is related to $\ell n\Psi$ suggests we look for solutions of Eq. 1 in the WKB form

$$\Psi = A\ e^{S/\nu} \ . \tag{15}$$

This when inserted in Eq. 1 yields

$$\frac{\partial A}{\partial t} + \frac{1}{\nu}\frac{\partial S}{\partial t} A - \nu\nabla^2 A - 2\nabla A\cdot\nabla S - \frac{(\nabla S)^2 A}{\nu} - (\nabla^2 S)\ A$$

$$- B\cdot\nabla A - \frac{(B\cdot\nabla S)\ A}{\nu} = \frac{PA}{2\nu}, \tag{16}$$

where we have written $P = 2\nu C$ since this is clearly to be understood as a "pressure" in Eq. 7. We choose S so that the terms in $1/\nu$ vanish, i.e.,

$$\frac{\partial S}{\partial t} - (\nabla S)^2 = \underset{\sim}{B}\cdot\nabla S + \frac{P}{2} \ . \tag{17}$$

The remainder of Eq. 16 is then

$$\frac{\partial A}{\partial t} - \nu\nabla^2 A - 2\nabla A\cdot\nabla S - (\nabla^2 S)A = \underset{\sim}{B}\cdot\nabla A \ . \tag{18}$$

For small ν the primary dependence on the arguments will occur through that of S. Indeed we expect A to be but a slowly varying function of position. From Eq. 4 we note that multiplying Ψ by a function of t alone has <u>no</u> effect on $\underset{\sim}{U}$. On the other hand, for S we have a quite general formal solution. Indeed, we note that Eq. 17 is the equation for Hamilton's principle function for a particle moving in a force field described by $\underset{\sim}{B}$ and P. Thus, for S we have the solution

$$S(\underset{\sim}{r},t;\underset{\sim}{r}_0,t_0) = -\int_{t_0}^{t} L(\underset{\sim}{r}(t^1),\dot{\underset{\sim}{r}}(t^1))dt^1 . \tag{19}$$

(Here L is the corresponding Lagrangian and $\underset{\sim}{r}(t^1)$ is the solution of the equations of motion which take on the value $\underset{\sim}{r}_0$ at t_0 and $\underset{\sim}{r}$ at t.)

Examples

1. Constant pressure gradient, $\underset{\sim}{B} = 0$

Choosing the x axis as the direction of the gradient and noting that Eq. 17 is then separable we see that

$$S = S_1\ (x) + S_2\ (y) + S_2\ (z) \tag{20}$$

where

$$\frac{\partial S_1}{\partial t} = \left(\frac{\partial S_1}{\partial x}\right)^2 + \frac{ax}{2} \tag{21}$$

and S_2 satisfies the equation with a = 0. For illustrative purposes we will work out this example in detail.

The Langrangian corresponding to Eq. 21 is

$$L = \frac{1}{4}\dot{x}^2 - \frac{ax}{2} \tag{22}$$

with equations of motion

$$\ddot{x}(t^1) = -a .$$

The solution which is x_0 at time 0 and x at time t is

$$x(t^1) = x_0 + C_1 t^1 - \frac{at^{1^2}}{2} \tag{23}$$

where

$$C_1 = \frac{x-x_0}{t} + \frac{at}{2} . \tag{24}$$

Using Eqs. 22 and 23 we find that Eq. 19 yields

$$S_1(x,t;x_0,0) = -\frac{(x-x_0)^2}{4t} + \frac{(x+x_0)at}{4} + \frac{a^2t^3}{48} . \tag{25}$$

Setting a = 0 gives

$$S_2(y,t;y_0,0) = -\frac{(y-y_0)^2}{4t} , \tag{26}$$

With S as determined here Eq. 18 for A is seen to be satisfied with A a function of time alone satisfying

$$\frac{\partial A}{\partial t} = -\frac{3}{2}\frac{A}{t} . \tag{27}$$

$$A = (\text{const})t^{-3/2} \tag{28}$$

where the constant is determined by Eq. 14. However, for most of our purposes it suffices merely to have found that A is position independent. Thus, the solution of the initial value problem for Eq. 17 in the present instance is

$$\underset{\sim}{U}(r,t) = -2\nu\nabla \, \ell n\Psi(\underset{\sim}{r},t) \tag{29}$$

where

$$\Psi(\underset{\sim}{r},t) = \int d\underset{\sim}{r}_0 \exp \frac{\left\{S'(\underset{\sim}{r},t;\underset{\sim}{r}_0,0) - \frac{1}{2}\int_o^{\underset{\sim}{r}_0} \underset{\sim}{U}(\underset{\sim}{r}',0)\bullet d\underset{\sim}{r}'\right\}}{\nu}$$

and

$$S'(\underset{\sim}{r},t;\underset{\sim}{r}_0 0) = -\frac{(\underset{\sim}{r}-\underset{\sim}{r}_0)^2}{4t} + \underset{\sim}{a} \cdot \frac{(\underset{\sim}{r}+\underset{\sim}{r}_0)t}{4} \qquad (31)$$

(Here we have generalized to the pressure gradient being in the direction of $\underset{\sim}{a}$ and omitted the t^3 term in Eq. 25--since it too plays no role in determining $\underset{\sim}{U}(\underset{\sim}{r},t)$.)

It may be remarked that the simplicity of the solution should have been expected from the quantum mechanical analog. What has been constructed is essentially the Green's function for a particle in a constant electric field. We therefore also would expect that other problems with known analogs are simple. For example: $\underset{\sim}{B} = 0$, P quadratic (harmonic oscillator); or $P = 0$, $\nabla \times \underset{\sim}{B} =$ constant, $\nabla \cdot \underset{\sim}{B} = 0$ (constant magnetic field); or any combination of these two and the problem just considered. We merely sketch the results.

2. $\underset{\sim}{B} = 0$, $P = \frac{1}{2} a_{ij} x_i' \; x_j'$

Introduce new coordinates such that P is diagonal, i.e.,

$$P = \frac{1}{2} \sum_i a_i^{\,2} \, x_i^{\,2} \qquad (32)$$

(Note: a_i can be either purely real or purely imaginary.) In these coordinates S is again separable--

$$S = \sum_i S_i \, (x_i) \qquad (33)$$

where

$$\frac{\partial S_i}{\partial \, t} = \left(\frac{\partial \, S_i}{\partial x_i}\right)^2 + \frac{a_i^{\,2} \, x_i^{\,2}}{4} \qquad (34)$$

Proceeding as before we obtain

$$S_i \, (x_i) = - \frac{a_i}{4\sin a_i t} \left\{(x_i^{\,2} + x_i^{\,0})^2 \cos a_i t - 2x_i x_i^{\,0}\right\} \qquad (35)$$

With these S_i we again readily find that A is a function (readily obtained) of t alone. Thus again Eq. 29 and 30 give the solution of the initial value problem where now S' is just the sum of the S_i.

This result is particularly simple in the case of isotropy (all $a_i = a$) Then

$$S'(r,t;r_0,0) = -\frac{a}{4\sin a\,t}\left\{(\underset{\sim}{r}^2 + \underset{\sim}{r}_0)^2 \cos at - 2\,\underset{\sim}{r}\cdot\underset{\sim}{r}_0\right\} \quad (36)$$

3. An Example With Non-Zero $\underset{\sim}{B}$

Consider the case when $\underset{\sim}{B}$ is a divergence free vector with constant curl--$\underset{\sim}{Q}$. Then

$$\underset{\sim}{B} = \frac{1}{2}\,\underset{\sim}{Q} \times \underset{\sim}{r} \quad (37)$$

To keep things simple we will also choose

$$P = \frac{\underset{\sim}{B}^2}{2} + \underset{\sim}{a}\cdot\underset{\sim}{r} \quad (38)$$

The appropriate Langrangian going with Eq. 17 is then of a familiar form

$$L = \frac{\dot{\underset{\sim}{r}}^2}{4} - \frac{\underset{\sim}{B}\cdot\dot{\underset{\sim}{r}}}{2} - \frac{\underset{\sim}{a}\cdot\underset{\sim}{r}}{2} \quad (39)$$

Again S is readily determined from Eq. 19. We find

$$S = -\frac{[\underset{\sim}{Q}\cdot(\underset{\sim}{r}-\underset{\sim}{r}_0)]^2}{Q^2\,4t} - \frac{1}{8}\frac{[\underset{\sim}{Q}\times(\underset{\sim}{r}-\underset{\sim}{r}_0)]^2}{Q}\frac{\sin Qt}{1-\cos Qt}$$

$$+ \frac{(\underset{\sim}{r}+\underset{\sim}{r}_0)\cdot\underset{\sim}{a}\,t}{8} + \frac{1}{8}\frac{\underset{\sim}{Q}\cdot[(\underset{\sim}{r}-\underset{\sim}{r}_0)\times\underset{\sim}{a}]}{Q}\left[\frac{t\sin Qt}{1-\cos Qt} - \frac{2}{Q}\right]$$

$$-\frac{\underset{\sim}{Q}\cdot(\underset{\sim}{r}\times\underset{\sim}{r}_0)}{Q} \quad (40)$$

Since again we have an S that is a quadratic function of the spatial coordinates, it follows that A is only a function of time and hence not needed. Indeed the solution of the initial value problem for Eq. 7 is again given by Eqs. 24 and 30 where now S' is as given by Eq. 40.

IV. CONCLUSIONS

We have seen that there exist generalizations of Burgers' equation which admit closed form solutions. The analog of the terms added to the original Burgers equation is the inclusion of forcing terms in the Navier-Stokes equation.

A number of directions for future work are suggested. From the connection with the Schrödinger equation we know there are other soluble models. For example, the case of a plane electromagnetic wave incident on a charged particle can be handled. It would be of some interest to apply the various approximate and statistical techniques which have been used on the Navier-Stokes equation to our soluble model equations.

In this connection it may be mentioned that there are other model equations due to Burgers(4) which, when suitably modified(5) yield simple solutions. Similar remarks hold for these equations.

REFERENCES

1. H. Bateman, "Some Recent Researches On The Motion of Fluids", Mon. Weather Per. 43, pp. 163-170 (1915).

   J. M. Burgers, "Application of a Model System to Illustrate Some Points of the Statistical Theory of Turbulence", Proc. Roy. Netherl. Academy of Sciences (Amsterdam) 43, p. 8, (1940).

2. E. Hopf, Comm. Pure and Applied Math. 3, pp. 201-230 (1950).

3. J. D. Cole, Quart. of Applied Math. 9, pp. 225-236 (1951).

4. J. M. Burgers, Vert. Nederl. Akad. Wetersch. Afd. Natuurk (Amsterdam) 17, 1, (1939).

5. K. M. Case and S. C. Chiu, "The Physics of Fluids" 12, 1799, (1969).

A CLOSURE HYPOTHESIS FOR THE HIERARCHY OF EQUATIONS FOR TURBULENT PROBABILITY DISTRIBUTION FUNCTIONS.

by

T. S. Lundgren

Department of Aerospace Engineering and Mechanics,
University of Minnesota, Minneapolis, Minnesota.

ABSTRACT

The hierarchy of equations for turbulent probability distribution functions is closed by relating the three point distribution function to lower order distribution functions. The theory is applied to isotropic, homogeneous turbulence at large wave number giving a nonlinear integral equation for the correlation function at small separation. The Kolmogorov spectrum is found in the inertial range and the Kolmogorov constant is determined.

INTRODUCTION

In an earlier paper (Lundgren (1967), also Monin (1967)), an infinite hierarchy of equations for turbulent probability distribution functions was derived. The lowest order equation is

$$\frac{\partial f}{\partial t} + \underline{v}_1 \cdot \frac{\partial f}{\partial \underline{r}_1} + \left(-\frac{1}{\rho}\frac{\partial \overline{p}_1}{\partial \underline{r}_1} + \nu \frac{\partial}{\partial \underline{r}_1}\cdot\frac{\partial}{\partial \underline{r}_1}\underline{u}_1\right)\cdot\frac{\partial f}{\partial \underline{v}_1}$$

$$= \frac{\partial}{\partial \underline{v}_1}\cdot\Big\{\frac{1}{4\pi}\int \frac{\partial}{\partial \underline{r}_1}\frac{1}{|\underline{r}_1-\underline{r}_2|}\left(\underline{v}_2\cdot\frac{\partial}{\partial \underline{r}_2}\right)^2 f_2'(1,2)\, d\underline{r}_2\, d\underline{v}_2 \tag{1}$$

$$- \lim_{\underline{r}_2\to\underline{r}_1} \nu \frac{\partial}{\partial \underline{r}_2}\cdot\frac{\partial}{\partial \underline{r}_2}\int \underline{v}_2 f_2'(1,2)\, d\underline{v}_2 \Big\}$$

where $f(\underline{r}_1, \underline{v}_1, t)\, d\underline{v}_1$ is the probability that the fluid velocity at point $\underline{r}_1$ is in the range $\underline{v}_1$, $\underline{v}_1 + d\underline{v}_1$. $f(\underline{r}_1, \underline{v}_1, t) \equiv f(1)$ is the one point probability distribution function. Here $\overline{p}_1$ is the mean pressure and $\underline{u}_1$ the mean velocity. The quantity $f_2'(1,2)$ which occurs in the integrals on the right hand side is defined by

$$f_2'(1,2) = f_2(1,2) - f(1)f(2)$$

where $f_2(1,2)$ is the two point probability distribution function defined such that $f_2(1,2)\,d\underline{V}_1\,d\underline{V}_2$ is the probability that the velocities at points 1 and 2 are in the ranges $\underline{V}_1, \underline{V}_1 + d\underline{V}_1$ and $\underline{V}_2, \underline{V}_2 + d\underline{V}_2$ respectively. The two point distribution function satisfies

$$\begin{aligned}
&\frac{\partial f_2'(1,2)}{\partial t} + \underline{V}_1 \cdot \frac{\partial f_2'(1,2)}{\partial \underline{r}_1} + \underline{V}_2 \cdot \frac{\partial f_2'(1,2)}{\partial \underline{r}_2} \\
&+ \left\{ -\frac{\partial \bar{p}_1}{\partial \underline{r}_1} + \nu \frac{\partial}{\partial \underline{r}_1} \cdot \frac{\partial}{\partial \underline{r}_1} \underline{u}_1 \right\} \cdot \frac{\partial f_2'(1,2)}{\partial \underline{V}_1} + \left\{ -\frac{\partial \bar{p}_2}{\partial \underline{r}_2} + \nu \frac{\partial}{\partial \underline{r}_2} \cdot \frac{\partial}{\partial \underline{r}_2} \underline{u}_2 \right\} \cdot \frac{\partial f_2'(1,2)}{\partial \underline{V}_2} \\
&+ \frac{\partial f(1)}{\partial \underline{V}_1} \cdot \left\{ -\frac{1}{4\pi} \int \left(\frac{\partial}{\partial \underline{r}_1} \frac{1}{|\underline{r}_1 - \underline{r}_3|} \right) \left(\underline{V}_3 \cdot \frac{\partial}{\partial \underline{r}_3} \right)^2 f_2'(2,3)\, d\underline{r}_3\, d\underline{V}_3 \right. \\
&\qquad \left. + \lim_{\underline{r}_3 \to \underline{r}_1} \nu \frac{\partial}{\partial \underline{r}_3} \cdot \frac{\partial}{\partial \underline{r}_3} \int \underline{V}_3 f_2'(2,3)\, d\underline{V}_3 \right\} \\
&+ \frac{\partial f(2)}{\partial \underline{V}_2} \cdot \left\{ \text{same as above with 1 and 2 interchanged} \right\} \qquad (2) \\
&+ \frac{\partial}{\partial \underline{V}_1} \cdot \left\{ -\frac{1}{4\pi} \int \left(\frac{\partial}{\partial \underline{r}_1} \frac{1}{|\underline{r}_1 - \underline{r}_3|} \right) \left(\underline{V}_3 \cdot \frac{\partial}{\partial \underline{r}_3} \right)^2 f_3'(1,2,3)\, d\underline{r}_3\, d\underline{V}_3 \right. \\
&\qquad \left. + \lim_{\underline{r}_3 \to r_1} \nu \frac{\partial}{\partial \underline{r}_3} \cdot \frac{\partial}{\partial \underline{r}_3} \int \underline{V}_3 f_3'(1,2,3)\, d\underline{V}_3 \right\} \\
&+ \frac{\partial}{\partial \underline{V}_2} \cdot \left\{ \text{same as above with 1 and 2 interchanged} \right\} \\
&= 0
\end{aligned}$$

where the quantity $f_3'(1,2,3)$ is defined by

$$f_3'(1,2,3) = f_3(1,2,3) - f(1)f_2(2,3) - f(2)f_2(1,3) - f(3)f_2(1,2) + 2\,f(1)f(2)f(3)$$

This in turn satisfies a higher order equation.

Equations (1) and (2) were derived from the Navier-Stokes equations for an incompressible fluid in the following way. The one point distribution function was defined by

$$f(1) = \langle \delta(\underline{v}_1 - \underline{v}(\underline{r}_1, t)) \rangle \tag{3}$$

where $\underline{v}$ is the instantaneous fluid velocity, $\delta(\underline{v}_1 - \underline{v}(\underline{r}_1, t))$ is a three dimensional Dirac delta function and $\langle \quad \rangle$ denotes an ensemble average. Eq. (1) is derived from the identity

$$\frac{\partial f(1)}{\partial t} = -\left\langle \frac{\partial \underline{v}(\underline{r}_1, t)}{\partial t} \cdot \frac{\partial}{\partial \underline{v}_1} \delta(\underline{v}_1 - \underline{v}(\underline{r}_1, t)) \right\rangle$$

by substituting for the time derivative from the Navier-Stokes equation and interpreting the result term by term. Similarly Eq. (2) is derived from the definition

$$f_2(1,2) = \langle \delta(\underline{v}_1 - \underline{v}(\underline{r}_1, t)) \, \delta(\underline{v}_2 - \underline{v}(\underline{r}_2, t)) \rangle. \tag{4}$$

This system of equations is similar to the B.B.G.K.Y. equations which are central to the field of statistical mechanics. These equations are derived from Liouville's equation, an equation for f_N in an N degree of freedom system, by integrating down to get the equations for lower order probability distribution functions. Since a fluid has an infinite number of degrees of freedom there is no Liouville equation. The nearest generalization is Hopf's (1952) equation for the probability functional for the whole velocity field. The hierarchy of equations produced above can also be derived from Hopf's equation. This was Monin's approach.

The probability distribution functions have a number of properties which follow directly from the formal definitions, Eq. (3) and Eq. (4). These are

i) The reduction property. Integration with respect to a velocity variable reduces the order by one. For instance

$$\int f(1) \, d\underline{v}_1 = 1$$

$$\int f_2(1,2)\, d\underline{v}_2 = f(1)$$

$$\int f_3(1,2,3)\, d\underline{v}_3 = f_2(1,2)$$

ii) The separation property. When two points are so widely separated that the fluid velocities are statistically independent the distribution functions decompose into products such as

$$\lim_{|\underline{r}_2-\underline{r}_1|\to\infty} f_2(1,2) = f(1)\, f(2)$$

$$\lim_{|\underline{r}_3-\underline{r}_1|\to\infty} f_3(1,2,3) = f_2(1,2)\, f(3)$$

An alternate way of expressing this is $f_2'(1,2)\to 0$, $f_3'(1,2,3)\to 0$, etc. whenever one point is distant from the others.

iii) The coincidence property. When two points coincide the velocities at these points must be the same with absolute certainty. That is $\lim_{\underline{r}_2\to\underline{r}_1} f_2(1,2) = 0$ unless $\underline{v}_2 = \underline{v}_1$. But since the reduction property must also be valid, we conclude

$$\lim_{\underline{r}_2\to\underline{r}_1} f_2(1,2) = f(1)\, \delta(\underline{v}_2 - \underline{v}_1)$$

$$\lim_{\underline{r}_2\to\underline{r}_1} f_3(1,2,3) = f_2(1,3)\, \delta(\underline{v}_2 - \underline{v}_1)$$

iv) The divergence condition. For an incompressible fluid the velocity field satisfies $\operatorname{div} \underline{v} = 0$. This leads to an infinite number of "continuity" equations

$$\frac{\partial}{\partial \underline{r}_1} \cdot \int \underline{v}_1\, f(1)\, d\underline{v}_1 = 0 \quad ,$$

$$\frac{\partial}{\partial \underline{r}_1} \cdot \int \underline{v}_1 f_2(1,2)\, d\underline{v}_1 = 0 \quad ,$$

$$\frac{\partial}{\partial \underline{r}_1} \cdot \int \underline{v}_1 f_3(1,2,3)\, d\underline{v}_1 = 0 \quad ,$$

and so on.

These properties will be automatically satisfied by solutions of the hierarchy which initially have these properties.

CLOSURE HYPOTHESIS

Since the complete hierarchy of equations presented in the last section is equivalent to the Navier-Stokes equation it is apparant that some kind of truncation or closure is necessary in order to get a tractable problem. In statistical mechanics, from which we draw inspiration, the closure is often made at the three point level. That is, the three particle distribution function is approximated in terms of lower order distribution functions. In the kinetic theory of liquids the superposition approximation or Kirkwood closure,

$$f_3(1,2,3) = \frac{f_2(1,2)\, f_2(1,3)\, f_2(2,3)}{f(1)\, f(2)\, f(3)}$$

is often used with distribution functions which have different properties than our functions (general reference, Rice and Gray (1965)). Another example, used for ionized gasses, is

$$f_3'(1,2,3) = 0 \quad ,$$

a linearized form of the Kirkwood closure. We propose to close at this level also, since the resulting equations contain most of the information desired about turbulence.

There are certain ground rules governing the choice of a closure.

Each of the four properties discussed in the introduction (reduction, separation, coincidence and divergence) must be satisfied by the three point distribution function in order for the resulting lower order equations to be consistent. Consider the Kirkwood closure. It is readily seen that the separation property is satisfied. However none of the other three properties hold, not even the reduction property. Therefore the Kirkwood closure is not a good candidate. The linearized Kirkwood closure must also be ruled out. It satisfies the separation, reduction and divergence properties but not the coincidence property.

We propose the following formal closure

$$f_3'(1,2,3) = F_3'(1,2,3) \tag{5}$$

or

$$f_3(1,2,3) - f(1)f_2(2,3) - f(2)f_2(1,3) - f(3)f_2(1,2) + 2f(1)f(2)f(3)$$
$$= F_3(1,2,3) - F(1)F_2(2,3) - F(2)F_2(1,3) - F(3)F_2(1,2) + 2F(1)F(2)F(3)$$

where F, F_2, F_3 are normal and joint normal distribution functions constructed with the correct velocity correlation tensor. These distribution functions are most easily expressed in terms of their characteristic functions Φ, Φ_2, Φ_3 defined by

$$F(1) = \frac{1}{(2\pi)^3}\int e^{i\underline{p}_1\cdot\underline{v}_1}\,\Phi(1)\,d\underline{p}_1$$

$$F_2(1,2) = \frac{1}{(2\pi)^6}\int e^{i\underline{p}_1\cdot\underline{v}_1 + i\underline{p}_2\cdot\underline{v}_2}\,\Phi_2(1,2)\,d\underline{p}_1\,d\underline{p}_2$$

$$F_3(1,2,3) = \frac{1}{(2\pi)^9}\int e^{i\underline{p}_1\cdot\underline{v}_1 + i\underline{p}_2\cdot\underline{v}_2 + i\underline{p}_3\cdot\underline{v}_3}\,\Phi_3(1,2,3)\,d\underline{p}_1\,d\underline{p}_2\,d\underline{p}_3$$

$$\Phi(1) = \exp\left\{-\frac{1}{2}\,\underline{p}_1\cdot\underline{R}(1,1)\cdot\underline{p}_1\right\}$$

$$\Phi_2(1,2) = \exp\left\{-\frac{1}{2}\underline{p}_1\cdot\underline{R}(1,1)\cdot\underline{p}_1 - \frac{1}{2}\underline{p}_2\cdot\underline{R}(2,2)\cdot\underline{p}_2 - \underline{p}_1\cdot\underline{R}(1,2)\cdot\underline{p}_2\right\}$$

$$\begin{aligned}\Phi_3(1,2,3) = \exp\Big\{&-\frac{1}{2}\underline{p}_1\cdot\underline{R}(1,1)\cdot\underline{p}_1 - \frac{1}{2}\underline{p}_2\cdot\underline{R}(2,2)\cdot\underline{p}_2 \\ &-\frac{1}{2}\underline{p}_3\cdot\underline{R}(3,3)\cdot\underline{p}_3 - \underline{p}_1\cdot\underline{R}(1,2)\cdot\underline{p}_2 \\ &-\underline{p}_1\cdot\underline{R}(1,3)\cdot\underline{p}_3 - \underline{p}_2\cdot\underline{R}(2,3)\cdot\underline{p}_3\Big\}\end{aligned}$$

where the correlation tensors $\underline{R}(i,j)$ are defined by

$$\begin{aligned}\underline{R}(i,j) &= \langle \underline{v}'(\underline{r}_i,t)\,\underline{v}'(\underline{r}_j,t)\rangle \\ &= \int (\underline{v}_i - \underline{u}(\underline{r}_i,t))(\underline{v}_j - \underline{u}(\underline{r}_j,t))\, f_2(i,j)\, d\underline{v}_i\, d\underline{v}_j\end{aligned}$$

The functions F, F_2, F_3 have all of the required properties if $\underline{R}$ has zero divergence and tends to zero as its two arguments are separated to infinity. The proposed closure satisfies exactly the separation, reduction and divergence properties but only approximately the coincidence property. This latter property would be exactly satisfied if the one point distribution function were normal and the two point distribution function joint normal and the error is small as long as deviations from normality are small. That is, it is a better approximation for nearly normal turbulence.

With this closure the resulting equations produce a complete theory of turbulence with an interesting structure. Equation (2) is a linear equation for f_2' which must be solved with appropriate initial and boundary conditions as if f and $\underline{R}$ were known functions. The resulting solution is a functional of these quantities, which are then determined by substituting the solution into Eq. (1), giving an equation for $f(1)$, and into the defining equation for $\underline{R}$,

$$\underline{R}(1,2) = \int (\underline{v}_1 - \underline{u}(\underline{r}_1,t))(\underline{v}_2 - \underline{u}(\underline{r}_2,t)) f_2(1,2)\, d\underline{v}_1\, d\underline{v}_2 \tag{6}$$

which gives a nonlinear integral equation for $\underline{R}$. Both of these equations are coupled in general. The equation for f is like a Boltzmann or Fokker-Planck equation.

ISOTROPIC HOMOGENEOUS TURBULENCE

Formulation. The general problem described in the last paragraph is much simpler for isotropic, homogeneous turbulence not only because of the simplified equations but also because in the infinite domain it is possible to use Fourier transforms. Even so, we cannot solve the complete problem but will restrict to large wave numbers.

For isotropic homogeneous turbulence the correlation tensor depends only on the difference between its two arguments, and time,

$$\underline{R}(1,2) = \underline{R}(\underline{r}_2 - \underline{r}_1)$$

further, because of incompressibility, it is expressible in terms of a single scalar function $h(|\underline{r}_2-\underline{r}_1|)$, the longitudinal correlation function, as

$$\underline{R}(\underline{\rho}) = \underline{R}(-\underline{\rho}) = U^2\left[\left(h(\rho) + \tfrac{1}{2}\rho h'(\rho)\right)\underline{I} - \tfrac{1}{2}\rho h'(\rho)\,\hat{\underline{\rho}}\hat{\underline{\rho}}\right]$$
$$U^2 = R(0)\ , \quad h(0) = 1 \tag{7}$$

where $\frac{3}{2}U^2$ is the turbulent energy per unit mass and $\hat{\underline{\rho}}$ denotes a unit vector. Both h and U also depend on time. The characteristic functions for the normal distributions are more simply expressed now as

$$\Phi(1) = \exp\left\{-\tfrac{1}{2}p_1^2 U^2\right\}$$
$$\Phi_2(1,2) = \exp\left\{-\tfrac{1}{2}p_1^2U^2 - \tfrac{1}{2}p_2^2U^2 - \underline{p}_1\cdot\underline{R}(\underline{r}_2-\underline{r}_1)\cdot\underline{p}_2\right\} \tag{8}$$

Introduce the new variable

$$\tilde{f}_2(\underline{v}_1, \underline{v}_2, \underline{r}_2-\underline{r}_1) = f_2'(1,2) - F_2'(1,2)$$

again suppressing the time dependence. This has the properties

$$\tilde{f}_2(\underline{v}_1, \underline{v}_2, \underline{\rho}) = \tilde{f}_2(\underline{v}_2, \underline{v}_1, -\underline{\rho})$$

$$\int \tilde{f}_2 \, d\underline{v}_2 = 0 \tag{9}$$

$$\frac{\partial}{\partial \underline{\rho}} \cdot \int \underline{v}_2 \tilde{f}_2 \, d\underline{v}_2 = 0$$

Substituting the closure into Eq. (2) (using $\underline{u} = 0$, $\bar{p}$ = const.) gives

$$\begin{aligned}
&\frac{\partial \tilde{f}_2}{\partial t} + (\underline{v}_2 - \underline{v}_1) \cdot \frac{\partial \tilde{f}_2}{\partial \underline{\rho}} \\
&+ \frac{\partial f(1)}{\partial \underline{v}_1} \cdot \Big\{ -\frac{1}{4\pi} \int \Big(\frac{\partial}{\partial \underline{\rho}} \frac{1}{|\underline{\xi} - \underline{\rho}|} \Big) \Big(\underline{v}_3 \cdot \frac{\partial}{\partial \underline{\xi}} \Big)^2 \tilde{f}_2(\underline{v}_2, \underline{v}_3, \underline{\xi}) \, d\underline{\xi} \, d\underline{v}_3 \\
&\qquad + \nu \frac{\partial}{\partial \underline{\rho}} \cdot \frac{\partial}{\partial \underline{\rho}} \int \underline{v}_3 \tilde{f}_2(\underline{v}_3, \underline{v}_2, \underline{\rho}) \, d\underline{v}_3 \Big\} \\
&+ \frac{\partial f(2)}{\partial \underline{v}_2} \cdot \Big\{ -\frac{1}{4\pi} \int \Big(\frac{\partial}{\partial \underline{\rho}} \frac{1}{|\underline{\xi} + \underline{\rho}|} \Big) \Big(\underline{v}_3 \cdot \frac{\partial}{\partial \underline{\xi}} \Big)^2 \tilde{f}_2(\underline{v}_1, \underline{v}_3, \underline{\xi}) \, d\underline{\xi} \, d\underline{v}_3 \\
&\qquad + \nu \frac{\partial}{\partial \underline{\rho}} \cdot \frac{\partial}{\partial \underline{\rho}} \int \underline{v}_3 \tilde{f}_2(\underline{v}_1, \underline{v}_3, \underline{\rho}) \, d\underline{v}_3 \Big\} \\
&= -J_1 - J_2
\end{aligned} \tag{10}$$

where

$$\begin{aligned}
J_1 = {} & \frac{\partial (f(1) - F(1))}{\partial \underline{v}_1} \cdot \Big\{ -\frac{1}{4\pi} \int \Big(\frac{\partial}{\partial \underline{r}_1} \frac{1}{|\underline{r}_1 - \underline{r}_3|} \Big) \Big(\underline{v}_3 \cdot \frac{\partial}{\partial \underline{r}_3} \Big)^2 F_2'(2,3) \, d\underline{r}_3 \, d\underline{v}_3 \\
& + \lim_{\underline{r}_3 \to \underline{r}_1} \nu \frac{\partial}{\partial \underline{r}_3} \cdot \frac{\partial}{\partial \underline{r}_3} \int \underline{v}_3 F_2'(2,3) \, d\underline{v}_3 \Big\}
\end{aligned} \tag{11}$$

$$+ \frac{\partial (f(2) - F(2))}{\partial \underline{v}_2} \cdot \Big\{ \text{same as above with 1 and 2 interchanged} \Big\}$$

$$J_2 = \frac{\partial F_2'(1,2)}{\partial t} + (\underline{V}_2 - \underline{V}_1)\cdot\frac{\partial F_2'(1,2)}{\partial(\underline{r}_2-\underline{r}_1)}$$

$$+\frac{\partial}{\partial \underline{V}_1}\cdot\Big\{-\frac{1}{4\pi}\int\Big(\frac{\partial}{\partial \underline{r}_1}\frac{1}{|\underline{r}_1-\underline{r}_3|}\Big)\Big(\underline{V}_3\cdot\frac{\partial}{\partial \underline{r}_3}\Big)^2 F_3(1,2,3)\,d\underline{r}_3\,d\underline{V}_3$$

$$+\lim_{\underline{r}_3\to\underline{r}_1} \nu \frac{\partial}{\partial \underline{r}_3}\cdot\frac{\partial}{\partial \underline{r}_3}\int \underline{V}_3\big(F_3(1,2,3)-F(2)\,F_2(1,3)\big)\,d\underline{V}_3\Big\} \qquad (12)$$

$$\frac{\partial}{\partial \underline{V}_2}\cdot\left\{\begin{matrix}\text{same as above with}\\ \text{1 and 2 interchanged}\end{matrix}\right\}$$

J_2 is the left hand side of Eq. (2) when the distribution functions are joint normal. It would of course be zero if "joint normal" were a solution of Eq. (2). Equation (10) is therefore driven by deviations from normality.

The problem is to solve Eq. (10) with an appropriate initial value, substitute the resulting solution into Eq. (1) to get an equation for $f^{(1)}$ and into the single scalar equation

$$\int \underline{V}_1\cdot\underline{V}_2\,\tilde{f}_2\,d\underline{V}_1\,d\underline{V}_2 = 0 \qquad (13)$$

to get an equation for h. As a consequence of the conditions required of the closure, if the initial value has properties (9) then the solution will also.

The problem can be posed more usefully in terms of the Fourier integral of $\tilde{f}_2$. Let

$$g(\underline{V}_1,\underline{V}_2,\underline{k}) = \frac{1}{(2\pi)^3}\int e^{-i\underline{k}\cdot\underline{\rho}}\,\tilde{f}_2(\underline{V}_1,\underline{V}_2,\underline{\rho})\,d\underline{\rho}$$

which satisfies the conditions (9) in the form

$$g(\underline{V}_1,\underline{V}_2,\underline{k}) = g(\underline{V}_2,\underline{V}_1,-\underline{k}) = g(\underline{V}_2,\underline{V}_1,\underline{k})^*$$

$$\int g\,d\underline{V}_1 = 0 \qquad (14)$$

$$\hat{\underline{k}}\cdot\int \underline{V}_1\,g\,d\underline{V}_1 = 0$$

where * denotes complex conjugate and $\hat{\underline{k}}$ is a unit vector in the direction of $\underline{k}$. The Fourier transform of Eq. (10) is

$$\frac{\partial g}{\partial t} + i\underline{k}\cdot(\underline{v}_2-\underline{v}_1)g$$
$$-i\underline{k}\cdot\frac{\partial f(1)}{\partial \underline{v}_1}\int(\hat{\underline{k}}\cdot\underline{v}_3)^2 g(\underline{v}_2,\underline{v}_3,\underline{k})^* d\underline{v}_3 - \nu k^2\frac{\partial f(1)}{\partial \underline{v}_1}\cdot\int\underline{v}_3 g(\underline{v}_2,\underline{v}_3,\underline{k})^* d\underline{v}_3$$
$$+i\underline{k}\cdot\frac{\partial f(2)}{\partial \underline{v}_2}\int(\hat{\underline{k}}\cdot\underline{v}_3)^2 g(\underline{v}_1,\underline{v}_3,\underline{k})d\underline{v}_3 - \nu k^2\frac{\partial f(2)}{\partial \underline{v}_2}\cdot\int\underline{v}_3 g(\underline{v}_1,\underline{v}_3,\underline{k})d\underline{v}_3$$
$$= -\frac{1}{(2\pi)^3}\int e^{-i\underline{k}\cdot\underline{\rho}}(J_1+J_2)\,d\underline{\rho} \qquad (15)$$
$$\equiv ik\,G(\underline{v}_1,\underline{v}_2,\underline{k})$$

Large Wave Number. The problem stated, while linear, is still difficult because of the time dependent coefficients $f(1)$ and $f(2)$. If these were independent of time one could solve the initial value problem by using Laplace transforms. We will simplify things at this stage by restricting to large values of the wave number, k, which occurs in Eq. (15) only as a parameter, premultiplying every term but the time derivative. When k is large we may neglect the time derivative after a short initial transient and treat the quasi-steady problem. The solution will depend on time then through $f(1)$ and $f(2)$ which may be expected to have characteristic time L/U, L a characteristic length. (L would be expected to be the integral scale.) The neglected time derivative may then be estimated and is small if $kL \gg 1$. We will retain a small remnant of the time derivative for analytical convenience, writing

$$(\hat{\underline{k}}\cdot(\underline{v}_2-\underline{v}_1)-i\delta)\,g(\underline{v}_1,\underline{v}_2,\underline{k})$$
$$-\hat{\underline{k}}\cdot\frac{\partial f(1)}{\partial \underline{v}_1}\int(\hat{\underline{k}}\cdot\underline{v}_3)^2 g(\underline{v}_2,\underline{v}_3,\underline{k})^* d\underline{v}_3 + i\nu k\frac{\partial f(1)}{\partial \underline{v}_1}\cdot\int\underline{v}_3 g(\underline{v}_2,\underline{v}_3,\underline{k})^* d\underline{v}_3$$
$$+\hat{\underline{k}}\cdot\frac{\partial f(2)}{\partial \underline{v}_2}\int(\hat{\underline{k}}\cdot\underline{v}_3)^2 g(\underline{v}_1,\underline{v}_3,\underline{k})d\underline{v}_3 + i\nu k\frac{\partial f(2)}{\partial \underline{v}_2}\cdot\int\underline{v}_3 g(\underline{v}_1,\underline{v}_3,\underline{k})dv_3$$
$$= G(\underline{v}_1,\underline{v}_2,\underline{k}) \qquad (16)$$

where we intend to take the limit as $\delta \to 0$.

To solve Eq. (16) we will derive equations for the integrals which occur in it. To this end, define

$$\mathcal{S}(\underline{v}_1, \underline{k}) = \int (\hat{\underline{k}} \cdot \underline{v}_3)^2 g(\underline{v}_1, \underline{v}_3, \underline{k})\, d\underline{v}_3 \quad ,$$

which has properties

$$\begin{aligned} &\mathcal{S}(\underline{v}_1, -\underline{k}) = \mathcal{S}(v, \underline{k})^* \\ &\int \mathcal{S}(\underline{v}_1, \underline{k})\, d\underline{v}_1 = 0 \\ &\int \underline{v}_1 \cdot \hat{\underline{k}}\, \mathcal{S}(\underline{v}_1, \underline{k})\, d\underline{v}_1 = 0 \end{aligned} \tag{17}$$

Now divide Eq. (16) by $\hat{\underline{k}} \cdot (\underline{v}_2 - \underline{v}_1) - i\delta$, multiply by $(\underline{v}_2 \cdot \hat{\underline{k}})^2$ and integrate on $\underline{v}_2$. This gives

$$\begin{aligned} &\mathcal{S}(\underline{v}_1, \underline{k}) - \hat{\underline{k}} \cdot \frac{\partial f(1)}{\partial \underline{v}_1} \int \frac{(\underline{v}_2 \cdot \hat{\underline{k}})^2 \mathcal{S}(\underline{v}_2, \underline{k})^*}{\hat{\underline{k}} \cdot (\underline{v}_2 - \underline{v}_1) - i\delta}\, d\underline{v}_2 + \int \frac{(\underline{v}_2 \cdot \hat{\underline{k}})^2 \hat{\underline{k}} \cdot \partial f(2)/\partial \underline{v}_2}{\hat{\underline{k}} \cdot (\underline{v}_2 - \underline{v}_1) - i\delta}\, d\underline{v}_2\, \mathcal{S}(\underline{v}_1, \underline{k}) \\ &= \int \frac{(\underline{v}_2 \cdot \hat{\underline{k}})^2 G(\underline{v}_1, \underline{v}_2, \underline{k})}{\hat{\underline{k}} \cdot (\underline{v}_2 - \underline{v}_1) - i\delta}\, d\underline{v}_2 \quad , \end{aligned} \tag{18}$$

two integrals involving $\int \underline{v}_3\, g\, d\underline{v}_3$ vanishing by isotropy and the divergence condition. Upon using conditions (17) plus $\int f(1)\, d\underline{v}_1 = 1$, $\int G\, d\underline{v}_1 = 0$ and $\int \underline{v}_1 \cdot \hat{\underline{k}}\, G\, d\underline{v}_1 = 0$, Eq. (18) simplifies to

$$\begin{aligned} &-\hat{\underline{k}} \cdot \frac{\partial f(1)}{\partial \underline{v}_1} \int \frac{\mathcal{S}(\underline{v}_2, \underline{k})^*}{\hat{\underline{k}} \cdot (\underline{v}_2 - \underline{v}_1) - i\delta}\, d\underline{v}_2 + \int \frac{\hat{\underline{k}} \cdot \partial f(2)/\partial \underline{v}_2}{\hat{\underline{k}} \cdot (\underline{v}_2 - \underline{v}_1) - i\delta}\, d\underline{v}_2\, \mathcal{S}(\underline{v}_1, \underline{k}) \\ &= \int \frac{G(\underline{v}_1, \underline{v}_2, \underline{k})}{\hat{\underline{k}} \cdot (\underline{v}_2 - \underline{v}_1) - i\delta}\, d\underline{v}_2 \end{aligned} \tag{19}$$

This may be written as a one dimensional integral equation by introducing

$$\mathscr{S} = \int \delta(\underline{v}_1 \cdot \hat{\underline{k}} - u_1)\, \mathcal{S}(\underline{v}_1, \underline{k})\, d\underline{v}_1 \quad ,$$

formed by integrating $\mathcal{S}$ over a plane perpendicular to $\hat{\underline{k}}$ - a plane mean. Also define

$$\mathcal{F}(u_1) = \int \delta(\underline{v}_1 \cdot \hat{\underline{k}} - u_1)\, f(1)\, d\underline{v}_1 \quad .$$

We will use

$$\int \delta(\underline{v}_1 \cdot \hat{\underline{k}} - u_1)\, \hat{\underline{k}} \cdot \partial f(1)/\partial \underline{v}_1\, d\underline{v}_1 = \partial \mathcal{F}(u_1)/\partial u_1 \quad .$$

Now multiply Eq. (19) by $\delta(\underline{v}_1 \cdot \hat{\underline{k}} - u_1)$ and integrate, getting

$$-\frac{\partial \mathcal{F}(u_1)}{\partial u_1}\int \frac{\mathcal{S}(u_2)^*}{u_2-u_1-i\delta}\,du_2 + \int \frac{\partial \mathcal{F}(u_2)/\partial u_2}{u_2-u_1-i\delta}\,du_2\,\mathcal{S}(u_1)$$

$$= \int \frac{\delta(\underline{v}_1\cdot\hat{\underline{k}}-u_1)\,G(\underline{v}_1,\underline{v}_2,\underline{k})}{\hat{\underline{k}}\cdot(\underline{v}_2-\underline{v}_1)-i\delta}\,d\underline{v}_1 \tag{20}$$

$$\equiv \varphi(u_1)$$

The function $\mathcal{S}$ may be recovered by eliminating $\int \mathcal{S}(u_2)^*/(u_2-u_1-i\delta)\,du_2$ between Eq. (19) and Eq. (20). Finally Eq. (20) may be reduced to a pair of singular integral equations by taking the limit as $\delta \to 0$, using the Plemelj formula (Muskhelishvili (1953)),

$$\lim_{\delta\to 0}\int \frac{\mathcal{S}(u_2)^*}{u_2-u_1-i\delta}\,du_2 = \pi i\,\mathcal{S}(u_1)^* + P\int \frac{\mathcal{S}(u_2)^*}{u_2-u_1}\,du_2$$

where P means principal value, and taking real and imaginary parts of the result. This gives

$$\mathrm{Im}\,\mathcal{S}(u_1)\,P\int \frac{\partial \mathcal{F}(u_2)/\partial u_2}{u_2-u_1}\,du_2 + \frac{\partial \mathcal{F}(u_1)}{\partial u_1}P\int \frac{\mathrm{Im}\,\mathcal{S}(u_2)}{u_2-u_1}\,du_2 = \mathrm{Im}\,\varphi(u_1) \tag{21}$$

$$\mathrm{Re}\,\mathcal{S}(u_1)\,P\int \frac{\partial \mathcal{F}(u_2)/\partial u_2}{u_2-u_1}\,du_2 - \frac{\partial \mathcal{F}(u_1)}{\partial u_1}P\int \frac{\mathrm{Re}\,\mathcal{S}(u_2)}{u_2-u_1}\,du_2 = \mathrm{Re}\,\varphi(u_1) \tag{22}$$

These independent singular integral equations may be solved by complex function theory methods described by Muskhelishvili (1953). The side conditions $\int \mathcal{S}(u_1)\,du_1 = \int u_1\mathcal{S}(u_1)\,du_1 = 0$ are needed for uniqueness.

The solution is required in order to get $\mathcal{S}$ for substitution into Eq. (16) for g. It is also needed directly for one of the integrals in Eq. (2). However, since we have restricted to large wave number it gives only part of the contribution to this integral, not a dominant part. For this reason we will not pursue the solution further here.

The other kind of integral which occurs in Eq. (16), $\int \underline{v}_3\, g\, d\underline{v}_3$, is needed for the viscous integral in Eq. (1). Again it does not give a dominant contribution. However, it is also required to get an equation for the correlation function, an equation which we do wish to pursue. To get an integral equation for this quantity, divide Eq. (16) by $\hat{\underline{k}}\cdot(\underline{v}_2-\underline{v}_1)-i\delta$, multiply by $\underline{v}_2$ and integrate on $\underline{v}_2$. This gives

$$\int \underline{v}_2 g(\underline{v}_1,\underline{v}_2,\underline{k})\,d\underline{v}_2 - \hat{\underline{k}}\cdot\frac{\partial f(1)}{\partial \underline{v}_1}\int \frac{\underline{v}_2 S(\underline{v}_2,\underline{k})^*}{\hat{\underline{k}}\cdot(\underline{v}_2-\underline{v}_1)-i\delta}\,d\underline{v}_2 + \int \frac{\underline{v}_2\,\hat{\underline{k}}\cdot\partial f(2)/\partial \underline{v}_2}{\hat{\underline{k}}\cdot(\underline{v}_2-\underline{v}_1)-i\delta}\,d\underline{v}_2\, S(\underline{v}_1,\underline{k})$$

$$+ i\nu k\,\frac{\partial f(1)}{\partial \underline{v}_1}\cdot\int \frac{\int \underline{v}_3 g(\underline{v}_3,\underline{v}_2,\underline{k})\,d\underline{v}_3}{\hat{\underline{k}}\cdot(\underline{v}_2-\underline{v}_1)-i\delta}\,\underline{v}_2\,d\underline{v}_2 + i\nu k\int \frac{\underline{v}_2\,\partial f(2)/\partial \underline{v}_2}{\hat{\underline{k}}\cdot(\underline{v}_2-\underline{v}_1)-i\delta}\,d\underline{v}_2\cdot\int \underline{v}_3 g(\underline{v}_1,\underline{v}_3,\underline{k})\,d\underline{v}_3$$

$$= \int \frac{\underline{v}_2\, G(\underline{v}_1,\underline{v}_2,\underline{k})}{\hat{\underline{k}}\cdot(\underline{v}_2-\underline{v}_1)-i\delta}\,d\underline{v}_2 \qquad (23)$$

By isotropy the integrals involving S,

$$\int \frac{\underline{v}_2\, S^*(\underline{v}_2,\underline{k})}{\hat{\underline{k}}\cdot(\underline{v}_2-\underline{v}_1)-i\delta}\,d\underline{v}_2$$

and

$$\int \frac{\underline{v}_2\,\hat{\underline{k}}\cdot\partial f(2)/\partial \underline{v}_2}{\hat{\underline{k}}\cdot(\underline{v}_2-\underline{v}_1)-i\delta}\,d\underline{v}_2$$

are both proportional to $\hat{\underline{k}}$. Therefore, subtracting from Eq. (23) its projection onto the $\hat{\underline{k}}$ direction eliminates these two integrals. Using $\int \hat{\underline{k}}\cdot\underline{v}_3\, g\, d\underline{v}_3 = 0$, this gives

$$\int \underline{v}_2 g(\underline{v}_1,\underline{v}_2,\underline{k})\,d\underline{v}_2 + i\nu k\,\frac{\partial f(1)}{\partial \underline{v}_1}\cdot\int \frac{\int \underline{v}_3 g(\underline{v}_3,\underline{v}_2,\underline{k})\,d\underline{v}_3}{\hat{\underline{k}}\cdot(\underline{v}_2-\underline{v}_1)-i\delta}(\underline{v}_2-\underline{v}_2\cdot\hat{\underline{k}}\hat{\underline{k}})\,d\underline{v}_2$$

$$+ i\nu k\int \frac{(\underline{v}_2-\underline{v}_2\cdot\hat{\underline{k}}\hat{\underline{k}})\,\partial f(2)/\partial \underline{v}_2}{\hat{\underline{k}}\cdot(\underline{v}_2-\underline{v}_1)-i\delta}\,d\underline{v}_2\cdot\int \underline{v}_3 g(\underline{v}_1,\underline{v}_3,\underline{k})\,d\underline{v}_3 \qquad (24)$$

$$= \int \frac{(\underline{v}_2-\underline{v}_2\cdot\hat{\underline{k}}\hat{\underline{k}})\, G(\underline{v}_1,\underline{v}_2,\underline{k})}{\hat{\underline{k}}\cdot(\underline{v}_2-\underline{v}_1)-i\delta}\,d\underline{v}_2$$

The equation can be further simplified by using

$$\int \frac{(\underline{v}_2-\underline{v}_2\cdot\hat{\underline{k}}\hat{\underline{k}})\,\partial f(2)/\partial \underline{v}_2}{\hat{\underline{k}}\cdot(\underline{v}_2-\underline{v}_1)-i\delta}\,d\underline{v}_2 = -(\underline{I}-\hat{\underline{k}}\hat{\underline{k}})\int \frac{\mathcal{F}(u_2)}{u_2-\hat{\underline{k}}\cdot\underline{v}_1-i\delta}\,du_2$$

which follows by integration by parts and isotropy, and

$$\int \frac{\int \underline{v}_3\, g(\underline{v}_2,\underline{v}_3,\underline{k})^* d\underline{v}_3}{\hat{\underline{k}}\cdot(\underline{v}_2-\underline{v}_1)-i\delta}(\underline{v}_2-\underline{v}_2\cdot\hat{\underline{k}}\hat{\underline{k}})\,d\underline{v}_2$$

$$=\frac{1}{2}(\underline{I}-\hat{\underline{k}}\hat{\underline{k}})\int \frac{\int \underline{v}_2\cdot\underline{v}_3\, g(\underline{v}_2,\underline{v}_3,\underline{k})^* d\underline{v}_3}{\hat{\underline{k}}\cdot(\underline{v}_2-\underline{v}_1)-i\delta}\,d\underline{v}_2 \tag{25}$$

which follows by isotropy and $\hat{\underline{k}}\cdot\int \underline{v}_3 g\, d\underline{v}_3 = 0$ since these imply $\int \underline{v}_3 g(\underline{v}_2,\underline{v}_3,\underline{k})d\underline{v}_3 = (\underline{v}_2-\underline{v}_2\cdot\hat{\underline{k}}\hat{\underline{k}})\,M(\underline{v}_2,\underline{v}_2,\hat{\underline{k}})$. We can then write

$$\int \underline{v}_2 g(\underline{v}_1,\underline{v}_2,\underline{k})d\underline{v}_2\left(1-i\nu k\int\frac{\mathcal{F}(u_2)}{u_2-\hat{\underline{k}}\cdot\underline{v}_1-i\delta}du_2\right)$$

$$+\, i\nu k\frac{\partial f(1)}{\partial \underline{v}_1}\cdot(\underline{I}-\hat{\underline{k}}\hat{\underline{k}})\frac{1}{2}\int\frac{\int \underline{v}_2\cdot\underline{v}_3\, g(\underline{v}_2,\underline{v}_3,\underline{k})^* d\underline{v}_3}{\hat{\underline{k}}\cdot(\underline{v}_2-\underline{v}_1)-i\delta}d\underline{v}_2 \tag{26}$$

$$=\int\frac{(\underline{v}_2-\underline{v}_2\cdot\hat{\underline{k}}\hat{\underline{k}})\,G(\underline{v}_1,\underline{v}_2,\underline{k})}{\hat{\underline{k}}\cdot(\underline{v}_2-\underline{v}_1)-i\delta}d\underline{v}_2$$

an equation for $\int \underline{v}_2 g\, d\underline{v}_2$ alone. This may be reduced to a scalar equation by scalar multiplication by $\underline{v}_1$ and to a one dimensional equation by defining

$$\mathcal{N}(u_1)=\iint \underline{v}_1\cdot\underline{v}_2\, g(\underline{v}_1,\underline{v}_2,\underline{k})\,d\underline{v}_2\,\delta(\underline{v}_1\cdot\hat{\underline{k}}-u_1)\,d\underline{v}_1 ,$$

multiplying by $\delta(\underline{v}_1\cdot\hat{\underline{k}}-u_1)$ and integrating on $\underline{v}_1$. The result

$$\mathcal{N}(u_1)\left(1-i\nu k\int\frac{\mathcal{F}(u_2)}{u_2-u_1-i\delta}du_2\right.$$

$$-\,i\nu k\,\mathcal{F}(u_1)\int\frac{\mathcal{N}(u_2)^*}{u_2-u_1-i\delta}du_2 \tag{27}$$

$$=\int\frac{(\underline{v}_1\cdot\underline{v}_2-\underline{v}_1\cdot\hat{\underline{k}}\,\underline{v}_2\cdot\hat{\underline{k}})\,G(\underline{v}_1,\underline{v}_2,\underline{k})\,\delta(\underline{v}_1\cdot\hat{\underline{k}}-u_1)}{\hat{\underline{k}}\cdot(\underline{v}_2-\underline{v}_1)-i\delta}d\underline{v}_1 d\underline{v}_2 ,$$

yields a pair of singular integral equations for $\mathrm{Re}\,\mathcal{N}$ and $\mathrm{Im}\,\mathcal{N}$ upon taking the limit as $\delta\to 0$. These are more difficult than the pair given by Eq. (21) and Eq. (22) because $\mathrm{Re}\,\mathcal{N}$ and $\mathrm{Im}\,\mathcal{N}$ do not decouple, but occur in both equations. In fact it is not known at this time how to solve these equations.

There is, however, a justifiable approximation which can be made at this stage. The viscous terms in Eq. (27) or Eq. (26) may be neglected if $\nu k/U \ll 1$. While k is supposed to be large, this means that it must not be too large. Later we will scale $1/k$ with the Kolmogorov length,

$$\ell = \nu^{3/4}/\varepsilon^{1/4}$$

where ε is the dissipation per unit mass. If we define a length L by $\varepsilon = U^3/L$ then the inequality above may be written

$$\frac{\nu k}{U} = \frac{k\ell}{\ell U/\nu} = \frac{k\ell}{(UL/\nu)^{1/4}} \ll 1 \quad .$$

Therefore if $k\ell = O(1)$ the inequality will be satisfied if the turbulent Reynolds number UL/ν is large. Upon dropping these viscous terms Eq. (26) gives

$$\int \underline{v}_2\, g(\underline{v}_1, \underline{v}_2, \underline{k})\, d\underline{v}_2 = \int \frac{(\underline{v}_2 - \underline{v}_2\cdot\hat{\underline{k}}\,\hat{\underline{k}})\, G(\underline{v}_1, \underline{v}_2, \underline{k})}{\hat{\underline{k}}\cdot(\underline{v}_2 - \underline{v}_1) - i\delta}\, d\underline{v}_2 \tag{28}$$

the desired result.

<u>An Equation for the Correlation Function</u>. When this last result is used with the condition to determine the correlation function, Eq. (13), we get

$$0 = \int \frac{(\underline{v}_1\cdot\underline{v}_2 - \underline{v}_1\cdot\hat{\underline{k}}\,\underline{v}_2\cdot\hat{\underline{k}})\, G(\underline{v}_1, \underline{v}_2, \underline{k})}{\hat{\underline{k}}\cdot(\underline{v}_2 - \underline{v}_1) - i\delta}\, d\underline{v}_1\, d\underline{v}_2 \tag{29}$$

where, the reader is reminded, G is a functional of the correlation function defined by Eq. (15) and Eqs. (11) and (12). The dependence on the correlation function is more easily extracted if characteristic functions are used. Toward this end define a characteristic function for G by

$$G(\underline{v}_1, \underline{v}_2, \underline{k}) = \frac{1}{(2\pi)^6} \int e^{i\underline{p}_1\cdot\underline{v}_1 + i\underline{p}_2\cdot\underline{v}_2}\, \mathrm{H}(\underline{p}_1, \underline{p}_2, \underline{k})\, d\underline{p}_1\, d\underline{p}_2 \tag{30}$$

Substitute this into Eq. (29) getting

$$0 = (\underline{I} - \hat{\underline{k}}\hat{\underline{k}}) : \int \frac{\partial}{\partial \underline{p}_1} \frac{\partial}{\partial \underline{p}_2} H(\underline{p}_1, \underline{p}_2, \underline{k}) \left\{ \frac{1}{(2\pi)^6} \int \frac{e^{i\underline{p}_1 \cdot \underline{v}_1 + i\underline{p}_2 \cdot \underline{v}_2}}{\hat{\underline{k}} \cdot (\underline{v}_2 - \underline{v}_1) - i\delta} dv_1 dv_2 \right\} d\underline{p}_1 d\underline{p}_2$$

In Appendix A it is shown that as $\delta \to 0$ the bracketed quantity above tends to

$$i\,\delta(\underline{p}_{1\perp})\,\delta(\underline{p}_{2\perp})\,\delta(\underline{p}_1 \cdot \hat{\underline{k}} + \underline{p}_2 \cdot \hat{\underline{k}})\,H(\underline{p}_2 \cdot \hat{\underline{k}})$$

where $\delta(\underline{p}_{1\perp})$ and $\delta(\underline{p}_{2\perp})$ are two dimensional delta functions whose arguments are the projections of $\underline{p}_1$ and $\underline{p}_2$ onto the plane perpendicular to $\hat{\underline{k}}$ and $H(\underline{p}_2 \cdot \hat{\underline{k}})$ is Heaviside's step function. The resulting equation for the correlation function is

$$0 = (\underline{I} - \hat{\underline{k}}\hat{\underline{k}}) : \int_0^\infty \left\{ \frac{\partial}{\partial \underline{p}_1} \frac{\partial}{\partial \underline{p}_2} H(\underline{p}_1, \underline{p}_2, \underline{k}) \right\}_{\substack{\underline{p}_1 = -p\hat{\underline{k}} \\ \underline{p}_2 = p\hat{\underline{k}}}} dp \tag{31}$$

The operations required to compute Eq. (31) are straight forward but lengthy. One must substitute the joint normal functions given by Eq. (8) into the function G defined by Eq. (15) and Eqs. (11) and (12), carry out the integrations needed to compute H from the Fourier inverse of Eq. (30) and then substitute this into Eq. (31). After differentiating with respect to $\underline{p}_1$ and $\underline{p}_2$ substitute $\underline{p}_1 = -p\hat{\underline{k}}$, $\underline{p}_2 = p\hat{\underline{k}}$ and integrate on p. These operations have been carried out. The resulting equation is

$$\int L_1(\underline{\rho}, \underline{k}) \cos \underline{k} \cdot \underline{\rho}\, d\underline{\rho} + \int L_2(\underline{\rho}, \underline{k}) \sin \underline{k} \cdot \underline{\rho}\, d\underline{\rho} = 0 \tag{32}$$

where

$$L_1(\underline{\rho}, \underline{k}) = -k \left\{ 2U^2 - Z_{ii} + \frac{\hat{\underline{k}} \cdot \underline{Z} \cdot \underline{Z} \cdot \hat{\underline{k}}}{\hat{\underline{k}} \cdot \underline{Z} \cdot \hat{\underline{k}}} \right\}$$

$$- \frac{\sqrt{\pi}}{(\hat{\underline{k}} \cdot \underline{Z} \cdot \hat{\underline{k}})^{1/2}} \left\{ \nu \nabla^2 Z_{ii} - \hat{\underline{k}} \cdot \nu \nabla^2 \underline{Z} \cdot \hat{\underline{k}} \right\}$$

$$- \frac{\sqrt{\pi}}{2(\hat{\underline{k}}\cdot\underline{\underline{z}}\cdot\hat{\underline{k}})^{3/2}} \Big\{ (\hat{\underline{k}}\cdot\nu\nabla^2\underline{\underline{z}}\cdot\hat{\underline{k}} - \hat{\underline{k}}\cdot\nu\nabla^2\underline{\underline{z}}\big|_0\cdot\hat{\underline{k}})(2U^2 + 3\hat{\underline{k}}\cdot\underline{\underline{z}}\cdot\hat{\underline{k}} - Z_{ii})$$

$$-2(\hat{\underline{k}}\cdot\nu\nabla^2\underline{\underline{z}} - \hat{\underline{k}}\cdot\nu\nabla^2\underline{\underline{z}}\big|_0)\cdot\underline{\underline{z}}\cdot\hat{\underline{k}}\Big\} \tag{33}$$

$$- \frac{3\sqrt{\pi}}{4(\hat{\underline{k}}\cdot\underline{\underline{z}}\cdot\hat{\underline{k}})^{5/2}} \Big\{ (\hat{\underline{k}}\cdot\nu\nabla^2\underline{\underline{z}}\cdot\hat{\underline{k}} - \hat{\underline{k}}\cdot\nu\nabla^2\underline{\underline{z}}\big|_0\cdot\hat{\underline{k}})(\hat{\underline{k}}\cdot\underline{\underline{z}}\cdot\underline{\underline{z}}\cdot\hat{\underline{k}} - (\hat{\underline{k}}\cdot\underline{\underline{z}}\cdot\hat{\underline{k}})^2)\Big\}$$

and

$$L_2(\underline{\rho}, \underline{k}) = \frac{1}{2\pi}\int \frac{d\underline{s}}{s^2}\left(\mathrm{I} + \mathrm{II} + \mathrm{III} + \mathrm{IV}\right)$$

$$\mathrm{I} = -\frac{\partial}{\partial \underline{s}}\frac{\partial}{\partial \underline{s}} : \Big\{\underline{Q}(\underline{\rho}, \underline{s})\big[\underline{\underline{z}}(\underline{s}-\underline{\rho}) - \underline{\underline{z}}(\underline{s})\big]\Big\}\cdot \frac{\underline{\underline{z}}(\underline{\rho})\cdot\hat{\underline{k}}\,\hat{\underline{k}}\cdot\hat{\underline{s}}}{(\hat{\underline{k}}\cdot\underline{\underline{z}}(\underline{\rho})\cdot\hat{\underline{k}})^2}$$

$$\mathrm{II} = \frac{\partial}{\partial \underline{s}}\frac{\partial}{\partial \underline{s}} : \Big\{\underline{Q}(\underline{\rho}, \underline{s})\,\underline{\underline{z}}(\underline{s}-\underline{\rho})\Big\}\cdot \frac{(\underline{\underline{I}} - \hat{\underline{k}}\hat{\underline{k}})\cdot\hat{\underline{s}}}{\hat{\underline{k}}\cdot\underline{\underline{z}}(\underline{\rho})\cdot\hat{\underline{k}}} \tag{34}$$

$$\mathrm{III} = -\frac{\partial}{\partial \underline{s}}\frac{\partial}{\partial \underline{s}} : \Big\{\underline{\underline{z}}(\underline{s}-\underline{\rho})\cdot\underline{\underline{z}}(\underline{s}) - \hat{\underline{k}}\cdot\underline{\underline{z}}(\underline{s}-\underline{\rho})\,\hat{\underline{k}}\cdot\underline{\underline{z}}(\underline{s})\Big\}\,\frac{\hat{\underline{k}}\cdot\hat{\underline{s}}}{\hat{\underline{k}}\cdot\underline{\underline{z}}\cdot\hat{\underline{k}}}$$

$$\mathrm{IV} = \frac{\partial}{\partial \underline{s}}\frac{\partial}{\partial \underline{s}} : \Big\{\underline{Q}(\underline{\rho}, \underline{s})\,\underline{Q}(\underline{\rho}, \underline{s})\Big\}\Bigg[\frac{\hat{\underline{k}}\cdot\hat{\underline{s}}\;\hat{\underline{k}}\cdot\underline{\underline{z}}(\underline{\rho})\cdot\underline{\underline{z}}(\underline{\rho})\cdot\hat{\underline{k}}}{(\hat{\underline{k}}\;\;\underline{\underline{z}}(\underline{\rho})\cdot\hat{\underline{k}})^3}$$

$$+\frac{1}{2}\,\frac{(-\hat{\underline{k}}\cdot\underline{\underline{z}}(\underline{\rho})\cdot\hat{\underline{s}} + 2\hat{\underline{k}}\cdot\hat{\underline{s}}\,U^2 - \hat{\underline{k}}\cdot\hat{\underline{s}}\,Z_{ii}(\underline{\rho}) + 2\hat{\underline{k}}\cdot\hat{\underline{s}}\,\hat{\underline{k}}\cdot\underline{\underline{z}}(\underline{\rho})\cdot\hat{\underline{k}}}{(\hat{\underline{k}}\cdot\underline{\underline{z}}(\underline{\rho})\cdot\hat{\underline{k}})^2}\Bigg]$$

In these expressions

$$\underline{Q}(\underline{\rho},\underline{s}) = \hat{\underline{k}}\cdot\underline{\underline{z}}(\underline{s}-\underline{\rho}) - \hat{\underline{k}}\cdot\underline{\underline{z}}(\underline{s})$$

$$\underline{\underline{z}}(\underline{\rho}) = U^2\underline{\underline{I}} - \underline{\underline{R}}(\underline{\rho})$$

and

$$\nu\nabla^2\underline{\underline{z}}\Big|_0 = -\nu\nabla^2\underline{\underline{R}}\Big|_0 = \frac{1}{3}\varepsilon\underline{\underline{I}} \tag{35}$$

where ε is the viscous dissipation per unit mass (Batchelor (1960), Hinze (1959)). These are obviously too many terms to cope with and there are even more terms that have not been written down. We have neglected the time derivatives in J_2 (Eq. (12)) and all of J_1 (Eq. (11)). These terms become negligible along with many of those written, when large k is exploited.

The asymptotic behavior of these integrals as k becomes large depends on the behavior of L_1 and L_2 for small ρ, the dominant terms being those which are most singular as $\rho \to 0$. This is well treated in Lighthill's book (Lighthill (1960)). Since $\underline{\underline{z}} \to 0$ as $\rho \to 0$ it is fairly certain that the dominant term from $L_1(\underline{\rho},\underline{k})$ is

$$-\frac{\sqrt{\pi}\,(\hat{\underline{k}}\cdot\nu\nabla^2\underline{\underline{z}}\cdot\hat{\underline{k}} - \hat{\underline{k}}\cdot\nu\nabla^2\underline{\underline{z}}|_0\cdot\hat{\underline{k}})\,U^2}{(\hat{\underline{k}}\cdot\underline{\underline{z}}\cdot\hat{\underline{k}})^{3/2}} \tag{36}$$

However the integrals in $L_2(\underline{\rho},\underline{k})$ make it more difficult to assess these terms.

A more orderly way to pick out the dominant terms is by a formal asymptotic expansion. We note that the parameters U, ν, ε occur in Eq. (32), where ε is defined by Eq. (35). It is assumed that these are independent parameters, that is, ε is not determined by the other two. While $\varepsilon \sim U^4/\nu$ would be dimensionally correct this is not what is observed. It is known experimentally that the length L defined by $\varepsilon = U^3/L$ is of the order of the energy containing eddies and is independent of viscosity. Introduce the Kolmogorov length and velocity scales (Batchelor (1960), Hinze (1959))

$$\ell = \nu^{3/4}/\varepsilon^{1/4}$$

$$v = \nu^{1/4}\varepsilon^{1/4}$$

Scale s, ρ, $1/k$ with ℓ and scale $\underline{Z}$ with v^2. Then $\nu \nabla^2 \underline{Z}$ $\sim \nu v^2/\ell^2 = \varepsilon$ so that $\nu \nabla^2 \underline{Z}$ and $\nu \nabla^2 \underline{Z}|_0$ are of the same order. Further, the parameter U will occur in combination v/U. If we introduce the length L by $U^3/L = \varepsilon$ then

$$\frac{v}{U} = \frac{1}{(UL/\nu)^{1/4}}$$

is small when the Reynolds number is large. It follows that Eq. (36) is the dominant term from $L_1(\underline{\rho}, \underline{k})$ while the dominant term from $L_2(\underline{\rho}, \underline{k})$ is

$$\frac{1}{2\pi} \int \frac{d\underline{s}}{s^2} \left(\frac{\frac{\partial}{\partial \underline{s}} \frac{\partial}{\partial \underline{s}} : \{ \underline{Q}(\underline{\rho}, \underline{s}) \underline{Q}(\underline{\rho}, \underline{s}) \}}{(\hat{\underline{k}} \cdot \underline{Z}(\underline{\rho}) \cdot \hat{\underline{k}})^2} U^2 \hat{\underline{k}} \cdot \hat{\underline{s}} \right) \tag{37}$$

The relative order of the neglected terms is $(UL/\nu)^{-1/2}$. (The terms neglected from Eq. (28) were of relative order $(UL/\nu)^{-1/4}$). Also, the dominant terms from L_1 and L_2 are of the same order, justifying scaling of $\underline{Z}$ with v^2, rather than with v^2 times some power of the Reynolds number. The integral equation is

$$\int \cos \underline{k} \cdot \underline{\rho} \left\{ \frac{\sqrt{\pi}}{2} \frac{\hat{\underline{k}} \cdot \nu \nabla^2 \underline{Z}(\underline{\rho}) \cdot \hat{\underline{k}} - \hat{\underline{k}} \cdot \nu \nabla^2 \underline{Z}(\underline{\rho})|_0 \cdot \hat{\underline{k}}}{(\hat{\underline{k}} \cdot \underline{Z}(\underline{\rho}) \cdot \hat{\underline{k}})^{3/2}} \right\} d\underline{\rho}$$

$$+ \int \sin \underline{k} \cdot \underline{\rho} \left\{ \frac{1}{4\pi} \int \frac{\hat{\underline{k}} \cdot \hat{\underline{s}}}{s^2} \frac{\frac{\partial}{\partial \underline{s}} \frac{\partial}{\partial \underline{s}} : \underline{Q}\,\underline{Q}}{(\hat{\underline{k}} \cdot \underline{Z}(\underline{\rho}) \cdot \hat{\underline{k}})^2} d\underline{s} \right\} d\underline{\rho} \tag{38}$$

$$= 0$$

where $\underline{Q} = \hat{\underline{k}} \cdot \underline{Z}(\underline{s} - \underline{\rho}) - \hat{\underline{k}} \cdot \underline{Z}(\underline{s})$. We have retained the dimensional form rather than define new dimensionless symbols. These integrals are not strictly Fourier integrals because $\hat{\underline{k}}$ occurs in the integrand, not merely in the sinusoidal part.

While we began by thinking in terms of an asymptotic expansion for large k, Eq. (38) has in fact been obtained by taking the limit as $(UL/\nu) \to \infty$ with $k\ell$ fixed, as a parameter expansion. This should be regarded as the first term in an inner expansion for the longitudinal correlation function $h(\rho)$, related to $\underline{\underline{Z}}$ by

$$\underline{\underline{Z}}(\underline{\rho}) = U^2((1-h(\rho)-\tfrac{1}{2}\rho h'(\rho))\underline{\underline{I}} + \tfrac{1}{2}\rho h'(\rho)\hat{\underline{\rho}}\hat{\underline{\rho}}) \tag{39}$$

$$h(0) = 1 , \qquad \varepsilon = -15\nu U^2 h''(0) .$$

When functions which should be solutions of this equation are substituted, the integrals are found to be divergent, in the ordinary sense, at large ρ . This is easily seen to occur because in taking the limit as $UL/\nu \to \infty$ we dropped natural convergence factors, the limit being improperly taken inside the integral. This difficulty may be eliminated by introducing artificial convergence factors $e^{-\lambda\rho}$ in each integral and then taking the limit $\lambda \to 0$ after the integrals are evaluated. We will not actually do this. It is equivalent to consider the integrands as generalized functions as described by Lighthill (1960). We will evaluate certain integrals using tables provided in his book.

Equation (38) is actually in its simplest looking form. One may appreciate the complexity which will result when $\underline{\underline{Z}}$, given by Eq. (39), is substituted. Some of the intermediate integrations can be performed. In the first integral, substitute spherical coordinates for $\underline{\rho}$ with $\hat{\underline{k}}$ the polar direction, $\hat{\underline{k}}\cdot\hat{\underline{\rho}} = \cos\theta$, and carry out the integration on the azimuthal angle. In the second integral, in the inner integration on $\underline{s}$, substitute

$$\hat{\underline{s}} = \hat{\underline{\rho}}\cos\alpha + \sin\alpha\,(\hat{\underline{i}}\cos\beta + \hat{\underline{j}}\sin\beta)$$

$$\hat{\underline{i}} = \frac{\hat{\underline{\rho}}\times\hat{\underline{k}}}{\sin\theta}$$

$$\hat{\underline{j}} = \hat{\underline{\rho}}\times\hat{\underline{i}} = \frac{\hat{\underline{\rho}}\cos\theta - \hat{\underline{k}}}{\sin\theta}$$

$$d\underline{s} = s^2 ds \sin\alpha\, d\alpha\, d\beta$$

and carry out the integration on β . Then substitute spherical coordinates for $\underline{\rho}$ and integrate over the azimuthal angle, denoting

$\cos\theta = x$, $\cos\alpha = y$. The resulting integral equation may be written

$$\int \cos k\ell x \, M_1(\rho, x)\, \rho^2 d\rho\, dx + \int \sin k\rho x \, M_2(\rho, x)\, \rho^2 d\rho\, dx = 0 \quad (40)$$

with

$$M_1(\rho, x) =$$

$$-\frac{\pi^{3/2}}{2}\left\{\frac{-\frac{2}{3}\varepsilon + 2\nu U^2\left[\left(-\frac{2h'(\rho)}{\rho} - 3h''(\rho) - \frac{1}{2}\rho h'''(\rho)\right) + \left(-\frac{2h'(\rho)}{\rho} + 2h''(\rho) + \frac{1}{2}\rho h'''(\rho)\right)x^2\right]}{U^3\left[1 - h(\rho) - \frac{1}{2}\rho h'(\rho) + \frac{1}{2}\rho h'(\rho) x^2\right]^{3/2}}\right\} \quad (41)$$

$$M_2(\rho, x) = \frac{\int_0^\infty P_1(s, x)\, ds - \int_0^\infty P_2(s, x)\, ds}{\left[1 - h(\rho) - \frac{1}{2}\rho h'(\rho) + \frac{1}{2}\rho h'(\rho) x^2\right]^2} \quad (42)$$

where

$$P_1(s, x) = \frac{1}{4} h'(s)^2 \left[T_{11}(\rho, s)\, x + T_{12}(\rho, s)\, x^3\right] + \frac{1}{4} s h'(s) h''(s) \left[T_{21}(\rho, s)\, x + T_{22}(\rho, s)\, x^3\right]$$

$$\begin{aligned} P_2(s, x) = \int_{-1}^{1} dy \Big\{ & \tfrac{1}{4} h'(\sigma) h'(s) \left[S_{11}(\rho, s, y)\, x + S_{12}(\rho, s, y)\, x^3\right] \\ & + \tfrac{1}{4} h'(\sigma)\, s h''(s) \left[S_{21}(\rho, s, y)\, x + S_{22}(\rho, s, y)\, x^3\right] \\ & + \tfrac{1}{4} \sigma h''(\sigma) h'(s) \left[S_{31}(\rho, s, y)\, x + S_{32}(\rho, s, y)\, x^3\right] \\ & + \tfrac{1}{4} \sigma h''(\sigma)\, s h''(s) \left[S_{41}(\rho, s, y)\, x + S_{42}(\rho, s, y)\, x^3\right] \Big\} \end{aligned}$$

and $\sigma = (s^2 + \rho^2 - 2s\rho y)^{1/2}$. The T functions and S functions, simple functions of ρ, s and ρ, s, y respectively, are recorded in Appendix B.

The Kolmogorov Spectrum. The integral equation presented in the last section has a simple solution in the nonviscous limit. This limit may be motivated as follows. Equation (38) was derived as an inner expansion, "inner" variables being those scaled with ℓ and ν. Introduce outer variables by scaling with L and U. Then upon taking the limit as $UL/\nu \to \infty$, the viscous term drops out, being of order $(UL/\nu)^{-1}$. Thus the "nonviscous" equation appears as the "outer" expansion of the "inner"expansion a limit which is significant for the matching (VanDyke (1964)) of this "inner" problem to the "outer", or inviscid initial value problem, which is too difficult to do yet. Physically this inviscid limit means that a viscous region of width ℓ near $\rho = 0$ has been shrunk to zero, or, from another point of view we are investigating the large ρ limit as seen on the scale ℓ.

The form of the nonviscous solution may be found by a modification of Kolmogorov's (1941) dimensional argument. We note that both U and ε occur in Eq. (40) (with $\nu = 0$) but that U occurs only in the combination $U^2(1-h)$, which must therefore be a function of ρ and ε alone. The only such function with dimension "velocity squared" is $(\varepsilon\rho)^{2/3}$. The solution must therefore be of the form

$$h = 1 - a\,\frac{\varepsilon^{2/3}}{U^2}\,\rho^{2/3} = 1 - a(\rho/L)^{2/3} \tag{43}$$

where a is a dimensionless constant. It is easily verified by direct substitution that the solution _is_ of this form and the constant has been determined by carrying out the lengthy integrations. While the integrals can all be found in standard tables or in Lighthill's book, one integral was done by computer to reduce the hand algebra. We find

$$a = .847 \tag{44}$$

The function h is related to the energy spectrum function, $E(k)$ by

$$\frac{1}{(2\pi)^3}\int e^{-i\underline{k}\cdot\underline{\rho}}\,\underline{R}(\underline{\rho})\,d\underline{\rho} = \frac{E(k)}{4\pi k^2}\left(\underline{I} - \hat{\underline{k}}\hat{\underline{k}}\right) \tag{45}$$

.

The function $E(k)$ has the property

$$\frac{3}{2}U^2 = \int_0^\infty E(k)\,dk$$

whence its interpretation as an energy density in wave number space. Substitution of Eq. (7) into Eq. (45) gives the formula

$$E(k) = \frac{k}{\pi}\int_0^\infty U^2 (3\rho h + \rho^2 h')\sin k\rho \, d\rho$$

by which E may be computed from h (Batchelor (1960), p. 49). In particular the behavior of h at small ρ determines E for large k. If $\nu \to 0$ first, the behavior of h for small ρ is given by Eq. (43) and the asymptotic behavior of $E(k)$ by the generalized integral

$$E(k) \sim -\frac{11}{3}\, a\,\varepsilon^{2/3}\,\frac{k}{\pi}\int_0^\infty \rho^{5/3}\sin k\rho\, d\rho$$

which is found in Lighthill's book. The result,

$$E(k) = C\,\varepsilon^{2/3} k^{-5/3}$$

is the celebrated Kolmogorov spectrum. (Batchelor (1960), p. 122). We find the Kolmogorov constant

$$C = \frac{11}{3}\cdot\frac{1}{\pi}\sin\frac{\pi}{3}\,\Gamma\left(1+\frac{5}{3}\right)a$$
$$= 1.521\,a$$
$$= 1.29$$

The only other theoretical prediction of the Kolmogorov constant is that of Kraichnan (1965) who got C = 1.77 from his Lagrangian history direct interaction theory. The most reliable experimental measurements are those in a tidal channel by Grant, Stewart and Moilliet (1961). They obtained C = 1.44 with a standard deviation of $\pm$.06 The complete range of values was from about 1.2 to 1.8.

DISCUSSION

A theory of turbulence has been proposed in this paper by a formal closure of the hierarchy of equations for turbulent probability distribution functions. By "formal" we mean the closure did not follow from any strong physical reasoning, though there was some guidance from statistical mechanics and certain necessary properties were satisfied. It resembles, in a broad sense, a closure proposed by Millionshtchikov (1941) and developed by Proudman and Reid (1954) and Tatsumi (1956). In this closure scheme, called the "cumulant discard hypothesis", the velocity moment equations are closed by relating fourth moments to second moments as if the distribution function were joint normal. In the present paper we also close by assuming a higher order quantity, $f_3'(1,2,3)$, is related to lower order quantities as if the distribution functions were joint normal. That such formal hypotheses may lead to difficulties is evidenced by the work of Ogura (1963) who showed that the "cumulant discard hypothesis" is unphysical, a consequence being negative values for the energy spectrum function. This shows that turbulence is <u>not</u> insensitive to the type of closure. (A more complete discussion of the "cumulant discard hypothesis" is given by Saffman (1968)).

Such difficulties have not yet appeared in the present theory, in fact the very good results obtained here for small scale phenomena suggest that the theory may be more generally valid. There are a number of attractive features. One is that it is not restricted to isotropic homogeneous turbulence. At its lowest level of information it gives an equation for the one point distribution function, something like a Boltzmann equation, coupled with an equation for the correlation tensor, higher order information being obtained, probably with less reliability, from Eq. (10) for the two point distribution function and from the closure itself. With some additional approximations such a Boltzmann-like equation can give a theory of turbulent exchange coefficients (a mixing length theory) similar to a theory of transport coefficients (a mean free path theory) in the kinetic theory of gases. The additional equations for the correlation tensor would be equivalent to conditions to determine the "mixing length", which is usually obtained empirically in such theories.

APPENDIX A

Evaluate the integral

$$I = \lim_{\delta \to 0} \frac{1}{(2\pi)^6} \int \frac{e^{i\underline{p}_1 \cdot \underline{v}_1 + i\underline{p}_2 \cdot \underline{v}_2}}{\hat{\underline{k}} \cdot (\underline{v}_2 - \underline{v}_1) - i\delta} \, d\underline{v}_1 \, d\underline{v}_2$$

Let

$$V_1 = u_1 \hat{\underline{k}} + \underline{V}_{1\perp}$$

$$V_2 = u_2 \hat{\underline{k}} + \underline{V}_{2\perp}$$

where $V_{1\perp}$ and $V_{2\perp}$ are projections of $\underline{V}_1$ and $\underline{V}_2$ onto plane perpendicular to $\hat{\underline{k}}$. Carry out the integration over the plane perpendicular to $\hat{\underline{k}}$. This gives

$$I = \lim_{\delta \to 0} \frac{1}{(2\pi)^2} \int \frac{e^{i\underline{p}_1 \cdot \hat{\underline{k}} u_1 + i\underline{p}_2 \cdot \hat{\underline{k}} u_2}}{u_2 - u_1 - i\delta} \, du_1 \, du_2 \, \delta(\underline{p}_{1\perp}) \delta(\underline{p}_{2\perp})$$

where $\delta(\underline{p}_{1\perp})$ and $\delta(\underline{p}_{2\perp})$ are two dimensional delta functions. In the remaining integral hold u_1 fixed and introduce $u_2 - u_1 = z$ as a new variable

$$\frac{1}{(2\pi)^2} \int \frac{e^{i(\underline{p}_1 \cdot \hat{\underline{k}} + \underline{p}_2 \cdot \hat{\underline{k}}) u_1} \, e^{i\underline{p}_2 \cdot \hat{\underline{k}} z}}{z - i\delta} \, du_1 \, dz$$

and carry out the integration on u_1 first, getting

$$\delta(\underline{p}_1 \cdot \hat{\underline{k}} + \underline{p}_2 \cdot \hat{\underline{k}}) \frac{1}{2\pi} \int \frac{e^{i\underline{p}_2 \cdot \hat{\underline{k}} z}}{z - i\delta} \, dz$$

Now take the limit as $\delta \to 0$, using the Plemelj formula to obtain

$$\delta(\underline{p}_1\cdot\hat{\underline{\kappa}}+\underline{p}_2\cdot\hat{\underline{\kappa}})\,\frac{1}{2\pi}\left(\pi i+P\int\frac{e^{i\underline{p}_2\cdot\hat{\underline{\kappa}}z}}{z}dz\right).$$

But

$$P\int_{-\infty}^{\infty}\frac{e^{i\underline{p}_2\cdot\hat{\underline{\kappa}}z}}{z}dz$$

$$=i\int_{-\infty}^{\infty}\frac{\sin\underline{p}_2\cdot\hat{\underline{\kappa}}z}{z}dz$$

$$=i\,\mathrm{sgn}\,\underline{p}_2\cdot\hat{\underline{\kappa}}\,\pi$$

where *sgn* denotes the algebraic sign. With this result we have

$$I=i\delta(\underline{p}_{1\perp})\,\delta(\underline{p}_{2\perp})\,\delta(\underline{p}_1\cdot\hat{\underline{\kappa}}+\underline{p}_2\cdot\hat{\underline{\kappa}})\,H(\underline{p}_2\cdot\hat{\underline{\kappa}}).$$

APPENDIX B

The T and S functions for Eq. (42).

$$T_{11}(\rho,s)=\frac{6\pi}{s\rho^4}\left(-\frac{2}{3}\rho^2 s^3-\frac{12}{5}s^5\right)\qquad s<\rho$$

$$=\frac{6\pi}{s\rho^4}\left(-\frac{16}{15}\rho^5\right)\qquad s>\rho\ .$$

$$T_{12}(\rho,s)=24\pi s^4/\rho^4\qquad s<\rho$$

$$=0\qquad s>\rho\ .$$

$$T_{21}(\rho,s)=\frac{\pi}{s\rho^4}\left(-\frac{12}{5}s^5-\frac{8}{3}s^3\rho^2\right)\qquad s<\rho$$

$$= \frac{\pi}{s\rho^4}\left(-\frac{16}{15}\rho^5\right) \qquad s > \rho$$

$$T_{22}(\rho, s) = \frac{1}{6} T_{12}(\rho, s)$$

$$\begin{aligned} S_{11} = & -6\left(\frac{s}{\sigma} - \frac{\rho}{\sigma}y\right) 2\pi y \\ & + \left[-\frac{\rho}{\sigma}\left(\frac{s}{\sigma} - \frac{\rho}{\sigma}y\right)^2 - 7\frac{\rho}{\sigma} - 4\frac{s\rho}{\sigma^2}\left(\frac{s}{\sigma} - \frac{\rho}{\sigma}y\right)\right] \pi(1-y^2) \\ & + \left[2\left(1 + \frac{s^2}{\sigma^2}\right)\left(\frac{s}{\sigma} - \frac{\rho}{\sigma}y\right) + \frac{s}{\sigma}\left(\frac{s}{\sigma} - \frac{\rho}{\sigma}y\right)^2 + 7\frac{s}{\sigma}\right] 3\pi y(1-y^2) \end{aligned}$$

$$\begin{aligned} S_{12} = & \left[2\frac{\rho^2}{\sigma^2}\left(\frac{s}{\sigma} - \frac{\rho}{\sigma}y\right)\right] 2\pi y \\ & + \left[-\frac{\rho}{\sigma}\left(\frac{s}{\sigma} - \frac{\rho}{\sigma}y\right)^2 - 7\frac{\rho}{\sigma} - 4\frac{s\rho}{\sigma^2}\left(\frac{s}{\sigma} - \frac{\rho}{\sigma}y\right)\right]\left(2\pi y^2 - \pi[1-y^2]\right) \\ & + \left[2\left(1 + \frac{s^2}{\sigma^2}\right)\left(\frac{s}{\sigma} - \frac{\rho}{\sigma}y\right) + \frac{s}{\sigma}\left(\frac{s}{\sigma} - \frac{\rho}{\sigma}y\right)^2 + 7\frac{s}{\sigma}\right]\left(2\pi y^3 - 3\pi y(1-y^2)\right) \end{aligned}$$

$$\begin{aligned} S_{21} = & \left[-\left(\frac{s}{\sigma} - \frac{\rho}{\sigma}y\right)\right] 2\pi y \\ & + \left[-3\frac{\rho}{\sigma} - 2\frac{s}{\sigma}\frac{\rho}{\sigma}\left(\frac{s}{\sigma} - \frac{\rho}{\sigma}y\right) + \frac{\rho}{\sigma}\left(\frac{s}{\sigma} - \frac{\rho}{\sigma}y\right)^2\right] \pi(1-y^2) \\ & + \left[3\frac{s}{\sigma} + \frac{s^2}{\sigma^2}\left(\frac{s}{\sigma} - \frac{\rho}{\sigma}y\right) - 2\left(\frac{s}{\sigma} - \frac{\rho}{\sigma}y\right) - \frac{s}{\sigma}\left(\frac{s}{\sigma} - \frac{\rho}{\sigma}y\right)^2\right] 3\pi y(1-y^2) \end{aligned}$$

$$\begin{aligned} S_{22} = & \left[\frac{\rho^2}{\sigma^2}\left(\frac{s}{\sigma} - \frac{\rho}{\sigma}y\right)\right] 2\pi y \\ & + \left[-3\frac{\rho}{\sigma} - 2\frac{s\rho}{\sigma^2}\left(\frac{s}{\sigma} - \frac{\rho}{\sigma}y\right) + \frac{\rho}{\sigma}\left(\frac{s}{\sigma} - \frac{\rho}{\sigma}y\right)^2\right]\left(2\pi y^2 - \pi(1-y^2)\right) \\ & + \left[3\frac{s}{\sigma} + \frac{s^2}{\sigma^2}\left(\frac{s}{\sigma} - \frac{\rho}{\sigma}y\right) - 2\left(\frac{s}{\sigma} - \frac{\rho}{\sigma}y\right) - \frac{s}{\sigma}\left(\frac{s}{\sigma} - \frac{\rho}{\sigma}y\right)^2\right] \times \\ & \left[2\pi y^3 - 3\pi y(1-y^2)\right] \end{aligned}$$

$$\mathcal{S}_{31} = \left[-\left(\frac{s}{\sigma}-\frac{\rho}{\sigma}y\right)\right] 2\pi y$$
$$+\left[-3\frac{\rho}{\sigma}+4\frac{s\rho}{\sigma^2}\left(\frac{s}{\sigma}-\frac{\rho}{\sigma}y\right)+\frac{\rho}{\sigma}\left(\frac{s}{\sigma}-\frac{\rho}{\sigma}y\right)^2\right]\pi(1-y^2)$$
$$+\left[4\frac{s}{\sigma}-\frac{\rho}{\sigma}y-2\frac{s^2}{\sigma^2}\left(\frac{s}{\sigma}-\frac{\rho}{\sigma}y\right)-\frac{s}{\sigma}\left(\frac{s}{\sigma}-\frac{\rho}{\sigma}y\right)^2\right]3\pi y(1-y^2)$$

$$\mathcal{S}_{32} = \left[-2\frac{\rho^2}{\sigma^2}\left(\frac{s}{\sigma}-\frac{\rho}{\sigma}y\right)\right] 2\pi y$$
$$+\left[-3\frac{\rho}{\sigma}+4\frac{\rho s}{\sigma^2}\left(\frac{s}{\sigma}-\frac{\rho}{\sigma}y\right)+\frac{\rho}{\sigma}\left(\frac{s}{\sigma}-\frac{\rho}{\sigma}y\right)^2\right]\left(2\pi y^2-\pi(1-y^2)\right)$$
$$+\left[4\frac{s}{\sigma}-\frac{\rho}{\sigma}y-2\frac{s^2}{\sigma^2}\left(\frac{s}{\sigma}-\frac{\rho}{\sigma}y\right)-\frac{s}{\sigma}\left(\frac{s}{\sigma}-\frac{\rho}{\sigma}y\right)^2\right]\left(2\pi y^3-3\pi y(1-y^2)\right)$$

$$\mathcal{S}_{41} = \left[-\frac{\rho}{\sigma}+2\frac{s\rho}{\sigma^2}\left(\frac{s}{\sigma}-\frac{\rho}{\sigma}y\right)-\frac{\rho}{\sigma}\left(\frac{s}{\sigma}-\frac{\rho}{\sigma}y\right)^2\right]\pi(1-y^2)$$
$$+\left[\frac{s}{\sigma}-\frac{s^2}{\sigma^2}\left(\frac{s}{\sigma}-\frac{\rho}{\sigma}y\right)-\left(\frac{s}{\sigma}-\frac{\rho}{\sigma}y\right)+\frac{s}{\sigma}\left(\frac{s}{\sigma}-\frac{\rho}{\sigma}y\right)^2\right]3\pi y(1-y^2)$$

$$\mathcal{S}_{42} = \left[-\frac{\rho^2}{\sigma^2}\left(\frac{s}{\sigma}-\frac{\rho}{\sigma}y\right)\right] 2\pi y$$
$$+\left[-\frac{\rho}{\sigma}+2\frac{s\rho}{\sigma^2}\left(\frac{s}{\sigma}-\frac{\rho}{\sigma}y\right)-\frac{\rho}{\sigma}\left(\frac{s}{\sigma}-\frac{\rho}{\sigma}y\right)^2\right]\left(2\pi y^2-\pi(1-y^2)\right)$$
$$+\left[\frac{s}{\sigma}-\frac{s^2}{\sigma^2}\left(\frac{s}{\sigma}-\frac{\rho}{\sigma}y\right)-\left(\frac{s}{\sigma}-\frac{\rho}{\sigma}y\right)+\frac{s}{\sigma}\left(\frac{s}{\sigma}-\frac{\rho}{\sigma}y\right)^2\right]\left(2\pi y^3-3\pi y(1-y^2)\right)$$

REFERENCES

G. K. Batchelor, (1960) The Theory of Homogeneous Turbulence, Cambridge University Press.

H. L. Grant, R. W. Stewart and A. Moilliet, (1962) Turbulence spectra from a tidal channel. J. Fluid Mech. 13, 237.

J. O. Hinze, (1959) Turbulence, McGraw-Hill.

E. Hopf, (1952) Statistical hydromechanics and functional calculus, J. Ratl. Mech. Anal. 1, 87.

A. N. Kolmogorov, (1941) Dissipation of energy in locally isotropic turbulence. C. R. Acad. Sci. U.R.S.S. 32, 16.

R. H. Kraichnan, (1965) Preliminary calculation of the Kolmogorov turbulence spectrum, Phys. Fluids 8, 995; (1966) Errata 9, 1884.

M. J. Lighthill, (1960) Fourier Analysis and Generalized Functions, Cambridge University Press.

T. S. Lundgren, (1967) Distribution functions in the statistical theory of turbulence, Phys. Fluids 10, 969.

M. Millionshtchikov, (1941) On the theory of homogeneous isotropic turbulence, C. R. Acad. Sci. U.R.S.S. 32, 619.

A. S. Monin, (1967) Equations of turbulent motion, P.M.M. 31, 1057.

N. Muskhelishvili, (1953) Singular Integral Equations, Noordhoff.

Y. Ogura, (1963) A consequence of the zero fourth order cumulant approximation in the decay of isotropic turbulence, J. Fluid Mech. 16, 33.

I. Proudman and W. H. Reid, (1954) On the decay of a normally distributed and homogeneous turbulent velocity field, Phil Trans. A., 247, 163.

S. A. Rice and P. Gray, (1965) The Statistical Mechanics of Simple Liquids, Interscience.

P. G. Saffman, (1968) Lectures on homogeneous turbulence in Topics in Nonlinear Physics, Ed. N. Zabusky, Springer-Verlag.

T. Tatsumi, (1957) The theory of decay process of incompressible isotropic turbulence, Proc. Roy. Soc. A, 239, 16.

M. VanDyke, (1964) Perturbation Methods in Fluid Mechanics, Academic Press.

Singular perturbation in some problems of Statistical Mechanics

by

M. Kac
Rockefeller University, New York

Two examples of singular perturbations which occur naturally in problems of Statistical Mechanics will be discussed.

The first concerns the discussion of the one-dimensional Ising model with interaction energy

$$E = -\alpha\gamma \sum_{1 \le i < j \le N} \exp\{-\gamma |i-j|\} \mu_i \mu_j \qquad (\mu_i = \pm 1)$$

in the limite as $\gamma \to 0$.

This problem is reduced to the study of the integral equation with the kernel

$$\cosh\sqrt{\nu\gamma}\; x \; \frac{W(x) P_\gamma(x|y)}{\sqrt{W(x)}\,\sqrt{W(y)}} \; \cosh\sqrt{\nu\gamma}\; y$$

where

$$\nu = \frac{\alpha}{\varkappa\tau}$$

$$W(x) = \frac{1}{\sqrt{2\pi}} \exp(-x^2/2)$$

and

$$P_\gamma(x/y) = \frac{1}{\sqrt{2\pi(1-\exp(-2\gamma))}} \exp\left\{\frac{(y-x\exp(-\gamma))^2}{2(1-\exp(-2\gamma)}\right\} .$$

The second example is taken from an approach to the problem of the disordered chain.

Here one is concerned with the maximum eigenvalue of the integral equation with the kernel $(0 \leq r,\ \rho < \infty)$

$$L_\varkappa(r,\rho) = \frac{2}{\Gamma(\varkappa/2)} f_\varkappa(r) f_\varkappa(\rho) + M_\varkappa(r,\rho)$$

where

$$f_\varkappa(r) = r^{\frac{\varkappa-1}{2}} \exp(-r^2)\sqrt{p\exp(-m\xi r^2) + q\exp(-M\xi r^2)}$$

$(p,q \geq 0,\ \ p + q = 1,\ \ \xi > 0,\ \ m,\ M > 0)$ and $M_\varkappa(r,\rho)$ a certain kernel which is perfectly well behaved for $\varkappa \to 0$.

The "singularity" of the problem is due only to the fact that $f_o(r)$ is not $L^2(0,\infty)$.

THE BOUNDING THEORY OF TURBULENCE AND ITS PHYSICAL SIGNIFICANCE IN THE CASE OF TURBULENT COUETTE FLOW

by

Friedrich H. Busse

Department of Planetary and Space Science

University of California

Los Angeles 90024

1. Introduction

The objective of a theory of turbulence is the deduction of expressions for the average properties of turbulent flows without the knowledge of the complex details of the instantaneous velocity field. The basic mathematical difficulty of the analysis of turbulence originates from the fact that the nonlinear terms of the Navier-Stokes equations prevent the separation of averaged and fluctuating quantities. A closed set of equations for the averaged quantities cannot be obtained without taking into account the details of the fluctuating velocity field. In order to deal with the unavoidable lack of information, two approaches have been used. Most theories of turbulence introduce additional heuristic assumptions which are based on more or less plausible physical arguments. The fact that these theories are only partially successful in predicting the observed properties of turbulent flows indicates that the lack of information cannot be bridged by simple assumptions. The bounding theory represents an alternative. The available information is used to derive bounds for averaged quantities. Thus the results of the theory reflect the unavoidable lack of information in a natural way. Howard (1963) has introduced this approach in his derivation of upper bounds on the heat transport by turbulent convection in a layer heated from below. It was shown more recently (Busse 1969a) that the same approach can be used to derive bounds in a wide variety of turbulent transport processes. In addition, the development of a multiple boundary layer technique (Busse 1969b) allows one to derive bounds in cases when the equation of continuity is taken into account. The results indicate a surprisingly close relation between the vector field optimizing the transport and the experimentally observed turbulent velocity field. This suggests that the physically realized manifold of solutions of the Navier-Stokes equations is distinguished among all possible solutions by the property of nearly extremal transport. The main purpose of the present paper is to apply the bounding method to the case of circular Couette flow. This case of turbulence offers a richer structure than the cases treated previously owing to the varying influence of the mean rotation and the geometrical difference between inner and outer boundary. Most of the analysis, however, can be reduced to mathematical problems solved earlier in the cited papers.

The basic idea of the bounding theory of turbulence is to consider a suitably defined 'trial'-class of vector fields which includes all velocity fields which are solutions of the Navier-Stokes equations. Among this class of vector fields the extrema of certain functionals can be determined by variational methods. Since all dynamically possible velocity fields are included in the class of vector fields, the extrema yield bounds for the physically realized values of the functionals. The success of this procedure depends on the fact that the trial-class of vector fields can be defined by more simple properties than the subclass of all possible solutions of the Navier-Stokes equations. In the following we shall use the equation of continuity, an energy balance, and the boundary conditions to define the trial-class of vector fields. Hence the fields represent possible velocity fields as far as kinematics and energetics are concerned. They do not necessarily satisfy the dynamic constraint of the Navier-Stokes equations. By requiring that the trial vector fields in the variational problem satisfy in addition part of the dynamical constraint, the bounds can be improved, of course. In principle the method offers a systematic procedure to determine the extrema of the respective functional among all solutions of the Navier-Stokes equations. To demonstrate the basic principles of the bounding theory we shall first discuss in §2 the case of plane Couette flow. In §3 the variational problem for the upper bound on the transport of angular momentum by turbulent circular Couette flow will be formulated. The bounding method has been applied to turbulent circular Couette flow before by Nickerson (1969), who considered the case when the equation of continuity is added as constraint for small values of the angular momentum transport. We shall focus our attention on the asymptotic case of large angular momentum transfer which is characterized by a multiple boundary layer structure. The corresponding solution of the Euler-Lagrange equation will be derived in §4. A discussion of the results follows in §5.

2. The mathematical problem in the case of plane Couette flow.

We consider the motion of a homogeneous incompressible fluid between two parallel rigid plates. The plates are infinitely extended and are moving relative to each other with the constant velocity V_0 in the direction described by the constant unit vector $\underset{\sim}{i}$. For the dimensionless description of the problem, we introduce the distance d between the plates as length scale, and d^2/ν as scale of the time where ν denotes the kinematic viscosity of the fluid. The Navier-Stokes equations for the velocity vector $\underset{\sim}{V}$ are

$$\nabla\times(\nabla\times\underset{\sim}{V}) + \nabla p + \underset{\sim}{V}\cdot\nabla\underset{\sim}{V} + \frac{\partial}{\partial t}\underset{\sim}{V} = 0 \quad , \qquad (2.1)$$

$$\nabla\cdot\underset{\sim}{V} = 0 \quad .$$

It is convenient to introduce a Cartesian system of coordinates with the origin in the center between the plates, the z-axis normal to the plates, and the x-axis in the direction of the unit vector $\underset{\sim}{i}$. The boundary conditions for the velocity vector $\underset{\sim}{V}$ are given by

$$\underset{\sim}{V} = \mp\frac{1}{2}\,Re\,\underset{\sim}{i} \qquad \text{at} \quad z = \pm\frac{1}{2} \quad , \qquad (2.2)$$

where the Reynolds number is defined by $Re \equiv V_0 d/\nu$. We assume that the velocity and the pressure are bounded everywhere, and that the averages of the velocity components and their products over planes z=const.

exist. The average over planes z = const. will be indicated by a bar above the respective quantities in contrast to the average over the total fluid layer which will be indicated by angular brackets $\langle \ldots \rangle$. Accordingly, the vector $\underset{\sim}{V}$ can be separated into two parts:

$$\underset{\sim}{V} = \underset{\sim}{U} + \check{\underset{\sim}{v}} \quad , \qquad \text{with} \quad \underset{\sim}{U} \equiv \overline{\underset{\sim}{V}} \quad .$$

By taking the average of equations (2.1) over planes z = const. and by subtracting the averaged equations from the original equations we obtain the following equations for $\underset{\sim}{U}$ and $\check{\underset{\sim}{v}}$.

$$\frac{\partial^2}{\partial z^2} \underset{\sim}{U} - \frac{\partial}{\partial t} \underset{\sim}{U} = \frac{\partial}{\partial z} \overline{\check{\underset{\sim}{u}} \check{w}} \quad , \tag{2.3}$$

$$- \frac{\partial}{\partial z} \overline{p} = \frac{\partial}{\partial z} \overline{\check{w}^2} \quad ,$$

$$\nabla \times (\nabla \times \check{\underset{\sim}{v}}) + \nabla (p - \overline{p}) + \frac{\partial}{\partial t} \check{\underset{\sim}{v}} + \check{\underset{\sim}{v}} \cdot \nabla \check{\underset{\sim}{v}} - \overline{\check{\underset{\sim}{v}} \cdot \nabla \check{\underset{\sim}{v}}} + \underset{\sim}{U} \cdot \nabla \check{\underset{\sim}{v}} \quad , \tag{2.4}$$

$$+ \check{w} \frac{\partial}{\partial z} \underset{\sim}{U} = 0 \quad ,$$

$$\nabla \cdot \check{\underset{\sim}{v}} = 0 \quad . \tag{2.5}$$

$\check{w}$ is the z-component of $\check{\underset{\sim}{v}}$, $\check{\underset{\sim}{u}}$ denotes the component of $\check{\underset{\sim}{v}}$ parallel to the plates. By multiplying equation (2.4) by $\check{\underset{\sim}{v}}$ and averaging it over the fluid layer we obtain

$$\frac{1}{2} \frac{d}{dt} \left\langle |\check{\underset{\sim}{v}}|^2 \right\rangle + \left\langle |\nabla \times \check{\underset{\sim}{v}}|^2 \right\rangle + \left\langle \check{\underset{\sim}{u}} \cdot \check{w} \frac{\partial}{\partial z} \underset{\sim}{U} \right\rangle = 0 \quad . \tag{2.6}$$

In deriving this relation we have used the fact that a number of integrals can be transformed into surface integrals which vanish, because $\check{\underset{\sim}{v}}$ vanishes at the boundaries $z = \pm \frac{1}{2}$, and because the contributions from the remaining surface become negligible as the average is extended over the infinite domain. We are interested in turbulent flow under stationary conditions long after any change in the relative motion of the plates has occurred. We define this case by the assumption that all averaged quantities like $\underset{\sim}{U}$ or $\overline{\underset{\sim}{\check{u}}\,\check{w}}$ do not depend on time. This assumption permits the integration of (2.3),

$$\frac{d}{dz}\underset{\sim}{U} = \overline{\underset{\sim}{\check{u}}\,\check{w}} - \langle \underset{\sim}{\check{u}}\,\check{w} \rangle - \mathrm{Re}\,\underset{\sim}{i} \qquad . \qquad (2.7)$$

The constant vector of integration has been determined by the boundary condition (2.2). Equation (2.7) and the fact that the averaged kinetic energy is time independent can be used to simplify relation (2.6),

$$\left\langle |\nabla\times\underset{\sim}{\check{v}}|^2 \right\rangle + \left\langle \left(\overline{\underset{\sim}{\check{u}}\check{w}} - \left\langle \underset{\sim}{\check{u}}\check{w} \right\rangle\right)^2 \right\rangle - \mathrm{Re}\left\langle \check{u}_x \check{w} \right\rangle = 0 \qquad . \qquad (2.8)$$

In the case of the laminar velocity field $\underset{\sim}{\check{v}} \equiv 0$, $\underset{\sim}{U} = -z\underset{\sim}{i}$, the dimensionless momentum transport between the plates is given by Re. In the case of turbulent flow the additional momentum transport in the direction of the relative motion of the plates is given by $\langle \check{u}_x \check{w} \rangle$. According to equation (2.8) this term is always positive. The goal of the bounding theory is to derive an upper bound for $\langle \check{u}_x \check{w} \rangle$ at a given value of the Reynolds number. Such an upper bound is given by the maximum $\mu(R)$ of $\langle u_x w \rangle$ among all vector fields $\underset{\sim}{v}$ which satisfy

$$\nabla\cdot\underset{\sim}{v} = 0 \qquad , \qquad (2.9)$$

$$\left\langle |\nabla\times\underset{\sim}{v}|^2 \right\rangle + \left\langle \left(\overline{\underset{\sim}{u}w} - \left\langle \underset{\sim}{u}w \right\rangle\right)^2 \right\rangle - R\left\langle u_x w \right\rangle = 0 \qquad , \qquad (2.10)$$

$$\underset{\sim}{v} = 0 \qquad \text{at} \qquad z = \pm\frac{1}{2} \qquad , \qquad (2.11)$$

and are bounded everywhere. Instead of solving this variational problem, it is more convenient to solve the following variational problem:

Given $\mu > 0$, find the minimum $R(\mu)$ of the functional

$$\mathcal{R}(\underset{\sim}{v},\mu) \equiv \frac{\left\langle |\nabla\times\underset{\sim}{v}|^2 \right\rangle}{\left\langle u_x w \right\rangle} + \mu\,\frac{\left\langle \left(\overline{\underset{\sim}{u}w} - \left\langle \underset{\sim}{u}w \right\rangle\right)^2 \right\rangle}{\left\langle u_x w \right\rangle^2} \qquad (2.12)$$

among all vector fields $\underset{\sim}{v}$ which satisfy (2.9), (2.11), are bounded everywhere and have $\langle u_x w \rangle > 0$. Because the functional (2.12) is homogeneous, the amplitude of the minimizing solution $\underset{\sim}{v}$ remains undetermined and can be chosen to satisfy

$$\langle u_x w \rangle = \mu \quad .$$

With this normalization the solution satisfies relation (2.10). Since it can be shown that $R(\mu)$ is a monotonic function of μ the two formulations of the variational problem are equivalent.

The variational problem stated above has been solved in I. It was shown in that paper that x-independent vector fields $\underset{\sim}{v}$ yield the minimum value $R(\mu)$, although no rigorous proof can be given for this property except in the case of low values of μ. In general, the minimizing solution $\underset{\sim}{v}$ of the variational functional will not satisfy the Navier-Stokes equations (2.4). This becomes evident when the following general representation of a divergence-free vector field is used for $\underset{\sim}{v}$,

$$\underset{\sim}{v} = \nabla\times(\nabla\times\underset{\sim}{k}v) + \nabla\times\underset{\sim}{k}\psi \equiv \underset{\sim}{\delta}v + \underset{\sim}{\varepsilon}\psi \quad .$$

$\underset{\sim}{k}$ is the unit vector in the z-direction. By using the same representation $\underset{\sim}{\check{v}}$ and multiplying equation (2.4) by $\underset{\sim}{\delta}\check{v}$ and by $\underset{\sim}{\varepsilon}\check{\psi}$, respectively, and averaging the result, we obtain two separate energy identities,

$$\begin{aligned} &\left\langle |\underset{\sim}{k}\times\nabla\nabla^2\check{v}|^2\right\rangle - \left\langle \underset{\sim}{\varepsilon}\check{\psi}\cdot(\underset{\sim}{\varepsilon}\check{\psi}+\underset{\sim}{\delta}\check{v})\cdot\nabla\underset{\sim}{\delta}\check{v}\right\rangle + \left\langle \delta_z\check{v}\ \underset{\sim}{\delta}\check{v}\cdot\frac{d}{dz}\underset{\sim}{U}\right\rangle = 0 \\ &\left\langle |\nabla\times(\nabla\times\underset{\sim}{k}\check{\psi})|^2\right\rangle + \left\langle \underset{\sim}{\varepsilon}\check{\psi}\cdot(\underset{\sim}{\varepsilon}\check{\psi}+\underset{\sim}{\delta}\check{v})\cdot\nabla\underset{\sim}{\delta}\check{v}\right\rangle + \left\langle \delta_z\check{v}\underset{\sim}{\varepsilon}\check{\psi}\cdot\frac{d}{dz}\underset{\sim}{U}\right\rangle = 0 \quad . \end{aligned} \tag{2.13}$$

The sum of these two energy relations yields relation (2.8), of course, when $\frac{d}{dz}\underset{\sim}{U}$ is replaced using relation (2.7). In contrast to relation (2.8) the first of relations (2.13) cannot be satisfied by x-independent fields v, ψ, since the second term of left hand side vanishes in this case and the third term is positive definite.

The bound on the momentum transport $\langle \check{u}_x \check{w} \rangle$ can be improved when the variational problem is formulated for a more restricted class of trial vector fields which still contains all possible turbulent velocity fields $\check{\underset{\sim}{v}}$. An obvious way to proceed in this direction would be to use the difference of the two energy relations (2.13) as a constraint in addition to the sum given by (2.10). As we have noticed above, the extremalizing solution will have to be three-dimensional in this case, and the appearance of cubic terms in the functional will yield complicated Euler-Lagrange equations. For this reason no attempt has been made yet to solve this problem.

3. The mathematical problem in the case of circular Couette flow.

We consider the motion of a homogeneous incompressible fluid between two coaxial cylinders of infinite length. The inner and the outer cylinder are rotating with the constant angular velocities $\hat{\Omega}_i$ and $\hat{\Omega}_o$, respectively. We shall use the radius r_o of the outer cylinder and $r_o{}^2/\nu$ as scales for length and time, respectively. The dimensionless Navier-Stokes equations can be written again in the form (2.1) since the turbulent flow is driven solely by the shear stresses acting at the boundary. We introduce a cylindrical system of coordinates (r,ϕ,y) with the y-axis along the axis of the cylinders in order to emphasize the analogy with the case of plane Couette flow. The boundary conditions for the velocity vector $\underset{\sim}{V}$ are given by

$$\underset{\sim}{V} = \Omega_o \underset{\sim}{\phi} \quad \text{at} \quad r = 1 , \qquad \underset{\sim}{V} = \Omega_i \eta \underset{\sim}{\phi} \quad \text{at} \quad r = \eta . \tag{3.1}$$

$\Omega_i = \hat{\Omega}_i \, r_o^2/\nu$ and $\Omega_o = \hat{\Omega}_o \, r_o^2/\nu$ represent the dimensionless expressions of the angular velocities. $\underset{\sim}{\phi}$ is the unit vector in the azimuthal direction, which we shall identify with the direction of rotation of the inner cylinder. We indicate the average over the surface r=const by a bar and separate the velocity field into two parts

$$\underset{\sim}{V} = \underset{\sim}{U} + \underset{\sim}{\check{v}} \qquad \text{with} \quad \underset{\sim}{U} = \overline{\underset{\sim}{V}} \quad .$$

As in the preceding section, it is assumed that quantities which depend only on r are time independent. The mean flow $\underset{\sim}{U}$ is determined by the equations

$$r \frac{d}{dr} \frac{U_\phi}{r} = \overline{\check{v}_r \check{v}_\phi} - h\langle \check{v}_r \check{v}_\phi h \rangle - hA$$

$$\frac{d}{dr} U_y = \overline{\check{v}_r \check{v}_y} - \frac{1-\eta^2}{2\eta \ln 1/\eta} \langle \overline{\check{v}_r \check{v}_y} / r \rangle / r \quad . \tag{3.2}$$

The function $h(r)$ and the parameter A are defined by

$$h(r) \equiv \frac{\eta}{r^2} \quad , \qquad A \equiv \frac{2\eta}{1-\eta^2} (\Omega_i - \Omega_o) \tag{3.3}$$

and the angular brackets denote the average over the annular region,

$$\langle \check{v}_r \check{v}_\phi h \rangle \equiv \frac{2}{1-\eta^2} \int_\eta^1 \overline{\check{v}_r \check{v}_\phi} \; h \; r dr \quad .$$

The energy balance for the fluctuating velocity field can be obtained in analogy to relation (2.8),

$$\left\langle |\nabla \check{v}_r|^2 + |\nabla \check{v}_\phi|^2 + |\nabla \check{v}_y|^2 + \frac{1}{r^2}(\check{v}_r^{\,2} + \check{v}_y^{\,2} - 2\check{v}_\phi \frac{\partial}{\partial \phi} \check{v}_r) \right\rangle$$
$$+ \left\langle (\overline{\check{v}_r \check{v}_\phi} - h\langle \check{v}_r \check{v}_\phi h \rangle)^2 \right\rangle + \left\langle (\overline{\check{v}_r \check{v}_y} - \frac{1-\eta^2}{2\eta r \ln 1/\eta} \langle \check{v}_r \check{v}_y / r \rangle)^2 \right\rangle \tag{3.4}$$
$$- A\langle \check{v}_r \check{v}_\phi h \rangle = 0$$

where the property

$$\langle h^2 \rangle = 1$$

has been used.

The quantity $2\pi\eta A$ represents the dimensionless torque exerted by the laminar flow $\check{\underset{\sim}{v}} \equiv 0$ on the outer cylinder per unit length. In the case of turbulent flow the additional torque $2\pi\eta\langle\check{v}_r\check{v}_\phi h\rangle$ is exerted by the Reynolds stress. According to relation (3.4) the additional torque has always the same sign as A. The goal of the bounding theory is to derive an upper bound on the absolute value of the turbulent part $2\pi\eta\langle\check{v}_r\check{v}_\phi h\rangle$ of the angular momentum transport.

As in the case of plane Couette flow, it is mathematically more convenient to ask for a lower bound $R(\mu)$ for the absolute value of the parameter A at a given value μ of the quantity $|\langle\check{v}_r\check{v}_\phi h\rangle|$ and to incorporate the constraint (3.4) into the definition of extremalized functional. Hence we shall consider the variational problem:

Given $\mu > 0$, find the minimum $R(\mu)$ of the functional

$$\mathcal{R}(\underset{\sim}{v},\mu) = \frac{\left\langle|\nabla v_r|^2 + |\nabla v_\phi|^2 + |\nabla v_y|^2 + \frac{1}{r^2}(v_r^2 + v_\phi^2 - 2v_\phi\frac{\partial}{\partial\phi}v_r)\right\rangle}{\langle v_r v_\phi h\rangle} \tag{3.5}$$

$$+ \frac{\mu}{\langle v_r v_\phi h\rangle^2}\left\{\left\langle(\overline{v_r v_\phi} - h\langle v_r v_\phi h\rangle)^2\right\rangle + \left\langle(\overline{v_r v_y} - \frac{1-\eta^2}{2\eta r \ln\eta^{-1}}\langle v_r v_y/r\rangle)^2\right\rangle\right\}$$

among all vector fields $\underset{\sim}{v}$ which are bounded everywhere and satisfy

$$\nabla\cdot\underset{\sim}{v} = 0 \quad , \tag{3.6a}$$

$$\underset{\sim}{v} = 0 \quad \text{at} \quad r = 1 \quad \text{and} \quad r = \eta \quad , \tag{3.6b}$$

$$\langle v_r v_\phi h\rangle > 0 \quad . \tag{3.6c}$$

As in the case of plane Couette flow we shall make the assumption that the minimum of the functional will be reached by vector fields which do not depend on the coordinate in the direction of the motion of the boundary. This assumption allows us to satisfy constraint (3.6a) by the introduction of the variables v and θ,

$$v_r = -\frac{\partial^2}{\partial y^2} v \equiv w \quad , \qquad v_y = \frac{1}{r}\frac{\partial^2}{\partial y \partial r} vr \ , \quad v_\phi = \theta \ .$$

The Euler-Lagrange equations for an extremum of the functional (3.5) in terms of the functions w and θ are

$$\left(\frac{\partial}{\partial r}\frac{1}{r}\frac{\partial}{\partial r} r + \frac{\partial^2}{\partial y^2}\right)^2 w - \left\{\mu(\overline{w\theta} - h) - \frac{h}{2}\left(R(\mu) + \mu\left\langle(\overline{w\theta} - h)^2\right\rangle\right)\right\}\frac{\partial^2}{\partial y^2}\theta = 0$$

$$\left(\frac{\partial}{\partial r}\frac{1}{r}\frac{\partial}{\partial r} r + \frac{\partial^2}{\partial y^2}\right)\theta - \left\{\mu(\overline{w\theta} - h) - \frac{h}{2}\left(R(\mu) + \mu\left\langle(\overline{wh} - h)^2\right\rangle\right)\right\} w = 0 \ . \qquad (3.7)$$

We have neglected the term $\overline{v_r v_y}$ since we anticipate that it vanishes for all solutions of (3.7). In addition, we have introduced the normalization condition

$$\langle w\theta h\rangle = 1 \qquad . \qquad (3.8)$$

Since the functional (3.5) is homogeneous, the solution w,θ of equations (3.7) multiplied by $\sqrt{\mu}$ will satisfy relation (3.4) with $R(\mu)$ in place of A. In the next section equations (3.7) will be solved in the asymptotic case of large μ.

4. Multi-α-solutions.

The variational problem (2.12) in the case of plane Couette flow has been solved in the paper (Busse 1970a) which we refer to as I. In this section we shall apply the same method of solution to the more general variational problem (3.5). The homogeneity of the problem with respect to the y-direction suggests that the solutions of the Euler-Lagrange equations (3.7) can be written in the form

$$w = w^{(N)} \equiv \sum_{n=1}^{N} (\phi_n(y) w_n(r) + \phi_n^*(y) w_n^*(r)) \quad ,$$

$$\theta = \theta^{(N)} \equiv \sum_{n=1}^{N} (\phi_n(y)\theta_n(r) + \phi_n^*(y)\theta_n^*(r)) \quad . \qquad (4.1)$$

With functions ϕ_n , ϕ_n^* satisfying the relations

$$\frac{d^2}{dy^2}\phi_n = - \alpha_n^2\phi_n \quad , \quad \frac{d^2}{dy^2}\phi_n^* = - \alpha_n^{*2}\phi_n^* \quad , \quad |\phi_n|^2 = |\phi_n^*|^2 = 1 \ . \tag{4.2}$$

The functions w_n , θ_n , ϕ_n and the wavenumber α_n , as well as their starred companions, will depend in general on the parameter N. This dependence will be indicated by an upper index (N) when it becomes necessary to distinguish different solutions. Solutions of the form (4.1) satisfy also the requirement that $\overline{v_y v_r}$ vanishes for a minimum of the functional (3.5). Quantities with a star and without a star have been introduced in (4.1) to represent the solutions $w^{(N)}$, $\theta^{(N)}$at the inner and outer boundary, respectively. We expect boundary layers since we are interested in solutions for large values of μ, in which case the second term in the definition (3.5) of the functional $\mathcal{R}$ will be dominant. To minimize this term $\overline{w\theta}$ has to approach the function h(r) as closely as is compatible with the first term on the right hand side of the definition (3.5). Since the boundary conditions

$$w = \frac{\partial}{\partial r} w = \theta = 0 \qquad \text{at} \quad r = \eta, 1 \tag{4.3}$$

have to be satisfied the dissipation term will increase as $\overline{w\theta}$ approaches h(r) close to the boundary. Thus the balance between the first and the second term in the definition (3.5) determines the minimum of the functional. When the dissipation term is rewritten in terms of the variables w and θ it becomes clear that it reaches a minimum when the dissipation associated with the radial scale is of the same order as the dissipation associated with axial scale. For this reason a solution seems to be preferred which corresponds to increasing values α_n as the boundary is approached. Hence we expect that $\overline{w\theta}$ will be composed of N boundary layers at both walls of the annulus. In each of the boundary layers at the outer wall $w_n\theta_n$ grows from the value determined by

$$w_n = \frac{d}{dr} w_n = \theta_n = 0 \qquad \text{at} \quad r = 1$$

until it reaches the value h_o assumed by $h(r)$ close to the outer bounding. Towards the interior $w_n \theta_n$ will then decay at the rate at which $w_{n-1} \theta_{n-1}$ is growing on the next larger boundary in order that the approximate relation

$$w_n \theta_n + w_{n-1} \theta_{n-1} \approx h_o \tag{4.4a}$$

is satisfied throughout all but the N-th boundary layer. Qualitatively the same description holds for the N boundary layers at the inner wall. The functions w_n^*, θ_n^* will be different from zero essentially only in the n-th and (n-1)-th boundary layer. They will take turns in carrying the momentum transport $\overline{w\theta}$ such that

$$w_n^* \theta_n^* + w_{n-1}^* \theta_{n-1}^* \approx h_o^* \equiv h(\eta) \tag{4.4b}$$

is satisfied except in the N-th boundary layer at the inner wall. A

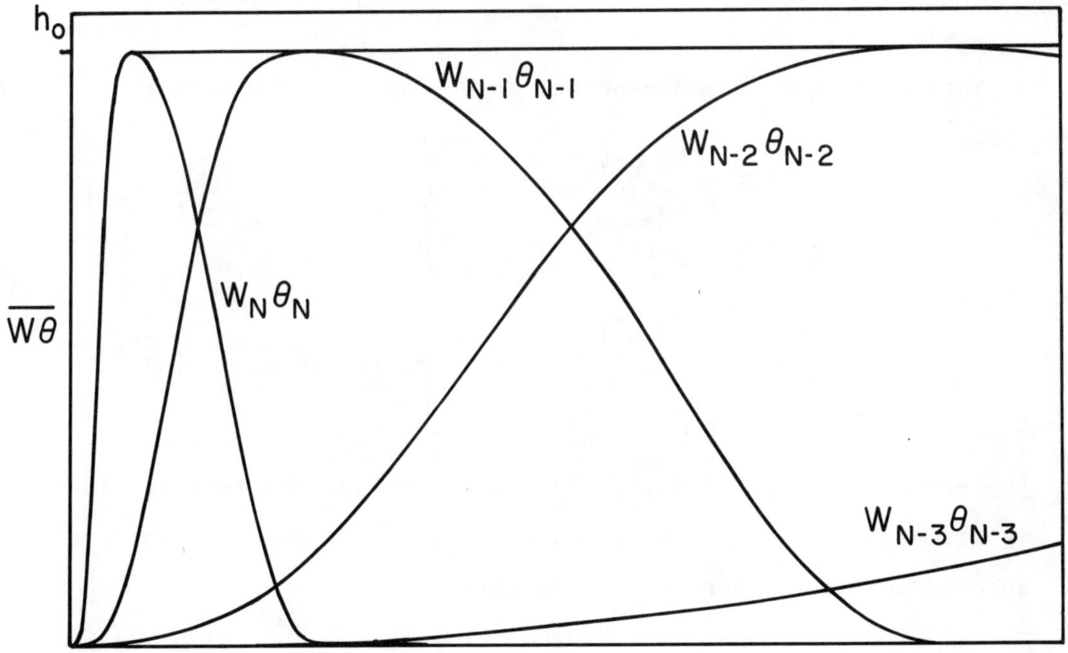

Figure 1

qualitative sketch of the multiple boundary layer structure is shown by Figure 1. In the interior of the annulus

$$w_1\theta_1 + w_1^*\theta_1^* \approx h(r) \tag{4.5}$$

holds. Assuming that the thicknesses of the respective boundary layers scale like μ^{-r_n} asymptotically, we introduce

$$\zeta_n \equiv \mu^{r_n}(1-r) \quad , \quad \zeta_n^* \equiv \mu^{r_n}(r-\eta) \tag{4.6}$$

as boundary layer coordinates at the outer and at the inner wall, respectively. Similarly, we express the asymptotic dependence on μ of the functions w_n, θ_n in the form

$$\left.\begin{aligned} w_n(r) &= \mu^{-s_n}\,\tilde{w}_n(\zeta_{n-1}) \\ \theta_n(r) &= \mu^{s_n}\,\tilde{\theta}_n(\zeta_{n-1}) \end{aligned}\right\} \quad \text{for } (1-r) \approx o(\mu^{-r_{n-1}})$$

$$\left.\begin{aligned} w_n(r) &= \mu^{-p_n}\,\hat{w}(\zeta_n) \\ \theta_n(r) &= \mu^{p_n}\,\hat{\theta}(\zeta_n) \end{aligned}\right\} \quad \text{for } (1-r) \approx o(\mu^{-r_n}) \tag{4.7}$$

We anticipate that the dependence on μ will be the same at the inner boundary

$$\left.\begin{aligned} w_n^*(r) &= \mu^{-s_n}\,\tilde{w}_n^*(\zeta_{n-1}^*) \\ \theta_n^*(r) &= \mu^{s_n}\,\tilde{\theta}^*(\zeta_{n-1}^*) \end{aligned}\right\} \quad \text{for } (r-\eta) \approx o(\mu^{-r_{n-1}})$$

$$\left.\begin{aligned} w_n^*(r) &= \mu^{-p_n}\,\hat{w}^*(\zeta_n^*) \\ \theta_n^*(r) &= \mu^{p_n}\,\hat{\theta}^*(\zeta_n^*) \end{aligned}\right\} \quad \text{for } (r-\eta) \approx o(\mu^{-r_n}) \tag{4.8}$$

The representations (4.6), (4.7), (4.8) include the interior if $r_n = 0$ is assumed for $n = 0$. Finally, we assume that the asymptotic dependence of the wavenumbers on μ is given by

$$\alpha_n^2 = \mu^{q_n}\, b_n^{\,2} \;, \qquad \alpha_n^{*2} = \mu^{q_n}\, b_n^{*2} \quad .$$

To determine r_n , p_n , s_n , q_n we consider the boundary layer approximation of the functional (3.5)

$$\hat{\mathcal{R}}(w^{(N)},\theta^{(N)},\mu) = \mu^{1-r_N} \frac{2}{1-\eta^2} \left\{ \int_0^\infty (h_o - \hat{w}_N\hat{\theta}_N)^2 d\zeta_N + \eta \circledast \right\}$$

$$+ \frac{2}{1-\eta^2} \left\{ \sum_{n=1}^{N} \mu^{3r_n-2p_n-q_n} \left(b_n^{-2} \int_0^\infty \hat{w}''^2 d\zeta_n + \eta \circledast \right) \right.$$

$$+ \sum_{n=2}^{N} \mu^{q_n-r_{n-1}-2s_n} \left(b_n^2 \int_0^\infty \tilde{w}_n^2 d\zeta_{n-1} + \eta \circledast \right) \tag{4.9}$$

$$\left. + \sum_{n=1}^{N} \mu^{r_n+2p_n} \left(\int_0^\infty \hat{\theta}_n'^{\,2} d\zeta_n + \eta \circledast \right) + \sum_{n=2}^{N} \mu^{q_n-r_{n-1}+2s_n} \left(b^2 \int_0^\infty \tilde{\theta}^2 d\zeta_{n-1} + \eta \circledast \right) \right\}$$

$$+ \mu^{q_1+s_1} (b_1^2 \langle \tilde{w}_1^2 \rangle + \circledast) + \mu^{q_1-s_1} (b_1^2 \langle \tilde{\theta}_1^2 \rangle + \circledast) .$$

The symbol $\circledast$ stands for the starred equivalent of the first term of the respective bracket. Since all parameters r_n , p_n , s_n , q_n appear with positive as well as with negative signs in the exponents of the expression (4.9), the functional $\hat{\mathcal{R}}$ reaches a minimum when all exponents are equal. This condition determines

$$r_n = \frac{1-4^{-n}}{2-4^{-N}} ; \qquad q_n = \frac{2-4\cdot 4^{-n}}{2-4^{-N}} ; \quad 2p_n = \frac{4^{-n}}{2-4^{-N}} ; \quad s_n = 0 . \tag{4.10}$$

Hence the μ dependence of the minimum $R^{(N)}(\mu)$ of the functional (4.9) can be written in the form

$$R^{(N)}(\mu) = \mu^{\frac{1}{2-4^{-N}}} F(N) . \tag{4.11}$$

The results can be used to obtain the boundary layer approximation of the Euler-Lagrange equations (3.7) for the variables $\hat{w}_n$, $\hat{\theta}_n$, $\tilde{w}_{n+1}$, $\tilde{\theta}_{n+1}$ and their starred equivalents. In the interior the equations reduce to

$$b_1^2\ \tilde{w}_1 + G_o\ \tilde{\theta}_1 = b_1^2\ \tilde{\theta}_1 + G_o\ \tilde{w}_1 = 0 \quad ,$$
$$b_1^{*2}\ \tilde{w}_1^* + G_o\ \tilde{\theta}_1^* = b_1^{*2}\ \tilde{\theta}_1^* + G_o\ \tilde{w}_1^* = 0 \quad , \tag{4.12}$$

with $G_o \equiv \mu^{r_N}(\tilde{w}_1\tilde{\theta}_1 + \tilde{w}_1^*\ \tilde{\theta}_1^* - h) - \frac{h}{2}\left\{ F(N) + \frac{2}{1-\eta^2}\left(\int_0^\infty (h_o - \hat{w}_N\hat{\theta}_N)^2 d\zeta_N + \eta \circledast\right)\right\}$.

The balance (4.12) requires

$$\tilde{w}_1 = \tilde{\theta}_1 \quad ; \quad \tilde{w}_1^* = \tilde{\theta}_1^* \quad ; \quad b_1 = b_1^* \quad . \tag{4.13}$$

Since r_N is positive the relation (4.5) is indeed satisfied asymptotically. The boundary layers at the outer wall are described by the equations

$$b_n^{-2}\hat{w}_n'''' - G_n\hat{\theta}_n = 0 \quad , \quad \hat{\theta}_n'' + G_n\ \hat{w}_n = 0 \quad , \tag{4.14}$$

$$b_{n+1}^2\tilde{w}_{n+1} - G_n\ \tilde{\theta}_{n+1} = 0 \quad , \quad b_{n+1}^2\ \tilde{\theta}_{n+1} - G_n\ \tilde{w}_{n+1} = 0 \quad , \tag{4.15}$$

with $G_n \equiv \mu^{r_N - r_n} (h_o - \hat{w}_n\hat{\theta}_n - \tilde{w}_{n+1}\tilde{\theta}_{n+1})$.

At the inner wall the same equations hold for the corresponding starred quantities. Again, we notice that the equations are consistent with the assumption (4.4) made above in all but the N-th boundary layer. In the N-th boundary layer $\tilde{w}_{N+1}\tilde{\theta}_{N+1}$ has to be replaced by zero, of course. The equations (4.14), (4.15) are identical to those considered in I. For further discussion we need only the results

$$b_n^{-2}\int_0^\infty \hat{w}_n''^2 d\zeta_n + b_{n+1}^2\int_0^\infty \tilde{w}_{n+1}^2 d\zeta_n = \int_0^\infty \hat{\theta}_n'^2 d\zeta_n + b_{n+1}^2\int_0^\infty \tilde{\theta}_{n+1} d\zeta_n = \left(\frac{b_{n+1}^4}{b_n}\right)^{1/3} h_o 3\beta \tag{4.16}$$

for $n = 1, \ldots, N-1$.

$$b_n^{-2}\int_0^\infty \hat{w}_N''^2 d\zeta_N = \int_0^\infty \hat{\theta}_N'^2 d\zeta_N = \frac{1}{4}\int_0^\infty (h_o - \hat{w}_N\hat{\theta}_N)^2 d\zeta_N = \left(\frac{h_o^2}{b_N}\right)^{1/3} h_o\sigma \quad . \tag{4.17}$$

Identical results hold for the starred quantities. The constants

$$\sigma = .337 \qquad \text{and} \quad 3\beta = 1.847$$

have been calculated by Howard (1963) and by Busse (1969b). Relations

(4.5), (4.13), (4.16), and (4.17) yield

$$F(N) = \frac{12\beta}{1-\eta^2}\sum_{n=1}^{N}\left\{h_o\left(\frac{b_{n+1}^4}{b_n}\right)^{1/3} + h_o^*\eta\left(\frac{b_{n+1}^{*4}}{b_n^*}\right)^{1/3}\right\} + 2b_1^2\langle h\rangle \qquad (4.18)$$

where the definitions

$$b_{N+1} \equiv h_o^{1/2}\,(\sigma/\beta)^{3/4} \quad , \quad b_{N+1}^* \equiv h_o^{*1/2}\,(\sigma/\beta)^{3/4}$$

have been introduced to simplify the notation. The final step of the analysis is the minimization of the expression (4.18) as a function of the parameters b_n, b_n^* (n=1, ..., N). The vanishing of the derivatives of $F(N)$ with respect to b_n, b_n^* requires

$$\left(\frac{b_{n+1}}{b_n}\right)^{4/3} = 4\left(\frac{b_n}{b_{n-1}}\right)^{1/3} , \quad \left(\frac{b_{n+1}^*}{b_n^*}\right)^{4/3} = 4\left(\frac{b_n^*}{b_{n-1}^*}\right)^{1/3} \quad \text{for } n = 2,\ldots,N$$

$$h_o\left(\frac{b_2}{b_1}\right)^{4/3} + h_o^*\left(\frac{b_2^*}{b_1}\right)^{4/3} = b_1\,\frac{\langle h\rangle(1-\eta^2)}{\beta}$$

with the solution

$$4b_1 = \left\{\frac{\beta 4^{7/3}}{(1-\eta^2)\langle h\rangle}\left(h_o\left(\frac{b_{n+1}}{4^{N-1}}\right)^{\frac{1}{1-4^{-N}}} + \eta\, h_o^*\left(\frac{b_{N+1}^*}{4^{N-1}}\right)^{\frac{1}{1-4^{-N}}}\right)\right\}^{\frac{1-4^{-N}}{2-4^{-N}}} \qquad (4.19)$$

$$b_{n+1} = 4^{n-1}\left\{\left(\frac{b_{N+1}}{4^{N-1}}\right)^{1-4^{-n}}(4b_1)^{4^{-n}-4^{-N}}\right\}^{\frac{1}{1-4^{-N}}} \quad \text{for } n = 1,\ldots,N$$

$$b_{n+1}^* = 4^{n-1}\left\{\left(\frac{b_{N+1}^*}{4^{N-1}}\right)^{1-4^{-n}}(4b_1)^{4^{-n}-4^{-N}}\right\}^{\frac{1}{1-4^{-N}}} \quad . \qquad (4.20)$$

Accordingly, the minimum of the functional (4.9) is given by

$$\hat{R}^{(N)}(\mu) = \mu^{\frac{1}{2-4^{-N}}}(2.4^N-1)2b_1^2\langle h\rangle \quad . \qquad (4.21)$$

We finish the analysis by proving that $\langle w^{(N)}\theta^{(N)}h\rangle = 1$ as assumed in (3.8). By multiplying the first equations in relations (4.14, (4.15) by $\hat{w}_n$ and $\tilde{w}_{n+1}$, respectively, adding them, and integrating them, we find using (4.16) and (4.4)

(4.22)

$$\mu^{-r_n} h_o \int_0^\infty (h_o - \tilde{w}_{n+1}\tilde{\theta}_{n+1} - \hat{w}_n\hat{\theta}_n) d\zeta_n = 3\mu^{-r_n} \beta h_o \left(\frac{b_{n+1}}{b_n}\right)^{1/3} \quad n = 1,\ldots,N-1.$$

By performing a similar operation in the case $n = N$ it is found

$$h_o \int_0^\infty (h_o - \hat{w}_N\hat{\theta}_N) d\zeta_N = b_N^{-2} \int_0^\infty \left\{ \hat{w}_N''^2 + (h_o - \hat{w}_N\hat{\theta}_N)^2 \right\} d\zeta_N = 5\left(\frac{h_o}{b_N}\right)^{1/3} h_o{}^\sigma .$$

Using the analogous expressions for the starred quantities and the result

$$G_o = -b_1^2$$

from (4.12) we obtain

$$\mu^{r_N} \langle h(h - \sum_{n=1}^N (w_n\theta_n + w_n^*\theta_n^*))\rangle = \frac{6\beta}{1-\eta^2} \sum_{n=1}^{N-1} \left\{ h_o \left(\frac{b_{n+1}^4}{b_n}\right)^{1/3} + \eta \otimes \right\} + b_1^2 \langle h \rangle$$

$$+ \frac{10\beta}{1-\eta^2} \left\{ h_o \left(\frac{b_{N+1}^4}{b_N}\right)^{1/3} + \eta \otimes \right\} - \frac{1}{2} F(N)$$

$$- \frac{4\beta}{1-\eta^2} \left\{ h_o \left(\frac{b_{N+1}^4}{b_N}\right)^{1/3} + \eta \otimes \right\}$$

$$= 0 \qquad ,$$

which completes the proof.

5. The Bound on the Angular Momentum Transport.

The asymptotic approximations $\hat{R}^{(N)}(\mu)$ for the minimum $R^{(N)}(\mu)$ of the function (3.5) become exact only in the limit of μ tending to infinity. Because of the monotonic dependence of $R^{(N)}(\mu)$ on μ it can be assumed, however, that the expression (4.21) provides a close approximation for moderately large values of μ. The maximum among the inverse functions $\mu^{(N)}(R)$ of $\hat{R}^{(N)}(\mu)$ yields an upper bound on the transport of angular momentum at a given value R of the parameter $|A|$. Denoting the ratio between the turbulent and the laminar torque by Nt

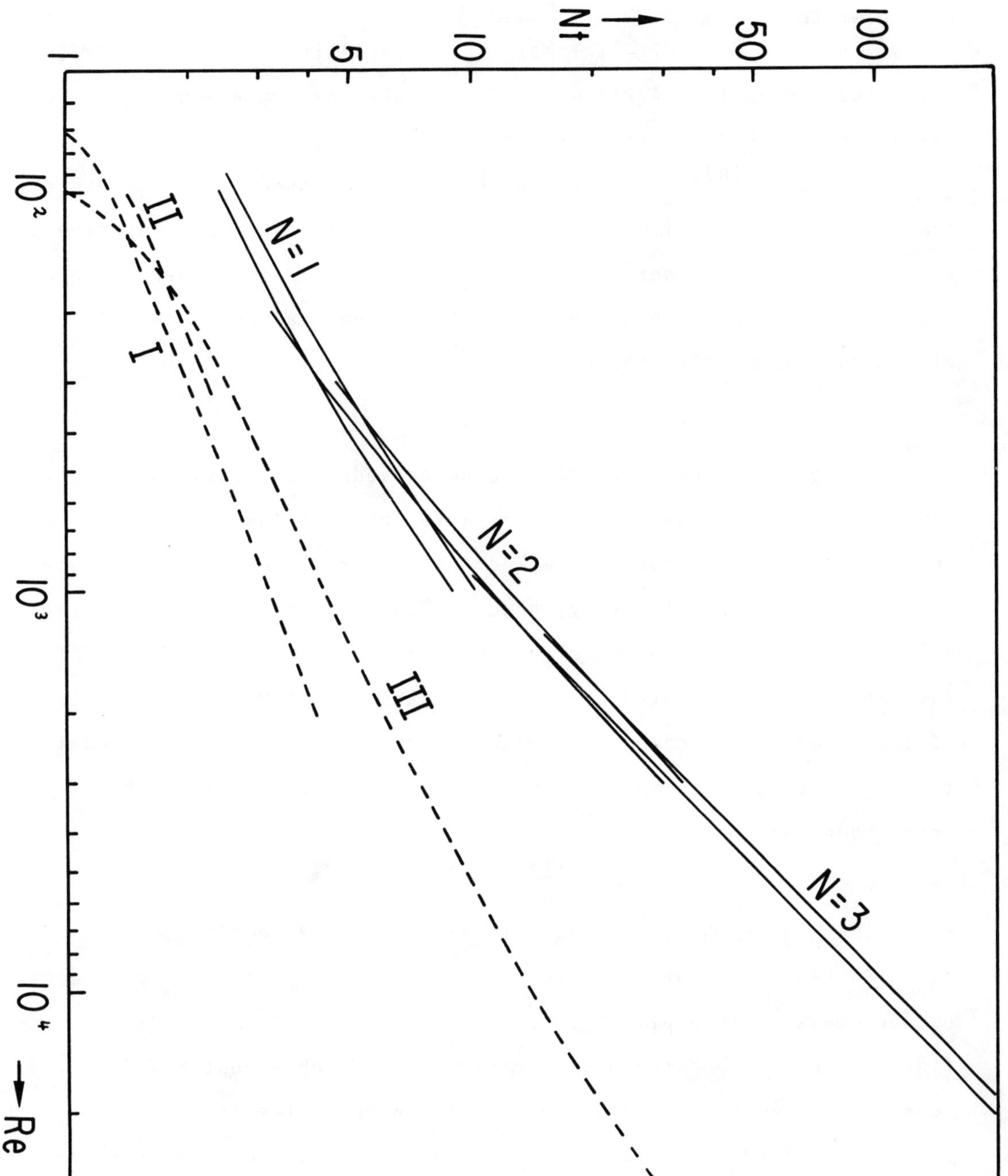

Figure 2: The upper bound for the ratio Nt of the turbulent torque to laminar torque for $\eta=.5$ (lower curve) and $\eta=.85$ (upper curve). Dashed lines indicate experimental results, I: $\eta=.5$ (Donnelly and Simon 1960), II: $\eta=.5$ (Debler et al. 1968), III: $\eta=.85$ (Wendt 1933).

we can write

$$Nt \leq 1 + \max_N (\mu^{(N)}(|A|))/|A| \quad . \qquad (5.1)$$

Using the expression (4.21) we find that the maximum at sufficiently low values of R is given by $\mu^{(1)}(R)$ and by one after the other of the functions $\mu^{(N)}(R)$ as R increases. The important consequence is that the absolute maximum is not given by a differentiable function but by a function with a series of kinks. In Figure 2 the upper bound on the ratio Nt has been drawn for two different values of η. Instead of the parameter $|A|$ the Reynolds number

$$Re = \frac{1}{2}|A|(1-\eta^2)(1-\eta)$$

has been used as parameter. The bound exceeds the experimentally measured values of the torque by a factor of 2 at lower values of the Reynolds number and does not reflect the observed power law at higher values of Re. Qualitatively, however, the bounds exhibit a similar dependence on the ratio η of the diameters of the cylinders. This fact,which has been noted earlier by Nickerson (1969), becomes evident if a more extensive comparison with the data by the authors cited in the figure caption is made. Asymptotically the expression for the upper bound becomes

$$Nt \leq Re \frac{2(1+\eta)}{\eta(\eta+\eta^{-1})^2}(\sigma^3\beta)^{-\frac{1}{2}} 4^{-14/3} \quad . \qquad (5.2)$$

Since the number N of boundary layers of the extremalizing vector field increases proportional to log μ , the asymptotic analysis does not become exactly applicable, even in the limit of μ tending to infinity. The ratio between the thicknesses of subsequent boundary layers tends toward 4 for the solution corresponding to the extremalizing value of N. Detailed estimates (Busse 1968), however, show that the error introduced by the finite value of this ratio does not change the results significantly. The boundary layer structure of the vector field optimizing the momentum transport has been sketched in Figure 3.

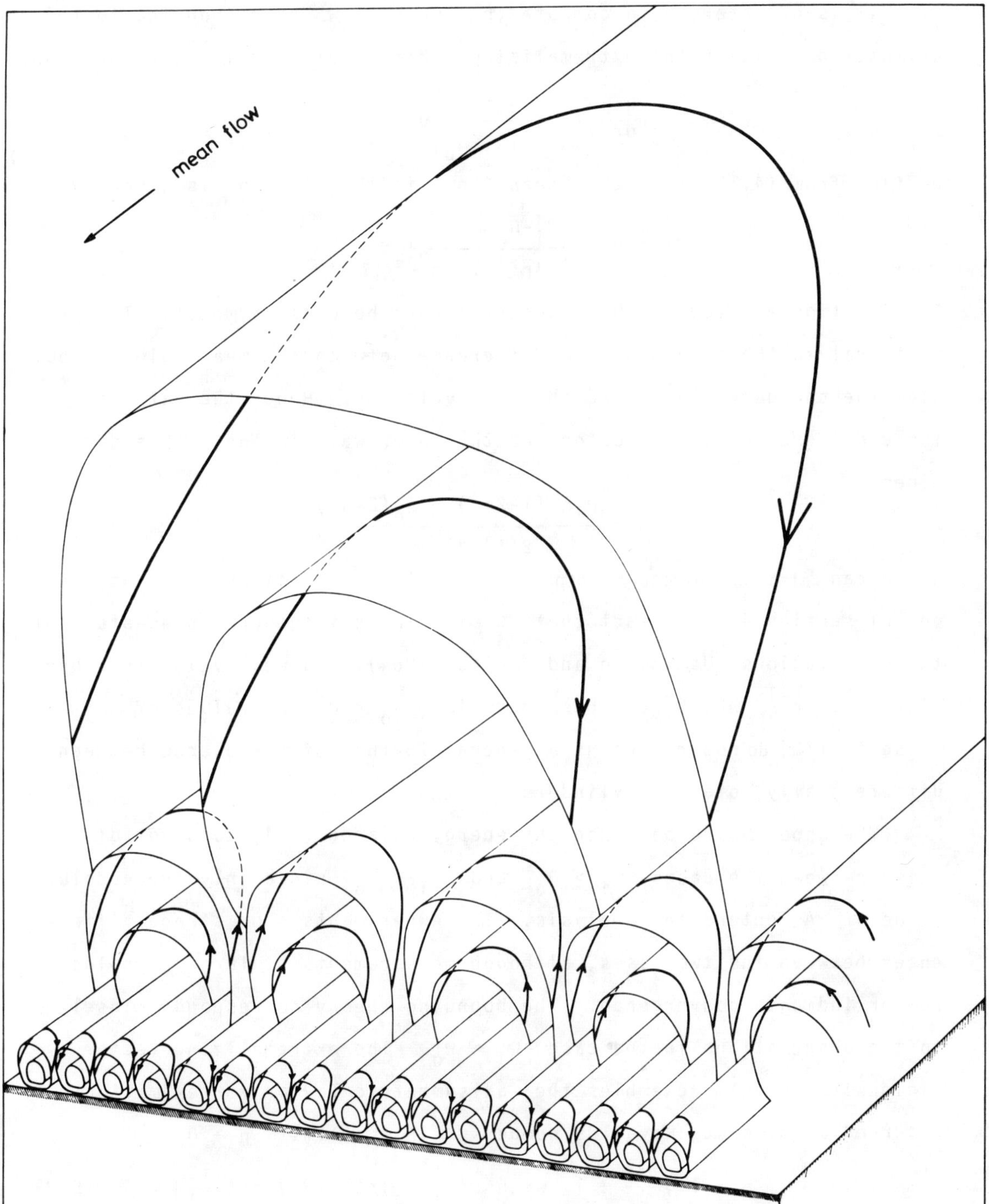

Figure 3

It is of interest to compute the "mean flow" corresponding to the asymptotic limit of the extremalizing vector field. Using the equation

$$r\frac{d}{dr}U^{\infty}/r = \frac{\Omega_i-\Omega_o}{|\Omega_i-\Omega_o|}\mu(w\theta-h) - h\ A$$

we find from (4.12) that the "mean flow" in the interior is given by

$$\frac{1}{r}\ U^{\infty} = \frac{\Omega_i-\Omega_o}{4r^2}\ \frac{\eta^2}{1-\eta^2} + C \qquad .$$

The friction exerted by the boundary layers becomes asymptotically proportional to the square of the difference between the mean velocity outside the boundary layers and the wall velocity. Hence the ratio of these differences at the outer and the inner wall is η^2. This determines

$$C = \frac{\Omega_i\eta^2(1-2\eta^2) + \Omega_o(2-\eta^2)}{2(1-\eta^4)}$$

which can also be directly computed from the expressions given at the end of section 4. The fact that C is finite in general suggests that the observations (Ustimenko and Zmeikov 1964) of a mean velocity proportional to r^{-1}, which have been made for $\Omega_o = 0$ and values of η^2 close to 1/2, do not represent a general feature of turbulence between differentially rotating cylinders.

The upper bound based on the energy balance (3.4) does not distinguish between the cases $\Omega_i > \Omega_o$ and $\Omega_i < \Omega_o$ since only the absolute value of A enters the analysis. The experiments show strong differences between the two cases, although measurements at large Reynolds number indicate convergence. The bounding theory corresponds closely to the observations in the case $\Omega_i > \Omega_o$. The extremalizing vector field with $N = 1$ resembles the **axisymmetric** Taylor vortex motion observed at moderate Reynolds numbers. In the limit $\mu \to 0$,

$$\eta \to 1\ ,\ \frac{\Omega_i}{\Omega_o} \to 1 \quad \text{with} \quad (\Omega_i-\Omega_o)/(\Omega_i+\Omega_o) = 4\,\frac{1-\eta}{1+\eta} \qquad (5.3)$$

the solution $w^{(1)},\theta^{(1)}$ even becomes identical to the hydrodynamic solution describing the Taylor vortex. This is a consequence of the property that the functional (3.5) includes the functional used in the energy method of stability as a special case corresponding to $\mu = 0$. The identity between the energy stability limit and the linear stability result in the special limit (5.3) has been shown in an earlier paper (Busse 1970b).

6. Conclusion.

The bounding theory of turbulence outlined in this paper in the case of Couette flow serves a threefold purpose. Its primary objective is to provide bounds on quantities characterizing turbulent transport processes. The Euler-Lagrange equations derived in the course of the analysis represent a second feature of interest all by themselves. They are strongly nonlinear partial differential equations which resemble closely the Navier-Stokes equations for the fluctuating velocity field. Yet they are simple enough to allow an asymptotic analytic solution in terms of a multiple boundary layer structure. The third and most exciting feature is the tendency of the realized turbulence to approach the structure of the vector field which extremalizes the turbulent transport. This tendency is most clearly exhibited by turbulent convection in a layer heated from below. We refer to the comparison with the measurements by Deardorff and Willis (1967) made by Busse (1969b). The discrete structure of the extremalizing vector field seems to represent an unphysical feature in view of the traditional notion of turbulence as a random process characterized by smooth spectra and continuous functions. The experimentally observed turbulence seems to reflect both aspects. An interesting example of the discrete structure is the series of kinks in the dependence of the turbulent heat transport on the Rayleigh number which was first observed by Malkus (1954) and which resembles the kinks in the upper bound. It is likely that similar kinks could be observed in the transport of angular momentum by turbulent circular Couette flow if sufficiently accurate methods of measurement are employed.

REFERENCES

Busse, F. H. Report MPI/Astro 8/68, Max-Planck-Institut für Physik und Astrophysik, Munich, (1968)

Busse, F. H. Z. Angew. Math. Phys. 20, 1-14, (1969a)

Busse, F. H. J. Fluid Mech. 37, 457-477, (1969b)

Busse, F. H. J. Fluid Mech. 41, 219-240, (1970a) referred to as I

Busse, F. H. Z. angew. Math. Mech. [ZAMM] 50, T173-T174, (1970b)

Deardorff, J. W. and Willis, G. E. J. Fluid Mech. 28, 675-704, (1967)

Debler, W., Füner, E., and Schaaf, B. Publ. EM 68-2, Dept. Eng. Mech., University of Michigan, Ann Arbor (1968)

Donnelly, R. J., and Simon, N. J. J. Fluid Mech. 7, 401-418, (1960)

Howard, L. N. J. Fluid Mech. 17, 405-432, (1963).

Malkus, W. V.R. Proc. Roy. Soc. A225, 196, (1954)

Nickerson, E. C. J. Fluid Mech. 38, 807-815, (1969)

Ustimenko, B. P., and Zmeikov, V. N. High Temperature 2, 220-228,(1964) Translated from Teplofizika Vysokikh Temperatur 2, 250-254, (1964)

Wendt, F. Ingenieur-Archiv 4, 577-594, (1933)

NUMERICAL SIMULATION OF TURBULENCE

Steven A. Orszag*

Department of Mathematics

M.I.T., Cambridge, Mass. 02139

and

G. S. Patterson, Jr.

Department of Engineering

Swarthmore College, Swarthmore, Pa. 19081

Numerical simulations of three-dimensional homogeneous, isotropic turbulence at wind-tunnel Reynolds numbers ($R_\lambda \sim 20$-40) are reported. The results of the simulations are compared with the predictions of turbulence theories.

We give a preliminary report of numerical simulations of three-dimensional homogeneous, isotropic turbulence in an incompressible fluid. To the best of our knowledge, these are the first such numerical simulations of three-dimensional turbulence. Two-dimensional 'turbulence' has been simulated by Lilly (1971) and others. The results of our simulations are compared with the predictions of turbulence theories. We only outline the method we used to solve the Navier-Stokes equations as details are available in the authors' other publications referred to below. The Navier-Stokes equations are solved in terms of the Fourier components of the velocity field. Periodic boundary conditions are applied in all three space directions so that discrete Fourier expansions are appropriate. The periodicity interval is chosen for now to be 2π in each direction. The equations of incompressible flow are

*Alfred P. Sloan Research Fellow.

$$[\frac{\partial}{\partial t} + \nu k^2] u_\alpha(\vec{k},t) = -ik_\beta(\delta_{\alpha\gamma} - \frac{k_\alpha k_\gamma}{k^2}) \sum_{\vec{p}+\vec{q}=\vec{k}} u_\beta(\vec{p},t) u_\gamma(\vec{q},t) \quad , \tag{1}$$

$$k_\alpha u_\alpha(\vec{k},t) = 0, \tag{2}$$

where ν is the kinematic viscosity, $\delta_{\alpha\beta} = 1$ if $\alpha=\beta$, $\delta_{\alpha\beta} = 0$ if $\alpha\neq\beta$, $\vec{k}$ has integral components, and

$$\vec{v}(\vec{x}) = \sum \vec{u}(\vec{k}) \exp[i\vec{k}\cdot\vec{x}] \tag{3}$$

is the physical-space velocity field. We use the convention that repeated Greek subscripts are summed over their range $\alpha = 1,2,3$.

Equations (1) and (2) involve a doubly infinite system of equations [noting that (2) is implied by (1) if (2) holds initially] that clearly could not be implemented on a finite computer. We obtain a finite set of ordinary differential equations for a finite number of dynamical quantities by setting $\vec{u}(\vec{k},t) \equiv 0$ for $|\vec{k}| \geq \bar{K}$, where the cutoff $\bar{K}$ will be chosen below. The dynamical quantities retained in the simulation are $\vec{u}(\vec{k},t)$ for $|\vec{k}| < \bar{K}$. This truncation is clearly the most natural for isotropic turbulence.

The resulting truncated Navier-Stokes equations may be solved on existing computers (provided, of course, that $\bar{K}$ is small enough). Some care is necessary to prevent needless inefficiencies in the simulation. In particular, the right-hand side of (1) should not be evaluated directly (which would require order $\bar{K}^6$ operations) but rather should be evaluated using transform methods (Orszag 1969, 1971a) that require only order $\bar{K}^3 \log \bar{K}$ algebraic operations. Fortunately, there exists a very efficient transform method for use with the spherical truncation made above (Patterson & Orszag 1971). The latter algorithm is most easily applied when $\bar{K} = K\sqrt{8/9}$, where K is a power of 2, because fast Fourier transforms on 2K points are required. With this choice of $\bar{K}$, evaluation of the Fourier-transformed equations requires about as much computer time as finite-difference simulations with $(2K)^3$ grid points. On the other hand, it has been shown (Orszag

1971b) that the simulations with cutoff $\bar{K}$ are at least as accurate as finite-difference simulations involving $(4K)^3$ grid points. Consequently, our simulations are at least an order-of-magnitude more efficient as regards both computer time and storage requirements than finite-difference simulations.

The calculations reported in this paper involve cutoffs $K = 16$, $\bar{K} = \sqrt{242}$, and $K = 8$, $\bar{K} = \sqrt{61}$. [The ratios $\bar{K}/K$ are slightly larger than $\sqrt{8/9}$ for reasons explained by Orszag & Patterson 1971.] There is one slight complication that is best mentioned at this point. The wavenumber cutoffs $\sqrt{242}$ and $\sqrt{61}$ are uncomfortably small for comparison with existing solutions of the equations of turbulence theories at $R_\lambda \sim 20$-40. This difficulty is easily circumvented by choosing the periodicity interval of $\vec{v}(\vec{x})$ to be $2\pi/L$ rather than 2π. Then only wavevectors $\vec{k}$ with all components multiples of L enter the representation (3). If equations (1) are only considered for such $\vec{k}$ then the effective wavenumber cutoff may be boosted to $L\bar{K}$ with no additional work. In the following, we choose periodicity interval π, so that the calculations have a wavenumber cutoff $2\bar{K}$. Hence, the big calculations (K=16) have a cutoff at $2\sqrt{242}$, while the small calculations (K=8) have a cutoff at $2\sqrt{61}$. Note that the calculations with K=16 have accuracy at least that of finite-difference simulations with $(64)^3$ grid points.

The details of the numerical code used to implement equations (1) and (2) with the spherical cutoffs are rather intricate. With cutoff $2\sqrt{242}$, discrete Fourier transforms on $(32)^3$ points must be performed and these could not be done conveniently within the 65,536 word core storage of the CDC 6600 computer used for the calculations. A new efficient method for out-of-core fast Fourier transforms was developed by one of us (GSP) and will be explained in detail elsewhere. No such problems occur for the cutoff $2\sqrt{61}$ simulations. However, efficient coding is still quite involved. Some details of the codes will be described in later publications by one of us (GSP). With cutoff $2\sqrt{61}$ and leapfrog time-differencing, our code requires $2\frac{1}{2}$ s per time step (with machine language fast Fourier transform routines and Fortran codes elsewhere). With cutoff $2\sqrt{242}$, the calculation requires 30 s of central processor time per time step and the code is designed so that the peripheral processing time for buffering large arrays in and out of core should be fully covered by calculations; peripheral hardware misalignment has so far precluded complete covering of the data transfer operations. Typical decay calculations with cutoff $2\sqrt{242}$ involve 200 time steps, or about $1\frac{1}{2}$ h of CDC 6600 computer time.

We choose a random initial incompressible flow field from a Gaussian ensemble with prescribed isotropic energy spectrum E(k) [see the Appendix to Orszag 1969]. In the calculations reported here, E(k) is chosen to be either

$$E(k) = 16\sqrt{2/\pi}\; v_0{}^2 k_{max}^{-5} k^4 \exp[-2(k/k_{max})^2] \qquad (4)$$

with the values (cf. Kraichnan 1964)

$$k_{max} = 4 \times 2^{1/4} = 4.75683, \quad v_0 = 1,$$

or

$$E(k) = \begin{cases} 0.75 & \text{if } 4 \le k < 6 \\ 0 & \text{otherwise} \end{cases} . \qquad (5)$$

As mentioned above, we use leapfrog time-differencing for the non-linear terms, with constant time steps. The viscous terms are treated using Crank-Nicolson time differencing.

At specified times, averages are computed by averaging over shells in Fourier space. The average modal energy, energy,dissipation, and transfer spectra, defined as averages over bands of width $2\Delta k$ centered at k_0, are given by

$$\begin{Bmatrix} \bar{U}(k_0) \\ \bar{E}(k_0) \\ \bar{D}(k_0) \\ \bar{T}(k_0) \end{Bmatrix} = \frac{1}{2\Delta k} \sum_{k_0 - \Delta k \le |\vec{k}| < k_0 + \Delta k} \begin{Bmatrix} (4\pi k^2)^{-1} u_\alpha(\vec{k}) u_\alpha(-\vec{k}) \\ \frac{1}{2} u_\alpha(\vec{k}) u_\alpha(-\vec{k}) \\ \nu k^2 u_\alpha(\vec{k}) u_\alpha(-\vec{k}) \\ u_\alpha(\vec{k}) t_\alpha(-\vec{k}) \end{Bmatrix} \tag{6}$$

respectively, where $t_\alpha(\vec{k})$ denotes the right-hand side of (1). The averaging interval $\Delta k = 1$ is used exclusively below.

We also compute band-averaged error spectra, defined as the rms expected error in the spectra (6) if fluctuations are statistically independent. For a quantity $\bar{A}$ defined as the sum of N fluctuating quantities a_i $(i=1,\ldots,N)$, the rms expected error in $\bar{A}$ is easily shown to be

$$\delta(\bar{A}) = \left(\sum_{i=1}^{N} a_i^{\,2} - \frac{1}{N} \bar{A}^2 \right)^{1/2} . \tag{7}$$

The errors $\delta(\bar{A})$ are used to calculate the error bars in figures 1-10. Clearly, these statistical errors increase rapidly with decreasing wavenumber, because of the existence of relatively few low wavenumber modes.

Details of the parameter values for the runs reported here are given in table I. Runs 1,2,3 correspond to runs 1,2,3, respectively, of Herring & Kraichnan (1971); our runs 1,2,3,4 correspond to runs 1,2,4,5, respectively, of Kraichnan (1964). Both these latter references give numerical solutions of the equations of turbulence theories. The values of Δt listed in table I were selected after some experimentation to ensure that time differencing errors were negligible. Several other runs with cutoff $2\sqrt{242}$ and many with cutoff $2\sqrt{61}$ were made, but the results for runs 1-4 are representative.

For both spectra (4) [with $v_0 = 1$] and (5), the initial rms velocity, defined by

$$3v_{rms}^2(t) = \sum u_\alpha(\vec{k},t) u_\alpha(-\vec{k},t) , \tag{8}$$

should be 1 in the absence of statistical fluctuations and truncation errors, the latter due to finite periodicity-box volume. Principally

TABLE I

Run	Initial spectrum	Cutoff	ν	Δt	Final t
1	(5)	$2\sqrt{61}$	0	.0032	2.0
2	(4)	$2\sqrt{242}$	.02	.004	.84
3a	(4)	$2\sqrt{242}$	.01189	.004	.84
3b	(4)	$2\sqrt{242}$	.01189	.004	.52
4	(4)	$2\sqrt{242}$	.01	.004	.60

because of sampling effects, the actual velocity fields used in runs 1-4 do not satisfy $v_{rms}(0) = 1$ exactly or the requirement of exact statistical isotropy. A measure of the latter effect is the distribution of rms velocity among the three coordinate directions. Define $\bar{v}_\alpha$ as the rms velocity along coordinate axis α so that

$$3v_{rms}^2 = \bar{v}_\alpha \bar{v}_\alpha .$$

Data for the runs 1-4 for v_{rms}, $\bar{v}_\alpha$, $L(0)$, $\lambda(0)$, $R_L(0)$, $R_\lambda(0)$ are given in table II. Here the integral scale $L(t)$ and the Taylor microscale $\lambda(t)$ may be defined by

$$L(t) = \tfrac{1}{4}\pi[v_{rms}(t)]^{-2} \sum k^{-1} u_\alpha(\vec{k},t) u_\alpha(-\vec{k},t), \tag{9}$$

$$\lambda(t) = [15 v_{rms}^2(t) / \sum k^2 u_\alpha(\vec{k},t) u_\alpha(-\vec{k},t)]^{1/2} . \tag{10}$$

The associated Reynolds numbers are

$$R_L(t) = L(t) v_{rms}(t)/\nu , \quad R_\lambda(t) = \lambda(t) v_{rms}(t)/\nu . \tag{11}$$

TABLE II

Run	$v_{rms}(0)$	$\bar{v}_1(0)$	$\bar{v}_2(0)$	$\bar{v}_3(0)$	$L(0)$	$\lambda(0)$	$R_L(0)$	$R_\lambda(0)$
1	.9428	.930	.879	1.015	.4956	.4635	∞	∞
2	.9772	.966	.949	1.015	.5470	.4305	26.72	21.03
3a	.9784	.998	.925	1.010	.5444	.4234	44.80	34.84
3b	.9960	1.017	1.018	.951	.5264	.4167	44.10	34.91
4	.9962	.936	.977	1.071	.5273	.4197	52.54	41.81

Run 1 gives a test of the approach to an inviscid equipartition ensemble. According to well known arguments (e.g. Kraichnan 1958), if the viscosity is zero, wavevector excitations are truncated at $\bar{K}$, and the inviscid Navier-Stokes equations are mixing, the statistics of rather arbitrary initial ensembles should approach that of a Gaussian equipartition ensemble as $t\to\infty$. In particular, the band-averaged modal energy spectrum $\bar{U}(k,t)$ should approach a constant as $t\to\infty$. This tendency is clearly illustrated by the results plotted in figure 1. Since the initial large-eddy circulation time $L(0)/v_{rms}(0)$ is roughly 0.5, it follows that approach to equipartition occurs on a time scale $4L(0)/v_{rms}(0)$. Several other runs with the inviscid Navier-Stokes equations gave quantitatively similar results.

The skewness of the longitudinal velocity-derivative, defined by

$$S(t) = -\langle(\partial v_1/\partial x_1)^3\rangle/\langle(\partial v_1/\partial x_1)^2\rangle^{3/2} \tag{12}$$

where $\langle\cdot\rangle$ denotes space average, is an important normalized measure of the nonlinear cascade of turbulent energy. The evolution of $S(t)$ in runs 1-4 is shown in figure 2. The curve labelled run 3 is for run 3a; the skewness for run 3b cannot be distinguished from that of run 2 on the scale of the graph. The decay of $S(t)$ to zero as $t\to\infty$ in run 1 is further evidence of the approach to an equipartition ensemble. It is interesting that the skewness for run 1 follows so closely the skewnesses for the other runs for $t \leq 0.2$. Evidently the $S(t)$ vs t curve for $t \leq 0.2 \simeq 0.4L(0)/v_{rms}(0)$ is determined by nonlinear transfer, not by either the cutoff $\bar{K}$ or viscous dissipation.

Within statistical fluctuations the $S(t)$ curves for runs 2-4 are identical. This is strong evidence for Reynolds-number independence of small-scale-turbulence structure, even at the moderate Reynolds numbers investigated here. For runs 2-4, the skewness is about 0.47-0.48 throughout the interval $0.5\leq t\leq 0.8$. At $t=0.5$, the values of $R_\lambda(t)$ are 13.5, 20.7, 20.9, 25.1 for runs 2, 3a, 3b, 4, respective-

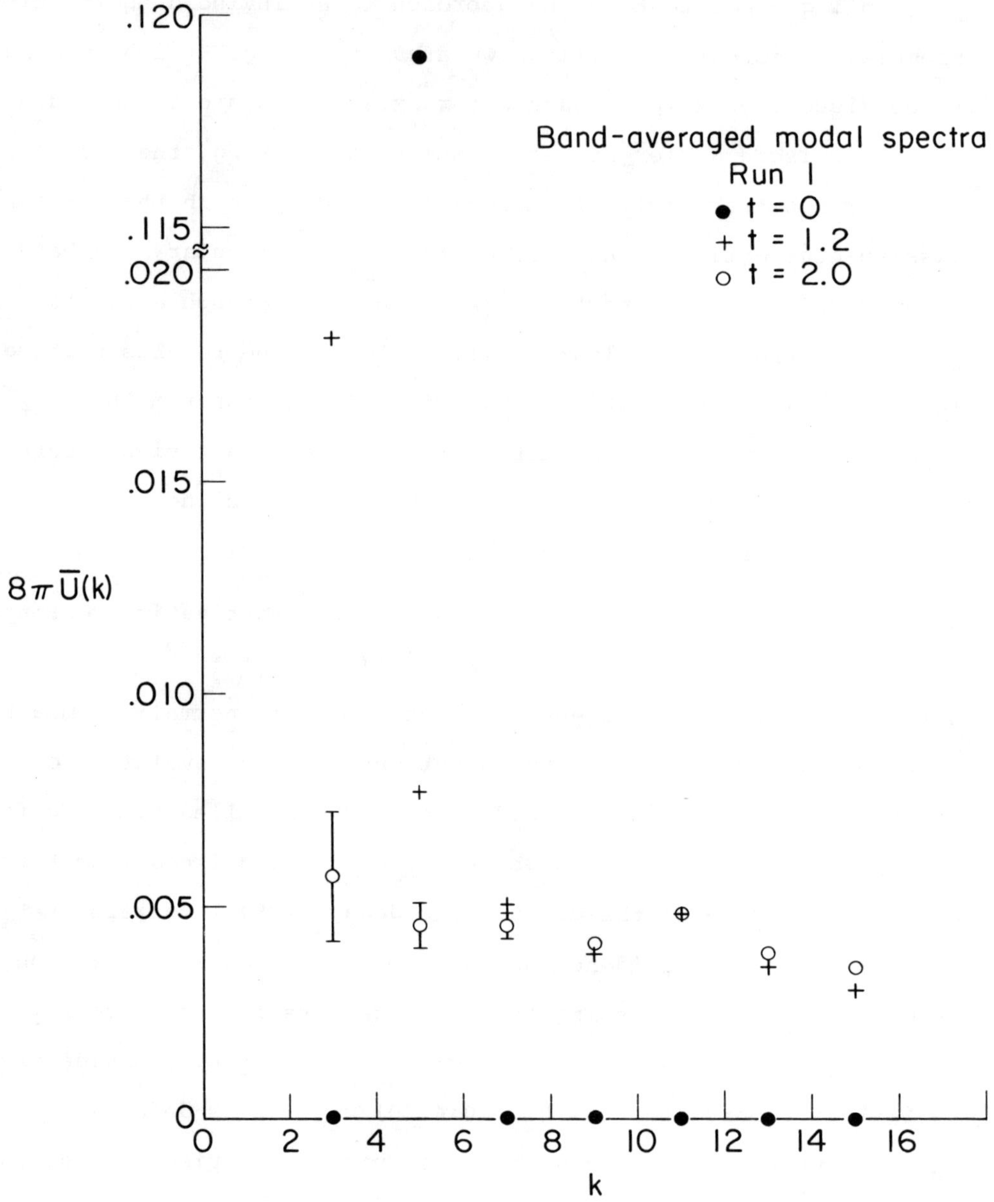
Band-averaged modal spectra
Run 1
● t = 0
+ t = 1.2
○ t = 2.0
.120
.115
.020
.015
.010
.005
0
8π Ū(k)
0
2
4
6
8
10
12
14
16
k

Figure 1

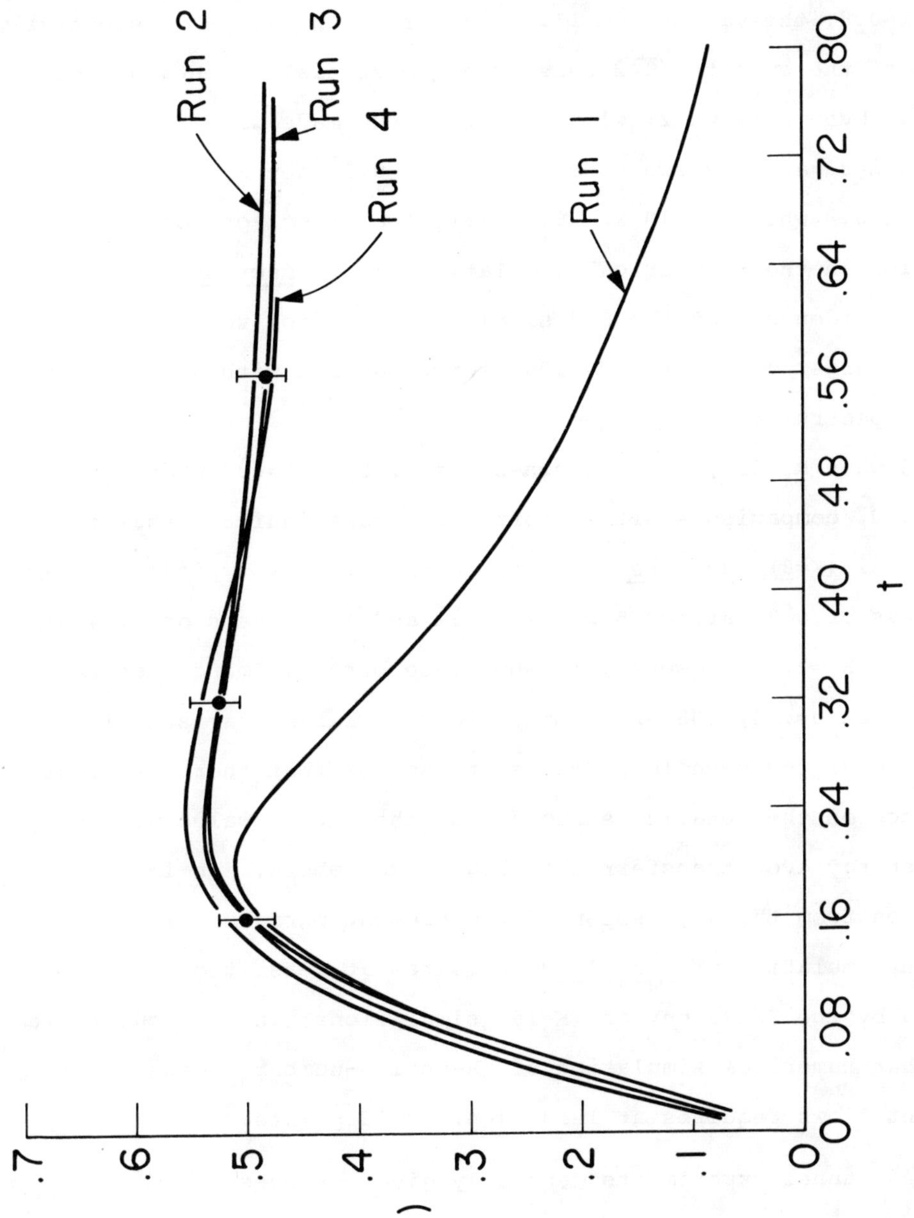

Figure 2

ly; at t=0.8, the values are 11.9, 17.7 for runs 2, 3a, respectively. The use of the cutoff $2\sqrt{242}$ runs is quite crucial to obtain these results. Even for run 2, with ν= .02, a run made with a cutoff of $2\sqrt{61}$ and K=8 gave a skewness of 0.38 at t=0.5 when R_λ = 12.6 and 0.39 at t=0.8 when R_λ= 10.9. Evidently the effect of too much truncation in these numerical simulations is to <u>decrease</u> the skewness. There is evidence that the K=8 simulations are not very accurate even with the larger viscosity ν = .04. A typical run with ν = .04 and initial spectrum (4) using the cutoff $2\sqrt{61}$ gives a skewness of 0.44 at t=0.5 when R_λ=6.7, and a skewness of 0.45 at t=0.8 when R_λ=5.7. With ν=.08, comparisons with cutoff $2\sqrt{242}$ runs indicate that the cutoff $2\sqrt{61}$ (K=8) runs <u>are</u> accurate. With ν=.08, a typical run gives a skewness of 0.36 at t=0.5 when R_λ=3.1 and a skewness of 0.34 at t=0.8 when R_λ=2.7. However, it should be borne in mind that the results with ν=.04, .08 are subject to very large statistical fluctuations; the Reynolds numbers are so low that there is great dependence on the detailed structure of the large scales with very little energy ever transferred to high wavenumbers. An important conclusion from the discussion of the present paragraph is that accurate simulation of <u>turbulence</u> requires at least the resolution afforded by the $2\sqrt{242}$ cutoff (K=16) simulations; in particular, it seems that numerical simulation of (Reynolds-number independent) turbulent flows requires at least $(64)^3$ grid points.

Wind-tunnel experiments generally give skewnesses in the range 0.3-0.4 (see, e.g., Batchelor 1953, Fig. 6.3). Admittedly, most of these measurements are made at somewhat higher values of R_λ than those used for the simulations here, but it is the authors' opinion that the discrepancies are genuine. The matter should be settled by the cutoff $2\sqrt{925}$ (K=32) simulations now being prepared for the CDC 7600 computer at the National Center for Atmospheric

Research; accurate simulations with $R_\lambda(0)\sim 75$-100 should then be possible.

In figure 3, the evolution of K(t), defined by

$$K(t) = [\varepsilon(t)/E(t)]L(t)/v_{rms}(t) \qquad (13)$$

where $\varepsilon(t)$ is the rate of energy dissipation and E(t) is the total energy, is plotted vs t for runs 2-4; there is no perceptible difference between runs 3a and 3b. K(t) is a measure of the dissipation per eddy-circulation-period and, according to the accepted tenets of equilibrium-range turbulence theory [e.g., Batchelor 1953, §6.1], K(t) [denoted A by Batchelor] should be a number of order unity which may vary but slightly with the time of decay and the initial conditions of the turbulence. It is by no means obvious that this condition is satisfied, although our Reynolds numbers may not be high enough to give a fair test of the hypothesis.

In figure 4, the decay of the band-averaged energy spectrum for run 3a is shown. In figures 5-9, we compare the results of our simulations 3a and 3b with the direct-interaction approximation (Kraichnan 1958,1964) and the test-field model with g=1.064 (Kraichnan 1971, Herring & Kraichnan 1971). It is evident that the direct-interaction approximation is superior as regards the various spectra and the dissipation rate $\varepsilon(t)$, but the test-field model gives a better representation of the skewness S(t). It has been found that the choice $g\simeq 1.2$ in the test-field model gives an excellent description of both the spectra and the skewness, though these results are not plotted here. Comparisons with the other runs have also been made. The nature of the errors of the direct-interaction turbulence theory remain roughly the same as for runs 3: significant underestimation of the skewness and very slight underestimation of the dissipation rate. For example, the behavior of $\varepsilon(t)$ for run 2 is plotted in figure 10; at this low Reynolds number,

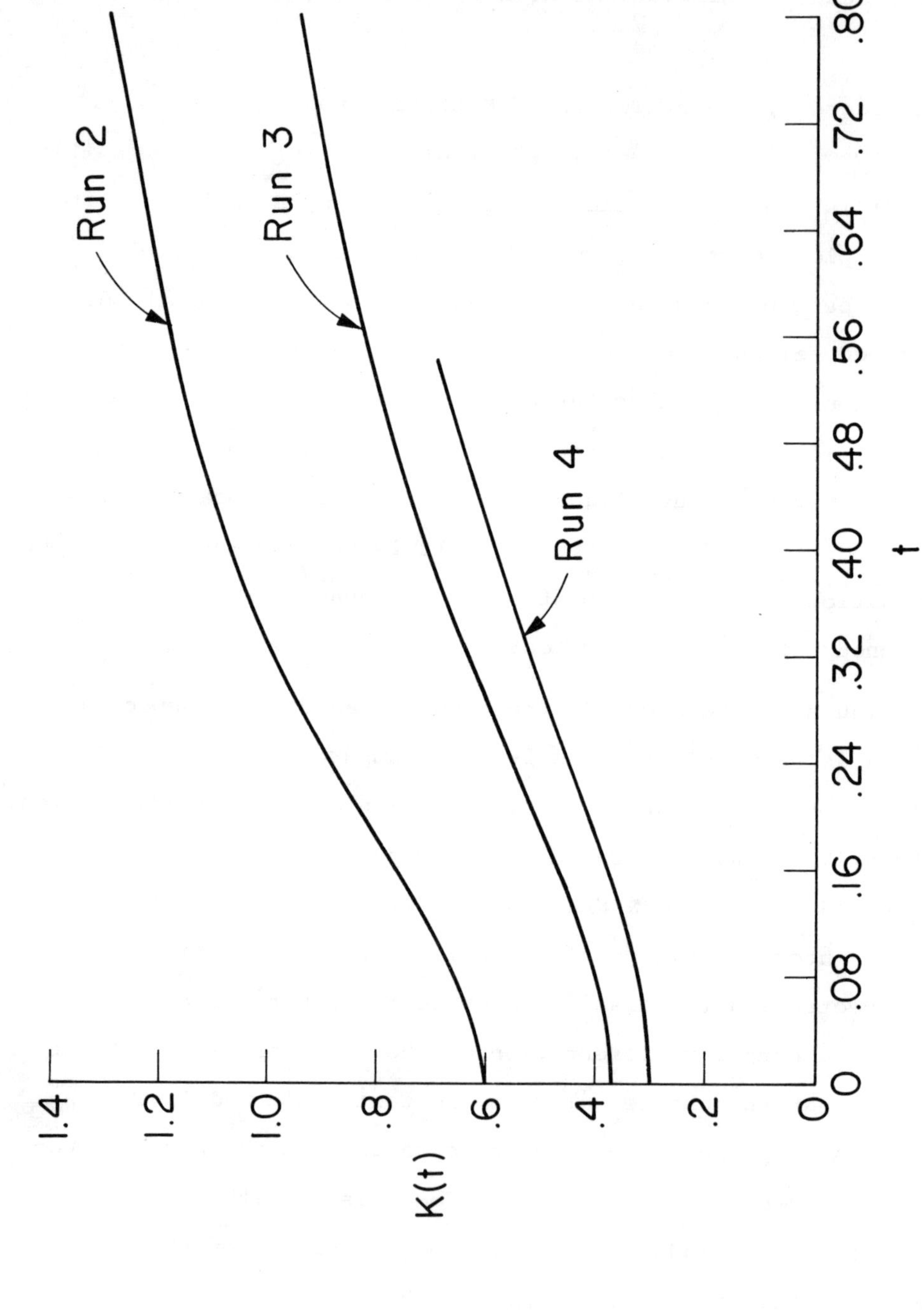

Figure 3

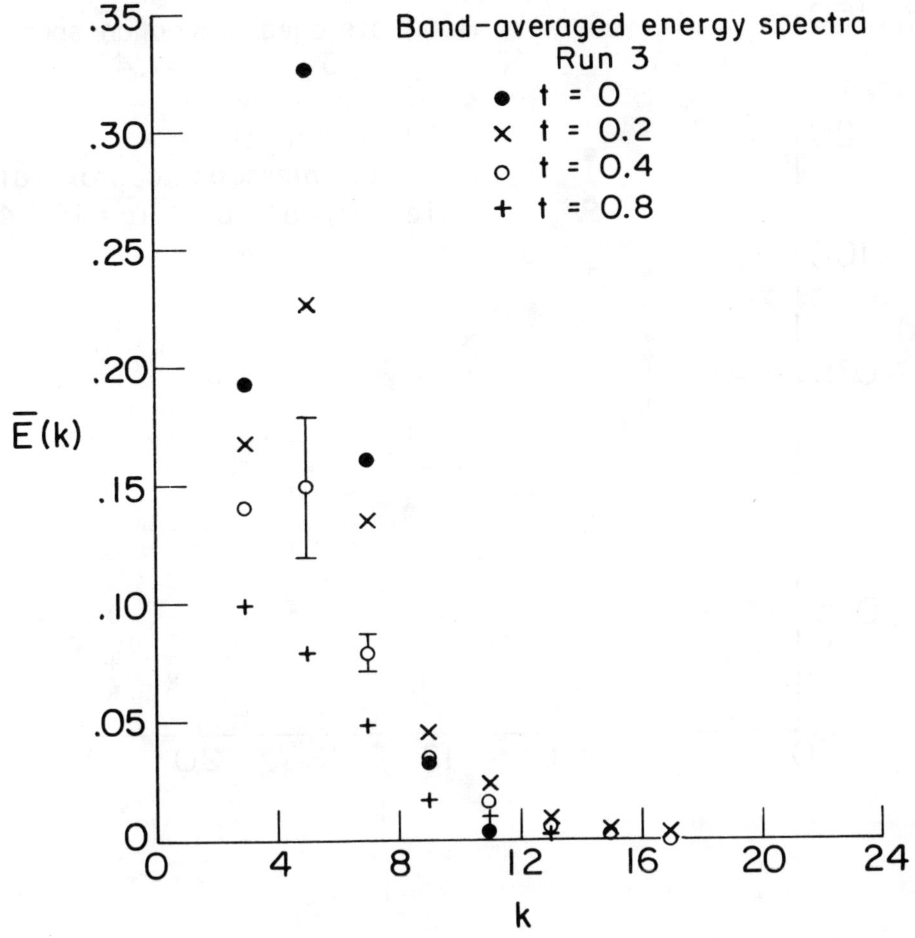
Band-averaged energy spectra
Run 3
● t = 0
× t = 0.2
○ t = 0.4
+ t = 0.8
.35
.30
.25
.20
.15
.10
.05
0
Ē(k)
0
4
8
12
16
20
24
k

Figure 4

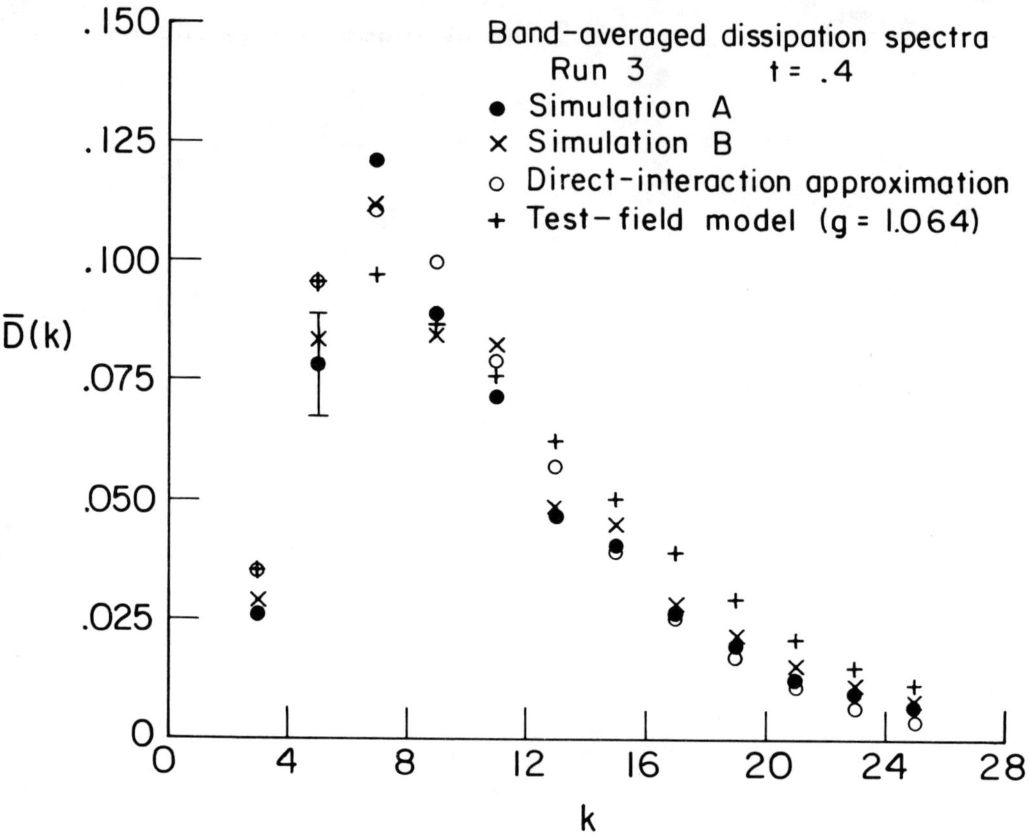
Band-averaged dissipation spectra
Run 3 t = .4
● Simulation A
× Simulation B
○ Direct-interaction approximation
+ Test-field model (g = 1.064)
$\bar{D}(k)$
.150
.125
.100
.075
.050
.025
0
0
4
8
12
16
20
24
28
k

Figure 5

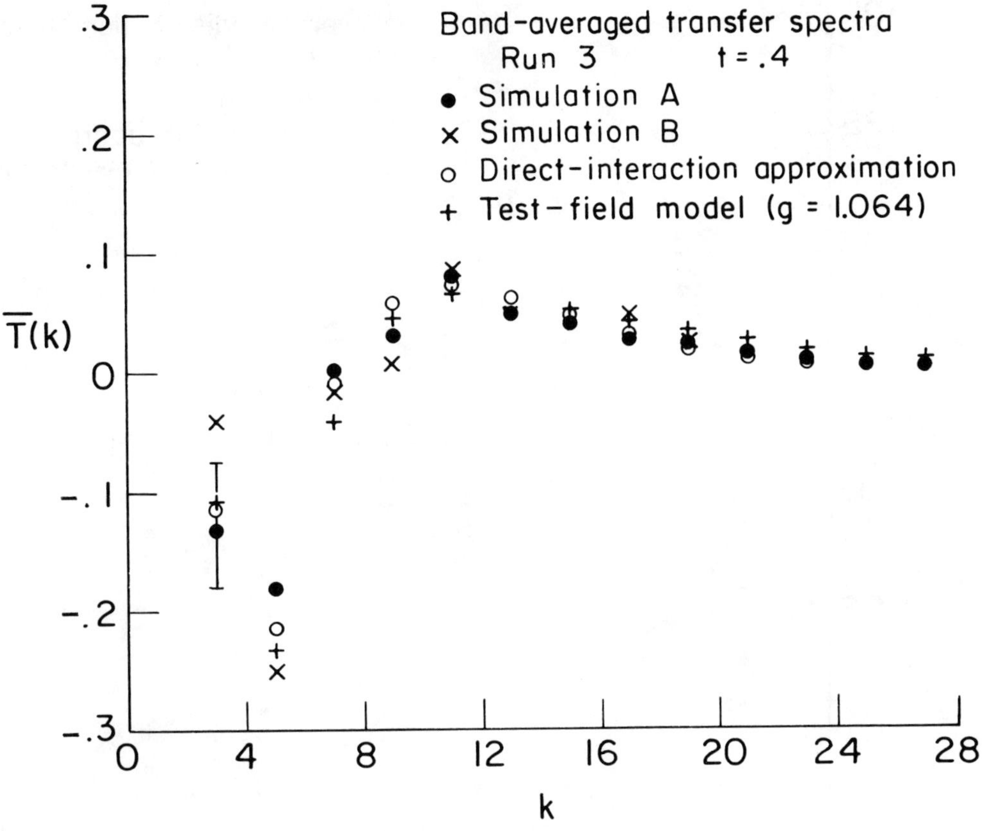
Band-averaged transfer spectra
Run 3 t = .4
Simulation A
Simulation B
Direct-interaction approximation
Test-field model (g = 1.064)
T̄(k)
.3
.2
.1
0
-.1
-.2
-.3
0
4
8
12
16
20
24
28
k

Figure 6

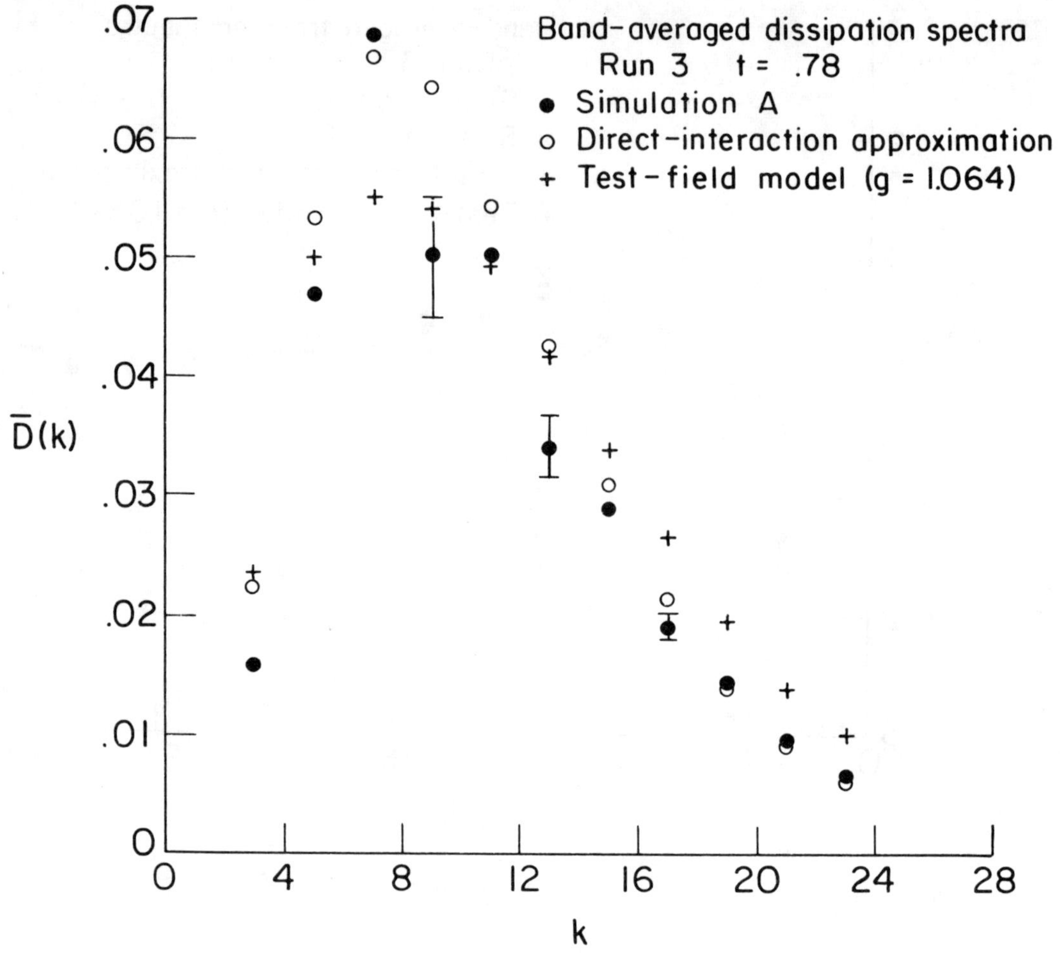
Band-averaged dissipation spectra
Run 3 t = .78
● Simulation A
○ Direct-interaction approximation
+ Test-field model (g = 1.064)
.07
.06
.05
.04
.03
.02
.01
0
D̄(k)
0
4
8
12
16
20
24
28
k

Figure 7

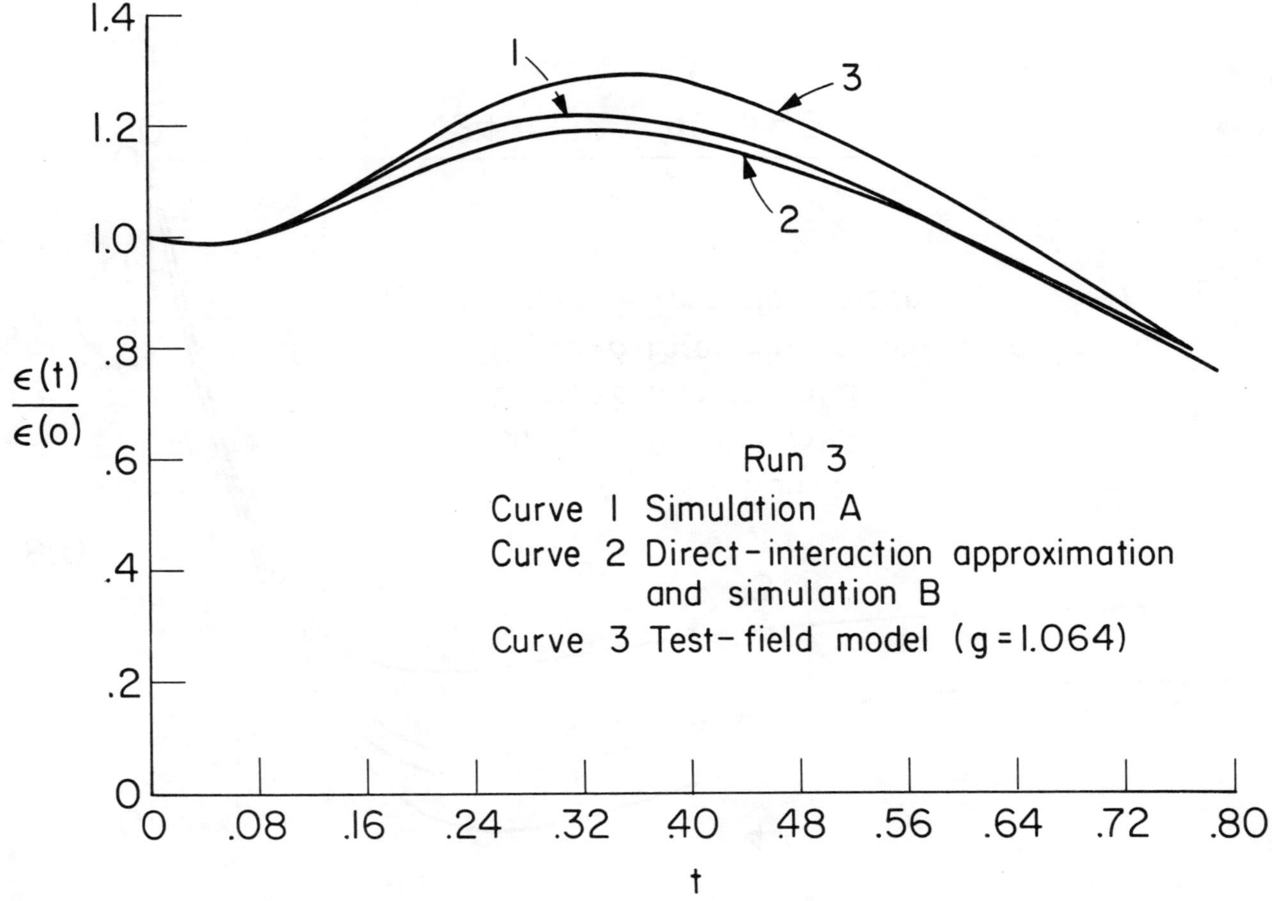
1.4
1.2
1.0
.8
.6
.4
.2
0
ε(t) / ε(o)
0
.08
.16
.24
.32
.40
.48
.56
.64
.72
.80
t
1
2
3
Run 3
Curve 1 Simulation A
Curve 2 Direct-interaction approximation and simulation B
Curve 3 Test-field model (g = 1.064)

Figure 8

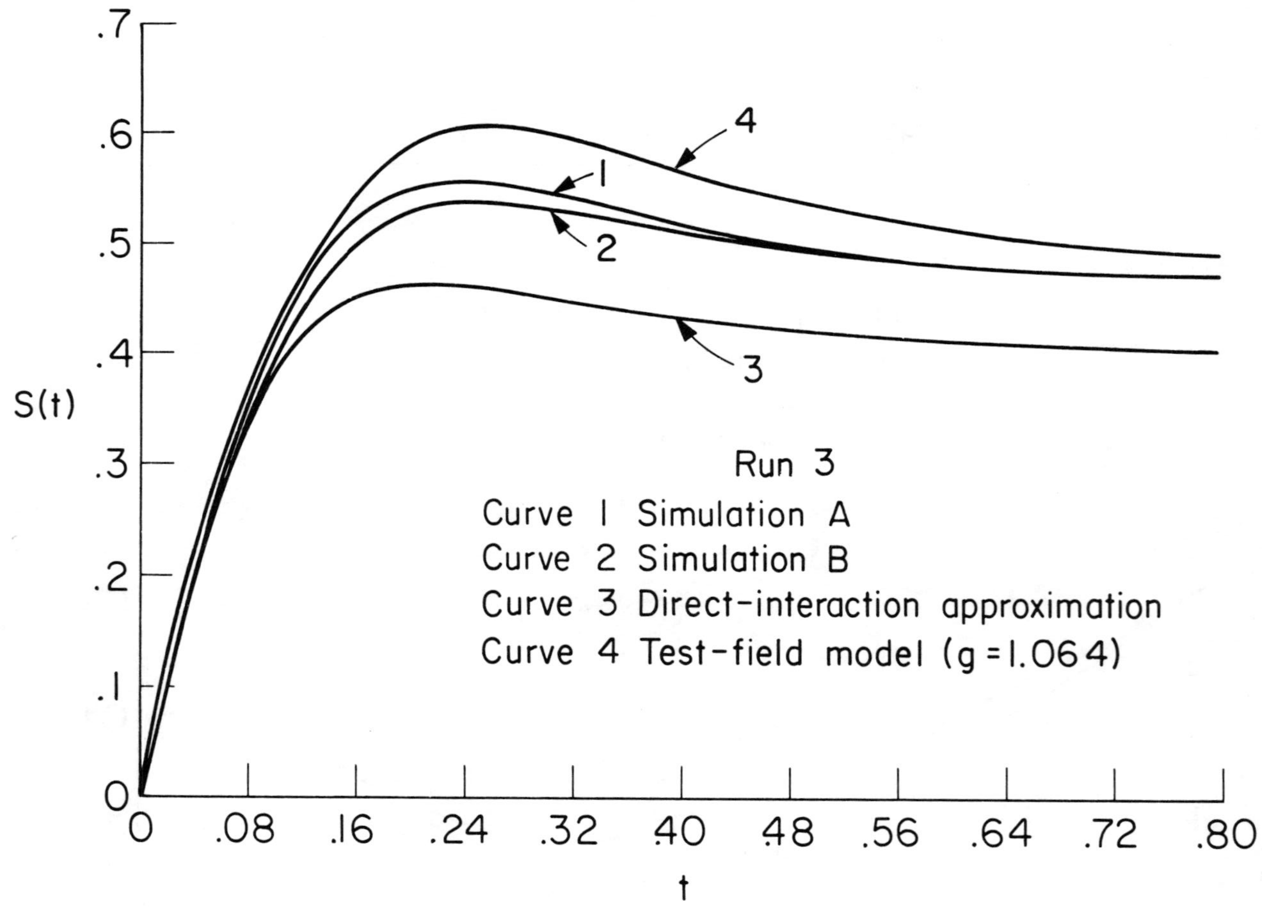
.7
.6
.5
.4
.3
.2
.1
0
S(t)
0
.08
.16
.24
.32
.40
.48
.56
.64
.72
.80
t
4
1
2
3
Run 3
Curve 1 Simulation A
Curve 2 Simulation B
Curve 3 Direct-interaction approximation
Curve 4 Test-field model (g = 1.064)

Figure 9

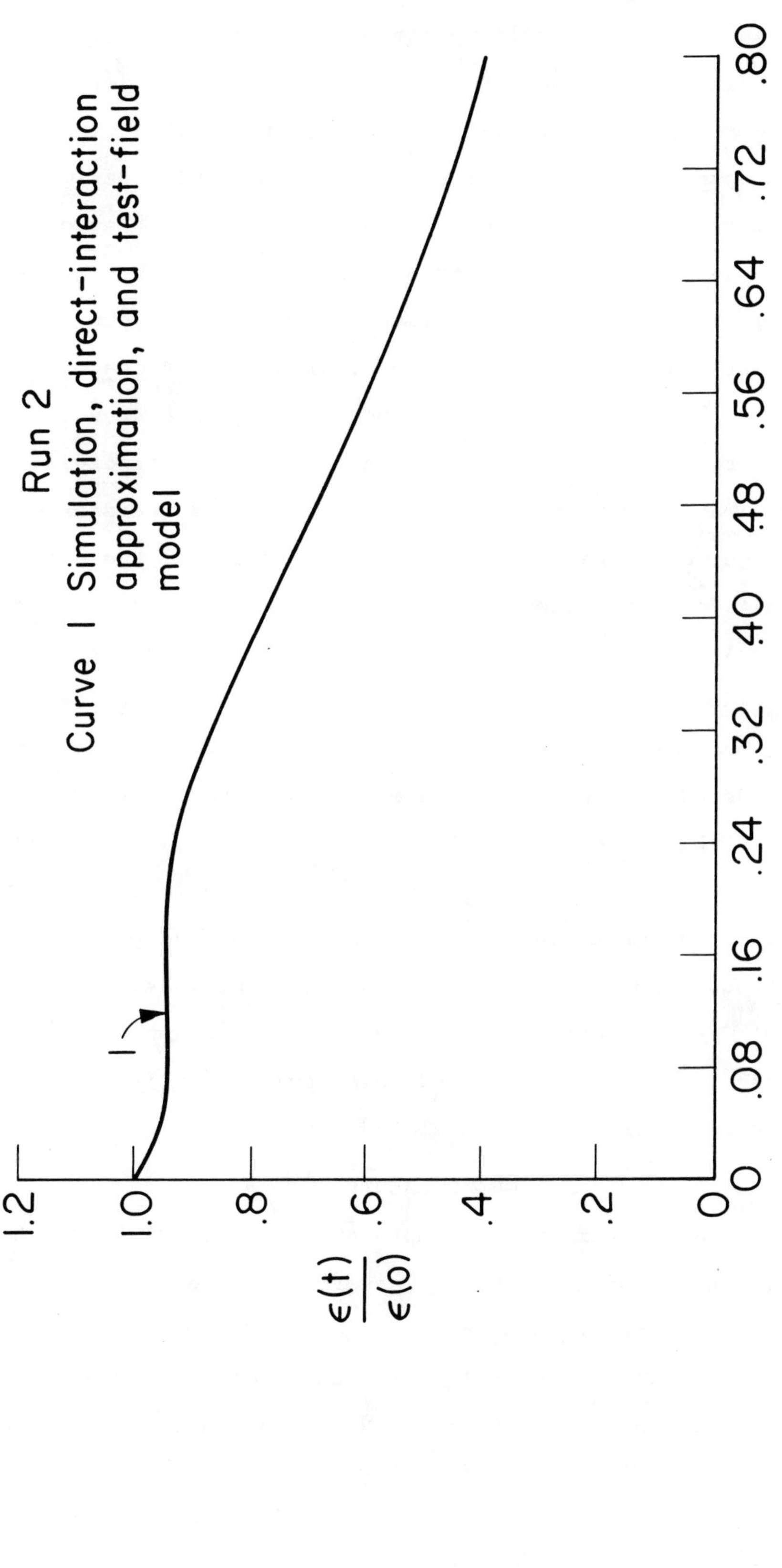
Run 2
Curve I Simulation, direct-interaction approximation, and test-field model
I
$\frac{\epsilon(t)}{\epsilon(0)}$
1.2
1.0
.8
.6
.4
.2
0
0
.08
.16
.24
.32
.40
.48
.56
.64
.72
.80
t

Figure 10

the theories and the numerical experiments are in close agreement for $\varepsilon(t)$. The nature of the errors in the test-field model is harder to assess. For run 2, the test-field model with g=1.064 gives a skewness consistently less than that of the numerical experiment. It seems that the test-field model has not yet achieved Reynolds-number independence at the Reynolds numbers considered here; some further tests of this feature are planned. In summary, our results to date suggest that the test-field model with g=1.064 overestimates the transfer of energy to high wavenumbers and gives a skewness that is too large at moderate Reynolds numbers. The direct-interaction approximation consistently underestimates the transfer of energy to large wavenumbers and, hence, underestimates the skewness and the magnitude of turbulent vortex stretching. However, considering the strongly nonlinear nature of turbulence at these Reynolds numbers, the adequately good agreement of both the direct-interaction approximation and the test-field models with the numerical experiments must be considered an impressive achievement of the theories.

In later publications, we will take up in considerably more detail the comparisons between the numerical simulations of turbulence and the predictions of turbulence theories, as well as present the results of turbulence simulations for the range of initial Reynolds numbers $R_\lambda(0) \sim 5-100$. Other interesting quantities, such as Eulerian and Lagrangian time correlations and velocity-probability-distribution functions, are currently being calculated from our data.

ACKNOWLEDGEMENTS

This work was performed while both authors were visitors at the National Center for Atmospheric Research, Boulder, Colorado. The success of our work owes much to the excellent operation and management of the CDC 6600 of NCAR's computer facility. The authors would like to thank Drs. J.R. Herring, R.H. Kraichnan, and C.E. Leith for stimulating discussion and encouragement, and Margaret Drake for much help with difficult buffering codes. Also, much of the computer time for the runs reported here was made available through the graciousness of Dr. A. Kasahara. One of us (SAO) would like to acknowledge support at M.I.T. by the Air Force Office of Scientific Research under Contract Number F44620-67-C-0007. The other author (GSP) would like to acknowledge partial support by the Sloan Foundation.

REFERENCES

Batchelor, G.K. 1953 The Theory of Homogeneous Turbulence. Cambridge: Cambridge University Press.

Herring, J.R. & Kraichnan, R.H. 1971 Comparison of Some Approximations for Isotropic Turbulence. Proceedings, Symp. on Stat. Models & Turbulence.

Kraichnan, R.H. 1958 Irreversible Statistical Mechanics of Incompressible Hydromagnetic Turbulence. Phys. Rev. 109, 1407-1422.

Kraichnan, R.H. 1964 Decay of Isotropic Turbulence in the Direct-Interaction Approximation. Phys. Fluids 7, 1030-1048.

Kraichnan, R.H. 1971 An Almost-Markovian Galilean Invariant Turbulence Model. J. Fluid Mech. 47, 513-524.

Lilly, D.K. 1971 Numerical Simulation of Developing and Decaying Two-Dimensional Turbulence. J. Fluid Mech. 45, 395-415.

Orszag, S.A. 1969 Numerical Methods for the Simulation of Turbulence. Phys. Fluids Suppl. II, 12, 250-257.

Orszag, S.A. 1971a Numerical Simulation of Incompressible Flows Within Simple Boundaries. I. Galerkin (Spectral) Representations. Stud. in Appl. Math., in press.

Orszag, S.A. 1971b Numerical Simulation of Incompressible Flows Within Simple Boundaries. II. Accuracy. J. Fluid Mech. , in press.

Patterson, G.S. & Orszag, S.A. 1971 Spectral Calculations of Isotropic Turbulence: Efficient Removal of Aliasing Interactions. Phys. Fluids, in press.

COMPARISON OF SOME APPROXIMATIONS FOR ISOTROPIC TURBULENCE

Jackson R. Herring
Goddard Space Flight Center
Greenbelt, Maryland 20771

and

Robert H. Kraichnan
Dublin, New Hampshire 03444

Several related turbulence approximations are studied with regard to dynamical properties and agreement of numerical predictions with laboratory and computer experiments. The approximations considered include the direct-interaction equations (Kraichnan 1964a) Herring's (1966) self-consistent-field theory, a generalization of Edwards' (1964) theory, the abridged Lagrangian-history, direct-interaction approximation (Kraichnan 1966a), the test-field model (Kraichnan 1971a), and an approximation, not previously described, in which one velocity field passively suffers convection by another. Most of the cited approximations are representable by stochastic model equations for the velocity amplitude. Explicit constructions are given for the stochastic models, in a form that can be approximated on a digital computer. These constructions are used to discuss the physical and mathematical differences between the model dynamics and actual Navier-Stokes dynamics. The numerical integrations of the various approximations are carried out for decaying isotropic turbulence, with several initial spectrum forms, over the range $20 \lesssim R_\lambda \lesssim 500$. Results are compared with laboratory and computer experiments by Stewart & Townsend, Ling & Huang, Van Atta & Chen, Grant, Stewart & Moilliet, and Orszag & Patterson.

1. INTRODUCTION AND SUMMARY

A number of turbulence approximations have been proposed with the common characteristics of yielding closed equations for the evolution of the energy spectrum of isotropic turbulence and of being related, with various degrees of closeness, to Reynolds-number expansions of the Navier-Stokes equation. In the present paper, we compare several of these approximations with each other and with laboratory and computer experiments on isotropic turbulence decay.

We emphasize approximations which are representable by a generalized Langevin-type model equation for the velocity amplitude (Leith 1971). In this class, we discuss the direct-interaction approximation (Kraichnan 1964a), a generalization of Edwards' (1964) theory, the test-field model (Kraichnan 1971a), and an approximation, not previously described, in which the actual advection process is replaced by one where a velocity field passively suffers advection by another velocity field. The existence of a model amplitude equation assures certain consistency properties, makes more accessible the analyticity properties of expansions of exact dynamics about the approximate dynamics (Kraichnan 1970), and makes clearer the differences in physics between approximation and Navier-Stokes equation.

In Section 2, we develop explicit constructions for the stochastic-model amplitude equations, so that they can be realized on computers if desired. The model representations of the direct-interaction approximation are central to this analysis. We find that the generalized Langevin models are intimately related to an alternative model equation for the direct-interaction approximation which was described some years ago. In Section 3, we compare the dynamical behaviors of the model amplitudes for the various theories with each other and with the correct Navier-Stokes dynamics. Also, we examine the relations between the models and two approximations, the self-consistent-field (Herring 1966) and the Lagrangian-history-direct-interaction (Kraichnan 1966a), which give broadly similar equations for covariances, but have no model-amplitude-equation representations.

Section 4 contains numerical results for isotropic turbulence decay obtained from the various approximations. The integrations cover moderate to large values of R_λ. The several approximations are compared with each other and with experiments, including current computer experiments by Orszag and Patterson (1971). The general conclusion from the numerical comparisons is that any of the approximations considered gives an adequate qualitative description of the evolution of the energy and vorticity spectrum for low and moderate R_λ. We find consistently that, of those considered, the Edwards approximation gives the least energy transfer, the test-field and Lagrangian-history-direct-interaction approximations give the most, while the direct-interaction and self-consistent-field approximations are intermediate. A surprising result is that the last-named two approximations, which give distinctly different results for systems of only a few degrees of freedom, yield almost indistinguishable results for isotropic-turbulence evolution. A parameter whose evolution distinguishes sensitively among the various approximations is the skewness of the longitudinal velocity derivative.

Only two of the approximations, the test-field model and the abridged Lagrangian-history-direct-interaction approximation, yield -5/3 inertial ranges, and only these are compared with the high-Reynolds-number data. The test-field model has an adjustable parameter (in contrast to all the other approximations). When this is chosen to yield the same inertial-range constant as the abridged Lagrangian-history direct-interaction approximation, the two approximations give very similar results for both small and large Reynolds numbers.

The reader who is uninterested in details of the stochastic-model construction may skip to Section 3, which discusses qualitative properties of the approximations and references the equations of Section 2 which are relevant to that discussion.

2. CONSTRUCTION OF STOCHASTIC MODELS

Leith (1971) and Kraichnan (1970) have pointed out that the direct-interaction approximation (Kraichnan 1964a) constitutes exact equations for the covariance function of a model system in which the velocity amplitude obeys a generalized Langevin equation. Let the Fourier-transformed Navier-Stokes equation in the infinite domain be written

$$(\partial/\partial t + \nu k^2)\, u_i(\mathbf{k}, t) = -i P_{ij}(\mathbf{k})\, k_m \sum_{\mathbf{p}+\mathbf{q}=\mathbf{k}} u_j(\mathbf{p}, t)\, u_m(\mathbf{q}, t), \tag{1}$$

where $P_{ij}(\mathbf{k}) = \delta_{ij} - k_i k_j/k^2$ and the wavevectors take all allowed values in a large cyclic box. $P_{ij}(\mathbf{k})$ expresses the action of the pressure forces. The Langevin representation of the direct-interaction equations for isotropic turbulence can be taken as

$$(\partial/\partial t + \nu k^2)\, u_i(\mathbf{k}, t) + \int_0^t \eta(k, t, s)\, u_i(\mathbf{k}, s)\, ds = q_i(\mathbf{k}, t), \tag{2}$$

where

$$\eta(k, t, s) = \pi k \iint_\Delta b_{kpq}\, G(p, t, s)\, U(q, t, s)\, p\, q\, dp\, dq, \tag{3}$$

$$q_i(\mathbf{k}, t) = -i P_{ij}(\mathbf{k})\, k_m \sum_{\mathbf{p}+\mathbf{q}=\mathbf{k}} \xi_j(\mathbf{p}, t)\, \xi_m(\mathbf{q}, t), \tag{4}$$

$$\langle \xi_i(\mathbf{k}, t)\, \xi_j^*(\mathbf{k}, t)\rangle = \langle u_i(\mathbf{k}, t)\, u_j^*(\mathbf{k}, t)\rangle . \tag{5}$$

Here U(k,t,s) is the covariance scalar defined by

$$(L/2\pi)^3 \langle u_i(\mathbf{k}, t)\, u_j^*(\mathbf{k}, s)\rangle = \frac{1}{2} P_{ij}(\mathbf{k})\, U(k, t, s), \tag{6}$$

where L is the cyclic box size ($L \to \infty$, eventually). G(k,t,s) is the average infinitesimal response scalar (Kraichnan 1964a). The field $\xi(\mathbf{k},t)$, which appears in (4), is statistically independent of $\mathbf{u}(\mathbf{k},0)$. We take $\xi(\mathbf{k},t)$ to be quasi-normally distributed (2nd and 4th order moments related as in a normal distribution). The integration $\iint_\Delta$ in (3) is over all of the p,q plane where k, p, and q can form a triangle, while

$$b_{kpq} = (p/k)(xy + z^3), \tag{7}$$

where x, y, z are the cosines of the interior angles opposite the triangle sides k,p,q, respectively.

The forcing term $q_i(\mathbf{k},t)$ and the damping-with-memory function $\eta(k,t,s)$ in the Langevin equation (2) depend on the velocity field only through ensemble averages. If the ensemble is infinite, they are unaffected by the value of the velocity field in any one typical realization, so that (2) is effectively a linear equation for $u_i(\mathbf{k},t)$ in such a realization. It follows immediately that the Green's function scalar of (2) satisfies

$$(\partial/\partial t + \nu k^2)\, G(k, t, t') + \int_{t'}^{t} \eta(k, t, s)\, G(k, s, t')\, ds = 0 \quad (t \geq t'), \tag{8}$$

$$G(k, t', t') = 1. \tag{9}$$

Also, the linearity gives

$$\langle u_j^*(\mathbf{k}, t')\, q_i(\mathbf{k}, t)\rangle = \int_0^{t'} G(k, t', s)\, \langle q_j^*(\mathbf{k}, s)\, q_i(\mathbf{k}, t)\rangle\, ds, \tag{10}$$

Using (10), (5), and the imposed quasi-normality of $\xi(\mathbf{k},t)$, we can obtain from (2) the energy-balance and covariance equations ($L \to \infty$)

$$(\partial/\partial t + 2\nu k^2)\, U(k, t, t) + 2\int_0^t \eta(k, t, s)\, U(k, s, t)\, ds$$

$$= 2\pi k \int_0^t ds \iint_\Delta a_{kpq}\, G(k, t, s)\, U(p, t, s)\, U(q, t, s)\, pq\, dp\, dq, \tag{11}$$

$$(\partial/\partial t + \nu k^2)\, U(k, t, t') + \int_0^t \eta(k, t, s)\, U(k, s, t')\, ds$$

$$= \pi k \int_0^{t'} ds \iint_\Delta a_{kpq}\, G(k, t', s)\, U(p, t, s)\, U(q, t, s)\, pq\, dp\, dq \quad (t \geq t'), \tag{12}$$

where $a_{kpq} = \frac{1}{2}(b_{kpq} + b_{kqp})$.

Equations (3), (8), (9), (11), (12) are the direct-interaction statistical equations. Equation (11) conserves the energy per unit mass. Setting

$$E(t) = \int_0^\infty E(k, t)\, dk$$

we have $dE(t)/dt = \epsilon(t)$, where

$$\epsilon(t) = 2\nu \int_0^\infty E(k, t)\, k^2\, dk$$

is the dissipation rate by viscosity and $E(k,t) = 2\pi k^2 U(k,t,t)$ is the kinetic energy spectrum function. The Langevin model (2) is conservative only in this average sense. Since $\eta(k,t,s)$ is a nonrandom damping while $q_i(\mathbf{k},t)$ is random, the energy in individual realizations clearly fluctuates, in contrast to the strict conservation property of (1).

An alternative model representation of the direct-interaction approximation, which retains the full conservation properties of (1), was described some years ago (Kraichnan 1958, 1961). We shall now describe a simplified version of that model and show its close relation to the Langevin model. We consider a collection of N systems, where eventually $N \to \infty$, and replace (1) by

$$(\partial/\partial t + \nu k^2)\, u_i^{(n)}(\mathbf{k}, t)$$

$$= -i P_{ij}(\mathbf{k})\, k_m \sum_{r,s=1}^{N} \phi_{nrs} \sum_{\mathbf{p}+\mathbf{q}=\mathbf{k}} u_j^{(s)}(\mathbf{p}, t)\, u_m^{(r)}(\mathbf{q}, t). \tag{13}$$

Here the superscripts label systems in the collection, and the N systems are coupled together into a supersystem according to the coefficients ϕ_{nrs}. The latter are fixed as follows. We first impose the symmetries

$$\phi_{nrs} = \phi_{srn}, \quad \phi_{nrs} = \phi_{nsr}. \tag{14}$$

where

$$b^1_{kpq} = \frac{1}{2}(1 + z^2)(1 - y^2), \quad b^2_{kpq} = b_{kpq} - b^1_{kpq},$$

$$2\, a^1_{kpq} = b^1_{kpq} + b^1_{kqp}, \quad a^2_{kpq} = a_{kpq} - a^1_{kpq}, \tag{19}$$

and

$$C_1 = N^{-2} \sum_{r,s} (\phi_{nrs})^2, \quad C_2 = N^{-2} \sum_{r,s} \phi_{nrs}\, \phi_{nsr},$$

$$C_3 = N^{-2} \sum_{r,s} \phi_{nrs}\, \phi_{srn}, \quad C_4 = N^{-2} \sum_{r,s} \phi_{nrs}\, \phi_{snr}. \tag{20}$$

$C_1 = 1$, by (15). If neither symmetry is imposed, $C_2 = C_3 = C_4 = 0$. If only the first (conservation) symmetry is imposed, $C_3 = 1$, $C_2 = C_4 = 0$. If only the second symmetry is imposed, $C_2 = 1$, $C_3 = C_4 = 0$. If both symmetries are imposed, all C's equal one and the original a_{kpq} and b_{kpq} are recovered. [Actually, the quantities $N^{-2}\Sigma\phi_{nrs}\phi_{rns}$ and $N^{-2}\Sigma\phi_{nrs}\phi_{rsn}$ also arise, but they equal C_4 for each of the four cases considered and therefore have been replaced by C_4 in writing (17) and (18).]

Let us consider the case where the second symmetry is imposed, but not the first. Then the statistical equations are (8), (9), (11), (12), but with $\eta(k,t,s) = 0$, thereby destroying energy conservation by eliminating the dynamical damping term on the left side of (11). But we can restore conservation in the supersystem, in the limit $N \to \infty$, by adding the term

$$\int_0^t \eta(k, t, s)\, u_i^{(n)}(k, s)\, ds$$

to the left-hand side of (13), with, now, $\eta(k,t,s)$ given by (3). This puts back the dynamical damping in the U and G equations. We have, then, an intimate relation between the two stochastic models (2) and (13) for the direct-interaction approximation. With the conservation symmetry imposed, the stochastic right-hand side of (13) does not act as a purely random forcing for $\mathbf{u}^{(n)}$. Instead, the correlations among the systems keep total energy conserved. But if we relax the symmetry, the ϕ's are sufficiently more random that these correlations are not maintained, and the right-hand side does act like a purely random forcing, like the right-hand side of (2). Then we can restore conservation only by introducing an explicit damping to counter the random input.

Thus we can realize the Langevin model of the direct-interaction approximation as a modification of (13) with $N \to \infty$. We replace (2) by a supersystem equation

$$(\partial/\partial t + \nu k^2)\, u_i^{(n)}(\mathbf{k}, t) + \int_0^t \eta(k, t, s)\, u_i^{(n)}(\mathbf{k}, s)\, ds = q_i^{(n)}(\mathbf{k}, t), \qquad (21)$$

where U(k,t,s) in (3) is given by (16), G(p,t,s) is the linearized Green's function of (21), and only the second symmetry of (14) is imposed. Alternatively, we can retain the form (4), (5) for $q_i^{(n)}(\mathbf{k},t)$ and take

$$\xi_i^{(n)}(\mathbf{k}, t) = \sum_{r=1}^{N} \phi_{nr}\, u_i^{(r)}(\mathbf{k}, t), \qquad (22)$$

where

$$\phi_{nr} = \pm N^{-1/2}, \qquad (23)$$

with the plus or minus taken completely at random for each (nr). This corresponds to

$$\phi_{nrs} = \phi_{nr}\, \phi_{ns}, \qquad (24)$$

a choice of the ϕ_{nrs} which is less random than that given by (15) and the second symmetry of (14), but one which yields the same U and G equations for $N \to \infty$. It should be noted that, even for $N \to \infty$, (22) does not yield strictly normal $\xi_i^{(n)}(\mathbf{k},t)$, because the $\mathbf{u}^{(n)}$ are dynamically coupled, as a whole, and, hence, not strictly statistically independent. However, the quasi-normality property assumed after (6) does hold for $N \to \infty$.

The stochastic models of form (13) or (21) can be realized on a computer by taking N finite, but large enough to give good statistics. This would supplement solutions of the U and G equations by illustrating how typical behavior of the actual velocity amplitudes differs between the two types of model and between models and Navier-Stokes equation. With finite N, the strict conservation in (13), with the conservation symmetry imposed, may give significantly different behavior than (21), which gives conservation only in the limit $N \to \infty$. It seems possible that (21) could exhibit runaway solutions for finite N, which could spoil the approximation to infinite-N behavior, after a sufficiently long time of evolution.

The models of form (13) or (21) can be constructed also for anisotropic, inhomogeneous flows. If only homogeneous turbulence is considered, models which yield the same U and G equations, but are more economical for computer realization, can be constructed by keeping a single system, rather than going to a supersystem, and introducing the random couplings among the triads of interacting wavevectors instead of among interacting systems. The limit $N \to \infty$ then is replaced by $L \to \infty$, which gives a dense packing of allowed modes in **k** space. In this formulation, (13) is replaced by (Kraichnan 1958).

$$(\partial/\partial t + \nu k^2)\, u_i(\mathbf{k}, t) = -i\, P_{ij}(\mathbf{k})\, k_m \sum_{\mathbf{p}+\mathbf{q}=\mathbf{k}} \phi_{\mathbf{kqp}}\, u_j(\mathbf{p}, t)\, u_m(\mathbf{q}, t), \tag{25}$$

where $\phi_{\mathbf{kqp}} = \pm 1$ and the symmetries imposed on the $\phi_{\mathbf{kqp}}$ in the various cases are the same as for the ϕ_{nrs} except that $\phi_{\mathbf{kqp}} = \phi_{-\mathbf{k},-\mathbf{q},-\mathbf{p}}$ is required in all cases to assure real amplitudes in **x** space. Equation (16) is replaced by an average over an appropriate neighborhood in **k** space.

Now let us consider the meaning of the second symmetry in (14). The quantity $k_m u_m(\mathbf{q},t)$ in (1) is the transform of the advection operator $\mathbf{u}\cdot\nabla$ in the Navier-Stokes equation. Suppose we call the first amplitude in $\mathbf{u}\cdot\nabla\mathbf{u}$ the convecting velocity and the second amplitude the convected velocity. Then it is easy to see that the effect of relaxing the second symmetry in (14) is to wipe out, in the equations for U and G, all contributions which involve a correlation between convecting and convected amplitude, together with all contributions which involve dynamical effects on the convecting velocity. These eliminations are expressed, in the U and G equations, by the replacement of a_{kpq} and b_{kpq} by a^1_{kpq} and b^1_{kpq}. The final results are as if both symmetries were retained in (14) but, instead of the Navier-Stokes equation, we started with

$$(\partial/\partial t - \nu\nabla^2)\,\mathbf{u} = -\mathbf{v}\cdot\nabla\mathbf{u} - \nabla p, \tag{26}$$

where **v** is a purely random velocity field, statistically independent of **u** at the initial instant, which is required to satisfy

$$\langle v_i(\mathbf{k}, t)\, v_i^*(\mathbf{k}, t')\rangle = \langle u_i(\mathbf{k}, t)\, u_i^*(\mathbf{k}, t')\rangle$$

We shall call the model without the second symmetry of (14) [C_2 and C_4 zero in (17), (18)] the unsymmetric model. The equivalent Langevin model of form (2) is obtained by replacing b_{kpq} with b^1_{kpq} in (3) and replacing $\xi_m(q,t)$ in (4) with $\xi'_m(q,t)$, where ξ and ξ' are statistically independent, but have the same covariance. The supersystem realization of form (21)-(22) is then obtained by adjoining to (22) a corresponding equation for $\xi_i'^{(n)}(\mathbf{k},t)$, involving coefficients

ϕ'_{nr} which are statistically independent of ϕ'_{nr}. Equation (24) then becomes

$$\phi_{nrs} = \phi'_{nr}\ \phi_{ns}. \tag{27}$$

The unsymmetric model of form (13) is analogous to a stochastic model that has been proposed for turbulence in a Vlasov plasma (Orszag & Kraichnan 1967).

At the risk of causing confusion, we should note that we can construct an alternative form of the fully symmetrized Langevin model (yielding a_{kpq} and b_{kpq}) by replacing (4) with the symmetrized version

$$q_i\,(\mathbf{k},\ t) = -\,i\,[P_{ij}\,(\mathbf{k})\,k_m + P_{im}\,(\mathbf{k})\,k_p] \sum_{\mathbf{p}+\mathbf{q}=\mathbf{k}} \xi_j\,(\mathbf{p},\ t)\,\xi'_m\,(\mathbf{q},\ t), \tag{28}$$

and requiring that both ξ and ξ' have covariances equal to that of $\mathbf{u}$, but otherwise allowing ξ and ξ' to have independent and wholly arbitrary isotropic statistics. Quasinormality is no longer required. This formulation was used by Kraichnan (1971a).

Now we shall turn to two approximations, the generalized Edwards theory and the test-field model, which lead to Langevin equations in which the forcing function is a white noise in time. For these models, we replace (2)-(4) by (Kraichnan 1971a)

$$[\partial/\partial\,t + \nu\,k^2 + \eta\,(k,\ t)]\ u_i\,(\mathbf{k},\ t) = q_i\,(\mathbf{k},\ t), \tag{29}$$

$$\eta\,(k,\ t) = \pi\,k \iint_{\Delta} b_{kpq}\,\theta_{pqk}\,(t)\,U\,(q,\ t)\,p\,q\,d\,p\,d\,q, \tag{30}$$

$$q_i\,(\mathbf{k},\ t) = -\,i\,P_{ij}\,(\mathbf{k}) \sum_{\mathbf{p}+\mathbf{q}=\mathbf{k}} [\theta_{kpq}\,(t)]^{1/2}\,w\,(t)\,\xi_j\,(\mathbf{p},\ t)\,\xi_n\,(\mathbf{q},\ t) \tag{31}$$

Here w(t) is a white-noise process,

$$\langle w\,(t)\,w\,(t')\rangle = 2\,\delta\,(t - t'), \tag{32}$$

$U(q,t) = U(q,t,t)$, and we continue to require (5). The quantity θ_{kpq} (t) is a characteristic memory time for the interaction of wavenumbers k,p, and q. Its presence takes the place of the explicit integration over history which occurs in (2).

Instead of (8), (11), and (12), the strict Langevin equation (29) gives

$$[\partial/\partial\,t + \nu\,k^2 + \eta\,(k,\ t)]\ G\,(k,\ t,\ t') = 0\ (t \geq t'), \tag{33}$$

$$U(k,\ t,\ t') = G(k,\ t,\ t')\ U(k,\ t')\ (t \geq t') \tag{34}$$

$$[\partial/\partial t + 2\nu k^2 + 2\eta(k,\ t)]\ U(k,\ t)$$
$$= 2\pi k \iint_{\Delta} a_{kpq}\ \theta_{kpq}(t)\ U(p,\ t)\ U(q,\ t)\ p\,q\,d p\,d q. \tag{35}$$

The generalized Edwards theory is completed by taking

$$\theta_{kpq}(t) = \int_0^t G(k,\ t,\ s)\ G(p,\ t,\ s)\ G(q,\ t,\ s)\ d s, \tag{36}$$

so that, in view of (34), the energy balance equation (33) resembles the direct-interaction equation (11).

The white-noise Langevin equation (29) can be substituted by a random-coupled model if we generalize (13) to

$$(\partial/\partial t + \nu k^2)\ u_i^{(n)}(\mathbf{k},\ t)$$

$$= -\, i\, P_{ij}(\mathbf{k})\, k_m \sum_{r,\,s=1}^{N} \phi_{nrs}(t) \sum_{\mathbf{p}+\mathbf{q}=\mathbf{k}} [2\tau^{-1}\,\theta_{kpq}(t)]^{1/2}\, u_j^{(s)}(\mathbf{p},\ t)\, u_m^{(r)}(\mathbf{q},\ t). \tag{37}$$

Here τ is a small time interval, and fresh choices of the form (14), (15) are made for the $\phi_{nrs}(t)$ at $t = 0, \tau, 2\tau, \ldots$. This system, like (13), is strictly conservative, except for the νk^2 term. The effect of the extra time dependence is that the quantities C_1, etc. in (20) are replaced by

$$2\, N^{-2}\, \tau^{-1}\, [\theta_{kpq}(t)\ \theta_{kpq}(s)]^{1/2} \sum_{r,\,s} \phi_{nrs}(t)\ \phi_{nrs}(s),\ \text{etc.}, \tag{38}$$

which now appear inside the time and wavenumber integrations in (3), (11), and (12). Equations (30), (33)-(35) are recovered in the limit $\tau \to 0$. The effects of relaxing the symmetries in (14) are as before. In particular, a realization of (29) can be constructed in analogy to the previous discussion for (2). In setting up an approximate realization on a computer, it is natural to take τ equal to the time step in the integration scheme.

Our final Langevin model is the test-field model (Kraichnan 1971a), which differs from all the others, discussed above, by exhibiting invariance to random Galilean transformation, a property which will be discussed in Section 3. This model differs from the generalized

Edwards' model only in that $\theta_{kpq}(t)$ is given by

$$\theta_{kpq}(t) = \int_0^t G^S(k, t, s)\, G^S(p, t, s)\, G^S(q, t, s)\, d s. \tag{39}$$

Here $G^S(k,t,s)$ is a Green's function which measures the rate at which the solenoidal part of a convected vector test field **w** exchanges excitation with the longitudinal part of the test field. This choice for $\theta_{kpq}(t)$ is made in order to have a memory time for energy transfer at high k which depends on intrinsic distortions of vortex structures and is unaffected by random, spatially uniform translation, in contrast to G(k,t,s), which is so affected. The model is completed by the following equations which determine $G^S(k,t,s)$, and $G^C(k,t,s)$, the corresponding Green's function for the longitudinal part of the test field:

$$[\partial/\partial t + \nu k^2 + \eta^S(k, t)]\, G^S(k, t, t') = 0,$$

$$[\partial/\partial t + \nu k^2 + \eta^C(k, t)]\, G^C(k, t, t') = 0, \tag{40}$$

$$\eta^S(k, t) = \pi k g^2 \iint_\Delta b^G_{kpq}\, \theta^G_{pkq}(t)\, U(q, t)\, p q\, d p\, d q,$$

$$\eta^C(k, t) = 2\pi k g^2 \iint_\Delta b^G_{kpq}\, \theta^G_{kpq}(t)\, U(q, t)\, p q\, d p\, d q, \tag{41}$$

$$\theta^G_{kpq}(t) = \int_0^t G^C(k, t, s)\, G^S(q, t, s)\, G^S(p, t, s)\, d s, \tag{42}$$

$$b^G_{kpq} = \frac{1}{2}(1 - y^2)(1 - z^2), \tag{43}$$

with $G^S(k,t,t) = G^C(k,t,t) = 1$. The quantity g is a dimensionless scaling parameter.

The Langevin equation (29) is supplemented, for the test-field model, by a Langevin equation of the same type for the test field (Kraichnan 1971a). Consequently, this model, also, can be substituted by random-coupled amplitude equations of the form (37). We shall not take space to write out the equations.

3. PROPERTIES OF THE APPROXIMATIONS

Before discussing the physical and mathematical properties of the several approximations, we shall list, all in one place, the defining equations, for the benefit of readers who skipped through Section 2.

Direct-Interaction Approximation (DIA)

Statistical equations: Energy balance (11); Green's function (8), (9); time displaced covariance (12); dynamical damping function (3), auxiliary equation (7). Langevin-model amplitude equation: (2), with auxiliary eqs. (3)-(7). Random-coupled-model amplitude equation: (13), with (14) and (15).

Unsymmetric Model (UNS)

In this approximation, the velocity field, in effect, passively suffers convection by a second field with the same covariance (26). Statistical equations: same as direct-interaction, except b_{kpq} and a_{kpq} are replaced throughout by b^1_{kpq} and a^1_{kpq} defined in (19). Langevin-model amplitude equation: same as direct-interaction, except b^1_{kpq} replaced b_{kpq} in (3) and, in (4), $\xi'_m(q,t)$ replaces $\xi_m(q,t)$, where ξ and ξ' are statistically independent, but each has the same covariance as $\mathbf{u}$. Random-coupled-model amplitude equation: (13), with (15) and only the first symmetry of (14).

Generalized Edwards Model (EDW)

Statistical equations: Energy balance (35); Green's function (33), (9); time-displaced covariance (34); dynamical damping function (30), (36). Langevin-model amplitude equation: (29), with auxiliary equations (30)-(32), (36). Random-coupled model-amplitude equation: (37), with (36).

Test-field Model (TFM)

Statistical Equations: Energy balance: (35), Green's function (33), (9); time-displaced covariance (34); dynamical damping function (30), (39)-(43). Langevin-model amplitude equation: (29), with auxiliary equations (30)-(32), (39), together with amplitude equations (not stated explicitly) for the convected test field. Random-coupled-model amplitude equations: (37), with (39), together with amplitude equations for the convected test field (not stated explicitly in Sec. 2).

We shall also summarize in the same way, the equations for the two additional approximations to be discussed, which do not have model-amplitude-equation representations.

Self-Consistent-Field Approximation (SCF)

Statistical equations: identical with direct interaction statistical equations except that now the time displaced covariance equation is (34) instead of (12). Langevin-model amplitude equation: none known to exist. Random-coupled-model amplitude equation: none known to exist.

Abridged Lagrangian-history direct-interaction approximation (ALH)

Statistical equations: energy balance equation and dynamical damping function are identical in form with direct-interaction equations (11), (3), (7), except that the Green's functions and time-displaced covariances which occur are now Lagrangian rather than Eulerian functions. Green's function and time-displaced covariance equations: the Eulerian functions determined in the other approximations are undetermined by this approximation. The Lagrangian Green's functions and time-displaced covariances are determined by integro-differential equations like (10) and (12) but more complicated (Kraichnan 1966a). Langevin-model amplitude equation: none known to exist. Random-coupled-model amplitude equation: none known to exist.

We shall start by discussing the direct-interaction approximation, and use it as a basis for pointing out the differences in behavior among the several alternative approximations. Some consistency properties of the DIA are immediately apparent from the random-coupled-model amplitude equation. First is conservation of energy, already noted. If the energy balance equation is written in the form

$$(\partial/\partial t + 2\nu k^2)\, E(k, t) = T(k, t), \tag{44}$$

then the conservation property is

$$\int_0^\infty T(k, t)\, dk = 0. \tag{45}$$

In addition to this overall property, the DIA, like the Navier-Stokes equation, yields a conservative interaction of each triad of wavevectors (detailed conservation).

A related property is the existence of an absolute-equilibrium equipartition solution,

$$U(k, t, t) = \text{constant independent of } k \text{ and } t, \tag{46}$$

if $\nu = 0$ and the system is conservatively truncated by removing all terms containing wave-numbers greater than some cutoff from the equations. This follows from the easily demonstrated Liouville theorem for the model equation (Kraichnan 1959). In absolute equilibrium, but not otherwise, the fluctuation-dissipation relation (34) holds. These statements about absolute equilibrium are valid for the Navier-Stokes system as well as the model.

A third property, which follows immediately from the existence of the model representation, is that the DIA for U(k,t,t') is the covariance scalar of some realizable probability distribution for $\mathbf{u}(\mathbf{k},t)$. In particular,

$$E(k, t) \geq 0,$$

$$[U(k, t, t')]^2 \leq U(k, t, t)\, U(k, t', t'). \tag{47}$$

The solutions $\mathbf{u}(\mathbf{k},t)$ of the Navier-Stokes equation, and therefore U(k,t,t') and G(k,t,t'), can be expanded formally in powers t and t'. Alternatively, a strength parameter λ can be inserted in the right-hand side of (1) and these quantities can be expanded in powers of λ. The latter is effectively a Reynolds number expansion. It can be verified from the statistical equations that, if all triple correlations vanish at t = 0, the DIA gives a U(k,t) which is identical with the exact U(k,t) through the term in t^3. U(k,t,t') and G(k,t,t') are even functions of λ provided that all odd initial correlations vanish. The DIA gives U(k,t,t') and G(k,t,t') correctly through the term in λ^2. The comparison of exact and DIA expansions for U and G in t or λ at higher orders depends on what is assumed for the initial statistics. The DIA results depend only on the initial covariances. For a multivariate-normal initial distribution, the most reasonable a priori choice, discrepancies between exact and DIA series appear at orders t^4 and λ^4.

In contrast to the behavior of U(k,t,t'), the unaveraged amplitude $\mathbf{u}(\mathbf{k},t)$ as given by the DIA Langevin-model amplitude equation, or $\mathbf{u}^{(n)}(\mathbf{k},t)$ as given by the DIA random-coupled amplitude equation, does not agree with the exact $\mathbf{u}(\mathbf{k},t)$ in a typical realization even through orders t or λ in the expansions. Thus the DIA models do not provide valid detailed approximations to any given Fourier amplitude in a given realization. Clearly, the DIA models can approximate the actual dynamics, to whatever extent they do, only in a statistical sense. Some further insight here is provided by the alternative random-coupled representation (25). The random coupling coefficients $\phi_{\mathbf{kpq}}$ scramble the dynamics in a way that makes it implausible that the model could reproduce the build-up of complicated correlations among large numbers of wavevector modes which occurs in solutions of the Navier-Stokes equations. However, since $\phi_{\mathbf{kpq}} = \pm 1$, the strengths of the couplings of elementary triads of wavevectors are the same as in the Navier-Stokes equation (where all $\phi_{\mathbf{kpq}} = 1$). This makes it plausible that, even at Reynolds numbers so high that accuracy through terms in λ^2 is irrelevant, the DIA may give valid approximations to the low-order statistical quantities U(k,t,t') and G(k,t,t'). A somewhat similar argument can be made on the basis of the Langevin equation (2). We neglect most of the correlations among the wavevector modes by replacing the right-hand side of the Navier-Stokes equation (1) with a purely random term $q_i(\mathbf{k},t)$. But, if the actual velocity distribution is not too far from Gaussian, $q_i(\mathbf{k},t)$ does have the approximate magnitude of the actual right-hand side, because of (4) and (5). The dynamical damping η (k,t,s) is just that required to fix up the worst consequence of ignoring correlations in the right-hand side: namely, violation of energy conservation. Again it is not unreasonable to expect an approximately correct U(k,t,t').

An important property of the exact dynamics not shared by the DIA is invariance of energy transfer to random Galilean transformations. It is clear for the Navier-Stokes equation that translation of each flow in an ensemble by a spatially uniform velocity, which varies randomly from realization to realization, can not affect energy transfer between wavenumbers. This is because it distorts nothing. However, if v is the rms value of this velocity, then for v large, the decay time of G(k,t,t') or U(k,t,t') as a function of t - t' is of order $(vk)^{-1}$, which is the convective dephasing time for simple translation of a sinusoidal disturbance of wavenumber k. This behavior of G(k,t,t') and U(k,t,t') is reproduced in the DIA. But, in the latter, G(k,t,t') and U(k,t,t') appear in the energy-balance equation in such a way that $(vk)^{-1}$ becomes, also, the effective memory time for build-up of energy transfer across wavenumber k. This last phenomenon is an artifact of the DIA and has the result that, at large Reynolds numbers, the DIA gives an inertial- and dissipation-range spectrum of the form (Kraichnan 1964c)

$$E(k) = C_{DI}\,(\epsilon\, v_0)^{1/2}\, k^{-3/2}\, \bar{f}\,(k/k_d), \tag{48}$$

where C_{DI} is a number of order unity, ϵ is the rate of energy dissipation per unit mass, v_0 is the rms turbulent velocity in any direction, and the characteristic dissipation wavenumber is $k_d = (\epsilon / v_0 \nu^2)^{1/3}$. Here $\bar{f}(0) = 1$, and $\bar{f}(x)$ falls off exponentially as $x \to \infty$. The appearance of v_0 in (48) reflects the spurious dependence of transfer on convection.

Now let us examine how the various properties discussed above for the DIA work out in the other approximations. All of the approximations we are studying yield both overall and detailed conservation of energy. All give equipartition, and the fluctuation-relaxation relation (34) in absolute equilibrium of the truncated, inviscid system. For the approximations representable by stochastic models, these properties follow immediately from (13) or (37). They may be verified directly from the U and G equations for the SCF and ALH approximations, which have no model representations. In the case of ALH, (34) is replaced by a similar relation involving Lagrangian functions (Kraichnan 1966a).

For the UNS model, the realizibility inequalities (47) follow, as for the DIA, from the existence of the amplitude representation (13). In the case of TFM and EDW, which have the amplitude representation (37), we must be careful because of the appearance of $[\theta_{kpq}(t)]^{1/2}$. If any $\theta_{kpq}(t)$ passes through the value zero during the evolution, (37) becomes meaningless thereafter, and (47) is no longer assured. It can be shown for the TFM that $\theta_{kpq}(t) > 0$ for $t > 0$, a fact which follows from the property $b^G_{kpq} \geq 0$ (Kraichnan 1971a), which implies $G^s(k,t,t') \geq 0$. Thus (47) does hold for this model. However, b_{kpq} takes negative as well as positive values, and it cannot be concluded in the same simple way that $G(k,t,t') \geq 0$ and that the EDW model guarantees (47). As we shall see, the numerical integrations suggest that (47) is well-maintained by the EDW model, and it seems likely that a more refined investigation could establish (47) analytically. For the SCF and ALH approximations, there is no known amplitude representation, and the evidence for (47) is solely from analytical solutions of the U and G equations in simple cases, together with numerical integrations in more realistic cases (Herring 1966, Kraichnan 1966a).

The extent to which the exact λ expansion of $U(k,t,t')$ and $G(k,t,t')$ is reproduced varies in a complicated way among the various approximations. By ignoring correlations between convected and convecting velocities, the UNS approximation throws out part of the order λ^2 contributions to both $U(k,t,t')$ and $G(k,t,t')$. The EDW, TFM, and SCF approximations make errors of order λ^2 in $U(k,t,t')$ except in the case of absolute equilibrium, where (34) holds for the Navier-Stokes dynamics. This is because the U(k,t,t') dynamical equation in these three approximations has the form (33), which does not have anything corresponding to the a_{kpq} term which appears in the DIA equation (12). However, the three approximations give U(k,t,t) correctly through order t^2, and SCF gives U(k,t,t) correct through order λ^2 for all t. The ALH approximation gives U(k,t), and the Lagrangian time-displaced covariance and Green's function, correct through order λ^2. [The full Lagrangian-history direct-interaction approximation, of which ALH is a simplification, yields Eulerian as well as Lagrangian two-time covariances, and gives all correct through order λ^2 (Kraichnan 1965).]

As with the DIA approximation, the UNS, TFM, and EDW model amplitude equations are inaccurate already at order λ, and can give, at best, a statistical resemblance to the behavior of the Navier-Stokes amplitudes. The TFM and EDW models have, in addition, the unphysical property of white-noise excitation, in contrast to the presumably smooth behavior of the Navier-Stokes amplitudes and all their time derivatives at finite Reynolds numbers. This property shows up in the statistical results as an unphysical cusp in U(k,t,t') at t = t'. According to (33) and (34), the logarithmic slope of U(k,t,t') at t = t' is $-\nu k^2 - \eta(k,t)$, whereas, in the Navier-Stokes dynamics, the slope must go smoothly to zero at t = t'. This trouble afflicts also the SCF approximation, which yields a logarithmic slope $-\nu k^2$. The UNS model, like the DIA, gives a smooth, zero-slope behavior at t = t', while the ALH approximation gives the proper slope behavior for the Lagrangian covariance which occurs in it (Kraichnan 1966a). A further consequence of (34) is that the decay time of U(k,t,t') as a function of (t - t') is $(\nu k^2)^{-1}$, for k where viscosity dominates. This means that the SCF, EDW, and TFM predict much shorter decorrelation times high in the dissipation range than would be expected from a physical picture where the very high wavenumbers arise from Fourier decomposition of relatively large flow structures (size $\gtrsim$ reciprocal of characteristic dissipation wavenumber) rather than exist as dynamically distinct entities.

The UNS, EDW, and SCF approximations share with the DIA a failure to give invariance of energy transfer to random uniform convection. In all these approximations, the decay time of U(k,t,t') and G(k,t,t') is qualitatively correct in the energy-containing and inertial ranges, and the trouble, as with the DIA, lies in the way these Eulerian functions enter the energy-balance equation. All four approximations lead to (48). The TFM and ALH approximations do exhibit the invariance to random uniform convection. They lead to the Kolmogorov spectrum

$$E(k) = C \epsilon^{2/3} k^{-5/3} f(k/k_s), \tag{49}$$

where $k_s = (\epsilon/\nu^3)^{1/4}$.

In the ALH approximation, the invariance is achieved by a modification of the DIA in which the memory integrals in (8), (11), and (12) are integrated backward in time along particle trajectories instead of at fixed points in laboratory coordinates (Kraichnan 1966a). The modification seems uniquely indicated, within the analytical framework that is used, and there are no adjustable constants.

In the TFM, a more ad hoc procedure is used, in the interest of maximum simplicity. The memory times for build up of energy-transfering correlations [the $\theta_{kpq}(t)$] are measured by the rate at which a convected test field is distorted by shearing so as to exchange excitation between its solenoidal and longitudinal parts. This is invariant to uniform convection, which has no shear, but we can expect that the mechanism chosen is, at best, only qualitatively valid as a measure of the distortion that produces energy transfer in the incompressible Navier-Stokes fluid. For this reason, the TFM is equipped with an adjustable scaling factor, g in (41), thereby differing from all the other approximations we consider. In the original derivation of the TFM (Kraichnan 1971a), the value $g = 1.064$ was chosen, in order to reproduce results of the DIA for the relaxation of small departures from absolute equilibrium in the interaction of the modes in a thin spherical shell, this being a case where the direct-interaction approximation was expected to be accurate. A more realistic procedure for the present study is to match some particular isotropic turbulence decay and see how well that value of g holds up for other cases. The TFM yields for C in (49) the value (Kraichnan 1971b)

$$C = 1.34\, g^{2/3}. \tag{50}$$

In Section 4, we shall exhibit integrations both for $g = 1.064$ (which gives $C = 1.40$) and for $g = 1.5$, a value chosen because it gives $C = 1.76$, which differs negligibly from the ALH value 1.77. The ALH approximation has previously been found to give an excellent fit (Kraichnan 1966a) to high-Reynolds-number data published by Grant, Stewart & Moilliet (1962).

We wish now to discuss further the relation between the DIA and SCF approximation. It was noted above that the SCF approximation can be regarded as a simplification of the DIA in which (12) is dropped and the fluctuation-dissipation relation (34) is assumed to hold out of, as well as in, absolute equilibrium. However, the original derivation of the SCF theory (Herring

1966) was independent of DIA. It involved truncation of an expansion of the probability distribution of the Fourier amplitudes about the product of exact univariate distribution functions. This expansion, which is explicitly an expansion in λ, is closely analogous to the expansion of triple moments in terms of exact covariances, whose corresponding truncation gives the DIA (Kraichnan 1966b). Recently, an alternative expansion of the probability distribution, using the diagram methods of the Brussels school, has again led to the SCF equations (Balescu & Senatorski, 1970).

Why then, does DIA give the full covariance U(k,t,t') accurate through terms in λ^2, while the SCF equations give this accuracy only for U(k,t,t)? The expansion of triple moments, on which the DIA is based, is intrinsically a many-time expansion. Moments of many-time distributions appear in the expansion, and they are reduced to covariances by appeal to an initial multivariate normal distribution (Kraichnan 1966b). On the other hand, both the original expansion from which the SCF equations arise, and the more recent treatment by Balescu and Senatorski, are based on the evolution of the single-time distribution of the Fourier amplitudes. Apparently this framework is too restricted to lead easily to approximations of sufficient symmetry in two time arguments to represent U(k,t,t') well. We shall see in Section 4 that, despite the different predictions for U(k,t,t'), the DIA and SCF equations yield results for E(k,t) that are extraordinarily similar.

The above discussion of the relation between the DIA and SCF approximation presumes that U(k, t, t') entering the SCF approximation is, in fact the same time displaced covariance that enters the DIA approximation. Strictly this is not true; in the SCF approximation, U (k,t,t') is that covariance obtained if, at t', the actual distribution function were replaced by a simple product of univariate distributions. An additional independent equation for the proper U(k,t,t') has been proposed by Balescu and Senatorski (1970) and it is this equation, or one playing its role, whose order in accuracy in λ should be examined. Nonetheless, the thrust of the criticism remains valid: working within the framework of a single time distribution function, it is difficult to propose a perturbation procedure uniformly valid for all difference times.

4. NUMERICAL INTEGRATIONS

The integrations described in this Section have the purposes of exhibiting qualiative features of the turbulence approximations, comparing them with each other, and confronting them with experiment. The third objective is hard to meet because it is impossible to know to what extent the initial state for the numerical integrations matches some point in the evolution of the laboratory flow. In the form they are given in this paper, all the approximations assume an initial state with zero triple correlations, and the integrations are carried out for relatively short times. The laboratory and field measurements, on the other hand, involve typically much longer times of evolutions, and from an initial state of complex character. In this circumstance, the most logical kind of comparison would appear to be between integrations and experiments both of which give self-preserving spectra, and to concentrate attention on the higher wave-numbers, which give more promise of universal behavior.

Orszag and Patterson (1971) have very recently succeeded in carrying out direct integrations of the Navier-Stokes equation for isotropic turbulent initial conditions at Reynolds numbers $R_\lambda \lesssim 40$. Here the initial conditions can be made a distribution which exactly matches that for the approximation integrations, apart from statistical fluctuations in the computer simulation. These computer experiments provide, for the first time, a clean basis of comparison for isotropic turbulence theory.

The initial energy spectrum in most of the evolution runs to be reported here had one or another of the following forms:

$$E(k, 0) = 16\,(2/\pi)^{1/2}\, v_0^2\, k_{peak}^{-5} k^4 \exp[-2(k/k_{peak})^2], \tag{51}$$

$$E(k, 0) = v_0^2\, \lambda^3\, (k\lambda)(1 + \sqrt{2}\, k\lambda) \exp(-\sqrt{2}\, k\lambda), \tag{52}$$

Where $k_{peak} = 4.76$.

$$E(k, 0) = v_0^2\, k_{peak}^{-2} \exp(-k/k_{peak}) k \tag{53}$$

The initial data for the several runs are given in the following table.

Run	Spectrum Form	v_0	λ	ν	$R_\lambda(0)$	Final t	k_{max}	k_{min}	k_0	k_1	N
1	See text	1.		0.	∞	1.2	16.	2.	1.		40
2	(51)	1.	.420	0.02	20.8	0.78	32.	2.	21.1	22.1	40
3	(51)	1.	.420	0.01189	34.5	0.78	32.	2.	21.1	22.1	40
4	(52)	1.	0.2	0.01	20.0	0.78	80.	0.	4.91	7.01	80
5	(53)	1.	0.17	0.01	19.0	0.78	80.	0.	4.91	7.01	80
6	(52)	1.	0.4	0.01	40.0	0.78	80.	0.	4.91	7.01	80
7	See text	1.95	.177	0.006475	53.4	0.39	80.	0.	4.91	7.01	80
8	See text	5.12	0.378	0.008	245	0.4	71.835	.1114	.3688	3.7901	80

Run 1 is an inviscid approach to equipartition, with the wavenumber range truncated to $2 < k < 16$, and an initial spectrum zero everywhere except $E(k,0) = .444$ for $4 < k < 6$. For Run 8,

$E(k,0) = 6.29k^{-5/3}$ for $2^{-19/6} < k < 2^{37/6}$, outside of which interval the system was truncated. The initial spectrum for Run 7 was a smooth curve passed through the experimental spectrum of Van Atta & Chen (1969).

The numerical techniques used for integrating the approximations are described in the Appendix. The parameters k_{max}, k_{min}, k_0, k_1, and N are defined there.

The computer experiments (compared here with Runs 1, 2, and 3) are reported in detail elsewhere (Orszag & Patterson 1971). These experiments were carried out in Fourier space, with cyclic boundary conditions, using fast-transform techniques. For Run 1, all wavevectors are included whose components are positive and negative even integers and such that $2 \leq k \leq 2\times(61)^{1/2}$. In Runs 2 and 3, the same mode spacing, and the range $2 \leq k \leq 2(242)^{1/2}$ are used. Each experiment involves only a single realization. In order to improve statistics, the experimental spectral data are all presented as averages over the bands $2 \leq k < 4$, $4 \leq k < 6$, etc. Those theoretical curves which are compared graphically to the numerical experiment have been similarly band averaged (indicated by over bar). The bars on the data points on some of the graphs indicate the standard deviation of the mean. The approximate number of wavevectors included in the experiments is 2000 for Run 1 and 16000 for Runs 2 and 3. For each wavevector, there are two independent degrees of freedom permitted by reality and incompressiblity. The theory integrations were done with continuous wavenumber integrations (infinite mode density), but, to get as close as possible to the experimental conditions, the same upper and lower wavenumner truncations were used. The inital values for Fourier amplitudes in the experiments were chosen, using a standard pseudorandom number generator, from a multivarate-normal distribution yielding the indicated spectrum.

Four laboratory and field experiments also are compared with the theory integrations. These are the low-Reynolds-number ($R_\lambda \sim 14$ to 20) wind-tunnel data of Stewart & Townsend (1951), low-Reynolds-number ($R_\lambda \sim 3$ to 30) water-tunnel data recently reported by Ling and Huang (1970), moderate-Reynolds-number ($R_\lambda \sim 35$ to 50) wind-tunnel data of Van Atta and Chen (1969), and the high-Reynolds-number ($R_\lambda \sim 3000$) tidal channel measurements of Grant, Stewart & Moilliet (1962).

The spectral quantities we compare are E(k,t), the vorticity spectrum $2k^2E(k,t)$, modal intensity U(k,t), and transfer spectrum T(k,t). The first three of these quantities represent the same information and, in each case, we choose the one which appears to give the clearest comparison. For the comparisons with the computer experiments, raw, unnormalized values are plotted. In the comparisons with the laboratory and field experiments, both the theoretical and experimental spectra are in nearly self-preserving forms at the times compared, and we normalize by the current Kolmogorov wavenumber and velocity scales $k_s = (\epsilon/\nu^3)^{1/4}$ and $v_s = (\nu\epsilon)^{1/4}$. In order to compare with the Stewart-Townsend and Grant-Stewart-Moilliet data, the energy spectrum is transformed into the one-dimensional spectrum

$$\phi_1(k) = \frac{1}{2}\int_k^\infty (1 - k^2/p^2)\, p^{-1} E(p)\, dp, \tag{54}$$

which is presented as $k^2 \phi_1(k)$.

We also examine the evolution of the following integral quantities: dissipation rate per unit mass $\epsilon(t)$, presented as $\epsilon(t)/\epsilon(0)$, skewness S(t) of longitudinal velocity derivative,

$$\begin{aligned} S(t) &= -\langle(\partial u_1(\mathbf{x}, t)/\partial x_1)^3\rangle / \langle(\partial u_1(\mathbf{x}, t)/\partial x_1)^2\rangle^{3/2} \\ &= (7.5)^{1/2}\left[\int_0^\infty k^2\, T(k, t)\, dk\right]\left[\int_0^\infty k^2\, E(k, t)\, dk\right]^{-3/2}, \end{aligned} \tag{55}$$

and the quantity $\Pi(t)$, defined as the maximum value of $\int_k^\infty T(k,t)dk$ (i.e., the area under the positive part of a T(k,t) versus k plot). We present $\Pi(t)$ as $\Pi(t)/\epsilon(0)$. S(t) is a nondimensional measure of strength of vorticity production.

Figures 1-18 compare the various approximations with the computer experiments, Figures 19-25 provide comparisons of the theories with each other and with the experiments of Ling & Huang and Stewart & Townsend, Figures 26-29 and Figures 30-32 compare the theories with, respectively, the Van Atta-Chen and Grant-Stewart-Moilliet data. In all these figures, the various approximations are identified by the acronyms introduced in Section 3. In the case of the test-field model,

TFM implies $g = 1.5$, while TFM' implies $g = 1.064$.

It is useful to discuss the comparisons as a whole, rather than take each up separately. We may note first that there are not large qualitative differences among the various approximations either in the inviscid approach to equipartition or in the curves of vorticity spectrum and transfer spectrum at low and moderate R_λ. There appears to be a consistent ranking of the approximations with regard to efficiency of energy transfer. EDW transfers the least energy, SCF and DIA transfer more energy, and the TFM transfers still more, except at the lowest R_λ values, where DIA and SCF transfer more energy than TFM.

The transfer inefficiency of EDW relative to DIA for the present decay calculations complements a similar behavior for steady-state turbulence noted some time ago. The reasons for the difference appear to be that DIA gives effectively longer memory times for build up of energy transfer, both because $U(k,t,t')$ is involved, which has (in DIA) a longer decay time than $G(k,t,t')$ for large k, and because, at lower k, the too-sharp cusp in the EDW $G(k,t,t')$ at $t' = t$, $t \neq 0$ gives a relatively fast initial fall off with $t - t'$, compared to DIA (Kraichnan 1964b). In decay calculations, there is an additional reason. The backward-into-time integration in the DIA energy-balance equation (11) makes the transfer at time t remember, in part, what the excitations were at times $s < t$, with the result that there is a tendency toward overshoot, in the transfer from initially strongly excited k's to initially weakly excited ones. This tendency is absent in the EDW energy-balance equation (34) because only current-time U's appear.

A remarkable feature of the integrations is the very close similarity of the DIA and SCF results for vorticity spectrum and transfer spectrum. In runs 4 through 7, this similarity is so great that the two approximations are indistinguishable on the spectrum plots. For this reason, only DIA is shown. The reason perhaps is a balancing of opposing tendencies. Because of (34), SCF gives to $U(k,t,t')$ unrealistically short decorrelation times, controlled by νk^2, at large k. This tends to decrease the memory times and, hence, the energy transfer by (11). At the same time, however, (34) makes transfer in SCF depend more strongly on past intensities than in DIA [$U(q,s,s)$ occurs instead of $U(q,t,s)$]. This produces bigger overshoot effects than in DIA (Herring 1966).

Figures 24 and 25 illustrate that the UNS model transfers substantially more efficiently than the DIA model, from which it differs by neglect of correlations between convected and convecting velocity and of reaction on convecting velocity. This result possibly is a hydrodynamic analog of Lenz' law. Inclusion of the reaction effects appear to brake the stretching of vortex tubes, if our statistical approximations are valid indicators.

The efficiency of transfer in TFM of course depends on the choice of g, rising as g decreases, because that increases the decay time of $G^S(k,t,s)$. The g value 1.5 evidently makes the energy transfer in TFM and ALH very similar quantitatively, both at low R_λ and in the high R_λ limit which indicated this choice. We have noted that ALH differs from DIA only in that the integrations back in time in the energy balance equation are along particle paths, so that Lagrangian time-correlations enter. The increased transfer efficiency of ALH and TFM ($g = 1.5$) over DIA, which is more pronounced at large R_λ, is associated with the matter of invariance to random uniform convection. The lack of this invariance in DIA gives relatively lower effective buildup times for energy transfer at the higher k's because the memory times entering the energy balance equation are convective decorrelation times. At very low R_λ, the updating of the convariance factors which enter the TFM energy balance equation, together with the rapid decay of the absolute levels of excitation, has a counter balancing effect, with the result that TFM actually transfer less energy overall than DIA.

The curves of S(t) appear to distinguish more sensitively among the different approximations than the other quantities presented. This is because the vorticity transfer, which enters S(t), is highly sensitive to the dynamics of the large wavenumbers, where the qualitative differences among the approximations are greatest. It is of interest that most of the approximations reproduce the overshooting transient in S(t) shown by the computer experiments. We surmise that the underestimation of the computer-experimental S(t) curves by DIA is a real effect, associated with lack of convection invariance. The TFM curves seem to fit the data excellently. The performance of DIA (and, equivalently, SCF) appears to improve on the spectrum plots of the computer experiments, where the high wavenumbers play a smaller role. In all the plots where TFM and DIA spectra are compared, TFM clearly transfers relatively more energy at high k, as evidenced by the higher tails of the $k^2E(k,t)$ and $T(k,t)$ curves. This again illustrates the effect of random-convection-invariance.

Runs 4 through 8 are examples where the initial spectrum quickly leads, for all the approximations shown, to spectrum forms which are nearly self-preserving, and independent of time, when normalized with Kolmogorov variables. The agreement of the approximations with the data of Stewart & Townsend, Van Atta & Chen, and Grant, Stewart & Moilliet seems as good as could be expected, subject to the ambiguities of interpretation mentioned at the beginning of this Section. In Run 8, we show only the random-convection-invariant approximations. The Reynolds number here is high enough that the theory curves are negligibly different from their infinite-Reynolds-number shape. The DIA and the other non-invariant approximations give a bad fit to the tidal-channel spectral data, in contrast to their performance at lower R_λ.

The spectrum (52) used in Runs 4 and 6 was chosen because Ling & Huang found that it fitted their water-tunnel data at $R_\lambda < 30$, and was self-preserving. Our results show that this spectrum does, in fact, quickly reach nearly self-preserving forms at $R_\lambda \sim 20$, for the several approximations, and that these evolved forms differ little from the initial spectrum. For DIA, the self-preserving form at $R_\lambda \sim 20$ also differs imperceptibly from that obtained in Run 5 with spectrum (53), which was found to lead to self-preservation in the original numerical studies of DIA (Kraichnan 1964a).

ACKNOWLEDGMENTS

We are grateful to S. A. Orszag and G. S. Patterson for making the results of their computer experiments available before publication. R. H. Kraichnan's work was supported by the Fluid Dynamics Branch, Office of Naval Research, under Contract N00014-67-C-0284.

REFERENCES

R. Balescu & A. Senatorski 1970. Ann. Phys. 58, 587.

S. F. Edwards 1964. J. Fluid Mech. 18, 239.

H. L. Grant, R. W. Stewart & A. Moilliet 1962. J. Fluid Mech. 12, 241.

J. Herring 1966. Phys. Fluids 9, 2106.

R. Kraichnan 1958. Second Symposium on Naval Hydrodynamics, edited by R. Cooper, Washington, Office of Naval Research, Publication ACR-38.

1959. Phys. Rev. 113, 1181.

1961. J. Math. Phys. 2, 124

1964a. Phys. Fluids 7, 1030.

1964b. Phys. Fluids 7, 1163.

1964c. Phys. Fluids 7, 1723.

1965. Phys. Fluids 8, 575.

1966a. Phys. Fluids 9, 1728.

1966b. Dynamics of Fluids & Plasmas, edited by S. I. Pai (Academic Press, New York).

1970. J. Fluid Mech. 41, 189.

1971a. J. Fluid Mech. 47, 513.

1971b. J. Fluid Mech. 47, 525.

C. E. Leith 1971. J. Atmos. Sci. 28, 145.

S. C. Ling & T. T. Huang 1970. Phys. Fluids 13, 2912.

S. A. Orszag & R. H. Kraichnan. 1967. Phys. Fluids 10, 1720.

S. A. Orszag & G. S. Patterson 1971. These proceedings.

R. W. Stewart & A. A. Townsend 1951. Phil. Trans. Roy. Soc. (London) A243, 359.

C. W. Van Atta & W. Y. Chen. 1969. J. Fluid Mech. 38, 743.

APPENDIX

This section gives some details of the numerical method used to integrate the integro-differential equations for U(k,t,t') and G(k,t,t'). For the DIA approximation these equations are (8), (11), and (12); the other methods are obtained from the above by either modifying the time structure of the right hand sides or changing the definition of the a-b coefficients or both.

The time integration of (8), (11), and (12) is essentially the same as described by Kraichnan (1964a). Consider (8) for example, and suppose that G(k,t,t') and U(k,t,t') are known for all k, and at the square array of mesh points $(t,t') = (t_n,t_m) = (n\Delta, m\Delta)$; $n,m = (0,1,2.\,.\,.\,N)$. Then $G(k,t_{N+1},t\ )$, $j = (0,.\,.\,.\,N)$ is found from the equation

$$G(k_1\ t_{N+1},\ t_j) = e^{-\nu\Delta k^2}\, G(k_1\ t_N,\ t_j) + (1 - e^{-\nu\Delta k^2})\,(F(t_N,\ t_j) + F(t_{N+1},\ t_j))/(2\,\nu\, k^2) \quad \text{(A1)}$$

Here F is the integral term in (8). This time integral is done here by Simpson's rule, supplemented, where needed, by the trapezoidal and 3/8 rule. Equation (A1) is Euler's modified method, stabilized by inclusion of exponential factors. It is an implicit method in that $F(t_N,t_i)$ involves $G(t_N,t_i)$, and must be solved by iteration. Such iteration is carried out to second order, starting with $F(t_{N+1},t_j)$ replaced by $F(t_N,t_j)$ in (A1). Finally (9) allows G to be computed in the domain (t_{N+1},t_{N+1}), given G and U in (t_N,t_N). The same procedure is used to solve (12), with (11) giving U along the time diagonal.

The wave number integration is performed as follows. First a convenient choice of points k_i, $i = (1,2,\ .\,.\,.\ M)$ is chosen at which the time integration is to be performed. The equations of motion for $U(k_i,t,t')$ and $G(k_i,t,t')$ ate then approximated by replacing U(p,t,t'), U(q,t,t') etc. in wave number convolutions by interpolated values derived from

$$\hat{U}(p,\ t,\ t') = u_0\ \phi_0^n(p) + \sum U(k_i,\ t,\ t')\ \phi_i^u(p) \quad \text{(A2)}$$

Here $U_0 = U(0,t,t')$. A similar approximation is made for G. In (A2), $\phi_i(k)$ are cardinal functions of the interpolation method $(\phi_i(k_j) = \delta_{ij})$.

The domain of the p-q integration in (8), (11), and (12) is a semi-infinite rectangular strip tilted by 45° to the p-q axes. The corners of the strip lie on the p-q axes at p=q=k. Two methods. of truncating the p-q integration are used. The simplest is to truncate the semi-infinite strip to a rectangular strip of length k_{max}, beyond which U (k) is negligible. Using a k_{max} too small so that U(k) is nonnegligible above cut-off, leads to energy non-conservation. The second method used is to discard (p-q) values for which $p > k_{max}$ or $q > k_{max}$. This method is energy-conserving, but has no directly assessible of truncation error.

The p-q integration was done by Gaussian quadrature using Gauss-Legendre weights and points distributed along the longitudinal and transverse strip directions. The order of integration was increased until no significant difference in estimates of the transfer function ocurred.

The conservative truncation method was used for runs 1, 2, and 3. These runs include a lower cut-off ($k_{min} = 2$) as well, so as to duplicate the conditions of the Orszag-Patterson experiment as exactly as possible. For the remaining runs, the non-conservative truncation was used. Results for run 8 were checked using the conservative method; errors in spectral quantities were below graphically detectable level.

The interpolating points k_i are distributed so as to well represent the energy spectrum. In general, this means a high density near the peak and a more sparse spacing at large k. For the calculations reported here the k_i were either uniformly spaced or generated by the formula,

$$k_i = k_0 \left(e^{i/k_i} - 1\right), \quad i = (1, 2, \ldots N) \tag{A3}$$

The interpolation procedure used in (A2) for all calculations here reported is based on cubic splines, with left end condition zero slope and free right end condition. In interpolating U and G, we regard them as functions of Z(k) instead of k. Here Z(k) is that function for which $Z_i = Z(k_i) = (0,1,2, \ldots N)$. The Z-transformation in general improves the interpolation since better estimates of derivatives are obtained from uniform than nonuniform spacing.

Data pertinent to the numerical integrations is given in the last 4 columns of table 1. Here, k_0 and k_1 appear in (A3), Δ is the time step, and N is the order (same in the longitudinal and transverse directions) of the Gauss-Legendre (p-q) integration. For the conservative (p-q) integrations – as well as those with a low wave number cut off – integrations over triangular areas must be subtracted from the rectangular integrations. The order of the Gaussian integration over these triangles is also N.

Discretizing errors (time step size Δ, interpolating point spacing, (p-q)-integration point density) and truncation errors from a finite k_{max} were determined by halving (or doubling) controlling parameters till significant differences in spectral and integral quantities disappeared. A measure of overall numerical accuracy is obtained from examining numerically the energy balance equation,

$$\mathcal{D} = \epsilon^{-1} \dot{E} - 1$$

Taking the output for E(t) and $\epsilon(t)$, numerically differentiating E(t), we find the following estimates for $\mathcal{D}$: for run 1, $\sim -10^{-4}$; runs 2 and 3 $\sim -10^{-3}$; runs 4, 5, and 6 $\sim -10^{-3}$ and run 7, 10^{-2}.

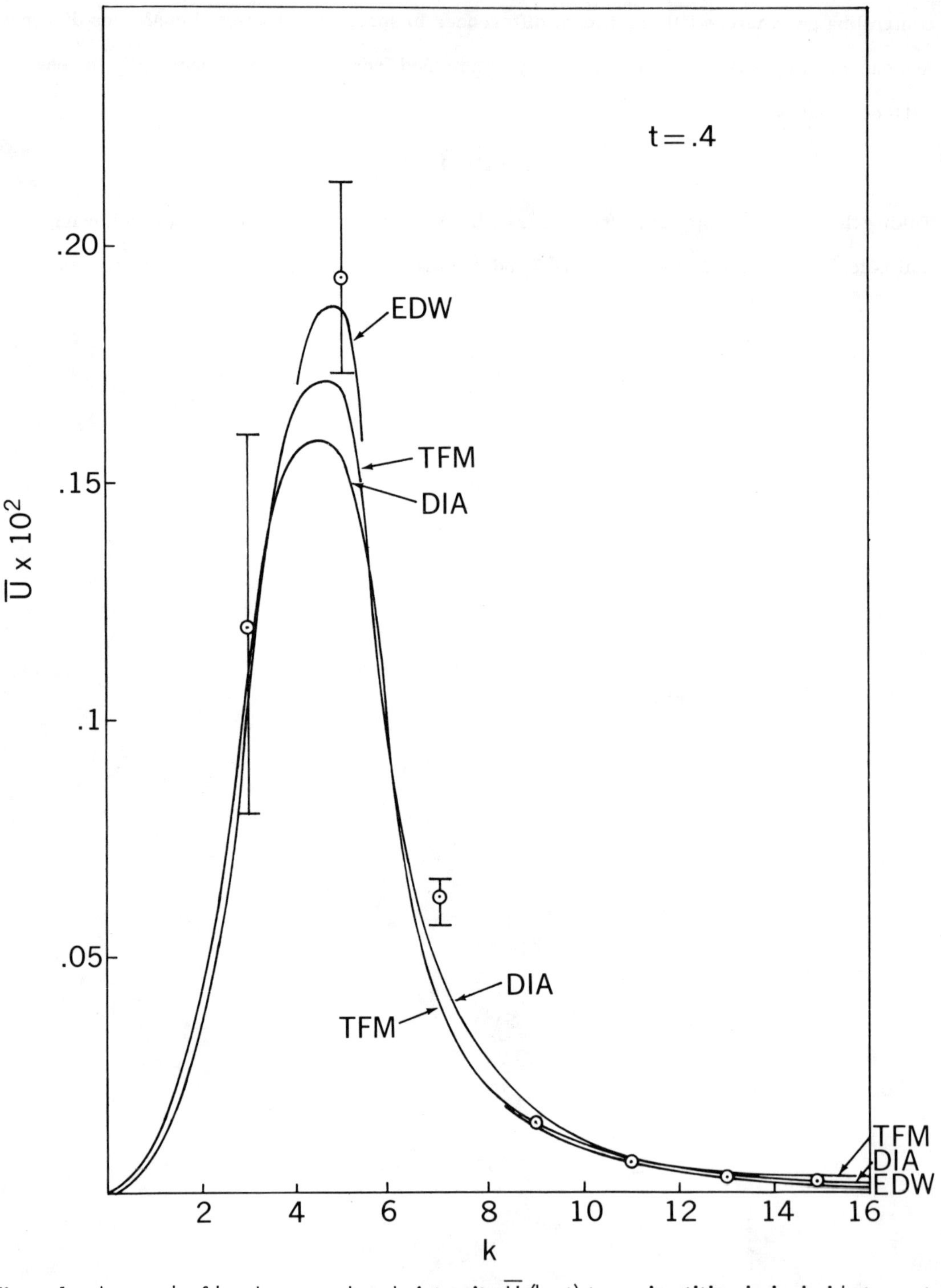

Figure 1. Approach of band averaged mode intensity $\overline{U}$ (k, t) to equipartition in inviscid, truncated system, Run 1. Curves show DIA, TFM, and EDW results at t = 0.4. Encircled points show band averages from computer-experiment.

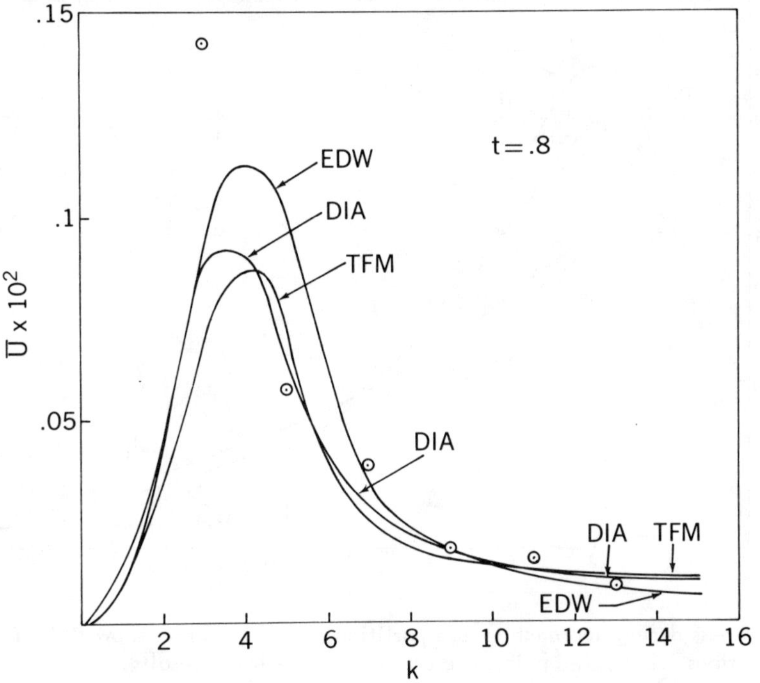

Figure 2. Approach of band averaged mode intensity $\overline{U}$ (k, t) to equipartition in inviscid, truncated system, Run 1. Curves show DIA, TFM, and EDW results at t = 0.8. Encircled points show band averages from computer-experiment.

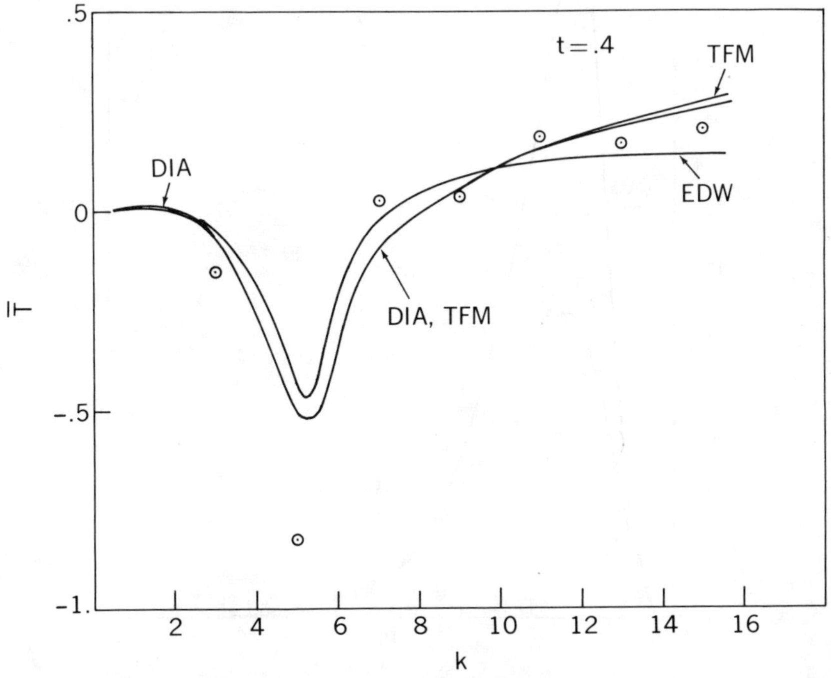

Figure 3. Band Average transfer function T (k, 0.4) during approach to equipartition. Data presented as in Figure 1.

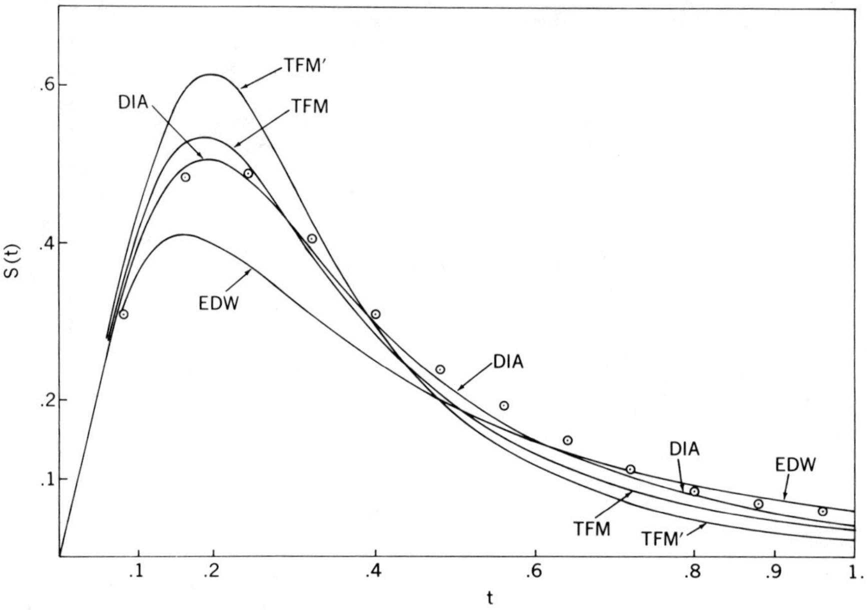

Figure 4. Skewness during approach to equipartition, Run 1. Curves show DIA, TFM, TFM' and EDW approximations. Encircled points are computer-experiment results.

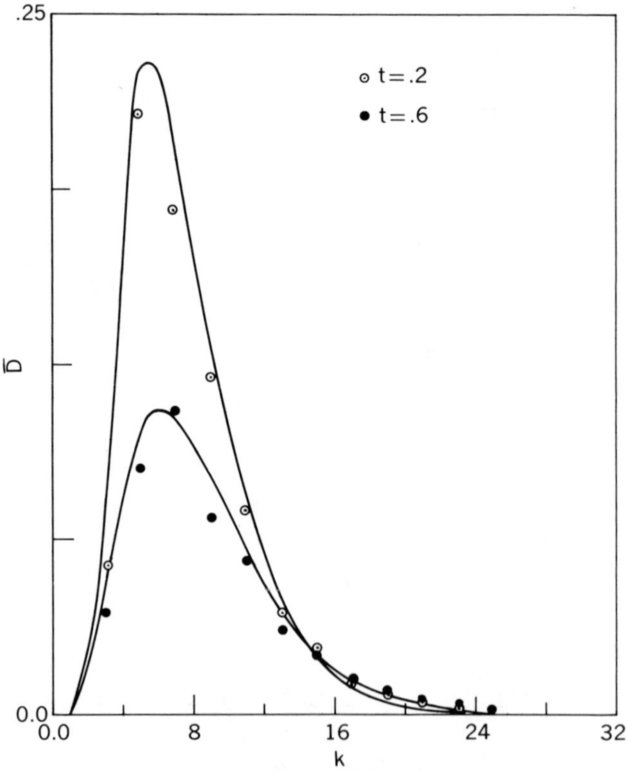

Figure 5. Evolution of band average $\bar{D} = \overline{2k^2 E(k, t)}$ for Run 2. Curves 1 and 2 are DIA results for t = 0.2 and 0.6. Symbols and are corresponding computer-experiments.

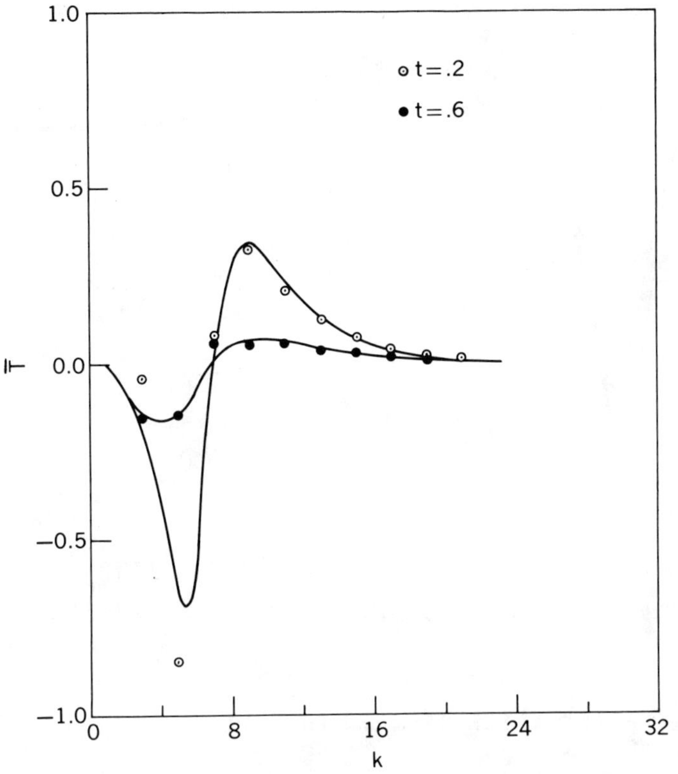

Figure 6. Evolution of band average of $\overline{T}$ for Run 2. Curves and points are as described in Figure 5.

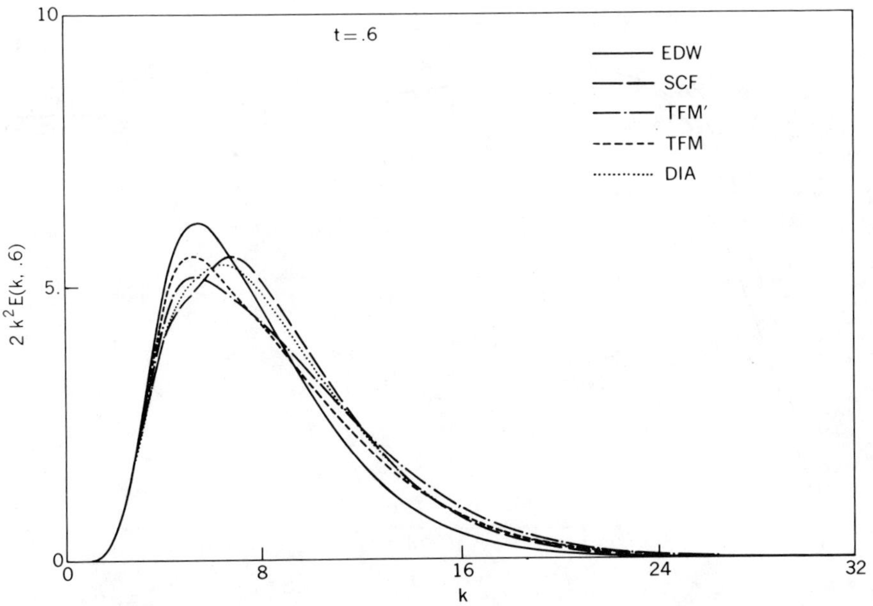

Figure 7. Comparison of DIA, SCF, TFM, and EDW results for $2k^2 E(k, t)$ for Run 2, t = 0.6. No band averaging has been preformed here.

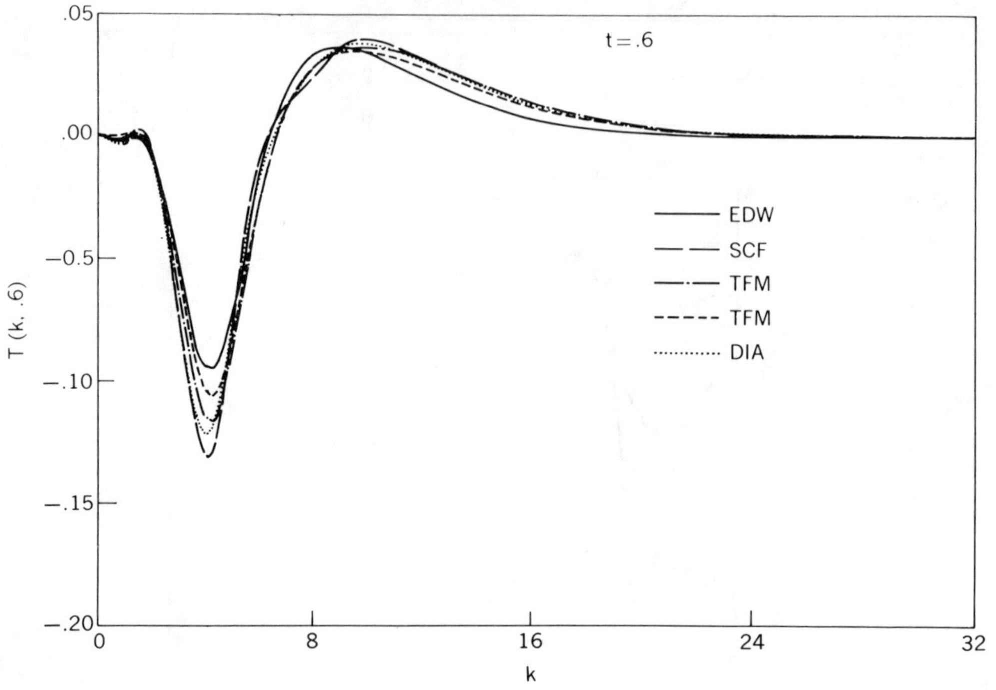

Figure 8. Comparison of DIA, SCF, TFM, and EDW results for T for Run 2. No band averaging has been preformed here.

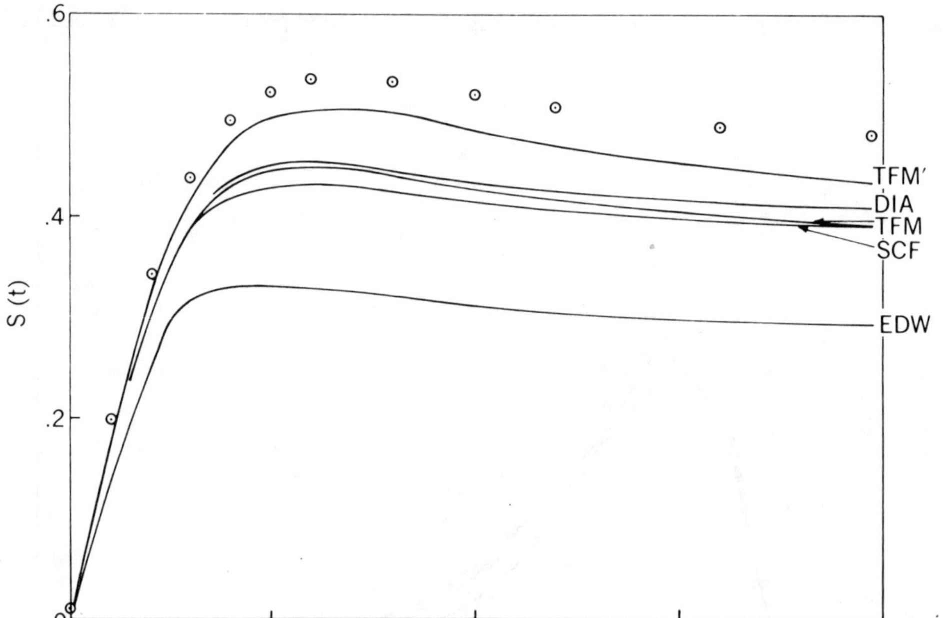

Figure 9. Evolution of S(t) for Run 2. Curves show results for the various approximations. Encircled points are computer experiment results.

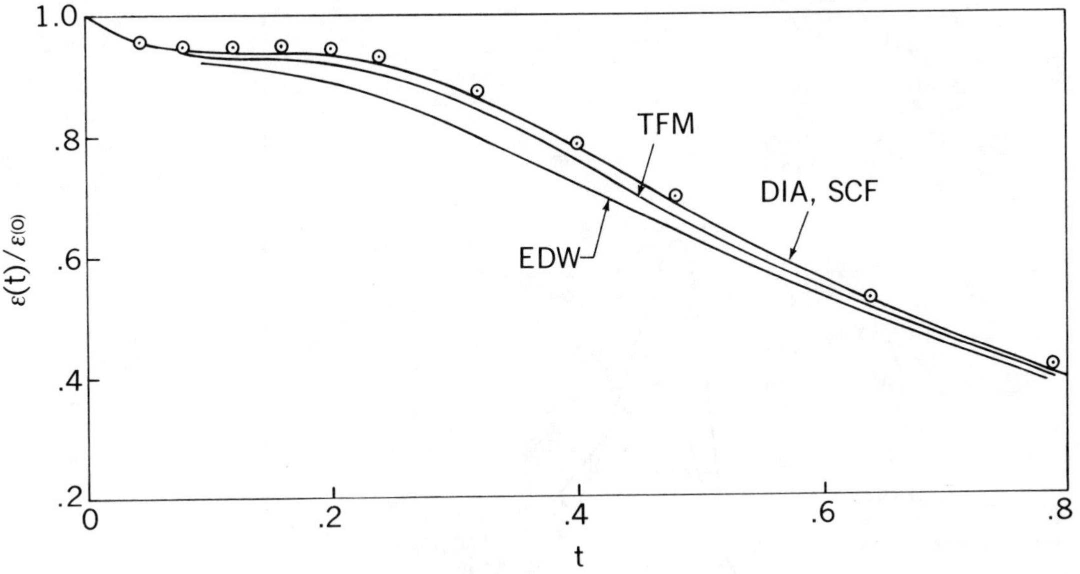

Figure 10. Evolution of $\epsilon(t)/\epsilon_{(0)}$ for Run 2. Encircled points are computer-experiment.

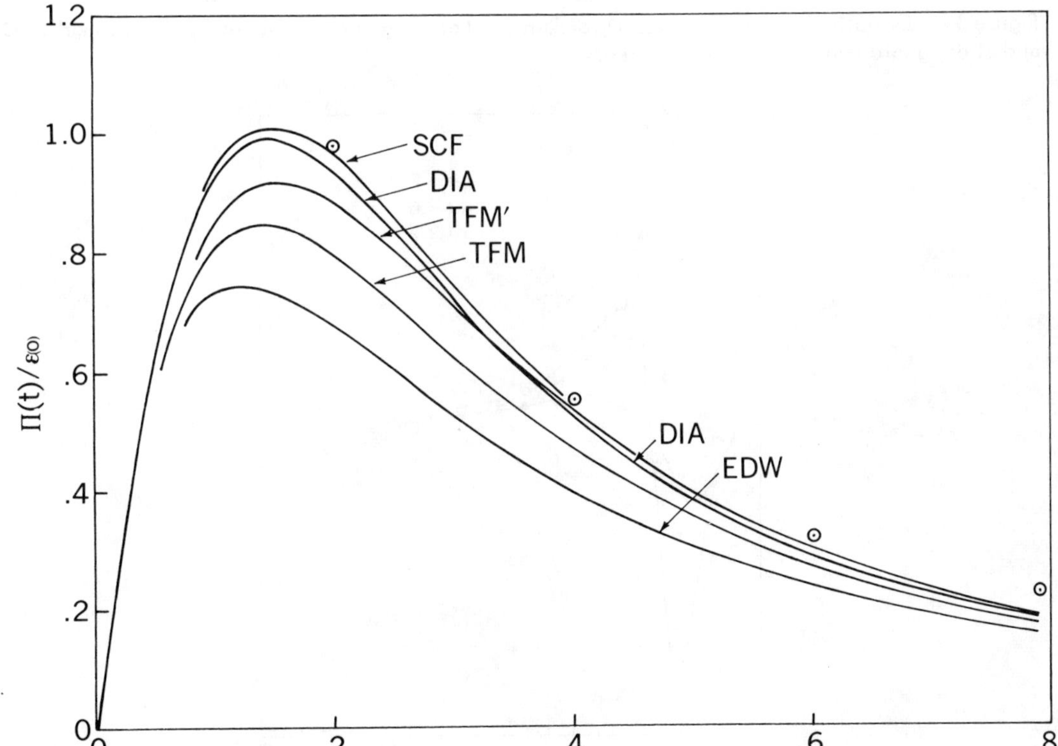

Figure 11. Evolution of $\Pi(t)/\epsilon_{(0)}$ for Run 2. Encircled points give band averaged computer-experiment results.

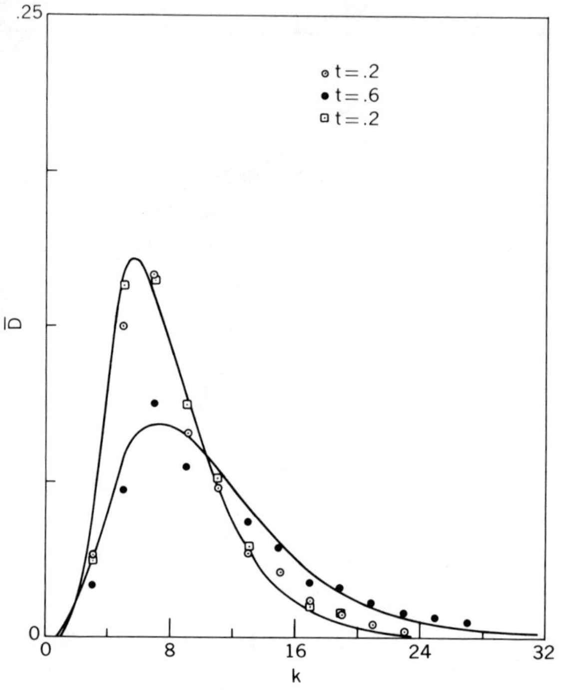

Figure 12. Evolution of band average $\overline{D}$ for Run 3. Labeling is as in Figure 5. Points marked ⊙ and ⊡ designate two computer experiments.

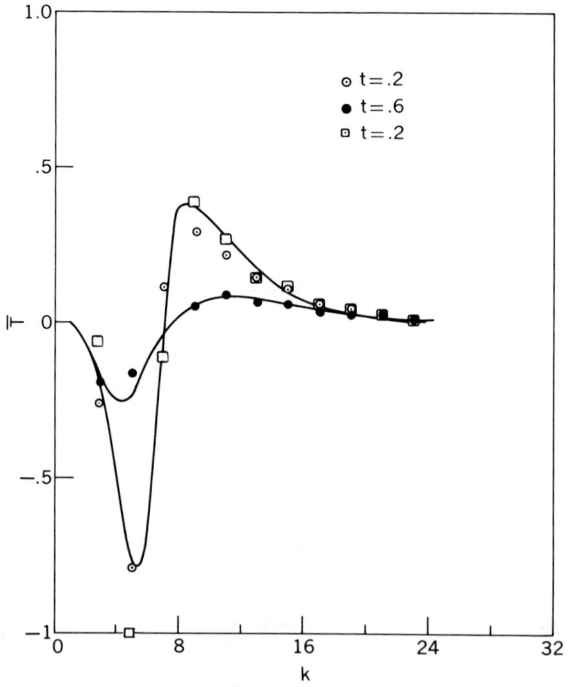

Figure 13. Evolution of band average $\overline{T}$ for Run 3. Curves and points as described in Figure 5. Points marked ⊙ and ⊡ designate two computer-experiments.

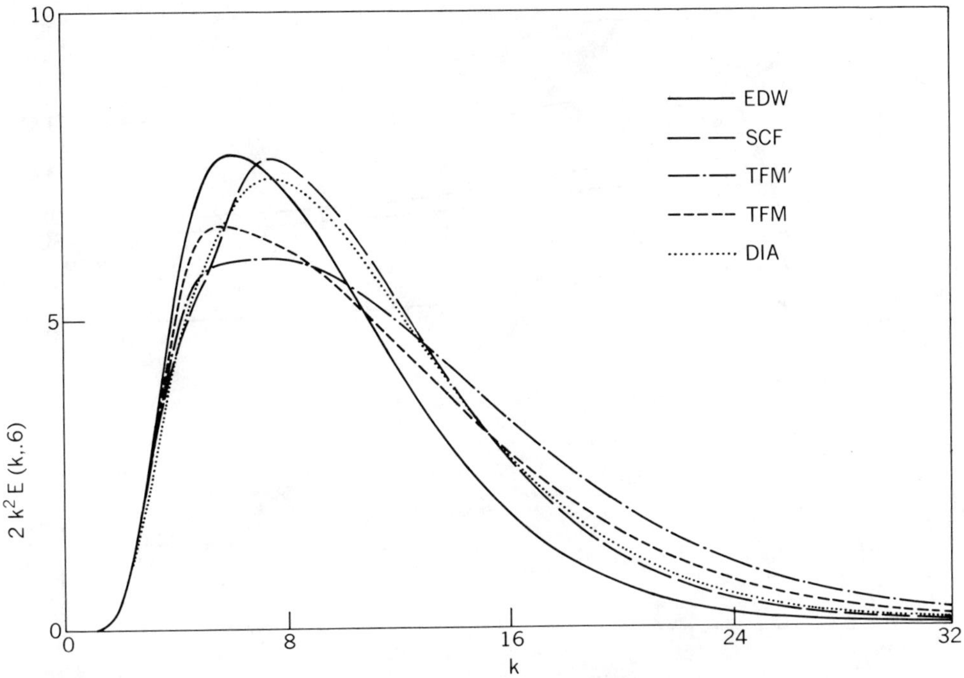

Figure 14. Comparison of DIA, SCF, TFM, and EDW results for $2k^2 E(k, t)$ for Run 3, $t = 0.6$. (No band averaging)

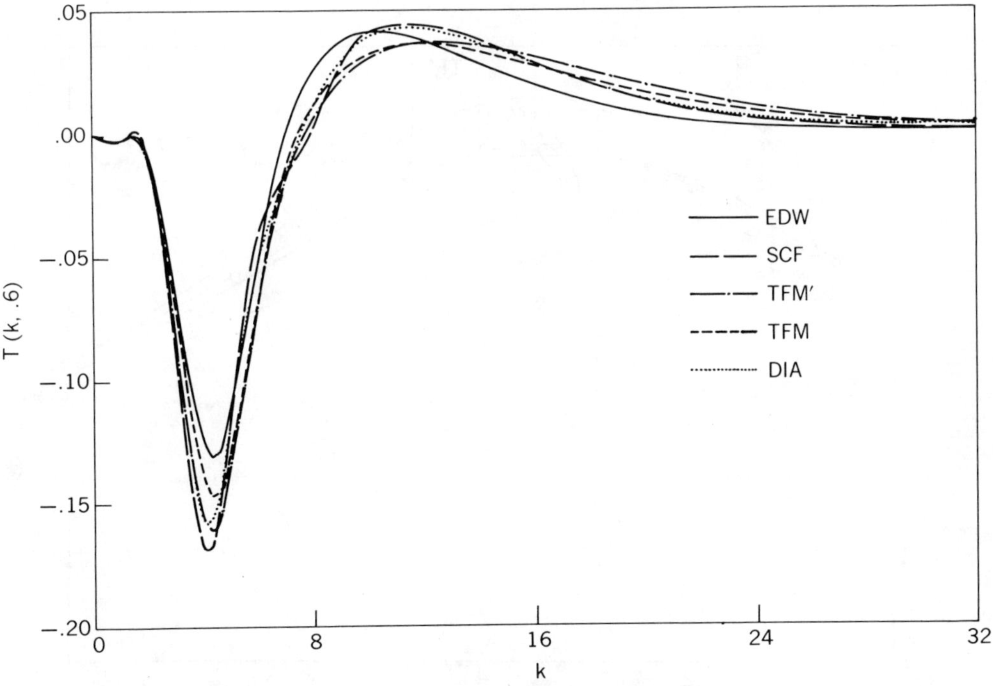

Figure 15. Comparison of DIA, SCF, TFM, and EDW results for $2k^2 E(k, t)$ for Run 3, $T(k, t)$. (No band averaging)

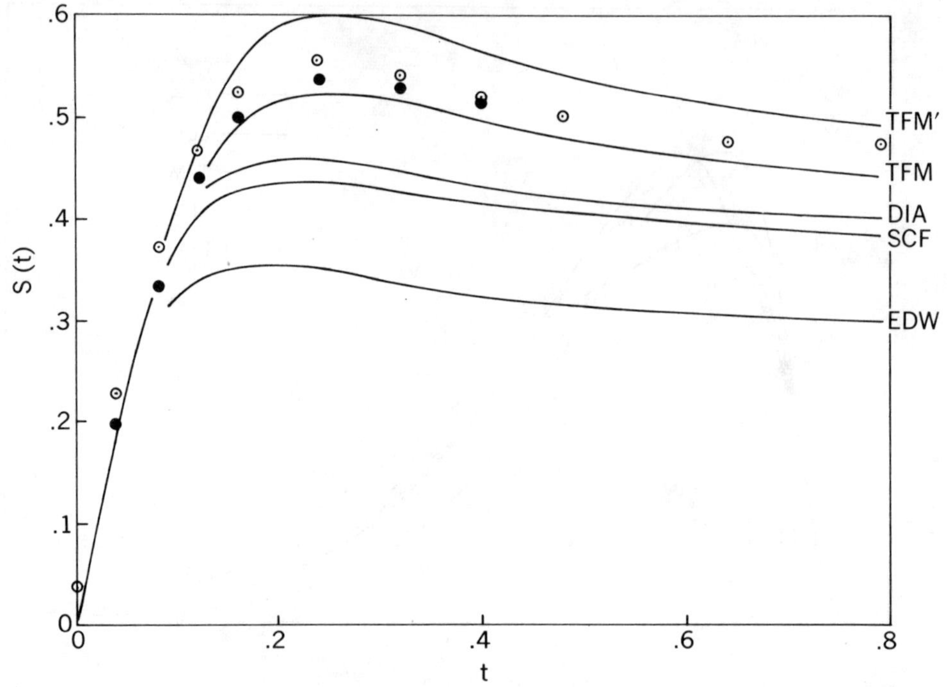

Figure 16. Evolution of S(t) for Run 3. Data for two computer-experiments are shown.

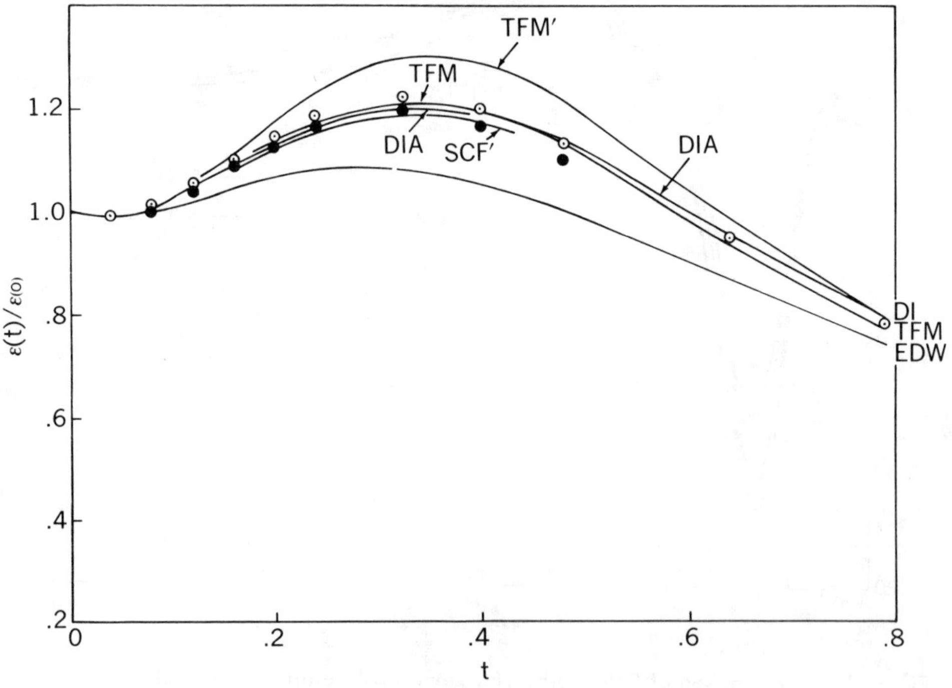

Figure 17. Evolution of $\epsilon(t)/\epsilon_{(0)}$ for Run 3.

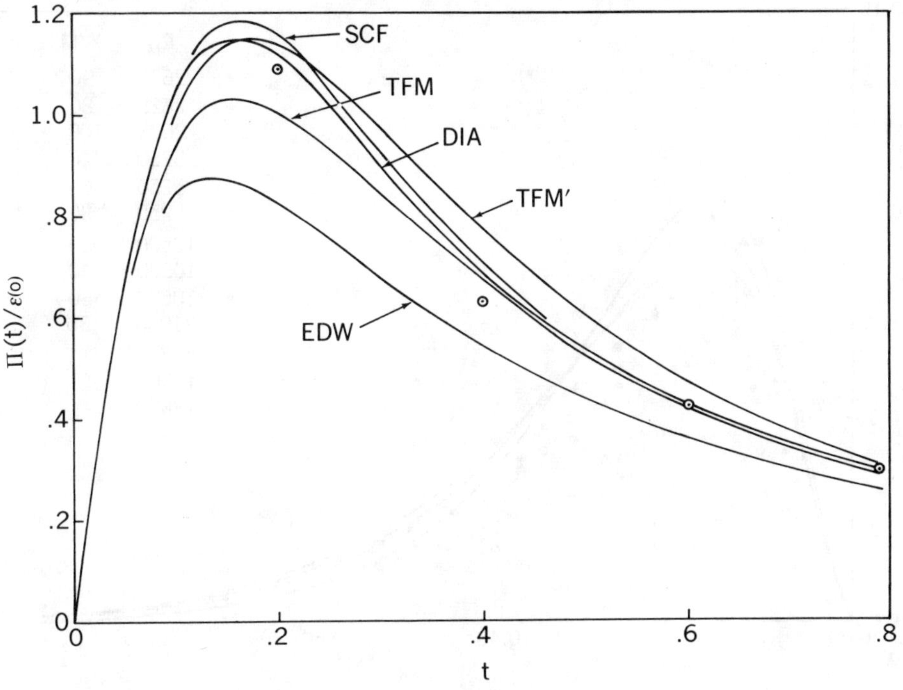

Figure 18. Evolution of (t)/$\epsilon_{(0)}$ for Run 3.

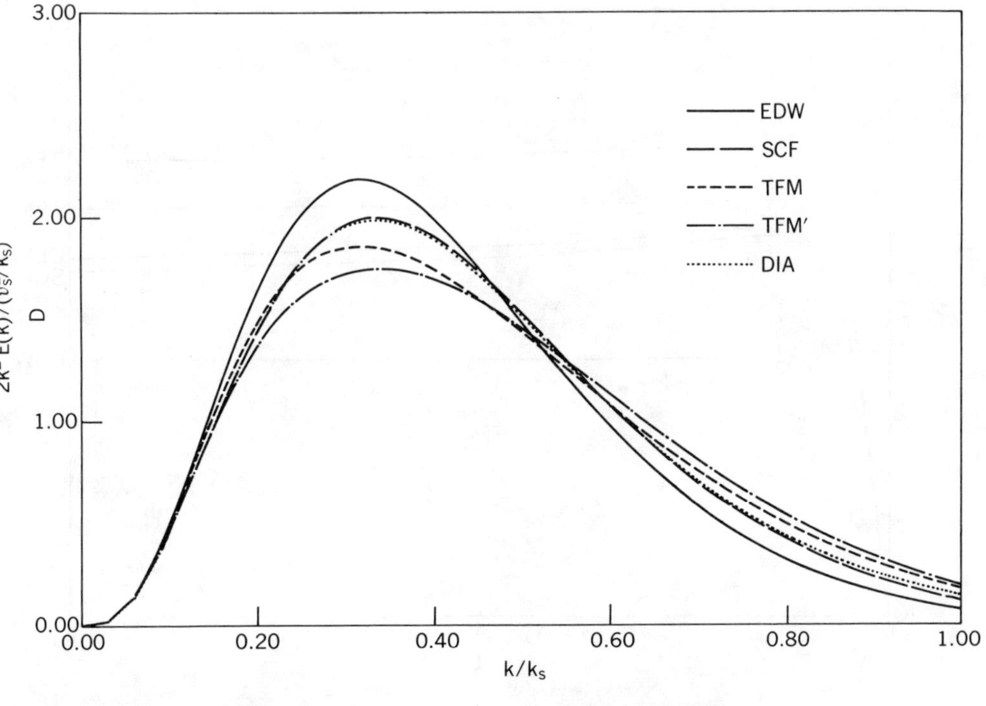

Figure 19. Evolved, normalized $2k^2 E(k, t)$ for Run 4.

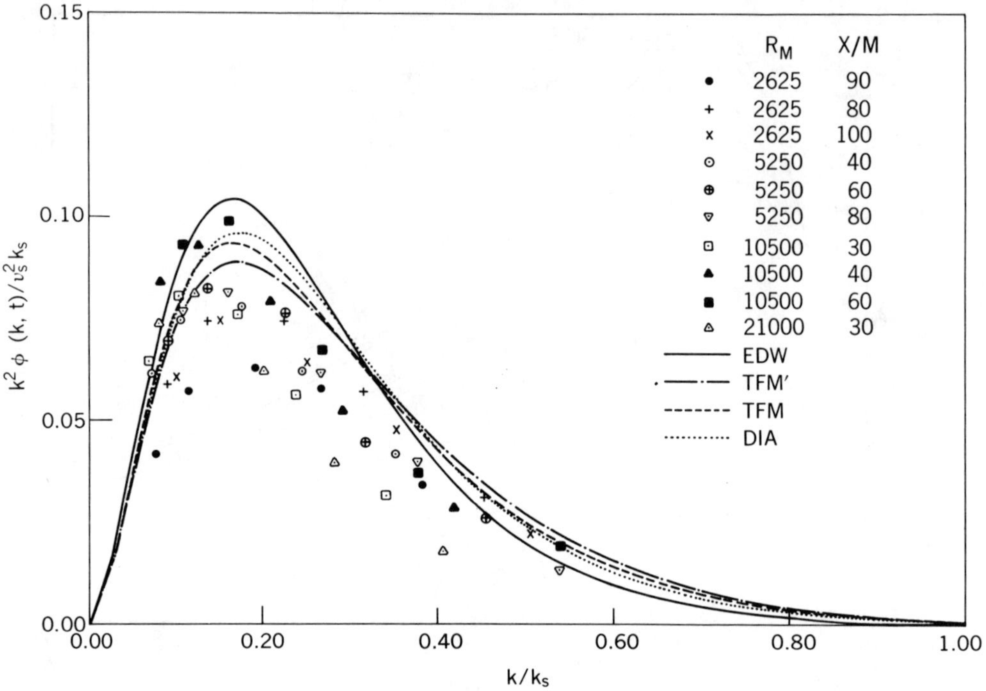

Figure 20. Evolved, normalized $2k^2 \phi\ (k,\ t)$ for Run 4. TFM' denotes TFM for g = 1.064. The symbols are laboratory data of Stewart and Townsend (1951).

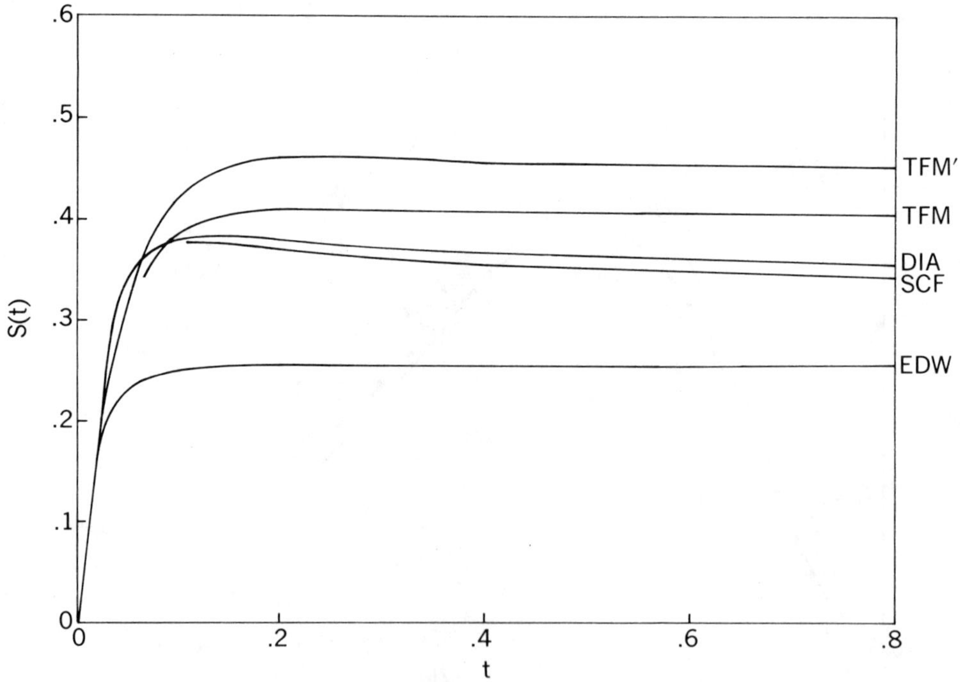

Figure 21. Evolution of S(t) for Run 4.

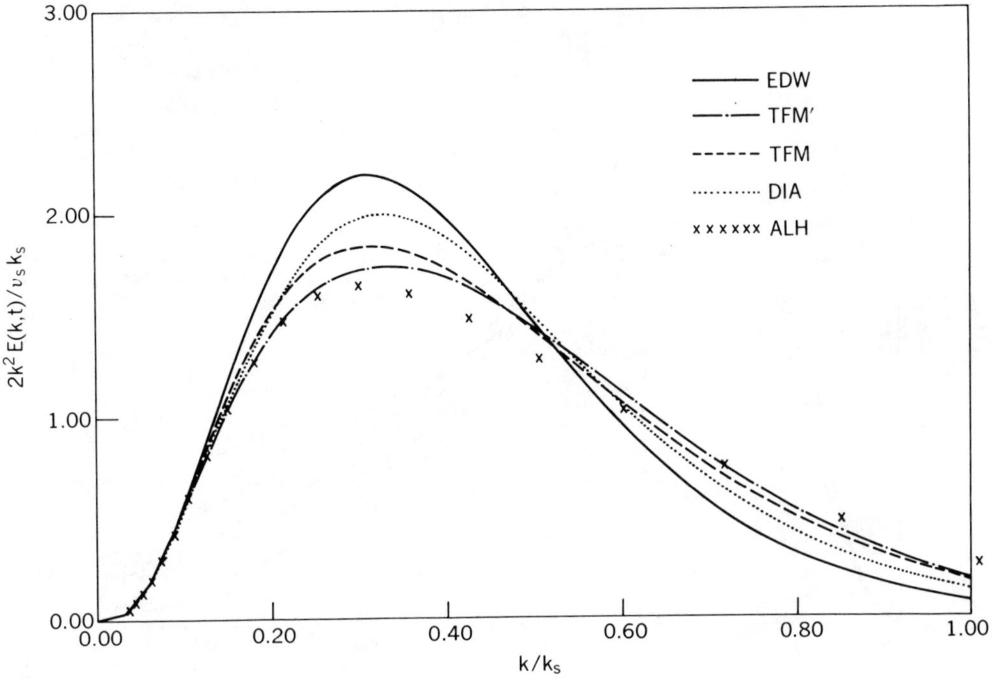

Figure 22. Evolved $2k^2 E(k, t)$ for Run 5.

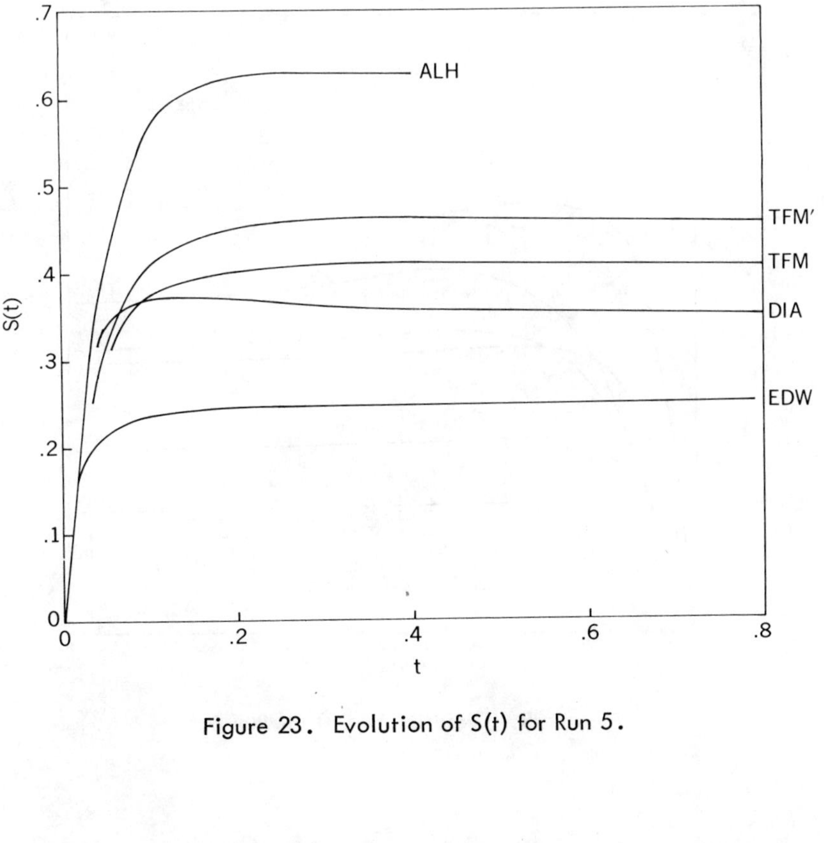

Figure 23. Evolution of S(t) for Run 5.

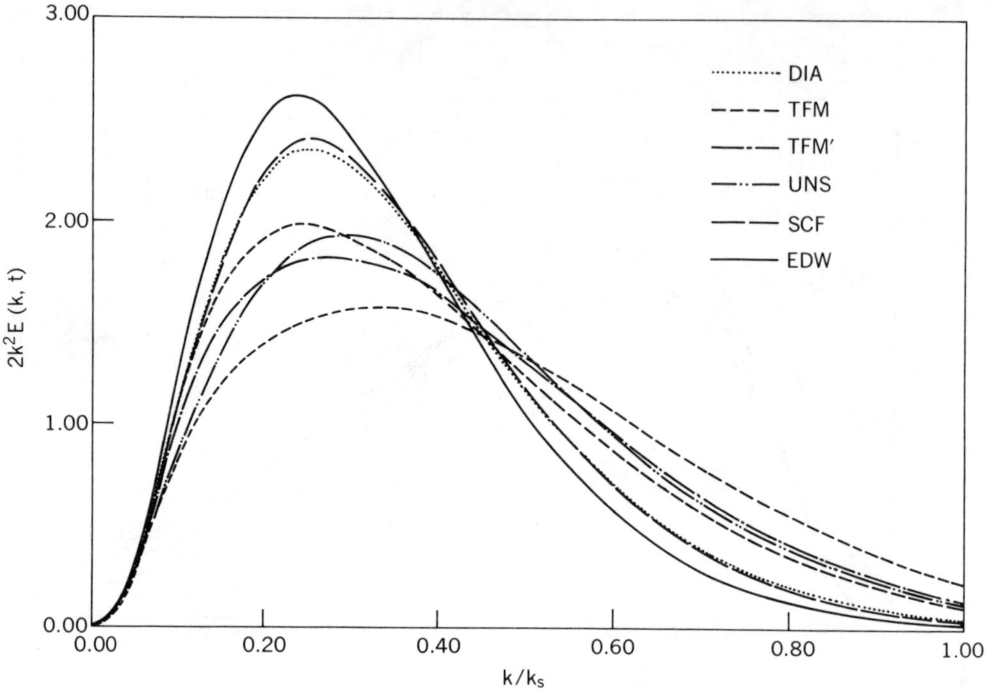

Figure 24. Evolved $2k^2 E(k, t)$ for Run 6.

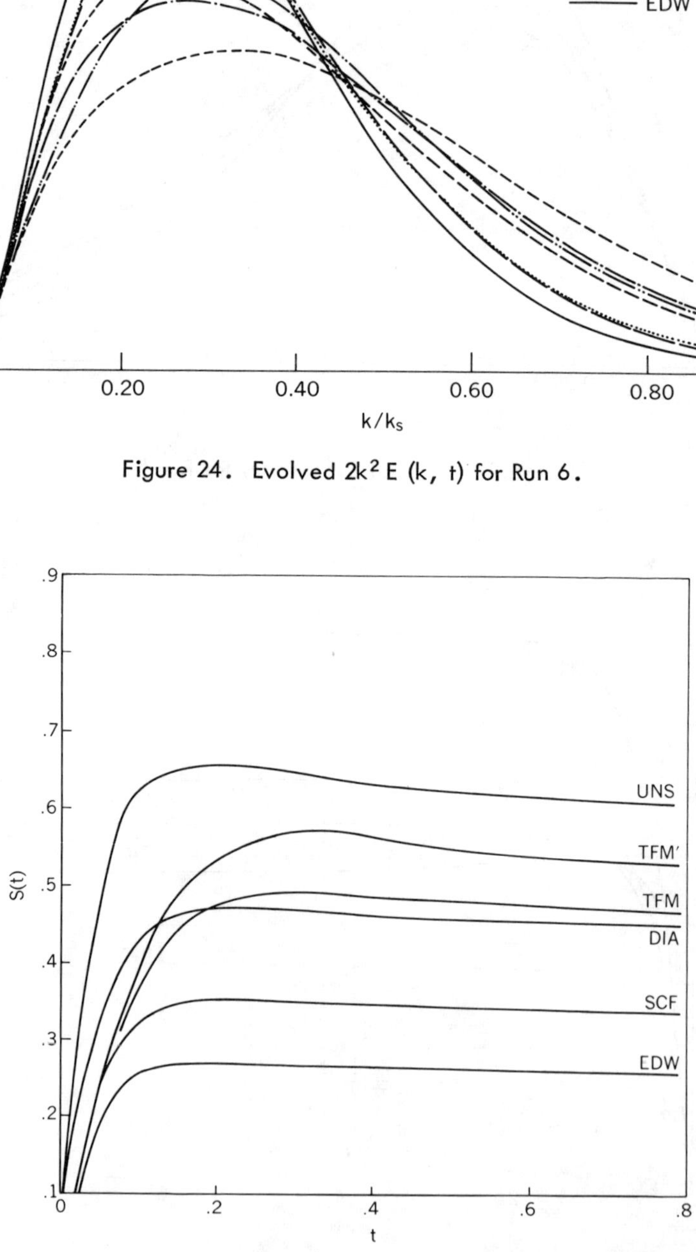

Figure 25. Evolution of S(t) for Run 6.

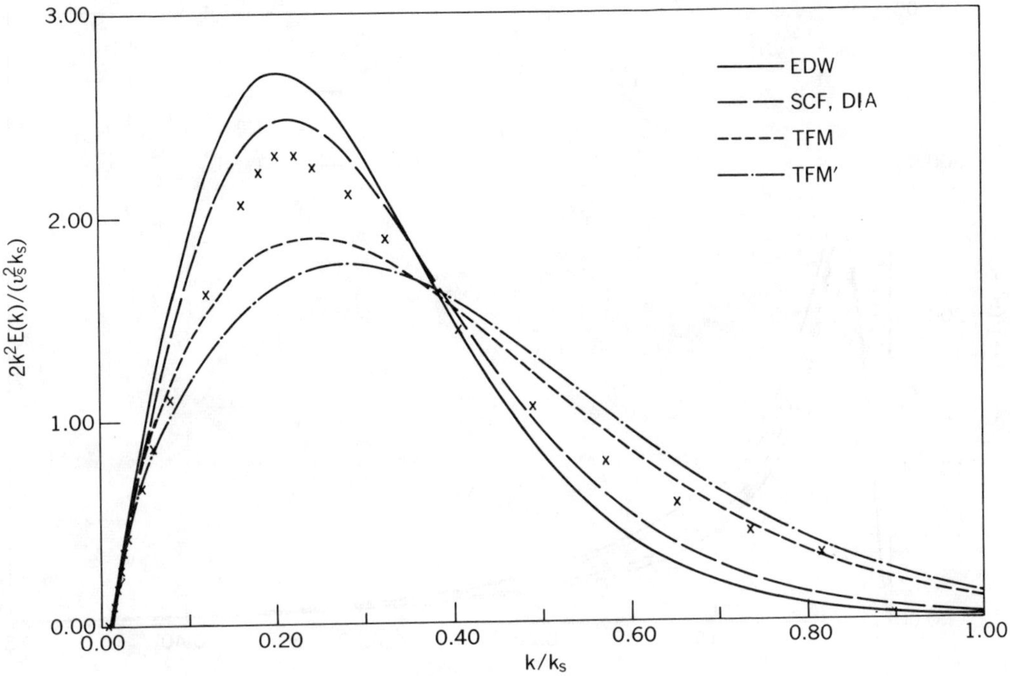

Figure 26. Evolved $2k^2 E (k, t)$ for Run 7.

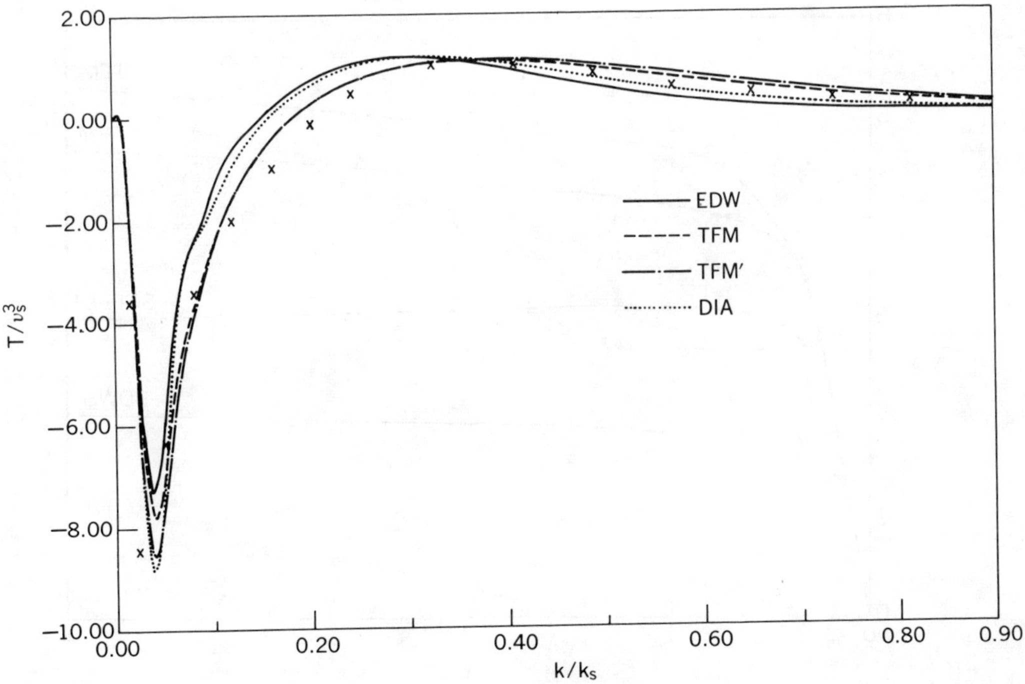

Figure 27. Evolved T (k, t) for Run 7.

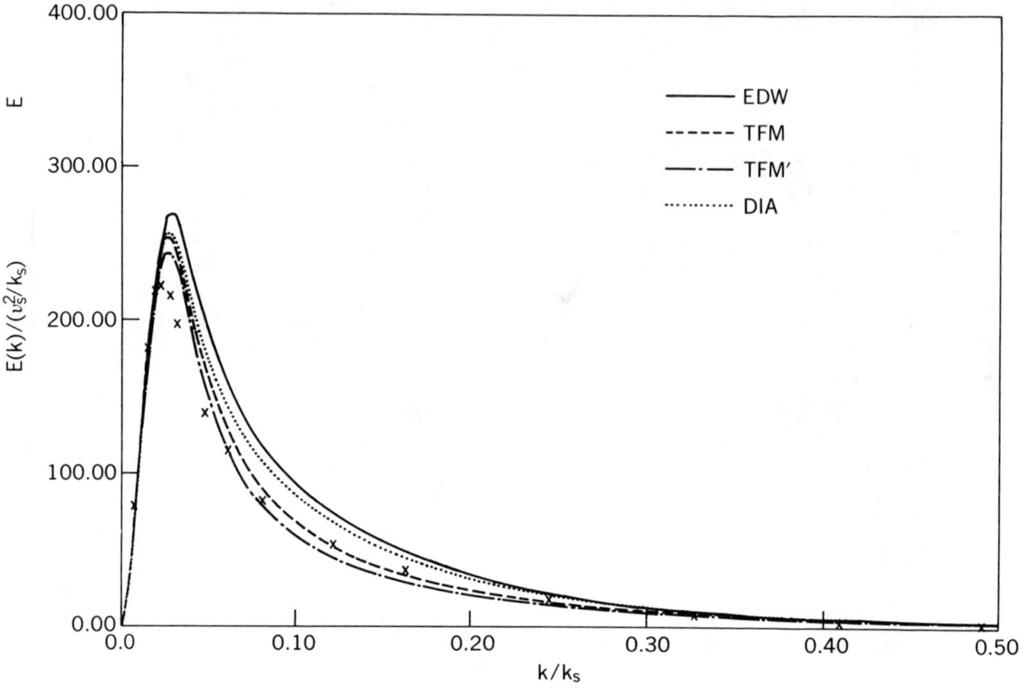

Figure 28. Evolved E (k, t) for Run 7.

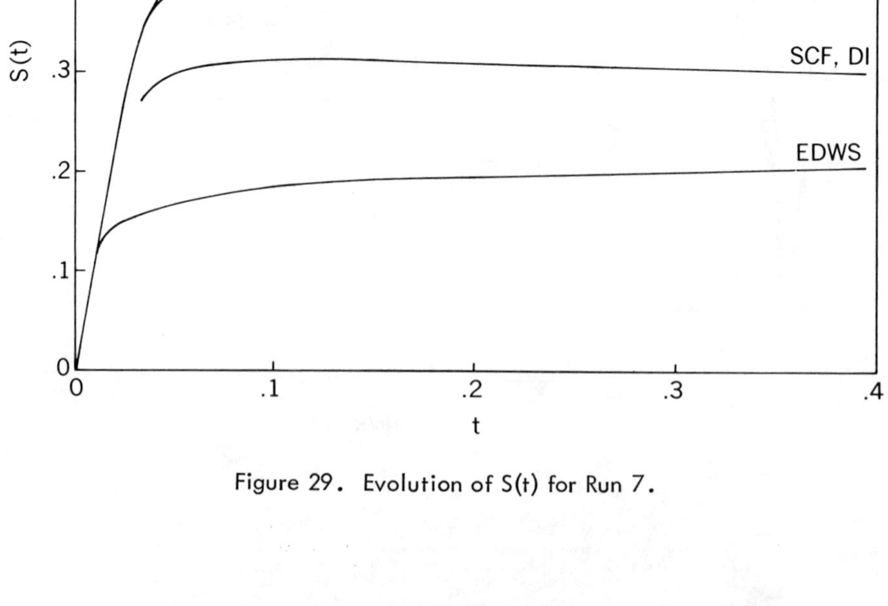

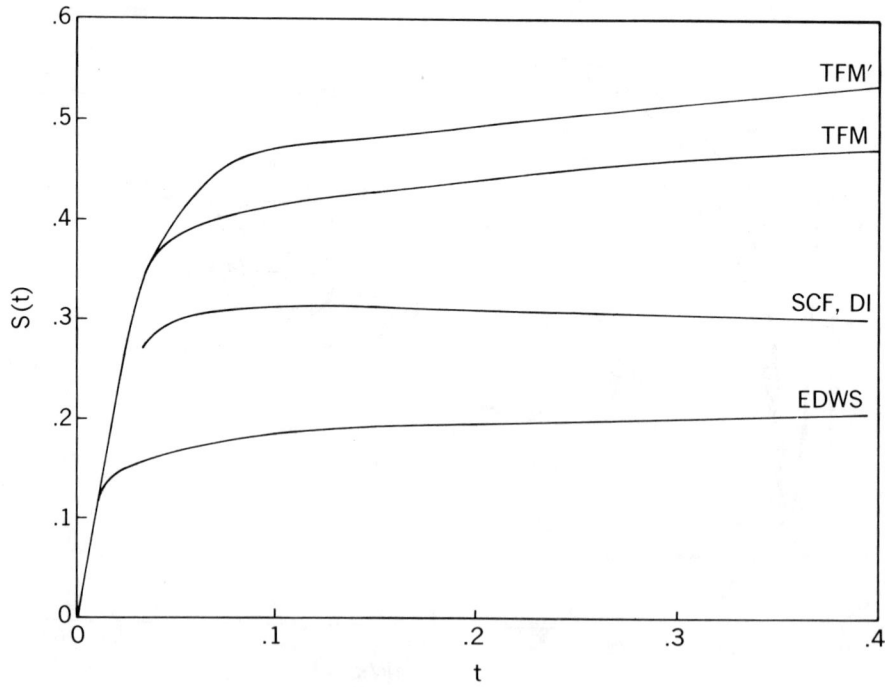

Figure 29. Evolution of S(t) for Run 7.

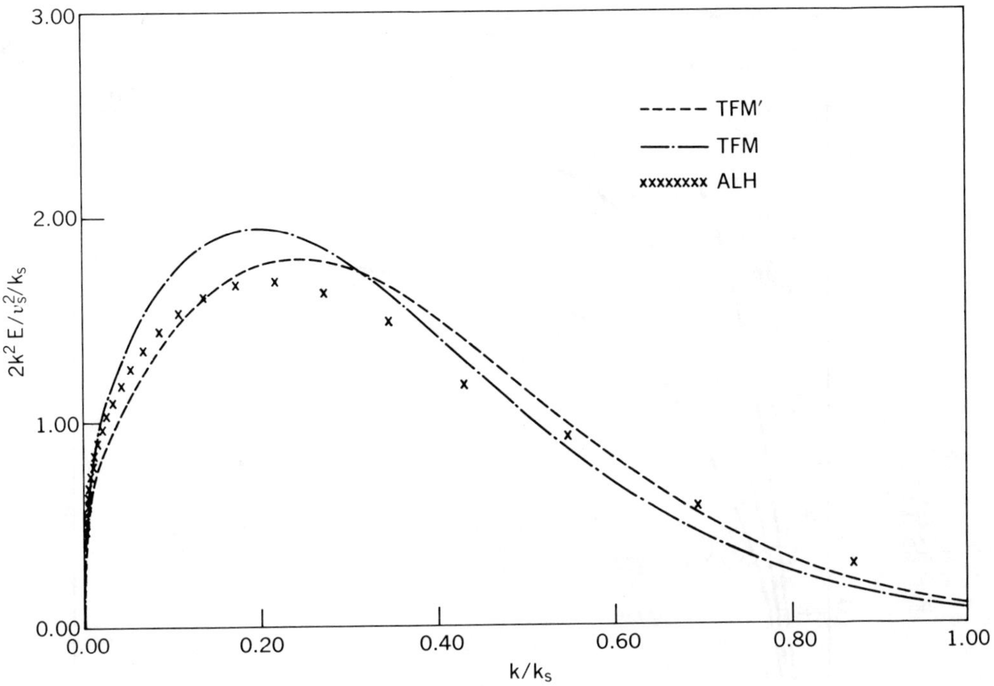

Figure 30. Evolved 2 k^2E (k,T) for Run 8.

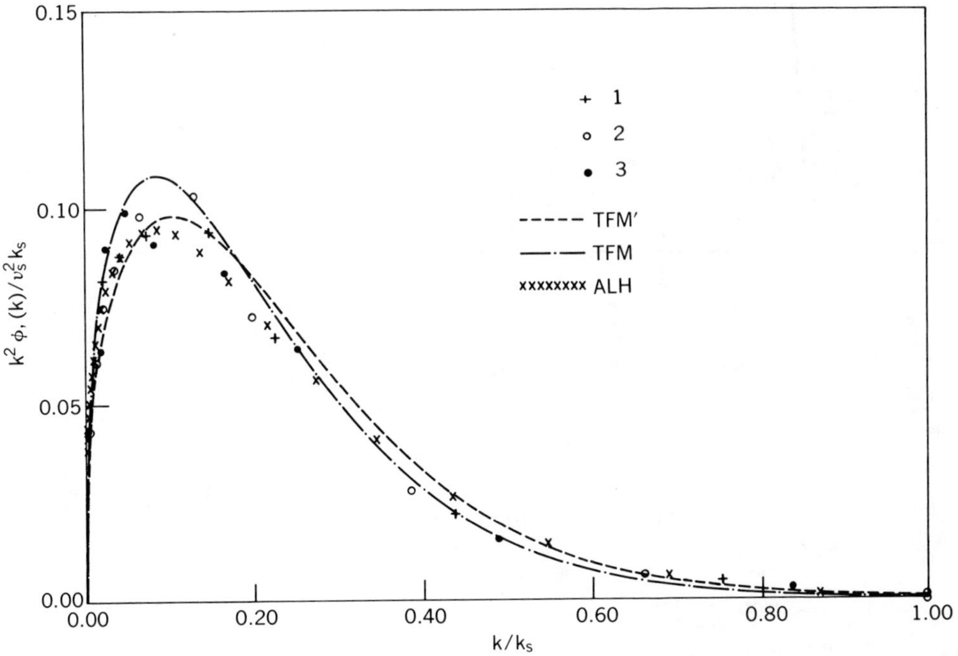

Figure 31. Evolved $k^2\phi_1$ (k) for Run 8, Compared With Tidal Channel Data of Grant, Stewart and Moilliet (1962).

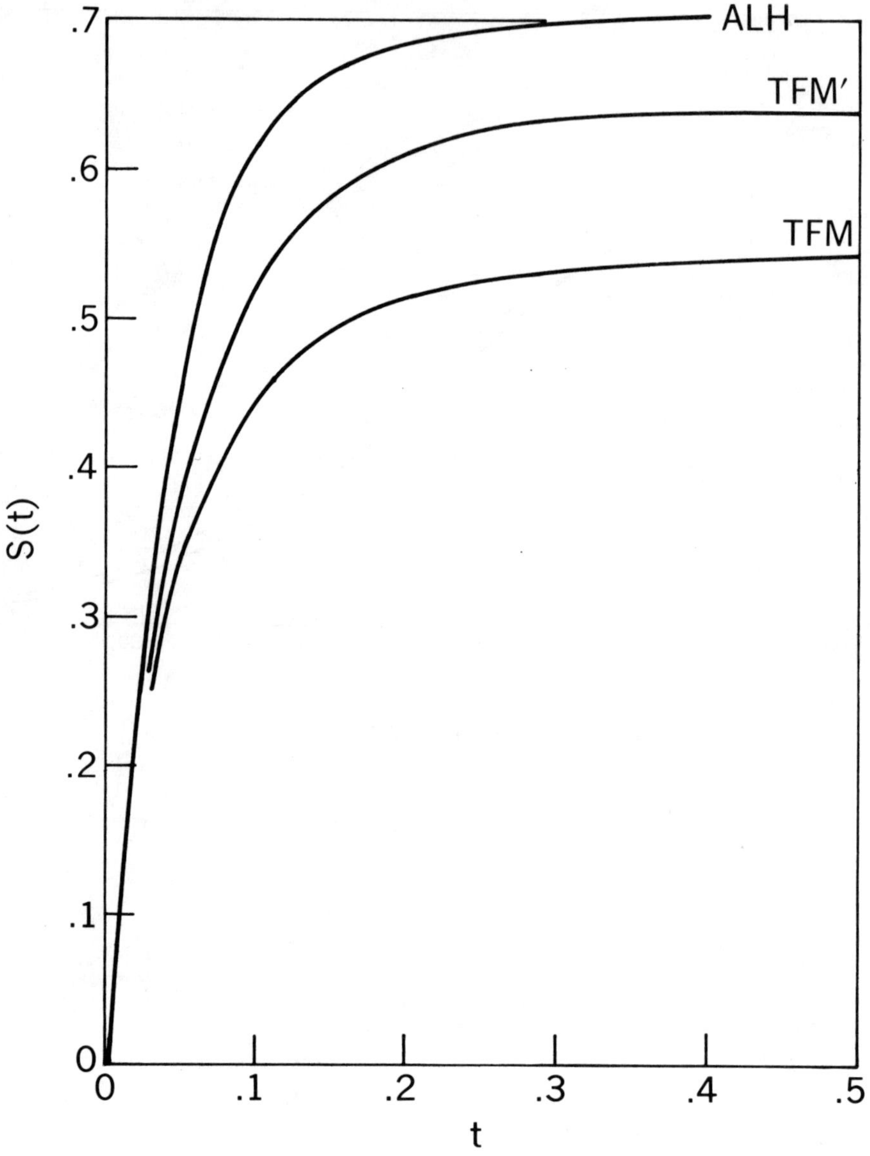

Figure 32. Evolution of S (t) for Run 8.

INVESTIGATING THE PREDICTABILITY OF TURBULENT MOTION

Edward N. Lorenz
Massachusetts Institute of Technology
Cambridge, Massachusetts

1. Introduction

Turbulence is sometimes cited as an ideal example of what in stochastic theory is termed a process. In the formal exposition of the theory, a process is identified with an ensemble of scalar or vector functions of time. A particular function of time which constitutes a member of the ensemble is termed a realization of the process. The realizations of a given process are supposed to have certain aspects in common; for instance, they may obey the same set of physical laws. An example of a realization would be a particular field of turbulent motion.

A stochastic process is one possessing some realizations which are identical to one another throughout the past but not in the future. In such a process the past of a realization usually restricts the realization to a subensemble whose future statistical properties differ from the properties of the total ensemble, but it does not determine the future of the realization uniquely. A stationary process is one where the statistical properties of the total ensemble do not vary with time.

Investigators who prefer to look upon turbulence as a stochastic process may be interested in predicting the future statistical properties of developing or decaying turbulence, or simply in determining the statistical properties of stationary turbulence. At the same time they may have little interest in predicting future states of particular realizations. Indeed, it is likely to be some assumed basic unpredictability of individual fields of turbulent motion which has made the application of stochastic theory attractive to these investigators.

There are nevertheless some instances where prediction of the behavior of particular fields of turbulent motion is of considerable interest and importance. This is notably true in the case of weather forecasting. The atmosphere is, after all, a turbulent fluid; the migratory cyclones and anticyclones which bring us much of our weather are among the more conspicuous turbulent elements.

Although one might offer a number of definitions of turbulence which would be reasonably satisfactory in real physical situations, we shall in this theoretical treatment regard turbulence as a process whose realizations are solutions of the Navier-Stokes equations (or some similar system of partial differential equations). This viewpoint is in keeping with much of the recent theoretical work. Our characterization is actually too general, since laminar motion also satisfies the equations. Perhaps a satisfactory definition would be an ensemble of nonperiodic solutions of the Navier-Stokes equations. Ensembles of solutions of simplified or otherwise modified forms of the Navier-Stokes equations will not qualify as turbulence; we shall instead regard them as models of turbulence.

It might then appear that turbulence so defined would be a deterministic rather than a stochastic process, since the Navier-Stokes equations are formally deterministic. We shall not concern outselves in this discussion with the possibility that, given an initial field of motion, the equations may not determine the motion uniquely at all future times. Let us simply note the possibility that two realizations which are nearly identical throughout the past may become unrecognizably different in the sufficiently distant future. This possibility can be verified as an actuality in some special cases. If there is slight uncertainty as to the present state of a realization, no system of prediction can choose rationally among the various possible states in the distant future, and the process, although perhaps formally deterministic, is for practical purposes stochastic. This effect is very important in limiting the range at which useful weather prediction is possible, since, even in the most populated regions of the globe, there are wide open spaces between weather stations, and therefore considerable uncertainty as to the state of the atmosphere at any particular time.

Even if we could observe turbulent motion without error, we could not predict its state in the distant future by any means presently available to us. We cannot find exact solutions of the Navier-Stokes equations except in very special cases; the best obtainable approximations are those yielded by stepwise numerical integration. To describe the observed or any predicted state with a computer we must replace a continuous field of motion by a finite set of numbers; the partial differential equations governing the motion must be replaced by a system of ordinary differential equations, and subsequently by a system of difference equations before stepwise numerical integration is possible. In short, we must use a model. Should there be no error initially, there will still be a slight error after completion of the first time step, and a larger one after the next. The errors which accumulate during the early steps will subsequently amplify just as if they had been present initially.

Predictability is therefore limited by the growth rate of errors. Yet it may be a serious oversimplification to talk about a single growth rate. In the atmosphere, at least, errors in different scales of motion seem to have their own growth rates. Doubling times are a few days for the largest systems, but only a few minutes for thunderstorms. The author (1969) has discussed in detail the possibility that the inevitable errors in the smaller scales, which soon become large, will then induce errors in somewhat larger scales, which will in turn become large and induce errors in still larger scales. This possible spreading of errors from smaller to larger scales makes it highly desirable to study turbulence with a model covering many octaves of the spectrum.

2. Realizations and ensemble statistics

The most straightforward way to investigate the predictability of turbulent motion would be to choose a "basic" initial state, and a "perturbed" initial state consisting of the basic state plus a small superposed "error", and then to examine the subsequent behavior of each state by solving the appropriate equations. One would thus be working directly with realizations. The amplification rate of the error would depend upon the choice of the basic state and also upon the form of the error, but the experiment could be repeated a number of times with different choices, whence a typical growth rate could be established.

As we have already seen, this procedure cannot be carried out in an exact manner, since we cannot solve the exact equations. We must be content to use a model. With this restriction, the procedure has indeed been carried out on many occasions.

Most investigations appear to have been performed with models which attempt to simulate the atmosphere, with varying degrees of sophistication. An early study by the author (1965) represented the state of the atmosphere by 28 numbers, and hence solved 28 simultaneous equations; a recent study by Smagorinsky (1969) uses more than 50,000 numbers. From the point of view of pure turbulence, however, the studies are typified by one recently performed by Lilly (1971), who dealt with two-dimensional turbulence, i.e., motion governed by the two-dimensional form of the incompressible Navier-Stokes equations. These equations may be reduced to a single partial differential equation representing, aside from the influences of viscosity and external forcing, the conservation of vorticity at points moving with the flow.

Some theoreticians refuse to acknowledge such motion as turbulence, since, if energy is fed into the largest scale of motion, it will not cascade to the smaller scales. However, energy fed into intermediate scales will spread to both larger and smaller scales, and much of the irregularity and apparent randomness characterizing three-dimensional turbulence will be found.

Lilly used in essence an infinite plane in which the motion was restricted to be periodic in each of two mutually perpendicular directions, with the same fundamental wave length in each direction. The complete flow was thus determined by the flow within a fundamental square. This flow was represented in the model by the values of the stream function at a uniform grid of 64×64 points. The partial derivatives occurring in the Navier-Stokes equations were represented by finite differences.

Herein lies the principal difference between turbulence and the model; scales of motion too small to be resolved by the grid are not explicitly acknowledged by the model, although their influence upon the larger scales may be partially included through a judicious formulation of the viscous effects. The same limitation also characterizes the most elaborate models of the atmosphere; thunderstorms and even considerably larger systems are commonly lost between grid points.

The principal factor limiting the number of grid points used in the computations was computer time rather than computer storage. With the fastest known computation schemes, which make use of the fast Fourier transform, the amount of computation per grid point per time step increases only slightly as the number of grid points increases. However, when the resolution is doubled (twice as many points in each direction), the time increment must be cut at least in half to avoid computational instability. Doubling the resolution therefore increases the labor by an order of magnitude. Extensive experiments covering ten or more octaves seem to be at best a thing of the future.

With 64 × 64 points, Lilly found that when the initial error was many orders of magnitude smaller than the basic flow upon which it was superposed, the energy spectrum of the error soon acquired a characteristic shape. Its subsequent growth, as long as it was still small compared to the basic flow, was quasi-exponential. The shape of the error spectrum differed somewhat from that of the basic flow, with the result that errors in the smaller scales reached their limiting values in advance of errors in the larger scales. It is difficult to say physically whether the errors in the smaller scales actually spread to the larger scales, or whether the errors in the larger scales simply took longer to mature.

A striking feature of the computations was the tendency for very large errors to occur at a very small number of points, at any particular time. Maximum errors exceeding ten standard deviations were not uncommon. The frequency distribution of the errors was far from Gaussian.

It is not obvious how Lilly's results would have been modified if he had been able to use higher resolution. The 64 × 64 grid gives some resolution down to wave number 32, and offers a fairly good description of wave number 16 (four grid points per wave length). Most of the energy of the growing error was contained in the well-represented wave numbers less than 16. However, the majority of the enstrophy was contained in wave numbers greater than 16, whence some significant information was probably missing.

As an alternative to dealing with realizations, which, as we have seen, cannot be too satisfactorily modeled, we may work directly with statistical properties of the process. From the equations which govern the realizations, we may derive systems of equations governing various ensemble statistics. These statistics may include the statistics of <u>differences</u> between fields of motion, i.e., of the errors.

Here we inevitably encounter a closure problem. Normally we wish to include ensemble averages of the velocity (or stream function, or vorticity) among the statistics to be considered; the time derivatives of these contain averages of quadratic quantities. Similarly the time derivatives of quadratic statistics include cubic statistics, etc. In order to obtain a finite closed system of equations we must introduce additional postulates. These generally take the form of specifying ensemble averages of higher-degree quantities in terms of averages of quantities of lower degree.

At this point the derived system of equations is far more complicated than the original system. However, certain reductions, which would not be possible if realizations were being used, can now frequently be made. An ensemble which is initially homogeneous will remain homogeneous if the forcing is homogeneous, and the number of dependent variables may be greatly reduced. Isotropy, another self-preserving property, allows further simplifications. But even then the derived system is generally more complicated than the original system, if third-degree statistics appear explicitly.

The greatest savings to be realized from working with statistics result from the further assumption that these statistics are rather slowly varying functions of their arguments, and may therefore be adequately described by their values for a limited number of arguments. The spectral density function, for example, may be satisfactorily depicted by three or four values per octave. As a consequence, the required amount of computation will not increase by an order of magnitude for every additional octave of resolution, but may simply be proportional to the number of octaves, or to some low power of this number.

A typical study of this sort has been performed by Leith (1971). For a closure scheme Leith used the eddy-damped Markovian approximation suggested by Orszag (1970), which effectively specifies fourth-degree statistics in terms of those of lower degree. Like Lilly, he dealt with two-dimensional turbulence which was spatially periodic in two directions; however, he was not confined to six octaves of the spectrum. Nevertheless, in view of his closure assumption, he was dealing with a model of turbulence rather than with turbulence itself.

Leith also found that initial error spectra of differing shapes, if small in amplitude, soon assumed a characteristic shape, and then grew with relatively little change in shape until they became comparable in magnitude to the spectrum of the basic flow. In some of his computations the basic flow was comparable in energy and horizontal scale to typical atmospheric flows, and he found error growth rates comparable to but somewhat more rapid than those which had generally been found from working with realizations of atmospheric motion.

3. A low-order model

We could eliminate the problem of finding a suitable closure scheme by returning to the use of realizations, but then we should reencounter the problem of representing sufficiently many scales of motion. The procedure which we shall summarize in the remainder of this work represents an attempt to overcome the latter problem, while not reintroducing the former. Naturally it may lead to other problems. We shall describe the procedure for the case of two-dimensional turbulence, and, as in the specific studies which we have described, we shall deal with spatially periodic fields of motion, which are completely specified by their behavior in a fundamental square.

The Navier-Stokes equations for two-dimensional incompressible flow reduce to the simple vorticity equation

$$\frac{\partial \nabla^2 \psi}{\partial t} = \frac{\partial \psi}{\partial y} \frac{\partial \nabla^2 \psi}{\partial x} - \frac{\partial \psi}{\partial x} \frac{\partial \nabla^2 \psi}{\partial y} + \nu \nabla^4 \psi + F \quad , \tag{1}$$

where t is time, x and y are rectangular Cartesian coordinates, ψ is a stream function for the flow (whence $\nabla^2\psi$ is the vorticity), ν is a coefficient of viscosity, and F is an external forcing function which prevents the motion from ultimately dying out. Because of the spatial periodicity the vorticity may be written

$$\nabla^2\psi = \sum_{\mathbf{J}} X_{\mathbf{J}} e^{iD^{-1}\mathbf{J}\cdot\mathbf{r}} , \tag{2}$$

where $2\pi D$ is the length of a side of the fundamental square, $\mathbf{r}$ is a two-dimensional vector with components (x, y), and $\mathbf{J}$ runs over all two-dimensional vectors both of whose components are integers. Reality of $\nabla^2\psi$ demands that $X_{-\mathbf{J}} = X_{\mathbf{J}}^*$, where the star denotes the complex conjugate. The mean kinetic energy E and the mean enstrophy V are then given by

$$E = \frac{1}{2} D^2 \sum_{\mathbf{J}} J^{-2} X_{\mathbf{J}} X_{\mathbf{J}}^* , \tag{3}$$

$$V = \frac{1}{2} \sum_{\mathbf{J}} X_{\mathbf{J}} X_{\mathbf{J}}^* , \tag{4}$$

where J denotes the magnitude of $\mathbf{J}$. The summations in (3) and (4) are redundant; i.e., identical products $X_{\mathbf{J}} X_{\mathbf{J}}^*$ and $X_{-\mathbf{J}} X_{-\mathbf{J}}^*$ are added together.

Since the coefficients $X_{\mathbf{J}}$ define the vorticity, and hence the stream function, they can serve equally well as dependent variables. From (1) and (2) it follows that

$$\frac{dX_{\mathbf{J}}}{dt} = \sum_{\mathbf{K},\mathbf{L}} C_{\mathbf{JKL}} X_{\mathbf{K}}^* X_{\mathbf{L}}^* - \nu D^{-2} J^2 X_{\mathbf{J}} + G_{\mathbf{J}} , \tag{5}$$

where $G_{\mathbf{J}}$ bears the same relation to F which $X_{\mathbf{J}}$ bears to $\nabla^2\psi$, and where

$$C_{\mathbf{JKL}} = \begin{cases} -\frac{1}{2}(K^{-2} - L^{-2})(\mathbf{K}\times\mathbf{L}) & \text{if } \mathbf{J}+\mathbf{K}+\mathbf{L} = 0 , \\ 0 & \text{if } \mathbf{J}+\mathbf{K}+\mathbf{L} \neq 0 . \end{cases} \tag{6}$$

Thus we may say that the vectors $\mathbf{J}$, $\mathbf{K}$, $\mathbf{L}$, or the variables $X_{\mathbf{J}}$, $X_{\mathbf{K}}$, $X_{\mathbf{L}}$, _interact_ if $\mathbf{J}+\mathbf{K}+\mathbf{L} = 0$.

The difficulty in handling equation (5) as it stands is that of handling an infinite system of equations. The customary simplification procedure is to omit all reference to vectors $\mathbf{J}$ and the corresponding variables $X_{\mathbf{J}}$ when either component of $\mathbf{J}$ exceeds some prechosen integer. This proves to be neither more or less restrictive than representing $\nabla^2\psi$ by its values at a prechosen grid of points; in either case the smaller scales of motion are not explicitly treated.

The procedure which we propose allows for representation of virtually all scales of motion. It is based upon the assumption that if the terms in the summations in (3) and (5) are arranged in a random order, one may be able to estimate the total sum after summing only a few terms, more or less as one estimates the outcome of an election after counting a few ballots.

As a preliminary step we choose a resolution factor α . We divide the spectrum into intervals, assigning the vector $\underset{\sim}{J}$ and the corresponding variable $X_{\underset{\sim}{J}}$ to the jth interval if $\alpha^j \leq J < \alpha^{j+1}$. The number N_j of vectors in the jth interval is then approximately $\pi(\alpha^2-1)\alpha^{2j}$. We let Q_{jkl} denote the number of triples of interacting vectors $\underset{\sim}{J}$, $\underset{\sim}{K}$, and $\underset{\sim}{L}$ belonging respectively to the jth, k^{th} , and l^{th} intervals. In defining N_j and Q_{jkl} we regard $\underset{\sim}{J}$ and $-\underset{\sim}{J}$ as separate vectors.

The mean kinetic energy may then be written

$$E = \frac{1}{2} D^2 \sum_{j=0}^{\infty} \sum_{\underset{\sim}{J}} J^{-2} X_{\underset{\sim}{J}} X_{\underset{\sim}{J}}^* , \tag{7}$$

with an analogous expression for V , where in the second summation $\underset{\sim}{J}$ runs over all vectors in the jth interval. Likewise equation (6) may be written

$$\frac{dX_{\underset{\sim}{J}}}{dt} = \sum_{k,l=0}^{\infty} \sum_{\underset{\sim}{K},\underset{\sim}{L}} C_{\underset{\sim}{J}\underset{\sim}{K}\underset{\sim}{L}} X_{\underset{\sim}{K}}^* X_{\underset{\sim}{L}}^* - \nu D^{-2} J^2 X_{\underset{\sim}{J}} + G_{\underset{\sim}{J}} , \tag{8}$$

where $\underset{\sim}{K}$ and $\underset{\sim}{L}$ run over all vectors in the kth and l^{th} intervals. Ideally α should be chosen small enough so that different variables in the same interval can be expected to exhibit similar statistical behavior.

As the principal step, we now select from each interval a relatively small number of vectors, and then omit reference to all vectors and the corresponding variables except those selected. We let M_j denote the number of selected vectors in the jth interval, while Q_{jkl} denotes the number of triples of interacting selected vectors $\underset{\sim}{J}$, $\underset{\sim}{K}$ and $\underset{\sim}{L}$ in the jth , k^{th} and l^{th} intervals respectively. Within each interval the behavior of the selected variables is supposed to be representative of that of all the variables. Obviously the success of the procedure, if success is attainable at all, will depend upon a judicious selection.

We now let the mean kinetic energy be represented by

$$E = \frac{1}{2} D^2 \sum_{j=0}^{\infty} c_j \sum_{\underset{\sim}{J}} J^{-2} X_{\underset{\sim}{J}} X_{\underset{\sim}{J}}^* , \tag{9}$$

with an analogous expression for V , where $\underset{\sim}{J}$ runs over all <u>selected</u> vectors in the jth interval. Likewise we replace equation (8) by

$$\frac{dX_{\underset{\sim}{J}}}{dt} = a_j \sum_{k,l=0}^{\infty} b_{jkl} \sum_{\underset{\sim}{K},\underset{\sim}{L}} C_{\underset{\sim}{J}\underset{\sim}{K}\underset{\sim}{L}} X_{\underset{\sim}{K}}^* X_{\underset{\sim}{L}}^* - \nu D^{-2} J^2 X_{\underset{\sim}{J}} + G_{\underset{\sim}{J}} , \tag{10}$$

where $\underset{\sim}{K}$ and $\underset{\sim}{L}$ run over all <u>selected</u> vectors in the kth and l^{th} intervals, and j denotes the interval in which $\underset{\sim}{J}$ lies. The factor c_j has been included in (9) to compensate as far as possible for the reduction of the number of terms in the second summation from N_j to M_j, while the factor b_{jkl} in (10) is intended to compensate for a reduction of the average number of terms in the summation over $\underset{\sim}{K}$ and $\underset{\sim}{L}$ from $N_j^{-1} Q_{jkl}$ to $M_j^{-1} P_{jkl}$. The additional factor a_j is included in (10) to compensate for the possibility that, given j, there may be integers k and l for which there are interacting variables in the jth, k^{th}, and l^{th} intervals, but no interacting selected variables. The remaining work in establishing the procedure consists of finding suitable values for c_j, b_{jkl}, and a_j.

Such values depend upon the manner in which the terms in the various summations in (7) and (8) combine. If the terms in a sum are mainly of one sign, one can estimate the total sum from the sum of a small number of terms, by multiplying the partial sum by the ratio of the total number of terms to the number of terms already summed. If however the terms tend to cancel, the sign of the total sum cannot be determined from the partial sum, but an expected magnitude can be obtained by multiplying the partial sum by the square root of the above mentioned ratio. We shall say that the terms combine systematically in the former case and randomly in the latter.

In the second summation in (7), all of the terms are nonnegative, and thus combine systematically. A suitable choice for c_j is therefore the ratio $M_j^{-1} N_j$.

Likewise, if the terms in the second summation in (8) should combine systematically, a suitable value for b_{jkl} would be

$$b'_{jkl} = (M_j^{-1} P_{jkl})^{-1} N_j^{-1} Q_{jkl} = P_{jkl}^{-1} Q_{jkl} c_j^{-1} . \tag{11}$$

If, however, the terms should combine randomly, a more appropriate value would be

$$b''_{jkl} = (b'_{jkl})^{1/2} = P_{jkl}^{-1/2} Q_{jkl}^{1/2} c_j^{-1/2} . \tag{12}$$

It is also desirable that b_{jkl} be chosen so that the mean kinetic energy E and the enstrophy V be conserved in the absence of viscosity and external forcing. From (9) and the analogous expression for V, and from (10) and (5), it follows that this will be the case if

$$c_j a_j b_{jkl} = c_k a_k b_{klj} = c_l a_l b_{ljk} . \tag{13}$$

The proper choice for b_{jkl} therefore depends upon a_j. For the time being, we shall note that there are interesting cases where a_j, while not necessarily equal to unity, is independent of j.

In these cases we note that (13) is satisfied if $b_{jkl} = b'_{jkl}$, but not if $b_{jkl} = b''_{jkl}$. Yet physically it seems more logical to assume that the terms in the second summation in (8) combine randomly, in which case the choice b''_{jkl} would be more appropriate. As a compromise, we choose b''_{jkl}, multiplied by factors which do not systematically increase or decrease when j, k, and l

are all increased by the same amount. Thus we let

$$b_{jkl} = b''_{jkl} \left(c_j^{-2} c_k c_l\right)^{1/6} \left(a_j^{-2} a_k a_l\right)^{1/3}$$

$$= P_{jkl}^{-1/2} Q_{jkl}^{1/2} c_j^{-5/6} \left(c_k c_l\right)^{1/6} a_j^{-2/3} \left(a_k a_l\right)^{1/3} \quad (14)$$

A satisfactory choice for a_j is more difficult to find. Whereas the selected interactions within a particular triple of intervals may be fairly representative of the remaining interactions, there is little reason to believe that interactions in those triples containing at least one selected interaction are representative of interactions in triples containing none. Yet some choice other than $a_j \equiv 1$ seems to be demanded. It would appear logical, given j, to let a_j depend upon the ratio of the number of pairs k, l for which the jth, k^{th}, and l^{th} intervals contain interacting variables to the number of pairs for which the intervals contain interacting selected variables. This is not possible, however, since the former number is infinite while the latter is generally finite.

A reasonably satisfactory solution is as follows. Given j, we let $\underset{\sim}{J}$ be a selected vector in the jth interval, and then let

$$I = \iint d(\log K)\, d(\log L) \quad , \quad (15)$$

with the limits of integration to be presently specified. We now let $I_0(j)$ be the value of I when K and L run over all pairs of values for which there exist vectors $\underset{\sim}{K}$ and $\underset{\sim}{L}$ (whose components need not be integers) of magnitude K and L which interact with $\underset{\sim}{J}$. Likewise we let $I_1(j)$ be the value of I when K and L satisfy the above conditions, with the further restriction that there be selected interactions in the triple of intervals to which $\underset{\sim}{J}$, $\underset{\sim}{K}$, and $\underset{\sim}{L}$ belong. With the assumption that the terms in the first summation in (8) combine randomly rather than systematically, our choice for a_j is $I_1^{-1/2} I_0^{1/2}$.

We briefly mention a few preliminary experiments which we have performed with this procedure. We have chosen a particular scheme with $\alpha = \sqrt{2}$, so that the intervals are half-octaves, with four selected vectors per interval. Each set of three consecutive intervals contains four triples of interacting selected variables; these are the only selected interactions.

We have attempted to reproduce some of the results which Lilly (1969, 1971) obtained with a 64 × 64 grid, by using similar forcing and similar resolution. We can cover six octaves of the spectrum with 48 variables, as opposed to Lilly's 4096. We obtain approximately the same mean kinetic energy and enstrophy. However, above the forcing wave number, Lilly obtained energy spectra conforming to the -3 power law, which incidentally agrees fairly well with what is found in the atmosphere, while our spectra generally fall off at least as rapidly as the -4 power.

Nevertheless, it appears that we can produce almost any desired spectrum through a suitable choice of forcing. With a spectrum resembling a typical atmospheric spectrum, our preliminary experiments have yielded growth rates of errors comparable to those apparently prevailing in the atmosphere.

ACKNOWLEDGMENT

This research was supported by the Atmospheric Sciences Section of the National Science Foundation under Grant GA-10276.

REFERENCES

Leith, C. E. J. Atmos. Sci. 28, 145-161 (1971)

Lilly, D. K. Phys. Fluids, Suppl. II 12, 240-249 (1969)

Lilly, D. K. NCAR Manuscript 71-90 (1971)

Lorenz, E. N. Tellus 17, 321-333 (1965)

Lorenz, E. N. Tellus 21, 289-307 (1969)

Orszag, S. A. Personal communication (1970)

Smagorinsky, J. Bull. Amer. Meteor. Soc. 50, 286-311 (1969)

USE OF C-M-W REPRESENTATIONS FOR
NONLINEAR RANDOM PROCESS APPLICATIONS

W. C. Meecham
W. C. Clever

University of California
Los Angeles

I. INTRODUCTION

The solution of the turbulence problem has thus far escaped the efforts of all researchers. However, a promising method of attack has been formulated recently, based upon a series expansion for nonlinear processes originally proposed by Norbert Wiener. The formulation expands the random variable in an infinite series with a set of ideal random functions as its basis. This expansion seemed ideally suited to the turbulence problem because it takes into account the nearly Gaussian statistics of real fluid turbulence. The first term has exactly Gaussian statistics. All of the statistical properties of the velocity field are made available by solving a set of integro-differential equations which represent the coefficients of the series.

The Wiener-Hermite expansion also appears attractive in a number of other ways. All approximations are realizable and thus negative energy spectra cannot occur. Also energy is conserved by the non-linear and pressure terms in the Navier Stokes equations, this energy is only dissipated by the viscous term, as it should be.

II. THE WIENER-HERMITE EXPANSION

The ideal random functions are made up from Hermite polynomial combinations of the white noise function. These Hermite polynomial combinations result in a set of statistically orthogonal functionals. The white noise function, a(x), is a random function constructed such that

$$\langle a(x_1)\, a(x_2) \rangle = \delta(x_1 - x_2) \tag{1}$$

The value of a(x) has a Gaussian distribution for each value of x and is statistically independent for different arguments.

Higher order functionals for the Wiener-Hermite expansion are constructed from Hermite polynomial combinations of a(x).

$$H^{(0)}(x) = 1$$

$$H^{(1)}(x) = a(x)$$

$$H^{(2)}(x_1,x_2) = a(x_1)\,a(x_2) - \delta(x_1-x_2) \qquad (2)$$

$$H^{(3)}(x_1,x_2,x_3) = a(x_1)\,a(x_2)\,a(x_3) + \sum_{x_1,x_2,x_3} \mathbb{P}\, a(x_1)\,\delta(x_2-x_3)$$

where $\sum_{x_1,x_2,x_3} \mathbb{P}$ means the sum of all distinct terms obtained by permuting the variables x_1, x_2, x_3. The functionals have been constructed such that they are mutually statistically orthogonal, in the sense that

$$\langle H^{(m)} H^{(n)} \rangle = 0 \quad \text{If } m \neq n$$

Averages of products of two or more of these Wiener-Hermite functionals can be obtained by taking the sum of all products of Dirac delta functions involving distinct "exogamous" pairings of the variables. Thus,

$$\langle H^{(2)}(x_1,x_2)\, H^{(2)}(x_3,x_4) \rangle = \delta_{13}\delta_{24} + \delta_{14}\delta_{23}$$

or

$$\langle H^{(1)}(x_1)\, H^{(2)}(x_2,x_3)\, H^{(3)}(x_4,x_5,x_6) \rangle = \delta_{14}\delta_{25}\delta_{36} + \delta_{16}\delta_{25}\delta_{34}$$
$$+\delta_{14}\delta_{26}\delta_{35} + \delta_{15}\delta_{26}\delta_{34} + \delta_{16}\delta_{24}\delta_{35} \quad \text{where } \delta_{ij} = \delta(x_i-x_j) \qquad (3)$$

The entire theory has been presented elsewhere; particularly the extension to three space dimensions. (Imamura and Meecham)

Wiener showed that his set of functionals was complete. Thus, any random function could be expanded in terms of an infinite series of the set of functionals $H^{(n)}(x_1 \ldots x_n)$. The expansion of a one dimensional velocity field of space and time would be written,

$$u(x,t) = \sum_{n=0}^{\infty} \int_{-\infty}^{\infty}\cdots\int K^{(n)}(x,x_1\cdots x_n,t)\, H(x_1\cdots x_n)\, dx_1\cdots dx_n \qquad (4)$$

where the kernel functions $K^{(n)}(x_1 \ldots x_n, t)$ are nonrandom functions of space and time. The first term, since it is a sum (integral sum) of Gaussian random variables, is exactly Gaussian. The expression can be thought of as a transform of the velocity field u(x,t) to a set of kernel functions $K^{(n)}$.

In Wiener's original formulation the set of Wiener-Hermite functionals, $H^{(n)}$ in (4), was considered independent of time. This meant the entire time dependence of u(x,t) had to be included in the kernel functions $K^{(n)}$. One could hardly expect such an expansion to converge well for all turbulent velocity fields, since the velocity field is continually referred to one instant of time. In an actual turbulent velocity field one would expect the fluid to have only a limited memory and the actual details at one instant of time should have a fading influence on the subsequent flow.

III. THE TIME VARIATION OF THE RANDOM BASE

The problems of a fixed base system were recognized by Wiener himself and he began the development of a time varying base system. His early work was corrected and extended by workers such as Bodner and Doi and Imamura. The main problem is to find out how to make the set of base functionals change in time so as to obtain the best convergence of the series. The expansion should allow the significant physical quantities to be retained in the first few kernels.

It will be assumed that the most general variation of the white noise base in time can be written as a power series involving the white noise basis itself. The differential equation for the variation of the basis is most easily written in Fourier transformed k-space.

The Fourier transform of H(x,t) need not really exist. However, the Fourier transforms of the kernels $K^{(n)}$ are assumed to exist and all of the results will be the same whether transforms are taken before or after averaging. The transforms need not be done in a finite box since all functions may be thought of in the sense of generalized functions.

The general expression for $\dot{H}$ in Fourier space is,

$$\frac{\partial}{\partial t}\tilde{H}(k) = \sum_{n=1}^{\infty}\int_{-\infty}^{\infty}\cdots\int L^{(n)}(k_1\cdots k_n)\,\tilde{H}^{(n)}(k_1\cdots k_n)\,\delta(k-k_1-\cdots\cdots-k_n)\,dk_1\cdots dk_n \qquad (5)$$

A necessary and sufficient condition for constant base statistics is,

$$\frac{\partial}{\partial t}<\tilde{H}(k_1)\cdots\cdots\tilde{H}(k_n)> = 0 \qquad (6)$$

This may be written,

$$\frac{\partial}{\partial t} < \tilde{H}(k_1) \cdots \tilde{H}(k_n) > = \sum_{k_1 \ldots k_n} \mathbb{P} < \tilde{H}(k_1) \cdots \tilde{H}(k_{n-1}) \frac{\partial}{\partial t} \tilde{H}(k_n) >$$

where $\sum_{k_1 \ldots k_n} \mathbb{P}$ denotes the sum of all distinct terms obtained by permuting the variables $k_1, \ldots, k_n$

Now using (5)

$$\frac{\partial}{\partial t} < \tilde{H}(k_1) \cdots \tilde{H}(k_n) > =$$

$$\sum_{m=1}^{\infty} \sum_{k_1 \ldots k_n} \mathbb{P} \int_{-\infty}^{\infty} \cdots \int L^{(m)}(q_1 \cdots q_m) < \tilde{H}^{(m)}(q_1 \cdots q_m) \tilde{H}(k_1) \cdots \tilde{H}(k_{n-1}) >$$

$$\delta(k_n - q_1 - \cdots q_m)\, dq_1 \cdots dq_m$$

$$= \sum_{m=1}^{n-1} \sum_{k_1 \ldots k_n} \mathbb{P} \int_{-\infty}^{\infty} \cdots \int L^{(m)}(q_1 \cdots q_m) < \tilde{H}^{(m)}(q_1 \cdots q_m) \tilde{H}(k_1) \cdots \tilde{H}(k_m) >$$

$$< \tilde{H}(k_{m+1}) \cdots \tilde{H}(k_{n-1}) > \delta(k_n - q_1 - \cdots - q_m)\, dq_1 \cdots dq_m$$

After averaging and integrating over $q_1, \ldots, q_m$

$$\frac{\partial}{\partial t} < \quad > = \sum_{m=1}^{n-1} \sum_{k_1 \ldots k_n} \mathbb{P} (2\pi)^m L^{(m)}(-k_1, \ldots, -k_m)\, \delta(k_1 + \cdots + k_m + k_n)\, \delta(k_{m+1} + k_{m+2}) \cdots \delta(k_{n-2} + k_{n-1})$$

Therefore a necessary and sufficient condition for constant base statistics is,

$$\sum_{k_1 \ldots k_{m+1}} \mathbb{P}\, L^{(m)}(k_1, \ldots, k_m)\, \delta(k_1 + \cdots + k_{m+1}) = 0 \qquad (7)$$

for all m and any $k_1, \ldots, K_{m+1}$

Note that none of the above will be changed if the $L^{(m)}$ are considered to be functions of t also. For m = 1 (7) becomes

$$\{ L(k_1) + L(k_2) \}\, \delta(k_1 + k_2) = 0$$

or

$$L(k) + L(-k) = 0$$

For m = 2 (7) becomes,

$$\left\{ L_s^{(2)}(k_1,k_2) + L_s^{(2)}(k_1,k_3) + L_s^{(2)}(k_2,k_3) \right\} \delta(k_1+k_2+k_3) = 0$$

$$L_s^{(2)}(k_1,k_2) = \frac{1}{2}\left[L^{(2)}(k_1,k_2) + L^{(2)}(k_2,k_1) \right] \tag{8}$$

IV. BURGERS' TURBULENCE

Burgers' turbulence provides a simplified method to test the applicability of the Wiener-Hermite expansion for the solution of the real fluid turbulence problem. Burgers' turbulence is governed by a one-dimensional analog of the Navier-Stokes equations, called Burgers' equation:

$$\frac{\partial}{\partial t} u(x,t) + u(x,t) \frac{\partial}{\partial x} u(x,t) = \nu \frac{\partial^2}{\partial x^2} u(x,t) \tag{9}$$

It differs from the Navier-Stokes equations in that it has no pressure term and no corresponding continuity equation. However, the equation retains the full nonlinearity and the viscous dissipation of the full Navier-Stokes equations.

In this equation the nonlinear term skews the velocity field into shocks or regions of large velocity gradient which are then dissipated by the viscous term on the right-hand side of Eq. (9). These shocks or near discontinuities give rise to a k^{-2} energy spectrum at equilibrium.

The principal difficulty in using Burgers' equation to demonstrate the use of the Wiener-Hermite expansion is that the statistics of the velocity field, unless it is driven by a Gaussian forcing term, are not generally near to Gaussian. And of course the Wiener-Hermite expansion should be best suited to solving problems involving a perturbation from Gaussianity.

Since the statistics for non-driven Burgers' turbulence are far from Gaussian, a truncated Wiener-Hermite expansion cannot be expected to work well for large times and for large Reynolds numbers. However by beginning with an exactly Gaussian velocity field and operating for only short times with moderate Reynolds numbers, one may hope for reasonably good results with a truncated Wiener-Hermite expansion for Burgers' turbulence. If encouraging results can be obtained for Burgers' equation, then the results would appear to be promising for real fluid turbulence.

It is more convenient to work in a Fourier transformed space; the result of taking Fourier transforms is (see Meecham and Siegel).

$$\left[\frac{\partial}{\partial t} + \frac{k^2}{Re}\right] \tilde{u}(k,t) = \frac{ik}{4\pi} \int_{-\infty}^{\infty}\int \tilde{u}(k_1)\,\tilde{u}(k_2)\,\delta(k-k_1-k_2)\,dk_1\,dk_2 \tag{10}$$

where

$$\tilde{u}(k,t) = \int_{-\infty}^{\infty} u(x,t)\,e^{ikx}\,dx \quad ; \quad u(x,t) = \frac{1}{2\pi}\int_{-\infty}^{\infty} \tilde{u}(k,t)\,e^{-ikx}\,dk$$

$$\tilde{H}(k,t) = \int_{-\infty}^{\infty} H(x,t)\,e^{ikx}\,dx$$

$Re = u_0' L_0/\nu$, with u_0' and L_0 respectively the initial RMS velocity and scale.

The expansion of the velocity field is written:

$$\tilde{u}(k,t) = 2\pi \sum_{n=1}^{\infty} \int_{-\infty}^{\infty}\cdots\int \tilde{K}(k_1,\ldots,k_n,t)\,\tilde{H}(k_1,\ldots,k_n,t)\,\delta(k-k_1-\cdots-k_n)\,dk_1\ldots dk_n \tag{11}$$

and where from (1) $[H \equiv a]$,

$$<\tilde{H}(k_1)\,\tilde{H}(k_2)> = 2\pi\,\delta(k_1+k_2) \tag{12}$$

V. EQUATIONS FOR THE KERNEL FUNCTIONS

The equations of motion of the kernels, $K^{(n)}$, can now be derived. They are obtained by multiplying (10) successively by $H^{(1)}(k_1)$, $H^{(2)}(k_1,k_2)\ldots$ and averaging after replacing $\tilde{u}(k,t)$ by the series. (11) The complete set of equations, with the time dependent base system varied in time according to (5) are as follows

$$\left[\frac{\partial}{\partial t} + \frac{(k_1+\cdots+k_n)^2}{Re}\right] \tilde{K}^{(n)}(k_1,\ldots,k_n,t)$$

$$- \frac{i(k_1+\cdots+k_n)}{2} \sum_{q=0}^{\infty}\sum_{p=0}^{n} \frac{(p+q)!\,(n-p+q)!}{(2\pi)^q\, n!\, q!} \times$$

$$\times \sum_{k_1 \dots k_n} \mathbb{P} \int_{-\infty}^{\infty} \cdots \int \tilde{K}^{(p+q)}(k_1 \dots k_p r_1 \dots r_q)\, \tilde{K}^{(n-p+q)}(k_{p+1} \dots k_n - r_1 \dots r_q)\, dr_1 \dots dr_q$$

$$= - \sum_{q=0}^{\infty} \sum_{p=0}^{n} \frac{(n-p+q+1)!\,(p+q)!}{n!\,(q+1)!} (2\pi)^{p-1} \times \tag{13}$$

$$\times \sum_{k_1 \dots k_n} \mathbb{P} \int_{-\infty}^{\infty} \cdots \int L_s^{(p+q)}(k_2 \dots k_p,\, r_1 \dots r_{q+1})\, \tilde{K}^{(n-p+q+1)}(k_{p+1} \dots k_n,\, r_1 \dots r_{q+1})\, \delta(r_1 + \dots + r_{q+1} - k_1 \dots - k_p)\, dr_1 \dots dr_{q+1}$$

The symbol $\sum_{k_1 \dots k_n} \mathbb{P}$ sums all distinct terms obtained by permuting the variables $k_1, \dots, k_n$. From these equations it can be seen that a time-dependent base system affects the growth rate of the various kernels. However, it can easily be shown that, if the infinite set of kernels is considered, derivatives in time of any set of moments of the form, $< \tilde{u}(k_1) \dots \tilde{u}(k_n) >$, do not depend upon whether the white noise base is varied in time. This fact is easily shown in the following way. The equation for $\frac{\partial}{\partial t} < \tilde{u}(k_1) \dots \tilde{u}(k_n) >$ can be written:

$$\frac{\partial}{\partial t} < \tilde{u}(k_1) \dots \tilde{u}(k_n) > = \sum_{k_1 \dots k_n} \mathbb{P} < \tilde{u}(k_1) \dots \tilde{u}(k_{n-1}) \frac{\partial}{\partial t} \tilde{u}(k_n) >$$

The symbol $\sum_{k_1 \dots k_n} \mathbb{P}$ sums all distinct terms obtained by permuting the variables $k_1, \dots, k_n$, and since

$$\frac{\partial}{\partial t} \tilde{u}(k_n) = - \frac{k_n^2}{R_e} \tilde{u}(k_n) + \frac{i k_n}{4\pi} \iint_{-\infty}^{\infty} \tilde{u}(q_1) \tilde{u}(q_2)\, \delta(k_n - q_1 - q_2)\, dq_1 dq_2$$

$$\frac{\partial}{\partial t} < \tilde{u}(k_1) \dots \tilde{u}(k_n) > = \frac{k_1^2 + \dots + k_n^2}{Re} < \tilde{u}(k_1) \dots \tilde{u}(k_n) >$$

$$+ \sum_{k_1 \dots k_n} \mathbb{P} \frac{i k_n}{4\pi} \iint_{-\infty}^{\infty} < \tilde{u}(k_1) \dots \tilde{u}(k_{n-1}) \tilde{u}(q_1) \tilde{u}(q_2) > \delta(k_n - q_1 - q_2)\, dq_1 dq_2$$

If the series,

$$\tilde{u}(k_1) = \sum_{j=1}^{\infty} \int_{-\infty}^{\infty} \cdots \int \tilde{K}^{(j)}(s_1 \ldots s_j)\, \dot{\tilde{H}}^{(j)}(s_1 \ldots s_j)\, \delta(k_1 - s_1 - \cdots - s_j)\, ds_1 \ldots ds_j$$

is substituted for the $\tilde{u}(k_i)$ on the right side of the above equation, it can be seen that no terms involving $\dot{H}$ appear. Therefore, at any time the value of $\frac{\partial}{\partial t} < u(k_1) \ldots u(k_n) >$ depends only upon the instantaneous values of the kernels $K^{(n)}$. In fact, by differentiating the equation again it can be seen that any order derivative in time,

$\frac{\partial^m}{\partial t^m} < u(k_1) \ldots (u(k_n) >$, does not depend upon $\dot{H}$. Therefore, a

Taylor series expansion shows that the evolution in time of any moment is not altered by a time dependent white noise base. However, this is only true if the infinite set of kernels and kernel equations is considered. When the Wiener-Hermite series is truncated, Burgers' equation is satisfied only approximately by the expansion of the velocity field. A moving base system can allow a closer approximation by permitting better series convergence.

VI. NUMERICAL RESULTS ON BURGERS' TURBULENCE

The early results using a fixed random base were encouraging but not completely satisfactory (Orzag and Bissonnette, Crow and Canavan). The kernel equations for a two term expansion were used for making the computations. The kernel equations are as follows:

$$\left\{ \frac{\partial}{\partial t} + \frac{k^2}{R_e} \right\} \tilde{K}(k) + \frac{k}{\pi} \int_{-\infty}^{\infty} \tilde{K}(k,q)\, \tilde{K}(q)\, dq = 0$$

$$\left\{ \frac{\partial}{\partial t} + \frac{(k_1+k_2)^2}{Re} \right\} \tilde{K}(k_1,k_2) - \frac{(k_1+k_2)}{2} \left[\tilde{K}(k_1)\tilde{K}(k_2) - \frac{2}{\pi} \int_{-\infty}^{\infty} \tilde{K}(k_1,q)\, \tilde{K}(k_2,-q)\, dq \right] = 0 \qquad (14)$$

Some of the results for the energy decay are shown in Fig. 1. The discrepancies were said to be due to the lack of convolution integrals, which lack restricts the energy cascade. The lack of convolution integrals (which allow distant Fourier modes to interact and which are present in Burgers' equation itself) permits satisfactory solutions only at moderately low Reynolds numbers.

This problem may be said to be related to the fact that the Wiener-Hermite expansion is not unique. The set of kernel equations (13) demonstrate this. If the random base system is allowed to vary in time, by prescribing a set of kernels $L^{(n)}$ which satisfy (7), the evolution in time of the kernels $K^{(n)}$ is changed.

Using the inviscid equipartition solution to Burgers' equation it can again be demonstrated that the entire infinite set of terms is required to represent even an exactly Gaussian velocity field, if the random base is not allowed to vary in time. Assuming the viscosity is zero below a given wave number results in the exactly Gaussian equipartition ensemble for Burgers' equation.

$$\nu(k) = 0 \qquad |k| \leq k_{max}$$
$$\nu(k) = \infty \qquad |k| > k_{max}$$

$$E(k,t) = C \qquad |k| \leq k_{max}$$
$$E(k,t) = 0 \qquad |k| > k_{max}$$

$$K^{(1)}(k) = [2\pi C]^{1/2} \qquad 0 \leq |k| \leq k_{max}$$
$$K^{(1)}(k) = 0 \qquad |k| > k_{max}$$

$$K^{(n)}(k) = 0 \qquad n \geq 2$$

However a two term expansion with a <u>fixed</u> random base does not preserve this solution. Equation (14) shows that $K^{(2)}$ will grow in time. Thus, even an exactly Gaussian velocity field requires an infinite number of terms to represent it exactly with a fixed base.

VII. A MOVING BASE SYSTEM FOR BURGERS' TURBULENCE

Thus, it is seen that a moving base system may be required if satisfactory results are to be expected from Burgers' turbulence in the general case. We must find how the base should be varied in time, or equivalently how the kernels $L^{(n)}$ should be prescribed. A convected base system such that

$$\left\{ \frac{\partial}{\partial t} + u(x,t) \frac{\partial}{\partial x} \right\} H(x,t) = 0$$

does not exist for Burgers' equation, because the base statistics cannot be preserved (Clever). Bodner suggested a moving base system which would preserve the equipartition criterion. Referring to Eq. (5), Bodner suggests the following moving base system.

$$L^{(2)}(k_1,k_2,t) = \frac{i}{6\pi}\tilde{K}^{(1)}(k,t)(k_1+2k_2)$$
$$L^{(n)} = 0 \qquad n \neq 2 \tag{15}$$

It was mentioned earlier that $L^{(1)}(k,t)$ may be omitted without loss of generality. Substituting (15) into (8), and using the fact that

$\frac{1}{2}\left[L^{(2)}(k_1,k_2) + L^{(2)}(k_2,k_1)\right]$, shows that the statistics of the white noise base system are preserved.

With this moving base system equation (13) yields the two partial differential equations to be solved for the kernels:

$$\left\{\frac{\partial}{\partial t} + \frac{k^2}{Re}\right\}\tilde{K}(k) + \frac{k}{\pi}\int_{-\infty}^{\infty}\tilde{K}(k,q)\tilde{K}(q)\,dq = \frac{1}{3\pi}\int_{-\infty}^{\infty}\tilde{K}(q)\tilde{K}(q,k-q)(2k-q)\,dq$$

$$\left\{\frac{\partial}{\partial t} + \frac{(k_1+k_2)^2}{Re}\right\}\tilde{K}(k_1,k_2) - \frac{(k_1+k_2)}{2}\left[\tilde{K}(k_1)\tilde{K}(k_2) - \frac{2}{\pi}\int_{-\infty}^{\infty}\tilde{K}(k_1,q)\tilde{K}(k_2,-q)\,dq\right] = -\frac{\tilde{K}(k_1+k_2)}{6}\left[\tilde{K}(k_1)(2k_2+k_1) + \tilde{K}(k_2)(2k_1+k_2)\right] \tag{16}$$

where

$$\tilde{K}(k) = \tilde{K}(-k) \text{ and } \tilde{K}(k_1,k_2) = \tilde{K}(-k_1,-k_2) \tag{16A}$$

The terms on the right-hand side of the equations are the new terms due to the moving base system. Inserting $K(k) = c$, $K(k_1,k_2) = 0$, and $(1/Re) = 0$ into the set of equations (16) shows that the equipartition solution is satisfied.

An assessment of the moving base system, described by Eqs. (16), will be made by comparing them with the results obtained using a fixed random base system of equations and with the results from a numerical experiment.

The most recent and comprehensive study of Burgers' turbulence, using a fixed base Wiener-Hermite expansion, was presented in a paper by Crow and Canavan. This paper was not overly optimistic in tone and placed much of its emphasis on the inadequacies of the fixed Wiener-Hermite expansion as applied to Burgers' turbulence. However, in light of the non-Gaussian character of non-driven Burgers' turbulence, their results seem to show that even a fixed base expansion can yield some useable results. Their results will now be described and compared with the results obtained with the time-dependent base system.

Crow and Canavan's results were obtained by assuming an initial energy spectrum and exactly Gaussian initial conditions. The fixed base set of kernel equations were solved numerically for the evolution in time of the first two kernels. The initial conditions used by Crow and Canavan[7] were,

$$E(k,0) = \frac{16}{3}\left(\frac{2}{\pi}\right)^{1/2} k^4 \exp(-2k^2)$$

so (17)

$$\tilde{K}^{(1)}(k,0) = 4\left(\frac{2}{3}\right)^{1/2}(2\pi)^{1/4} k^2 \exp(-k^2)$$

and

$$\tilde{K}^{(2)}(k_1,k_2,0) = 0$$

In addition, Crow and Canavan performed a corresponding set of numerical experiments on Burgers' equation in order to compare the results. The initial conditions for the numerical experiment were set up by constructing a random velocity field field based upon Eq. (4). This can be done by noting that since there is only one nonzero kernel,

$$u(x,0) = \int_{-\infty}^{\infty} K^{(1)}(x-y,0)\, H^{(1)}(y)\, dy$$

Thus by constructing a white noise function $H(x)$, a random velocity field (member of the ensemble) is obtained by a convolution integral with the inverse transform of the kernel function $\tilde{K}^{(1)}(k,0)$. The kernel in physical space at t=o is the inverse transform of (17),

$$K^{(1)}(x,0) = \left(\frac{2}{3}\right)^{1/2}\left(\frac{2}{\pi}\right)^{1/4}\left(1-\tfrac{1}{2}x^2\right)\exp(-x^2/4)$$

The ergodic hypothesis was assumed and spatial averages were used in place of ensemble averages since they were assumed to be equal.

Crow and Canavan computed the energy decay, $E = \int_{-\infty}^{\infty} E(k,t)dk$, with time. For large Reynolds numbers the agreement between the numerical experiment and the results from the solution of the fixed base set of equations for the first two kernels was not satisfactory. These results are presented in Fig. 1.

At Reynolds numbers of 5 and lower the solution of the fixed base set of equations shows a reasonable degree of agreement with the results obtained with the numerical experiment. However, at a Reynolds number of 5 (Fig. 2), although the total energy decay showed fair agreement with the actual solution obtained by numerical experiment, after a moderate time the energy $E_2 = \frac{2}{(2\pi)^2} \iint_{-\infty}^{\infty} |K^{(2)}(k_1,k_2,t)|^2 \delta(k-k_1-k_2)\, dk_1 dk_2$, or the non-Gaussian energy associated with the second kernel, became greater than the Gaussian energy, $E_1 = \frac{1}{2\pi} |K^{(1)}(k,t)|^2$. This fact argues that the fixed base expansion may not be converging rapidly enough and higher order kernels may be necessary for reliable results.

To test a moving base system, the set of kernel equations (16) representing Bodner's moving base system was used with the initial conditions given by (17). The results are presented in Fig. 3. From the results it appears that Bodner's moving base system, based on satisfying the inviscid equipartition solution, does indeed extend the range of Reynolds numbers where useful solutions are possible. The useful solutions now extend into a range where the nonlinear energy transfer significantly affects the total energy decay. A linear dissipation rate, which was obtained by assuming $\frac{d}{dt} E(k,t) = -2k^2 E(k,t)/Re$, is shown in the subsequent figures and demonstrates that considerable energy transfer has taken place.

Figure 4 shows the results of the energy decay for a Reynolds number of 5. These results for the energy decay appear to be in very good agreement with the numerical results of the numerical experiment conducted by Crow and Canavan - shown in Fig. 5. Also, now in contrast to the results obtained using a fixed random base, shown in Fig. 2, the energy in the first or Gaussian kernel is always larger than the energy in the second kernel. The figures show a close agreement between the distribution of energy among the Gaussian and non-Gaussian kernels for the moving base system and for the numerical experiment.

In the numerical experiment Crow and Caravan obtained the energy

E_1 by projecting the velocity field from the numerical experiment back upon the original random base to obtain the first kernel. Since

$$u(x) = \int_{-\infty}^{\infty} K^{(1)}(x-y)\,H(y)\,dy + \iint_{-\infty}^{\infty} K^{(2)}(x-y, x-z)\,H^{(2)}(y,z)\,dy\,dz + \cdots$$

$$\langle u(x)\,H(y)\rangle = K^{(1)}(x-y)$$

$$K^{(1)}(x) = \langle u(x+y)\,H(y)\rangle \qquad E_1 = \int_{-\infty}^{\infty} \left[K^{(1)}(x)\right]^2 dx$$

However, since the energy E_1 was obtained by projecting back upon the original random base, E_1 does not represent the entire Gaussian energy of the velocity field. After a time some of the Gaussian energy becomes represented in terms of the initial white noise base, by "non-Gaussian" terms. It was previously shown, when considering the inviscid equipartition ensemble, that the fixed base system of equations excites the "non-Gaussian" kernels, although the solution remains Gaussian for all time. Compared to the numerical experiment, two term Wiener-Hermite expansion could conceivably have a larger fraction of its total energy associated with the first kernel if the random base could be varied in an optimum manner. As time proceeds the convergence of the series becomes more closely connected with the manner in which the random base is varied in time.

When the moving base system of equations is used, a new convolution integral appears for $\dot{\tilde{K}}^{(1)}$. The convolution integral means that the evolution of $\tilde{K}^{(1)}$ is no longer effectively confined to low wave numbers, as with the fixed random base, thus disposing of the objection raised by Crow and Canavan.[7] This allows the dissipation to take place at a more realistic rate by allowing greater energy transfer to high wave numbers.

Although it appears that due to the non-Gaussian character of Burgers' turbulence the non-Gaussian energy must dominate the energy spectrum at high wave numbers, the distribution of energy need not be as bad as that given by a fixed base expansion. The manner in which the random base is varied in time also has an effect upon the distribution of energy among the various kernels (see Eq. 13), Indeed Bodner's moving base system (15) significantly changed the distribution of energy between Gaussian and non-Gaussian kernels.

It might be possible that some moving base would allow the kernels $K^{(n)}$ for $n > 2$ to be ignored and still have the dynamics of the system act properly to a moderately high range of wave numbers; a suggestion is offered below. It appears that the dynamics of the system of equations (16) does not represent Burgers' turbulence well at high wave numbers, and poor results occur if the inertial range of wave numbers extends too far into this range.

The Wiener-Hermite expansion with Bodner's time-dependent base system, allowing the equipartition solution with a truncated expansion, is able to produce some useful results when used on Burgers' turbulence. There is certainly a significant improvement over the fixed base results. The time-dependent base system is able to retard the transfer of energy from the Gaussian to the non-Gaussian kernels, which is so strongly present with the fixed base system. This means the non-Gaussian kernels are held to a smaller portion of the total energy of the velocity field. This is more consistent with the idea of an expansion about Gaussianity, although Burgers' turbulence, in contrast to real fluid turbulence, has statistics which may be far from Gaussian. What is still required, however, is the optimum method of varying the random base in time.

VIII. A PROPOSAL FOR AN OPTIMUM MOVING BASE FOR BURGERS' MODEL

Given a random velocity field (here, that of Burgers' model) which is nearly Gaussian, it is possible to find approximately the white noise process associated with it. We begin by finding that process. Refer to (4) where we now suppose that the white noise process H(x) also depends on time (the time variable is not written). We shall suppose that for the present purpose u(x,t) is known and invert to find H. We work to second order in the functionals and use definitions already given, to find for a statistically homogeneous process (time implicit).

$$\tilde{H}(k) = \tilde{K}^{-1}(k)\left\{\tilde{H}(k) - \iint \tilde{K}(k_1,k_2) \times \left[\tilde{H}^{(2)}(k_1,k_2)\,\delta(k-k_1-k_2)\,dk_1 dk_2\right] + \cdots\right\} \tag{18}$$

solving by iteration for $\tilde{H}$, to lowest order we find:

$$\tilde{H}(k) \cong \tilde{K}^{-1}(k)\Big\{\tilde{H}(k) - \iint \tilde{K}(k_1,k_2)\big[\tilde{K}^{-1}(k_1)\,\tilde{u}(k_1)\,\tilde{K}^{-1}(k_2)\,\tilde{u}(k_2) - (2\pi)^2\,\delta(k_1+k_2)\big]\,\delta(k-k_1-k_2)\,dk_1\,dk_2\Big\} \tag{19}$$

This is the white noise process associated with u at each instant of time. Now differentia e (19) to obtain $\dot{\tilde{H}}$; use the equation of motion (9) to find

$$\begin{aligned}\dot{\tilde{H}}(k) = {} & \tilde{K}^{-1}(k)\Big\{-\frac{ik}{4\pi}\int\big[\tilde{u}_1(k') + \tilde{u}_2(k')\big]\,\tilde{u}_1(k-k')\,dk' \\ & - \nu k^2\big[\tilde{u}_1(k) + \tilde{u}_2(k)\big]\Big\} - \tilde{H}(k)\,\tilde{K}^{-1}(k)\,\dot{\tilde{K}}(k) \\ & - \tilde{K}^{-1}(k)\int\Big\{\dot{\tilde{K}}(k_1,k_2)\,\tilde{H}^{(2)}(k_1,k_2) + \tilde{K}(k_1,k_2)\big[2\tilde{H}(k_2) \\ & \Big(-\frac{ik_1}{4\pi}\int\tilde{u}_1(k')\,\tilde{u}_1(k_1-k')\,dk' - \nu k_1^2\,\tilde{u}(k_1)\Big) \\ & - 2\tilde{H}(k_1)\,\tilde{H}(k_2)\,\dot{\tilde{K}}(k_1)\,\tilde{K}^{-1}(k_1')\big]\Big\}\,\delta(k-k_1-k_2)\,dk_1\,dk_2\end{aligned} \tag{20}$$

working to first order in the second order Wiener-Hermite term. Here u_1 and u_2 are the terms for $n = 1$ and 2 in (4) and similarly for their transforms $\tilde{u}_1$ and $\tilde{u}_2$. (The mean flow vanishes as previously, so the function $K^{(0)}$ vanishes in (4). If the time dependent $\tilde{H}$ given in (20) is to retain its statistics, the conditions (6) must be satisfied. Multiply (20) by $H(k_2)$ and average; add to this the corresponding result interchanging the wave number arguments and require the sum to vanish. We find after manipulation

$$\begin{aligned}\dot{\tilde{K}}(k) = {} & -ik\int\tilde{K}(k_1,k)\,\tilde{K}(k_1)\,dk_1 \\ & + 2i\int\tilde{K}(k_1,k-k_1)\,k_1\,\tilde{K}^{-1}(k_1)\,\tilde{K}(k_1-k) \\ & \times\tilde{K}(k)\,dk_1 - \nu k^2\,\tilde{K}(k)\end{aligned} \tag{21}$$

We use the symmetry property (16A). Next multiply (20) by $\tilde{H}(k_3)\,\tilde{H}(k_4)$ average, add permutation as in (6) to find the equation governing $\tilde{K}(k_1,k_2)$, after changes in the arguments,

$$\frac{\partial}{\partial t}\left[\tilde{K}(k_1)\tilde{K}(k_2)\tilde{K}(k_1,k_2)+\tilde{K}(k_2)\tilde{K}(k_3)\tilde{K}(k_2,k_3)+\tilde{K}(k_3)\tilde{K}(k_1)\tilde{K}(k_3,k_1)\right]=$$

$$\frac{i}{4\pi}\left[k_1\tilde{K}^2(k_2)\tilde{K}^2(k_3)+k_2\tilde{K}^2(k_3)\tilde{K}^2(k_1)+k_3\tilde{K}^2(k_1)\tilde{K}^2(k_2)\right]-\nu\Big[\tilde{K}(k_1)\tilde{K}(k_2)\tilde{K}(k_1,k_2)(k_3^2-k_1^2-k_2^2)+\tilde{K}(k_2)\tilde{K}(k_3)\times\tilde{K}(k_2,k_3)(k_1^2-k_2^2-k_3^2)+\tilde{K}(k_3)\tilde{K}(k_1)\times\tilde{K}(k_3,k_1)(k_2^2-k_3^2-k_1^2)\Big] \quad (22)$$

with $k_1+k_2+k_3=0$

It is remembered that $\tilde{K}(k_1,k_2)$ may be assumed symmetric in its arguments without loss of generality.

The energy spectrum function is related to the kernels as by Meecham and Siegel.

$$E(k)=(2\pi)^{-1}\left[|\tilde{K}(k)|^2+\frac{1}{\pi}\int|\tilde{K}(k_1,k-k_1)|^2\,dk_1\right] \quad (23)$$

If, for a nearly-Gaussian process, the first term is large we see from (22) that the time rate of change will be small if

$$k_1E(k_2)E(k_3)+k_2E(k_3)E(k_1)+k_3E(k_1)E(k_2)\approx 0$$

$$\text{with}\quad k_1+k_2+k_3=0$$

This condition is close to that of earlier work (Reid) which is known to lead to the correct equilibrium energy spectrum for Burgers' model: i.e. $E\sim k^{-2}$ for larger k (see Eq. (5,6), et. seq. of Ref. 11). This, and other characteristics of (21) and (22) suggest encouraging computation for the "optimum" base proposed here.

IX. RESULTS OF A MOVING BASE ON REAL FLUID TURBULENCE

After noticing the significant improvement when using a time-dependent base on Burgers' turbulence it was decided to see what the results would be on real homogeneous, isotropic fluid turbulence. In Fourier space the Navier-Stokes equations for homogeneous incompressible flow can be written,

$$\left\{ \frac{\partial}{\partial t} + \nu|\underset{\sim}{k}|^2 \right\} u_i(\underset{\sim}{k}) = i k_\beta \, \Delta_{ij}(\underset{\sim}{k}) \frac{1}{(2\pi)^3} \iiint_{-\infty}^{\infty} u_j(\underset{\sim}{q}) \, u_\ell(\underset{\sim}{k}-\underset{\sim}{q}) \, d\underset{\sim}{q}$$

and the Wiener-Hermite expansion can be written,

$$u_i(\underset{\sim}{k}) = K_{ij}(\underset{\sim}{k}) \, H_j(\underset{\sim}{k}) + \frac{1}{(2\pi)^3} \iiint_{-\infty}^{\infty} K_{ij\ell}(\underset{\sim}{k}-\underset{\sim}{q}, \underset{\sim}{q}) \, H_{j\ell}(\underset{\sim}{k}-\underset{\sim}{q}, \underset{\sim}{q}) \, d\underset{\sim}{q} + \cdots$$

Formally all of the results from Burgers' turbulence can be extended by simply adding subscripts to account for the three dimensions. Thus, the equation for the time-dependent base is written,

$$\frac{\partial}{\partial t} H_i(\underset{\sim}{k},t) = \iiint L^{(2)}_{i\alpha\beta}(\underset{\sim}{k}-\underset{\sim}{q}, \underset{\sim}{q}, t) \, H_{\alpha\beta}(\underset{\sim}{k}-\underset{\sim}{q}, \underset{\sim}{q}, t) \, d\underset{\sim}{q} + \cdots$$

and as for Burgers' model, constant base statistics imply (Doi and Imamura).

$$\Big\{ L^{(2)}_{ij\ell}(\underset{\sim}{k}_2, \underset{\sim}{k}_3) + L^{(2)}_{i\ell j}(\underset{\sim}{k}_3, \underset{\sim}{k}_2) + L^{(2)}_{j\ell i}(\underset{\sim}{k}_3, \underset{\sim}{k}_1) + L_{ji\ell}(\underset{\sim}{k}_1, \underset{\sim}{k}_3)$$

$$+ L^{(2)}_{\ell ij}(\underset{\sim}{k}_1, \underset{\sim}{k}_2) + L_{\ell ji}(\underset{\sim}{k}_2, \underset{\sim}{k}_1) \Big\} \, \delta(\underset{\sim}{k}_1 + \underset{\sim}{k}_2 + \underset{\sim}{k}_3) = 0$$

Bodner's choice was again made for the time-dependent base system. The choice of this base system was motivated by asking that it give a solution of the exactly Gaussian equipartition ensemble with a truncated two term expansion. The details of the moving base and the resulting kernel equations are given in Bodner's paper. The equations given can be simplified by the statistical isotropy of the velocity field, and three equations for three scalar functions result. As in Burgers' equation a $K^{(2)}K^{(2)}$ type integral results in the kernel equations. To facilitate computations this term was dropped on the grounds that it should be small for a nearly Gaussian process. The equations which result are energy conserving. They are,

$$\left\{\frac{d}{dt} + \nu k^2\right\} u(k) + \frac{1}{2(2\pi)^2} \int_0^\infty\!\!\int_{-1}^{1} \frac{q^2\, u(q)\, (1-y^2)}{(k^2 + 2kqy + q^2)}\, dy$$

$$\Big\{ \big[2k^2 (k^2 + 2kqy) + k^2 q^2 (y^2+1)\, F_1(k,q,y)$$

$$+ kqy\,(k^2 + kqy)\, F_2(k,q,y) \Big\}\, dq = \tag{22}$$

$$\frac{1}{2(2\pi)^2} \int_0^\infty\!\!\int_{-1}^{1} \frac{q^2 (1-y)^2\, u\left(\sqrt{k^2 + 2kqy + q^2}\right)}{(k^2 + 2kqy + q^2)}\, dy$$

$$\times \Big\{ (q^2 + kqy)(k^2 + 2kqy)\, F_1(q,k,y)$$

$$+ \left[2k^2 q^2 y^2 + k^2 (q^2 + kqy) \right] F_2(q,k,y) \Big\}\, dq$$

$$\left\{\frac{d}{dt} + \nu k^2\right\} F_1(q,k,y) = u(q) \left[u(k) - u\left(\sqrt{k^2 + 2kqy + q^2}\right) \right]$$

$$\left\{\frac{d}{dt} + \nu k^2\right\} F_2(q,k,y) = u\left(\sqrt{k^2 + 2kqy + q^2}\right) \left[u(k) - u(q) \right]$$

$$E_1(k) = \frac{k^2 \left[\dot{u}(k)\right]^2}{2(\pi)^2}$$

$$E_2(k) = \frac{k^4}{(2\pi)^4} \int_0^\infty\!\!\int_{-1}^{1} \frac{q^2 (1-y^2)}{k^2 + 2kqy + q^2} \Big\{ q^2 (1+y^2)\, F_2^2(k,q,y)$$

$$+ qy\,(k + qy)\, F_1(k,q,y)\, F_2(k,q,y) \Big\}\, dy\, dq$$

Some preliminary results of a numerical solution to these equations have been obtained. They are compared with the experimental results obtained by Ling and Huang. The results were for decaying turbulence downstream from a grid in a water channel. The experimental results indicated that the dimensionless energy spectra, energy transfer and correlation function were, within experimental error, independent of time.

$$f(r^*) = \frac{1}{1 + \left(\frac{r^*}{\alpha}\right)^2} \qquad r^* = \sqrt{5}\,\frac{r}{\lambda} \qquad \alpha = 3.2$$

$$E^*(k^*) = \frac{\alpha}{2}\left[\alpha k^* + (\alpha k^*)^2\right] e^{-\alpha k^*} \qquad k^* = \frac{\lambda k}{\sqrt{5}}$$

$$T^*(k^*) = \frac{\alpha^2}{2}\left\{2\alpha k^{*3} + \left(2 - \frac{\alpha^2}{2}\right) k^{*2} - \alpha k^* - 1\right\} k^* e^{-\alpha k^*}$$

$$E(k) = \frac{\langle u^2\rangle \lambda}{\sqrt{5}}\, E^*(k^*) \qquad T(k) = \frac{\langle u^2\rangle r\sqrt{5}}{\lambda}\, T^*(k^*)$$

Here λ is the dissipation length (Ling and Huang).

The numerical solution of Eqs. (22), using the above spectrum for E(k), was begun using exactly Gaussian initial conditions; a demanding restriction for the theory. This was done for the sake of simplicity in spite of the fact that the initial energy transfer was therefore zero. However, Fig. 6 shows that the energy transfer has reached its equilibrium form in a dimensionless time of order one.

The results for the energy decay are presented in Figure 7. An initial fluctuation Reynolds number of 10 was used, using $\sqrt{\langle u^2\rangle}$ for the velocity scale and a length scale equal to the inverse of the wave number where the energy spectrum was a maximum. There is insufficient energy decay and the dissipation length scale, λ , is increasing too rapidly. This is thought to be due to the fact that there is somewhat insufficient energy transfer from the low wave number region of the spectrum. Fig. 6 compares the experimental and theoretical transfer.

Fig. 7 shows that the moving base system which is used is not much better than the results obtained with a fixed random base. These results are in turn somewhat better than assuming a linear dissipation rate (with zero energy transfer). The Gaussian fraction of the energy spectrum is not much greater than with a fixed moving base set of equations. When the non-Gaussian energy becomes large, the neglect of the $K^{(2)}K^{(2)}$ probably becomes significant among other difficulties.

Although the results are not entirely satisfactory, they are an improvement. It is possible that a search for a better moving base could give some significant results for moderate Reynolds numbers.

LIST OF FIGURES

Fig. 1 Energy decay of Burgers' turbulence with a fixed base expansion.

Fig. 2 Decay of Burgers' turbulence using a fixed base expansion.

Fig. 3 Burgers' turbulence decay with a time-dependent random base.

Fig. 4 Decay of Burgers' turbulence using a time-dependent random base.

Fig. 5 Energy decay of Burgers' turbulence from a numerical experiment.

Fig. 6 Energy Transfer of Homogeneous Turbulence

Fig. 7 Energy Decay of Homogeneous Turbulence

REFERENCES

Bodner, S. E., Phys. Fluids, 12, 33 (1969).

Clever, W. C., Ph.D. Thesis, University of California, Los Angeles (1970).

Crow, S. C. and Canavan, G. H., J. Fluid Mech. 41, 387 (1970).

Doi, D. and Imamura, T., Progr. Theoret. Phys. (Kyoto), 41, 358 (1969).

Imamura, T., Meecham, W. C., and Siegel, A., J. Math Phys., 6, 695 (1965).

Ling, S. C. and Huang, T. T., Phys. Fluids, 13, 2912 (1970).

Meecham, W. C., J. Fluid Mech. 41, 179 (1970).

Meecham, W. C. and Siegel, A., Phys. Fluids 7, 1178 (1964).

Orzag, S. A. and Bissonnette, L. R., Phys. Fluids 10, 2603 (1967).

Reid, W. H., Appl. Sci. Res. A, 6, 85 (1956).

Wiener, N., Nonlinear Problems in Random Theory (Technology Press, Cambridge, Mass. (1958)).

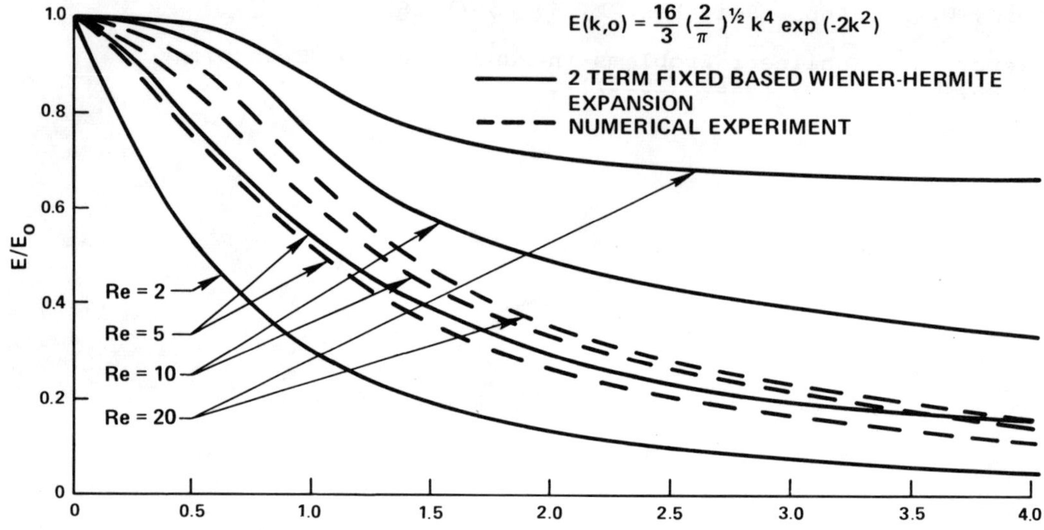

Figure 1. Energy Decay of Burgers' Turbulence with a Fixed Base Expansion

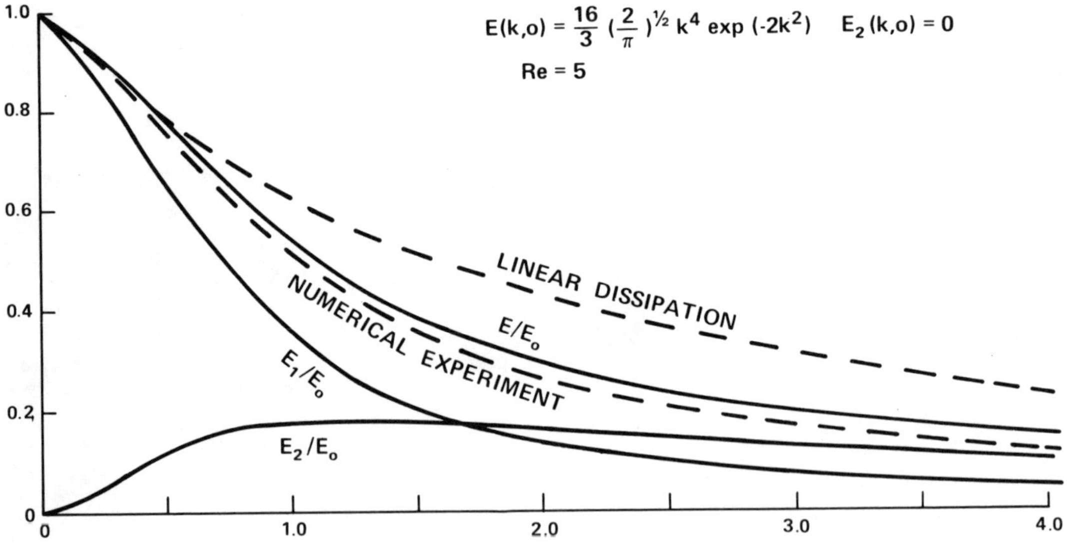

Figure 2. Decay of Burgers' Turbulence Using a Fixed Base Expansion

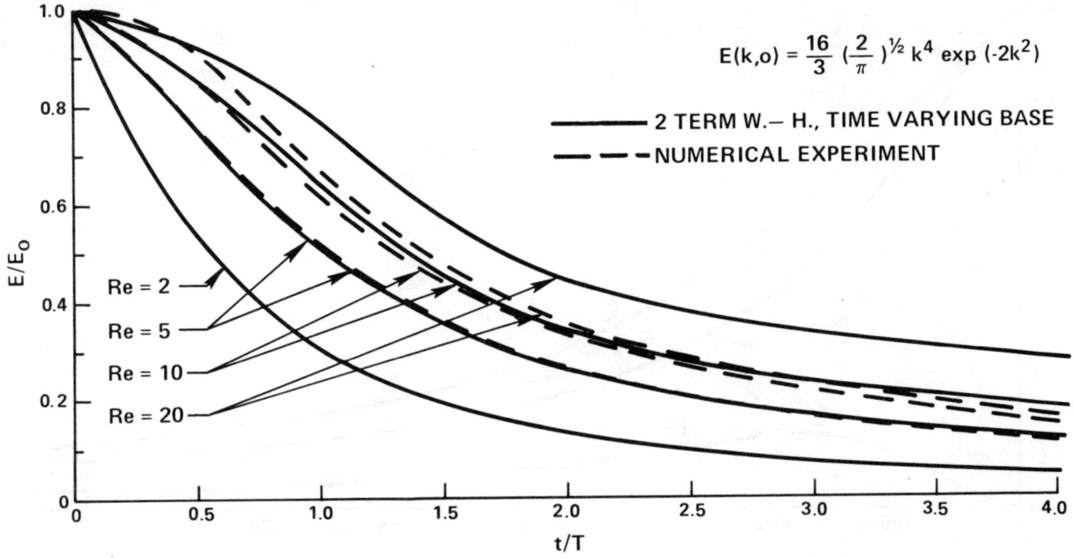

Figure 3. Burgers' Turbulence Decay with a Time Dependent Random Base

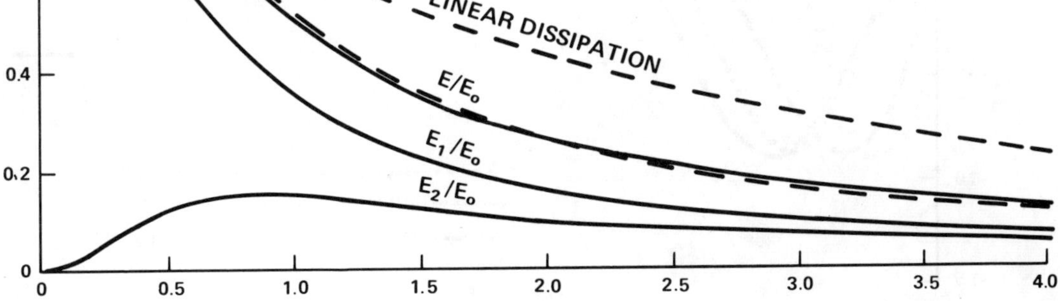

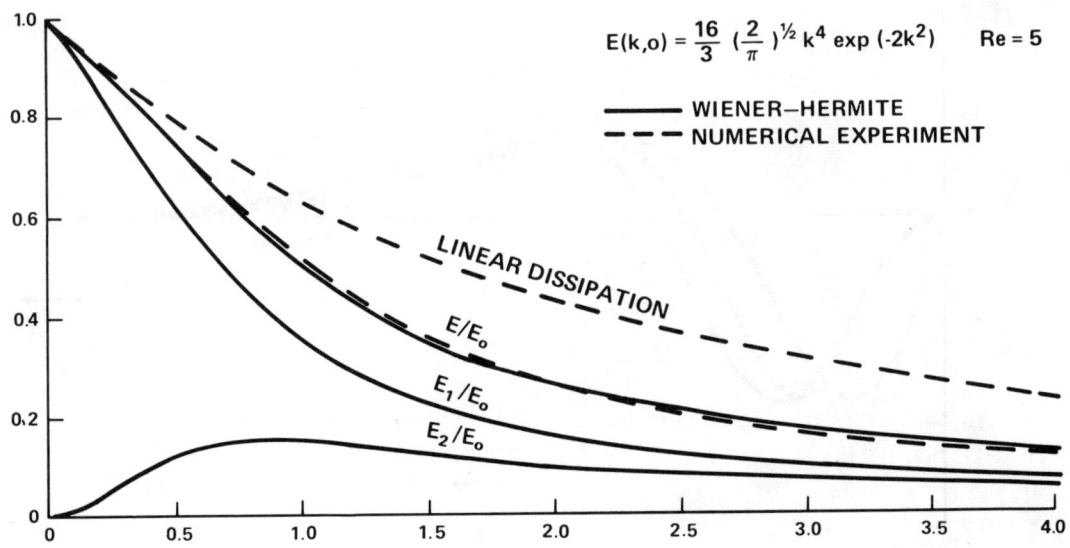

Figure 4. Burgers' Turbulence Decay Using a Time-Dependent Random Base

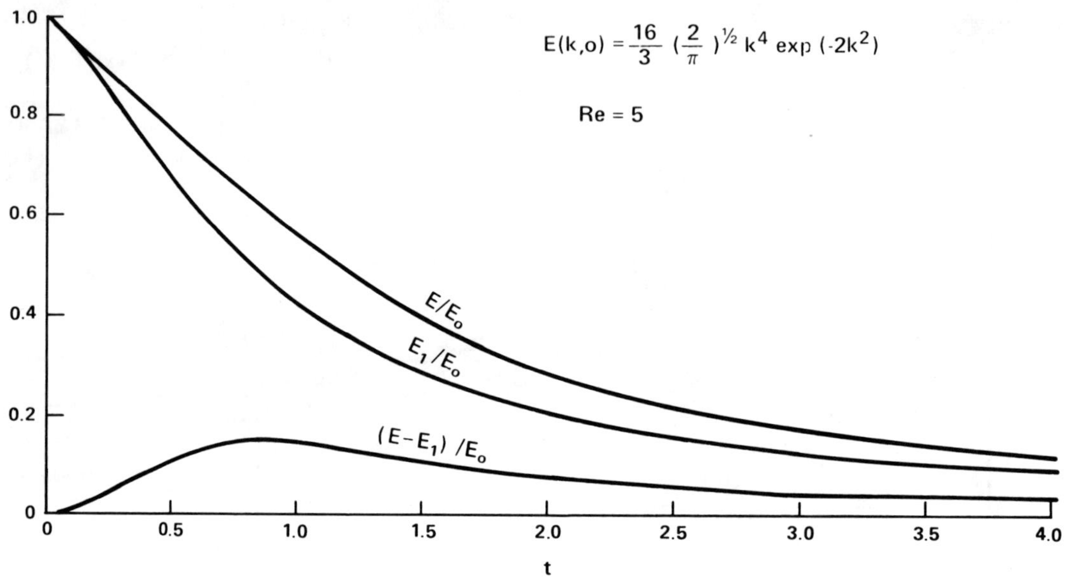

Figure 5. Energy Decay of Burgers' Turbulence from a Numerical Experiment

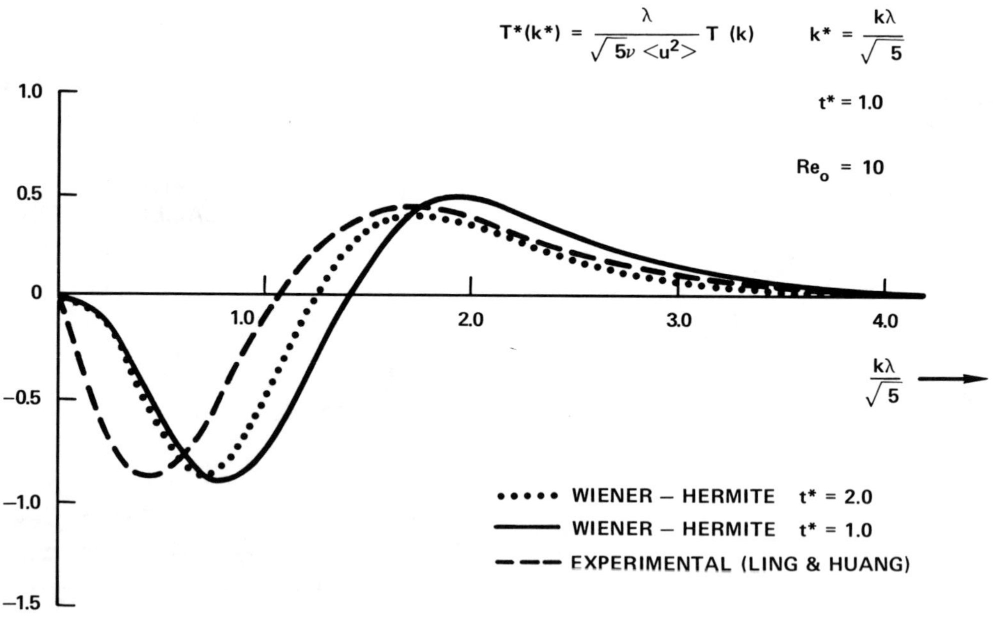

Figure 6. Energy Transfer of Homogeneous Turbulence

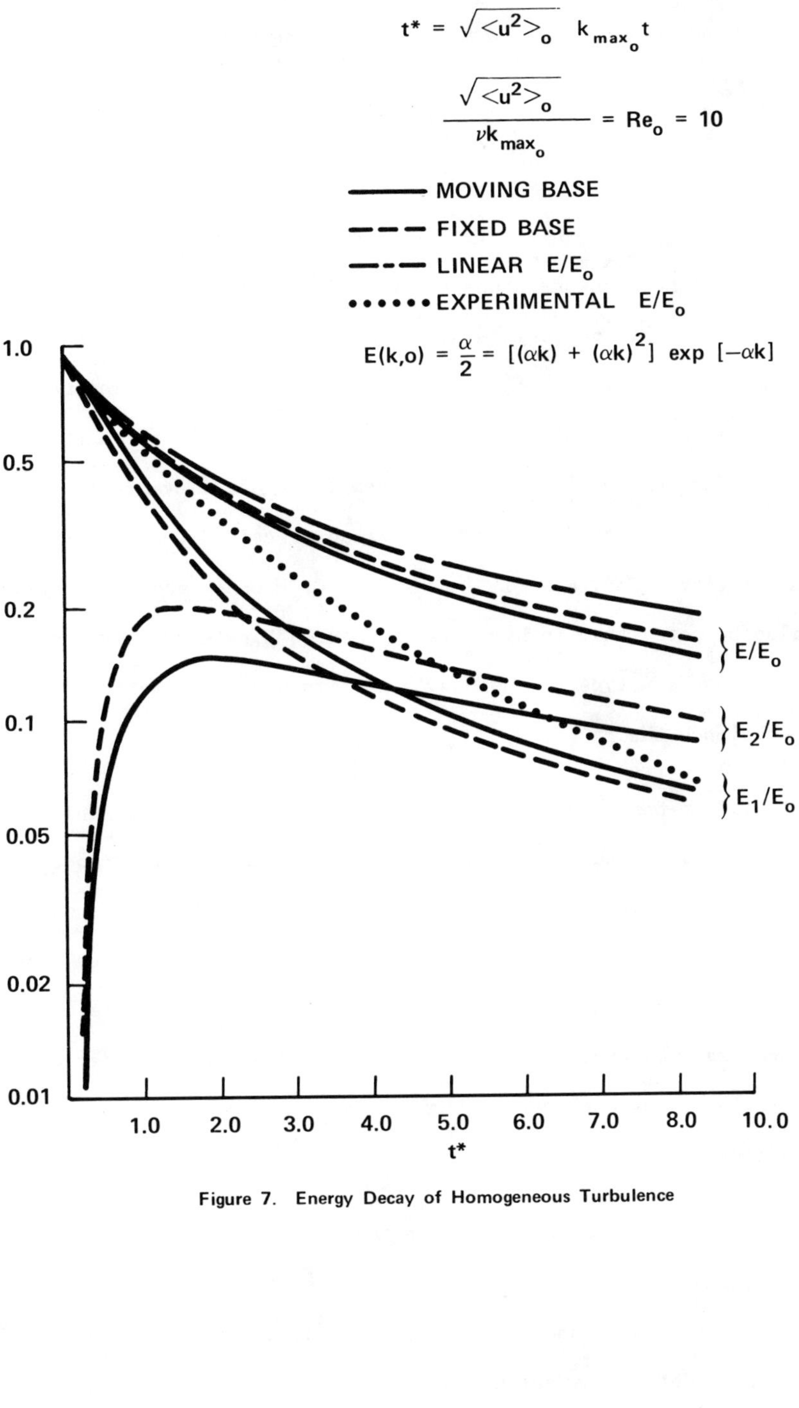

Figure 7. Energy Decay of Homogeneous Turbulence

HOMOGENEOUS CHAOS EXPANSIONS

by

G. Kallianpur
University of Minnesota

1. Introduction In an earlier article [10] we have given a general version of the theory of homogeneous chaos for Gaussian processes completing and extending the ideas of R. H. Cameron and W. T. Martin [1], K. Ito [7] and S. Kakutani [9], (See also J. Neveu [16]). Since this theory yields nonlinear expansions for a large and interesting class of functionals of Gaussian processes its study and further development would seem to be of relevance in turbulence theory. Indeed, such problems have already received the attention of specialists in this field. (See, e.g. T. Imamura, W. C. Meecham and A. Siegel, [6]). It may also be noted that in his M.I.T. lectures N. Wiener suggested the use of homogeneous chaos type expansions in certain areas of statistical mechanics and turbulence theory [19].

In this paper we take as our point of departure the principal results of [10] which are stated in Section 2. Since the stochastic processes occurring most often in turbulence theory are generalized random fields and processes depending on many parameters (e.g. time and space co-ordinates) the remaining sections are devoted to a detailed derivation of the homogeneous chaos for such processes. (Sections 3 and 4). In Section 6 we consider the nonlinear expansion of functionals of stationary Gaussian processes defined on topological homogeneous spaces.

Section 3 has many points of contact with the work of T. Hida and N. Ikeda [5] but, it seems to us that there are also some differences, particularly in the nature of the assumptions. The techniques of Section 3 also extend to certain kinds of non Gaussian generalized fields. In what follows we shall use the abbreviation RKHS for Reproducing kernel Hilbert space.

2. Definitions and earlier results.

We recall the principal results on homogeneous chaos expansions derived in [10] and introduce the necessary notation. We shall have to refer the reader to [10] for details since to present them here would needlessly overburden the paper.

Let $\{x(t),\ t \in T\}$ be a real Gaussian stochastic process defined on a probability space $(\Omega,\ \underline{A},\ P_0)$ where $\underline{A}$ is taken to be the completion with respect to P_0 of A^o, the minimal σ-field with respect to which all the random variables $x(t,\omega)(t \in T)$ are measurable. The mean function, $E[x(t)]$ will be assumed to be zero and the covariance $E[x(t)x(s)]$ will be denoted by $R(t,\ s)$. We shall assume that the infinite set T is a complete, separable metric space. Then R determines a Hilbert space $H(R)$, called the RKHS of R (or of the process) which has the following properties: $H(R)$ consists of real functions f on T such that $R(\cdot\ ,\ t) \in H(R)$ and $(f,\ R(\cdot\ ,\ t)) = f(t)$ for every t in T. Because of the assumptions on T, $H(R)$ is also separable and infinite dimensional. Let $L^2(\Omega,\ \underline{A},\ P_0)$ or, briefly, $L^2(\Omega,\ P_0)$ be the Hilbert space of square integrable random variables on Ω (with the usual identification of functions differing on null sets) and let $L_1(X)$ be the closed linear subspace of $L^2(\Omega,\ P_0)$ spanned by the finite real linear combinations $\sum_{i=1}^{n} c_i\, x_{t_i}$. In what follows we shall use the well known congruence ψ_1 between $L_1(X)$ and $H(R)$ under which the image of x_t is $R(\cdot\ ,\ t)$.

Let H_1 and H_2 be real separable Hilbert spaces with inner products $(\ ,\ )_1$ and $(\ ,\ )_2$ and let the tensor (or direct) product Hilbert space of H_1 and H_2, be denoted by $H_1 \otimes H_2$. Here we shall briefly recapitulate a few elementary facts about tensor products of Hilbert spaces which will be useful to us later and also introduce the notion of the symmetric tensor product.

For any two elements $h_1,\ h_2$ $(h_i \in H_i)$ let $h_i \otimes h_2$ denote their tensor product which is an element of $H_1 \otimes H_2$. Let $(\ ,\ )$ denote the inner product for $H_1 \otimes H_2$. If h_i, g_i are elements of H_i then $(h_1 \otimes h_2, g_1 \otimes g_2) = (h_1, g_1)_1 (h_2, g_2)_2$. The set of all finite linear combinations of the form $\sum_{k=1}^{n} c_k (h_1^k \otimes h_2^k)$ where c_k's are real, $h_1^k \epsilon H_1$ and $h_2^k \epsilon H_2$ are dense in $H_1 \otimes H_2$. Let $\{f_\alpha\}$ be a set that spans H_1, i.e., such that H_1 is the closed linear manifold spanned by the f_α's. Similarly let $\{g_\beta\}$ be a set that spans H_2. Then the set of elements $\{f_\alpha \otimes g_\beta\}$ spans $H_1 \otimes H_2$. Let $\{e_i\}_1^\infty$ be a complete orthonormal system (CONS) in H_1 and $\{f_j\}_1^\infty$ be a CONS in H_2. Then $\{e_i \otimes f_j, i, j = 1, 2, \ldots\}$ is a CONS in $H_1 \otimes H_2$. All these statements generalize in an obvious way to any finite number of Hilbert spaces H_i $(i = 1, \ldots, p)$. The tensor product Hilbert space in that case is written $H_1 \otimes \ldots \otimes H_p$. When all the spaces are the same, say, H, we introduce the shorter notion $\overset{p}{\otimes} H$.

Let $h_1 \otimes \ldots \otimes h_p$ be an element of $\overset{p}{\otimes} H$. Define

$$\sigma(h_1 \otimes \ldots \otimes h_p) = \frac{1}{p!} \sum_{\pi} h_{\pi_1} \otimes \ldots \otimes h_{\pi_p}$$

where $\pi = (\pi_1, \ldots, \pi_p)$ is a permutation of the integers $(1, \ldots, p)$. The symmetric tensor product Hilbert space $\sigma[\overset{p}{\otimes} H]$ is defined as the closed linear subspace of $\overset{p}{\otimes} H$ generated by elements of the form

$$\sum_{k=1}^{n} c_k \, \sigma(h_1^k \otimes \ldots \otimes h_p^k).$$

In fact, it can be shown that σ defined as above can be extended to define a projection operator on $\overset{p}{\otimes} H$ whose range is $\sigma[\overset{p}{\otimes} H]$. From now on we take $H = H(R)$, and $H = L_1(X)$.

We shall assume throughout that $\{e_i\}_1^\infty$ is a fixed but arbitrary complete orthonormal system (CONS) in $H(R)$. Let $\{\xi_i\}_1^\infty$ be the CONS in $L_1(X)$ (independent standard normal random variables) where each ξ_i corresponds under the congruence ψ_1 to e_i. Let $p \geq 1$ be any integer. It is now clear that the elements $e_{i_1} \otimes \ldots \otimes e_{i_p}$ as $i_1, \ldots, i_p$ range independently from 1 to infinity form a CONS for $\overset{p}{\otimes} H(R)$. It is more convenient to arrange the elements of this CONS in the following manner. Let $\lambda_1, \ldots, \lambda_r$ be the distinct integers in the sequence $(i_1, \ldots, i_p)$ with λ_1 occurring n_1 times, λ_2 occurring n_2 times, etc. Then $n_i > 0$ and $n_1 + \ldots + n_r = p$. The following lemma gives a description of CONS in $\sigma(\overset{p}{\otimes} H(R))$ and $\sigma(\overset{p}{\otimes} L_1(X))$ in terms of $\{e_i\}$ and $\{\xi_i\}$. Define

$$(2.1) \qquad e_{\lambda_1,\ldots,\lambda_r}^{n_1,\ldots,n_r} = \frac{\sqrt{p!}}{(n_1!\ldots n_r!)^{1/2}} \sigma(e_{i_1} \otimes \ldots \otimes e_{i_p}).$$

where

$$(2.2) \qquad \sigma(e_{i_1} \otimes \ldots \otimes e_{i_p}) = \frac{1}{p!} \sum_{(j)\sim(i)} e_{j_1} \otimes \ldots \otimes e_{j_p},$$

and (j) is a permutation $(j_1, \ldots, j_p)$ of $(i) = (i_1, \ldots, i_p)$. We define the elements $\xi_{\lambda_1,\ldots,\lambda_r}^{n_1,\ldots,n_r}$ in a similar manner.

<u>Lemma 2.1</u>.

(i) The elements $\{e_{\lambda_1,\ldots,\lambda_r}^{n_1,\ldots,n_r}, \sum_1^r n_i = p, \lambda_i = 1, 2, \ldots, n_i > 0\}$ form a CONS in $\sigma(\overset{p}{\otimes} H(R))$;

(ii) The elements $\{\xi_{\lambda_1,\ldots,\lambda_r}^{n_1,\ldots,n_r}, \sum_k^r n_i = p, \lambda_i = 1, 2, \ldots, n_i > 0\}$ form a CONS in $\sigma(\overset{p}{\otimes} L_1(X))$.

The linear map ψ_p which sends each $\xi_{\lambda_1,\ldots,\lambda_r}^{n_1,\ldots,n_r}$ into $e_{\lambda_1,\ldots,\lambda_r}^{n_1,\ldots,n_r}$ extends to a congruence (inner product preserving isomorphism) between $\sigma(\overset{p}{\otimes} L_1(X))$ and $\sigma(\overset{p}{\otimes} H(R))$. The congruences ψ_p $(p = 1, 2, \ldots)$ will be called canonical. It is shown in [10] (Theorem (2.1)) that $\sigma(\overset{p}{\otimes} H(R))$ is also a RKHS with a kernel obtained by an appropriate symmetrization of the kernel $\overset{p}{\otimes} R$ defined on $T^p \times T^p$ by $\overset{p}{\otimes} R(s_1,\ldots,s_p; t_1,\ldots,t_p) = \prod_{i=1}^p R(s_i, t_i)$. Let $H_p (p = 0, 1, \ldots)$ be Hilbert spaces with inner products $(\, , \,)_p$. Denote by $\sum_{p\geq 0} \oplus H_p$ their orthogonal (external) direct sum.

<u>Lemma 2.2</u>. Every element u of $\sum_{p\geq 0} \oplus \sigma[\overset{p}{\otimes} L_1(X)]$ can be expressed in the form

$$(2.3)\qquad u = \sum_{p\geq 0}\ \sum_{n_1+\ldots+n_r=p}\ \sum_{\lambda_1,\ldots,\lambda_r} (f_p, e^{n_1,\ldots,n_r}_{\lambda_1,\ldots,\lambda_r})_p\ \xi^{n_1,\ldots,n_r}_{\lambda_1,\ldots,\lambda_r}$$

where

$$(2.4)\qquad f_p \in \sigma\,[\overset{p}{\otimes} H(R)]\ , \qquad (p = 0, 1, \ldots)$$

are uniquely determined and

$$(2.5)\qquad \sum_{p\geq 0}\ \sum_{n_1+\ldots+n_r=p}\ \sum_{\lambda_1,\ldots,\lambda_r} (f_p, e^{n_1,\ldots,n_r}_{\lambda_1,\ldots,\lambda_r})^2_p < \infty .$$

Let f be the element in $\sum_{p\geq 0} \oplus\, \sigma[\overset{p}{\otimes} H(R)]$ given by

$$(2.6)\qquad f = \sum_{p\geq 0}\ \sum_{n_1+\ldots+n_r=p}\ \sum_{\lambda_1,\ldots,\lambda_r} (f_p, e^{n_1,\ldots,n_r}_{\lambda_1,\ldots,\lambda_r})_p\ e^{n_1,\ldots,n_r}_{\lambda_1,\ldots,\lambda_r} .$$

Then the map (linearly defined) $\psi(u) = f$ where u and f are respectively given by (2.3) and (2.6) defines a congruence between the two Hilbert spaces and we write

$$(2.7)\qquad \sum_{p\geq 0} \oplus\, \sigma[\overset{p}{\otimes} L_1(X)] \overset{\psi}{\cong} \sum_{p\geq 0} \oplus\, \sigma[\overset{p}{\otimes} H(R)].$$

We can now state the main result of [10] (Theorem 4.3) as

<u>Theorem 2.1</u>. For any $u \in L^2(\Omega, P_0)$ the (L^2-convergent) expansion

$$(2.8)\qquad u = \sum_{p\geq 0}\ \sum_{n_1+\ldots+n_r=p}\ \sum_{\lambda_1,\ldots,\lambda_r} (f_p, e^{n_1,\ldots,n_r}_{\lambda_1,\ldots,\lambda_r})_p \prod_{i=1}^{r} h_{n_i}(\xi_{\lambda_i})$$

holds, where

$$(2.9)\qquad f_p\ ,\ e^{n_1,\ldots,n_r}_{\lambda_1,\ldots,\lambda_r} \in \sigma[\overset{p}{\otimes} H(R)]$$

are uniquely determined by u and $h_n(x)$ is the nth normalized Hermite polynomial in x, i.e.,

(2.10) $$h_n(x) = (-1)^n (n!)^{-\frac{1}{2}} e^{x^2/2} \left(\frac{d}{dx}\right)^n e^{-x^2/2} .$$

Equivalently, we obtain the decomposition

(2.11) $$L^2(\Omega, P_0) \cong \sum_{p\geq 0} \oplus \sigma[\overset{p}{\otimes} L_1(X)].$$

Here $\cong$ is a congruence relation. The Hilbert space $\sigma[\overset{p}{\otimes} L_1(X)]$ is the so-called p^{th} homogeneous chaos. More precisely (following Kakutani's definition, see [10]) if $p \geq 1$ and $\hat{G}_p$ in the linear subspace in $L^2(\Omega, P_0)$ of all polynomials in $\{\xi_i\}_1^\infty$ of degree not exceeding p, and $G_p = \hat{G}_p \ominus \hat{G}_{p-1}$ is the set of all polynomials in $\hat{G}_p$ orthogonal to every polynomial in $\hat{G}_{p-1}$, then the closed linear subspace $\bar{G}_p$ is the pth homogeneous chaos. ($\bar{G}_0 = G_0 = \{1\}$, the linear subspace of constant random variables). If J denotes the congruence map in (2.11) then J sends the element $\prod_{i=1}^r h_{n_i}(\xi_{\lambda_i})$ into $\xi_{\lambda_1,\ldots,\lambda_r}^{n_1,\ldots,n_r}$. In particular this means that $\sigma[\overset{p}{\otimes} L_1(X)]$ may be identified with $\bar{G}_p$.

It is shown in [13] that the kernel Γ on $H(R) \times H(R)$ given by

$$\Gamma(m, m') = \exp\{(m, m')\} \qquad (m, m' \in H(R))$$

generates on RKHS $H(\Gamma)$ whose elements f are functions on or in H(R) defined by

(2.12) $$f(m) = \sum_{p\geq 0} \sum_{n_1+\ldots+n_r=p} \sum_{\lambda_1,\ldots,\lambda_r} a_{\lambda_1,\ldots,\lambda_r}^{n_1,\ldots,n_r} \prod_{i=1}^r \frac{(m, e_i)^{n_i}}{\sqrt{n_i!}}$$

where the coefficients $a_{\lambda_1,\ldots,\lambda_r}^{n_1,\ldots,n_r}$ are uniquely determined by f and where further the sum of squares of the coefficients is convergent. Let [,] denote the inner product in $H(\Gamma)$. For m in $H(R)$ let $m^{\otimes p}$ stand for $\underbrace{m \otimes\ldots\otimes m}_{p}$. Then the series $\sum_{p\geq 0} (p!)^{-1/2} m^{\otimes p}$ defines an element of $\sum_{p\geq 0} \oplus \sigma[\overset{p}{\otimes} H(R)]$ which we denote by $\exp[\hat{\otimes} m]$. We quote the following result from [10] (Lemma 4.3 and Theorem 4.5).

Theorem 2.2. For every m in H(R) $\exp[\hat{\otimes}\, m] \in \sum_{p\geq 0} \oplus\, \sigma[\overset{p}{\otimes} H(R)]$ and

$$(\exp[\hat{\otimes}\, m],\ \exp[\hat{\otimes}\, m']) = \exp\{(m,m')\}. \tag{2.13}$$

The relation (2.13) extends to a congruence map ψ^* between $H(\Gamma)$ and $\sum_{p\geq 0} \oplus\, \sigma[\overset{p}{\otimes} H(R)]$.

3. Homogeneous chaos of Gaussian generalized random fields.

We shall now apply Theorem 2.1 to obtain nonlinear expansions for L^2-functionals of generalized Gaussian processes. For the index set T of the previous section we take E, a nuclear, separable, Fréchet space over the reals. Let $\Omega = E'$ the topological dual of E. Elements of E' will be denoted by x and are generalized functions, i.e., continuous linear functionals on E. The value of x at φ in E will be denoted by $x(\varphi)$. As in the last section A^o is the σ-field of subsets of E' generated by all sets of the form $\{x \in E' : [x(\varphi_1), \ldots, x(\varphi_n)] \in B\}$ (n arbitrary, $\varphi_1,\ldots,\varphi_n$ in E and B a Borel set in R^n). Let P_0 be a Gaussian measure on $(E',\underline{A})$ with zero mean and covariance functional R. I.e., for every φ in E

$$\int_{E'} x(\varphi)\, P_0(dx) = 0 \quad \text{and} \tag{3.1}$$

$$R(\varphi, \psi) = \int_{E'} x(\varphi)\, x(\psi)\, P_0(dx)\,, \qquad (\varphi, \psi \text{ in } E). \tag{3.2}$$

The stochastic process $x(t)$ is now replaced by $x(\varphi)$. The measure P_0 or the functional x is called a Gaussian generalized process or generalized random field (GRF, for short). Examples of E are provided by (a) $\mathcal{D}_K$ the space of infinitely differentiable, real functions defined on R^p and vanishing outside a compact set K, and (b) $C^\infty(S^{p-1})$ the space of infinitely differentiable functions on the sphere S^{p-1} in R^p. The topologies on these spaces are defined as in L. Schwartz [17]. $\mathcal{D}_K$ (and similarly $C^\infty(S^{p-1})$) is then known to be a nuclear separable Fréchet space, a fact which essentially ensures the existence of probability measures on (E', A)

corresponding to pre-assigned finite dimensional distributions. (See Gelfand and Vilenkin [4]). From (3.2) it is immediately seen that R is a bilinear functional. The assumptions on the space E and the fact that P_0 is Gaussian actually yield the continuity of R. This can be seen as follows. ([11], Lemma 4.1). Write $\sigma^2(\varphi) = R(\varphi, \varphi)$. If φ_0 is any point of E and $\{\varphi_n\}$ in E is a sequence converging to φ_0, then the sequence of random variables $\{x(\varphi_n)\}$ on $(E', \underline{A}, P_0)$ converges in probability to $x(\varphi)$ since it converges a.s (indeed for every x) to it. Hence for every $\epsilon > 0$

$$(3.3) \qquad P_0(\{x \in E' : | x(\varphi_n) - x(\varphi) | > \epsilon\}) < \epsilon$$

for all sufficiently large n. The probability on the left hand side of (3.3) equals

$$(3.4) \qquad (\tfrac{2}{\pi})^{\frac{1}{2}} \int_{\epsilon/\sigma(\varphi_n - \varphi_0)}^{\infty} e^{-\frac{1}{2}t^2} \, dt.$$

We now assert that

$$(3.5) \qquad \sigma^2(\varphi_n - \varphi_0) \to 0.$$

If (3.5) is not true, for some $\epsilon' > 0$ there is a subsequence $\{\varphi_{n'}\}$ such that $\sigma(\varphi_{n'} - \varphi_0) \geq \epsilon' > 0$. Then the integral in (3.4) (with $n = n'$) is bounded below by $(\frac{2}{\pi})^{1/2} \int_{\epsilon/\epsilon'}^{\infty} e^{-t^2/2}\, dt$ which contradicts (3.3) if ϵ is chosen small enough. The continuity of R on $E \times E$ now follows from (3.5), the bilinearity of R and the Schwarz inequality.

Since R is continuous and E is a separable complete metric space it follows that $H(R)$ is a separable Hilbert space. Furthermore, it is easy to verify that the elements of $H(R)$ are generalized functions for, by the definition of an RKHS and the bilinearity of R, every m in $H(R)$ is seen to be a linear functional on E while continuity follows from the inequality $|m(\varphi)| \leq \|m\|_{H(R)} \cdot R(\varphi, \varphi)$. From now on the covariance R will be assumed to be non-degenerate, i.e.

(3.6) $$R(\varphi, \varphi) = 0 \text{ implies } \varphi = 0.$$

The map $i : E \to H(R)$ defined by

(3.7) $$i(\varphi) = R(\cdot, \varphi)$$

is obviously linear and 1-1 into $H(R)$ because of (3.6). Let us now define an inner product $(,)$ in E by $(\varphi, \psi) = (R(\cdot, \varphi), R(\cdot, \psi))_{H(R)}$. From (3.7) we have

(3.8) $$(i(\varphi), i(\psi))_{H(R)} = (\varphi, \psi)$$

for φ and ψ in E. If $\hat{E}$ denotes the completion of E with respect to $(,)$ from (3.8) i extends to a congruence between $\hat{E}$ and $H(R)$ enabling us to identify $\hat{E}$ with $H(R)$. We have thus proved the following fact.

Lemma 3.1.

(3.9) $$E \subset \hat{E} \subset E'.$$

If $(\xi_j)_1^\infty$ is a CONS in $L_1(X)$ the congruence between the latter and $\hat{E}$ shows that the sequence $\{\hat{\varphi}_j\}_1^\infty$ in $\hat{E}$ where $\hat{\varphi}_j$ corresponds to ξ_j is a CONS in $\hat{E}$. To display this correspondence more effectively we denote the $\xi_j(x)$ by $x(\hat{\varphi}_j)^\sim$. (We believe this notation is due to J. Kuelbs). It should be noted that, in general, $x(\hat{\varphi}_j)^\sim$ is not a linear functional.

Theorem 2.1 is now applicable to $L^2(E', \underline{A}, P_0)$. Writing $\hat{\varphi}_i$ for e_i and $\hat{\varphi}_{\lambda_1,\dots,\lambda_r}^{n_1,\dots,n_r}$ for $e_{\lambda_1,\dots,\lambda_r}^{n_1,\dots,n_r}$ in Theorem 2.1 we obtain the following result.

Theorem 3.1. Every u belonging to $L^2(E', \underline{A}, P_0)$ has the (L^2-convergent) expansion

$$(3.10) \qquad u(x) = \sum_{p\geq 0} \sum_{n_1+\dots+n_r=p} \sum_{\lambda_1,\dots\lambda_r} (F_p, \hat{\varphi}^{n_1,\dots,n_r}_{\lambda_1,\dots,\lambda_r}) \prod_{i=1}^{r} h_{n_i}(x(\hat{\varphi}_i)^{\sim})$$

where

$$F_p \text{ and } \hat{\varphi}^{n_1,\dots,n_r}_{\lambda_1,\dots,\lambda_r} \text{ belong to } \sigma[\overset{p}{\otimes}\hat{E}].$$

We wish to extend the validity of Theorem 3.1 to the case where E is an LF space ([18], p. 126) which is, in general, neither separable nor a Fréchet space. Let $E_n (n = 1, 2, \dots)$ be an increasing sequence of separable, nuclear Fréchet spaces and let $E = \bigcup_{n=1}^{\infty} E_n$ be the (countable) strict inductive limit of $\{E_n\}$. Then E is nuclear and hence the property that a probability measure can be determined on $(E', \underline{A})$ (E' = dual of E, $\underline{A}$ defined as before) by an assignment of consistent finite dimensional distributions continues to hold. An examination of the proof of Theorem 2.1 shows that Theorem 3.1 will be valid if we take for E a countable strict inductive limit of the E_n's provided we assure ourselves of the separability of the RKHS $H(R)$. This is done in the following lemma.

Lemma 3.2. Let P_0 be a Gaussian measure on $(E', \underline{A})$ with zero mean and covariance R. Then $H(R)$ (and hence also $\hat{E}$) is separable.

Proof. Let $M(R; E_n)$ be the closed linear manifold of $H(R)$ generated by $\{R(\cdot, \varphi), \varphi \in E_n\}$. Then $M(R; E_n)$ is a separable subspace of $H(R)$. This is shown as follows. Let $\{\psi^n_j\}_{j=1}^{\infty}$ be a countable set dense in E_n which exists since E_n is a separable, Fréchet space. Let $m \in M(R; E_n)$ and $\epsilon > 0$. By the definition of $M(R; E_n)$ there exists a $\varphi = \Sigma c_i\varphi_i$, a finite linear combination of $\varphi_i \in E_n$ such that

$$(3.11) \qquad \|m - R(\cdot, \varphi)\|_{H(R)} < \epsilon/2.$$

Let $\{\psi^n_{j'}\}$ be the subsequence of the dense set $\{\psi^n_j\}$ which converges to φ $(j' \to \infty)$. Now R restricted to $E_n \times E_n$ is continuous as has been shown above.

Hence $R(\psi_j^n, \psi_j^n) \to R(\varphi, \varphi)$ which in turn implies that
$\|R(\cdot, \varphi) - R(\cdot, \psi_{j'}^n)\|_{H(R)}^2 = R(\varphi, \varphi) - 2R(\varphi, \psi_{j'}^n) + R(\psi_{j'}^n, \psi_{j'}^n) \to 0$ as $j' \to \infty$.
Hence there exists a $\psi_{j'}^n$ belonging to the countable dense set in E_n such that

$$\|R(\cdot, \varphi) - R(\cdot, \psi_{j'}^n)\|_{H(R)} < \epsilon/2$$

which together with (3.11) shows that the countable set $\{R(\cdot, \psi_j^n)\}_{j=1}^{\infty}$ is dense (in $H(R)$-topology) in $M(R; E_n)$.

Now denote by $M(R)$ the closed linear subspace of $H(R)$ spanned by $M(R; E_n)$ $(n = 1, 2, \ldots)$. To prove the separability of $H(R)$ it suffices to show that

(3.12) $$M(R) = H(R).$$

Let $m \in H(R)$, $(m \neq 0)$ be $\perp$ $M(R)$. Then $m \perp m(R; E_n)$ for every n, and in particular $m \perp R(\cdot, \varphi)$ for all $\varphi \in E_n$ $(n = 1, 2, \ldots)$. This means that for φ in E_n

(3.13) $$m(\varphi) = (m, R(\cdot, \varphi))_{H(R)} = 0.$$

(3.13) implies that $m(\varphi) = 0$ for all φ in E since if φ is in E, $\varphi \in E_n$ for some n and (3.13) holds. This proves $m = 0$ and (3.12).

The important application we have in mind is $E = \mathcal{D}(\mathbb{R}^p)$ (denoted henceforth by $\mathcal{D}$), the Schwarz space of infinitely differentiable functions vanishing outside compacts of $\mathbb{R}^p$. If $K_n (n = 1, 2, \ldots)$ is an increasing sequence of compacts with $\cup K_n = \mathbb{R}^p$ then $\mathcal{D}$ is the countable strict inductive limit of $\mathcal{D}_{K_n}$ where the latter are the spaces introduced in (a) above. The topology on $\mathcal{D}$ that makes it so is not metrizable or separable ([17], pp. 64-66).

<u>Remark:</u> If it is possible to choose a sequence $\{\varphi_j\}$ contained in E which is a CONS in $\hat{E}$ (an assumption made for convenience by Hida and Ikeda in their work [5]) then the random variables $x(\hat{\varphi}_j)^\sim = x(\varphi_j)^\sim$ and the CONS $\hat{\varphi}_{\lambda_1, \ldots, \lambda_r}^{n_1, \ldots, n_r}$

can be expressed in terms of the φ_j.

In the following section we discuss applications of Theorems 2.1 and 3.1 with particular reference to square integrable functionals of Gaussian random fields (i.e. Gaussian processes of several parameters) or Gaussian GRF's. The general expansions derived in these two theorems make clear the importance of concretely describing the RKH spaces of the various important random fields. The examples of the next section are presented mainly from this point of view.

4. Discussion of some examples.

We shall first consider an interesting, and possibly useful example of a Gaussian GRF based on the p-parameter Wiener process. By the latter we mean a process $Y(t)$ $(t \in \mathbb{R}^p,\ p > 1)$ Gaussian and centered and such that

$$(4.1) \qquad E\, Y(t)\, Y(s) = \prod_{i=1}^{p} \min(t_i, s_i) \qquad (t_i \geq 0,\ \ s_i \geq 0)$$

where $t = (t_1, \ldots, t_p)$ and $s = (s_1, \ldots, s_p)$.

The covariance (4.1) is the tensor product (p-fold) of the covariance of the one-parameter Wiener process $W(t)$. $Y(t)$ is like $W(t)$ in the following sense. There is a class of generalized random processes called Gaussian white noise whose covariances are given by

$$(4.2) \qquad B(\varphi, \psi) = \int_{\mathbb{R}^n} \varphi(t)\, \psi(t)\, dt \qquad (t \in \mathbb{R}^n).$$

For each n, the corresponding random process is given by a measure on the space of Schwarz distributions on $\mathcal{D}(\mathbb{R}^n)$. We designate such a process as n-parameter white noise. Each of these processes is stationary with independent values at every point [4]. For $n = 1$, it is well known that the above white noise process is the derivative in Schwarz's sense of the one parameter Wiener process. More precisely, we have the following: Consider the generalized process

$$(4.3) \qquad W(\varphi) = \int_{\mathbb{R}^1} W(t)\, \varphi(t)\, dt$$

(Define $W(t) = 0$ if $t < 0$ although we may equally well consider the Wiener process defined for all real t). Let DW $(D = \frac{d}{dt})$ be the derivative of W, i.e. the distribution given by $DW(\varphi) = -W(D\varphi)$. Then DW is a generalized Gaussian process with covariance

$$(4.4) \qquad E\, DW(\varphi)\, DW(\psi) = \int_{R^1} \varphi(t)\, \psi(t) dt.$$

For $n = p$ (>1) we show that the p-parameter Gaussian white noise is the $\frac{\partial^p}{\partial t_1\, \partial t_2 \cdots \partial t_p}$ - derivative of $Y(t)$. This fact is contained in a result of Dudley's ([3], Theorem 4.1). We give here a direct and simple proof which uses an a.s. uniformly convergent orthogonal expansion for $Y(t)$. Write $D^p = \frac{\partial^p}{\partial t_1 \cdots \partial t_p}$.

<u>Lemma 3.3</u>. If $Y(t)$ is the p-parameter Wiener process then $D^p Y$ is (p-parameter) Gaussian white noise.

For ease of writing we give the proof for $p = 2$. First of all we observe that since our interest is only in the probability laws of the processes involved we may choose any convenient representation for $Y(t_1, t_2)$. Let η_{mn} be a double sequence of independent $N(0, 1)$ random variables on some probability space (Ω, Q). Functions $G_n \in L^2([0, \infty))$ $(n = 1, 2, \ldots)$ can then be found such that almost surely (we suppress the probability parameter ω)

$$(4.5) \qquad \sum_{m,n=1}^{\infty} \eta_{mn}\, G_m(t_1)\, G_n(t_2)$$

converges uniformly with respect to (t_1, t_2) in every bounded rectangle. A typical function G may be chosen in the following way (see Z. Ciesielski [2]). Let f be a member of the Haar family which is a CONS in $L^2[0, 1]$ and let $g(t) = (\frac{2}{\pi})^{1/2} (1 + t^2)^{-1/2} f(\frac{2}{\pi})$ arc tan t). Then $G(t) = \int_0^t g(u)du$ $(t \geq 0)$. With this choice the uniform convergence of (4.5) has been shown by G. Zimmerman in her thesis [22]. It is easy to verify that

$$(4.6) \qquad \sum_{n=1}^{\infty} G_n(t_1)\, G_n(s_1) = \min(t_1, s_1),$$ (the converge nce being uniform on

compacts) so that the process defined by (4.5) is easily seen to be a version of $Y(t_1, t_2)$. The generalized random field corresponding to the process Y is given by

$$(4.7) \qquad Y(\varphi) = \int_{\mathbb{R}^2} Y(t_1, t_2)\, \varphi(t_1, t_2)\, dt_1\, dt_2$$

where $\varphi \in D(\mathbb{R}^2)$. First consider φ of the form $\varphi(t_1, t_2) = \varphi_1(t_1)\, \varphi_2(t_2)$ where $\varphi_i \in D(\mathbb{R}^1)$. Using the uniform convergence over compacts of (4.5) we obtain (a.s.)

$$\iint_{\mathbb{R}^2} Y(t_1, t_2)\, \varphi_1(t_1)\, \varphi_2(t_2)\, dt_1\, dt_2$$
$$= \sum_{m,n} \eta_{mn} \left[\int G_m(t_1)\, \varphi_1(t_1)\, dt_1\right] \left[\int G_n(t_2)\, \varphi_2(t_2)\, dt_2\right]$$

and hence

$$(4.8) \qquad E[Y(\varphi)]^2 = \sum_{m,n} \left[\int G_m(t_1)\, \varphi_1(t_1)\, dt_1\right]^2 \left[\int G_n(t_2)\, \varphi_2(t_2)\, dt_2\right]^2$$
$$= \sum_{n=1}^{\infty} \left[\int G_n(t_1)\, \varphi_1(t_1)\, dt_1\right]^2 \cdot \sum_{n=1}^{\infty} \left[\int G_n(t_2)\, \varphi_2(t_2)\, dt_2\right]^2$$

The first factor on the right hand side of (4.8) is

$$\sum_{n=1}^{\infty} \iint G_n(t_1)\, G_n(s_1)\, \varphi_1(t_1)\, \varphi_1(s_1)\, dt_1\, ds_1$$
$$= \iint \min\,(t_1, s_1)\, \varphi_1(t_1)\, \varphi_1(s_1)\, dt_1\, ds_1$$
$$= B_W\,(\varphi_1, \varphi_1)$$

where B_W is the covariance functional of the one parameter Wiener process $W(t)$. Hence

$$(4.9) \qquad E[Y(\varphi)]^2 = B_W(\varphi_1, \varphi_1)\, B_W(\varphi_2, \varphi_2).$$

From (4.9) and (4.4) we get

$$(4.10) \qquad E[D^2 Y(\varphi)]^2 = E[Y(D^2\varphi)]^2 = B_W(\varphi_1', \varphi_1')\, B_W(\varphi_2', \varphi_2')$$

(the prime denoting derivative)

$$= B(\varphi_1, \varphi_1)\ B(\varphi_2, \varphi_2)$$

$$= \iint \varphi^2(t_1, t_2)\ dt_1\ dt_2.$$

Similarly, if $\psi \in D(R^2)$ with $\psi(t_1, t_2) = \psi_1(t_1)\ \psi_2(t_2)$, $\psi_i \in D(R^1)$ we get

(4.11) $$E\langle D^2Y, \varphi\rangle \langle D^2Y, \psi\rangle = \iint \varphi(t_1, t_2)\ \psi(t_1, t_2)\ dt_1\ dt_2.$$

Now the class of functions in $D(R^2)$ expressible as $\sum_{j=1}^{k} \varphi_1^j(t_1)\ \varphi_2^j(t_2)$ with $\varphi_1^j, \varphi_2^j \in D(R^1)$ is dense in $D(R^2)$ in the topology of $D(R^2)$ (see Yosida, [21] p. 66). This fact together with the continuity of the covariance functional of D^2Y shows that (4.11) holds for all φ, ψ in $D(R^2)$. This finishes the proof.

Writing $R(\varphi, \psi) = \int_{R^p} \varphi(t)\ \psi(t)\ dt$, $\varphi, \psi \in D(R^p)$ it is easy to verify that the RKHS $H(R) = L^2(R^p, \text{Leb.})$.

We now consider Gaussian generalized locally homogeneous random fields (GLHRF) which are extensions to generalized processes of the concept of a process with stationary increments. These fields, particularly isotropic GLHRF are encountered in the theory of isotropic turbulence and have been studied in detail by A. M. Yaglom [20]. Let $\mathcal{D}_1 \equiv \mathcal{D}_1(R^p)$ be the class of functions φ in $\mathcal{D}$ such that

(4.12) $$\int_{R^p} \varphi(t)\ dt = 0.$$

By a GLHRF we mean a random linear functional x on $\mathcal{D}_1$ whose covariance $R(\varphi, \psi)$ satisfies

(4.13) $$R(\varphi, \psi) = R(\tau_h\varphi, \tau_h\psi)$$

where $\tau_h\varphi(t) = \varphi(t + h)$ $(h \in R^p)$. The mean $E[x(\varphi)]$ will be assumed to be zero for all $\varphi \in \mathcal{D}_1$.

We wish now to find a general formula for the elements of the RKHS H(R). It is shown in [20] that

$$(4.14)\qquad R(\varphi,\psi) = \int_{R^p-\{0\}} \tilde{\varphi}(\lambda)\,\overline{\tilde{\psi}(\lambda)}\,F(d\lambda) + \mathfrak{a}\,\nabla\tilde{\varphi}(0).\overline{\nabla\tilde{\psi}(0)}.$$

Here $\tilde{\varphi}(\lambda)$ is the Fourier transform of $\varphi(t)$, $\nabla\tilde{\varphi}(0) = (\frac{\partial\tilde{\varphi}}{\partial\lambda_1}(0),\ldots,\frac{\partial\tilde{\varphi}}{\partial\lambda_p}(0))$, $\mathfrak{a} = (a_{jk})$ is a constant, p x p non negative Hermitian matrix and F is a measure on $R^p - \{0\}$ such that

$$(4.15)\qquad \int \frac{|\lambda|^2}{(1+|\lambda|^2)^{k+1}}\,F(d\lambda) < \infty \quad (|\lambda| = \text{length of } \lambda),$$

for some non negative k. (It should be noted that we are now considering complex-valued φ as well as $x(\varphi)$). From the discussion given in [20] (p. 294) it is easy to deduce the nature of H(R). It is found that $m \in H(R)$ is of the form

$$(4.16)\qquad m(\varphi) = \int_{R^p-\{0\}} \tilde{\varphi}(\lambda)\,\overline{\hat{m}(\lambda)}\,F(d\lambda) + \mathfrak{a}\nabla\tilde{\varphi}(0)\cdot\alpha$$

where the function $\hat{m}(\lambda)$ and the complex vector α are uniquely determined and $\int|\hat{m}(\lambda)|^2\,F(d\lambda) < \infty$.

Let us now consider the case where $\mathfrak{a} = 0$ and $k = 0$ in (4.15). It has been shown in [20] that the latter condition is satisfied if and only if the GLHRF is given by an ordinary (i.e. non generalized) LHRF $x(t)$ $(t \in R^p)$ such that $x(\varphi) = \int x(t)\,\varphi(t)\,dt$.

From (4.15) with $k = 0$ it follows that F is a finite measure so that $\int|\hat{m}(\lambda)|\,F(d\lambda) < \infty$. From (4.16) if $m \in H(R)$ we may write

$$m(\varphi) = \int [\int\{e^{i(t,\lambda)} - 1\}\,\varphi(t)\,dt]\,\overline{\hat{m}(\lambda)}\,F(d\lambda)$$

where (t,λ) = inner product R^p. Using $\hat{m} \in L^1(dF)$ and Fubini's theorem we then obtain

(4.17) $$m(\varphi) = \int \varphi(t)\, M(t)\, dt$$

where

(4.18) $$M(t) = \int [e^{i(t,\lambda)} - 1]\, \overline{\hat{m}(\lambda)}\, F(d\lambda).$$

$M(t)$ is obviously locally integrable. The formulas (4.17) and (4.18) thus describe the RKHS of all (non-generalized) LHRF's. The expressions (4.17) and (4.18) contain Molchan's characterization of the RKHS of Lévy's p-parameter Brownian motion [15]. The latter as a generalized field may be defined by

(4.19) $$x(\varphi) = \int \varphi(t)\, x(t)\, dt \qquad (\varphi \in \mathcal{D}_1)$$

where $x(t)$ is a centered Gaussian process with mean zero and covariance

(4.20) $$Ex(t)\, x(s) = \frac{1}{2}\,(|t| + |s| - |t-s|),$$

$t, s \in R^p$ and $|t|$ is the length of the vector t. [14].

5. Homogeneous chaos for vector-valued generalized random fields.

There are apparently problems in turbulence theory which make it necessary to consider nonlinear or homogeneous chaos expansions for functions of a vector-valued generalized random field (GRF). (See [6]). An example of a vector-valued Gaussian GRF is provided by the gradient of Lévy's multi-parameter Brownian motion mentioned by Ito in connection with isotropic turbulence [8]. Accordingly we consider now the extension of Theorem 3.1 to vector-valued GRF's. Let F be a finite dimensional linear inner product space and let $\mathcal{E}$ be the tensor product $F \otimes \mathcal{D}$ where $\mathcal{D} = \mathcal{D}(R^p)$. Since $\mathcal{D}$ and F are nuclear so is $\mathcal{E}$. Since $\mathcal{D}$ is complete and F is finite dimensional, the topological completion of the nuclear space, $F \hat{\otimes} \mathcal{D}$ is $F \otimes \mathcal{D}$ itself. Finally $F \otimes \mathcal{D}$ is the countable strict inductive limit of nuclear, separable Fréchet spaces. All these facts are to be found in [18] (Chapter 40 et-seq.) or easily deducible from the material given there. Hence Lemma 3.2 holds with E of that lemma replaced by $\mathcal{E}$. (In other words if K is the covariance functional of a zero mean Gaussian process P_0 on

$(\mathcal{E}', \underline{A})$ then the RKHS $H(K)$ is separable.) Consequently, (see remarks preceeding Lemma 3.2) Theorem 3.1 is true for any random variable belonging to $L^2(\mathcal{E}', \underline{A}, P_0)$. Observe that the space $\hat{\mathcal{E}}$ now replaces $\hat{E}$ in that theorem. Let x be a generic element of $\mathcal{E}'$. Some remarks are in order to explain why x is a vector-valued (d dimensional assuming d to be the dimension of F) generalized function. It is easy to see that the space $F \otimes \mathcal{D} = \mathcal{D}(R^p; F)$, the space of infinitely differentiable F-valued functions on R^p vanishing outside compacts of R^p. (See [18], pp. 412-414). Denote by $\underline{\varphi}$ an element of $\mathcal{D}(R^p; F)$. Fixing an orthonormal basis $\underline{e}_1,\dots,\underline{e}_d$ in F we may write

$$\underline{\varphi}(t) = \sum_{j=1}^{d} \varphi_j(t)\, \underline{e}_j ,$$

where (identifying the dual of F with itself) $\varphi_j(t) = \langle \underline{e}_j, \underline{\varphi} \rangle$, $\langle\ ,\ \rangle$ denoting inner product in F. Then each $\varphi_j \in \mathcal{D}$ and $\varphi_j\ \underline{e}_j$ is nothing but $\underline{e}_j \otimes \varphi_j$. Now we have

$$(5.1) \qquad x(\underline{\varphi}) = \sum_{j=1}^{d} x(\underline{e}_j \otimes \varphi_j).$$

For each $j = 1,\dots,d$ we define $x_j \in \mathcal{D}'$ which are the components of x as follows.

$$(5.2) \qquad x_j(\varphi) = x(\underline{e}_j \otimes \varphi) \qquad (\varphi \in \mathcal{D}).$$

From (5.1) we then have

$$(5.3) \qquad x(\underline{\varphi}) = \sum_{j=1}^{d} x_j(\varphi_j).$$

Conversely, starting from the x_j it is customary (see [20], p. 279) to consider the following one-dimensional generalized function depending on the complex parameter $\underline{\alpha} = (\alpha_1,\dots,\alpha_d) \in F$:

$$(5.4) \qquad T_\alpha(\varphi) = \sum_{j=1}^{d} \alpha_j\, x_j(\varphi) \qquad (\varphi \in \mathcal{D}).$$

(5.4) may be regarded as an alternative formulation of (5.1). Indeed given

(5.4) we can get back x in (5.1) if we define for every $\underline{\varphi} = \sum_{j=1}^{d} \underline{e}_j \otimes \varphi_j$, $x(\underline{\varphi}) = \sum_{j=1}^{d} T_{\underline{e}_j}(\varphi_j)$.

Example 1. Let P_0 be the Gaussian measure on $(\mathcal{E}', \underline{A})$ with the characteristic functional $e^{-1/2\, K(\underline{\varphi}, \underline{\varphi})}$, where

$$K(\underline{\varphi}, \underline{\psi}) = \int_{R^p} \langle \underline{\varphi}(t), \underline{\psi}(t) \rangle dt = \sum_{j=1}^{d} \int_{R^p} \varphi_j(t)\, \overline{\psi_j(t)}\, dt. \tag{5.5}$$

K is the covariance functional $K(\underline{\varphi}, \underline{\psi}) = E\, x(\underline{\varphi})\, \overline{x(\underline{\psi})}$. (We make the obvious changes if only real generalized processes are to be considered as was done in Section 3). It is clear from (5.5) that $E[x_j(\varphi)\, \overline{x_k(\psi)}] = 0$ if $j \neq k$, for all $\varphi, \psi \in \mathcal{D}$ and hence (because of Gaussianity) that the components x_j are independent. This example corresponds to the case discussed in [6] and is the vector-valued analogue of the GRF given by (4.4). It is easy to see that the RKHS is the direct sum

$$H(K) = \underbrace{L^2(R^p) \oplus \ldots \oplus L^2(R^p)}_{d}. \tag{5.6}$$

Example 2. Gradient of a LHGRF. Let L be a LHGRF whose covarance R is given by formulas (4.14) (with $\alpha = 0$) and (4.15). Let $D_j = \frac{\partial}{\partial t_j}$ and let x be the vector-valued GRF with components given by $x_j = D_j L$. It is well known that $x = (D_1 L, \ldots, D_p L)$ is a homogeneous GRF since L is a locally homogeneous GRF. If $\underline{\varphi} = \sum_{j=1}^{d} \varphi_j\, \underline{e}_j$ and $\underline{\psi} = \sum_{j=1}^{d} \psi_j\, \underline{e}_j$, from (4.17), for φ, ψ in $\mathcal{D}$,

$$E\, x_j(\varphi)\, \overline{x_k(\psi)} = E\, L(D_j\varphi)\, \overline{L(D_k\psi)} = \int \widetilde{D_j\varphi(\lambda)}\, \overline{\widetilde{D_k\psi(\lambda)}}\, F(d\lambda) = \int \lambda_j \lambda_k\, \widetilde{\varphi(\lambda)}\, \overline{\widetilde{\psi(\lambda)}}\, F(d\lambda)$$

where $\lambda = (\lambda_1, \ldots, \lambda_p)$. Hence from (5.3) if $\underline{\varphi} = \Sigma\, \varphi_j\, \underline{e}_j$ and $\underline{\psi} = \Sigma\, \psi_j\, \underline{e}_j$ we obtain

$$K(\underline{\varphi}, \underline{\psi}) = \int \sum_{j,k=1}^{p} \lambda_j \lambda_k\, \widetilde{\varphi_j(\lambda)}\, \overline{\widetilde{\psi_k(\lambda)}}\, F(d\lambda). \tag{5.7}$$

Writing $T_\alpha = \sum_{j=1}^{p} \alpha_j x_j$ instead of (5.7) we have for each pair of complex vectors $\underline{\alpha}$ and $\underline{\beta}$,

(5.8) $$E\, T_\alpha(\varphi)\, \overline{T_\beta(\psi)} = \int \widehat{\varphi(\lambda)}\, \overline{\widehat{\psi(\lambda)}}\, F(d\lambda;\, \alpha,\, \beta)$$

where the measure

(5.9) $$F(d\lambda;\, \alpha,\, \beta) = \sum_{j,k=1}^{p} \alpha_j\, \overline{\beta_k}\, \lambda_j\, \lambda_k \quad F(d\lambda).$$

If L is the Gaussian random field defined by Lévy's p-parameter Brownian motion (see (4.19) and (4.20)) $L(\varphi) = \int x(t)\, \varphi(t)\, dt$, then it can be shown that (5.8) and (5.9) reduce to the formula derived by Ito in [8].

6. Nonlinear expansions for stationary processes on homogeneous spaces.

Let us return to Theorem 2.1 and assume that the index or "time" set T is a separable homogeneous space under a group $G = \{g\}$, and that $\{x_t\}$ is a (zero mean) Gaussian, mean-continuous G-stationary process, i.e., such that

(6.1) $$R(gt,\, gs) = R(t,\, s) \quad \text{for all } g \text{ in } G\, .$$

The kernel R is then called G-invariant. The Hilbert space $H(R)$ then inherits the following easily verified (but important) properties given in Krein's illuminating paper [12]. For m in $H(R)$ of the form

(6.2) $$m = \sum_{i=1}^{n} c_i R(\cdot,\, t_i) \qquad (c_i\text{'s real})$$

define

(6.3) $$U_g m = \Sigma\, c_i\; R(\cdot,\, gt_i).$$

If m' is another element as in (6.2) then

(6.4) $$(U_g m,\, U_g m') = (m,\, m')$$

and U_g then extends to a unitary operator on $H(R)$. It is further easy to check that the family $\{U_g\}$ $(g \in G)$ is a group of unitary operators on $H(R)$, i.e.

(6.5) $$U_{g_1} U_{g_2} = U_{g_1 g_2} \qquad (U_g^* = U_{g^{-1}}).$$

Furthermore for every m in $H(R)$

(6.6) $$(U_g m)(t) = m(g^{-1}t).$$

Let $\{e_i\}_1^\infty$ be any CONS in $H(R)$. The G-invariance of R ensures that $\{e_i\}$ is an "ortho-invariant" basis (the term is Krein's). I.e. for each g $\{U_g e_i\}_{i=1}^\infty$ is a CONS in $H(R)$ (which is obvious) and

(6.7) $$U_g e_i = \sum_{j=1}^{\infty} \alpha_{ij}(g)\, e_j$$

where $((\alpha_{ij}(g)))$, $\alpha_{ij}(g) = (U_g e_i, e_j)$ is a unitary matrix. Using the congruence relations of Section 2 we first show how the operator U_g (and hence the group $\{U_g\}$) can be extended to the larger Hilbert space $\sum_{p\geqslant 0} \oplus\, \sigma[\overset{p}{\otimes} H(R)]$. Recall from Section 2 that $H(\Gamma)$ consists of functions f on $H(R)$ and has for its kernel $\Gamma(m, m') = \exp\{(m, m')\}$. The G-invariance of R induces the $\{U_g\}$-invariance of Γ since

(6.8) $$\Gamma(U_g m, U_g m') = \exp\{(U_g m, U_g m')\} = \exp\{(m, m')\} = \Gamma(m, m').$$

In view of (6.8) the definition of V_g on $H(\Gamma)$ can be carried out in exactly a similiar manner as that of U_g. For f in $H(\Gamma)$ of the form

(6.9) $$f = \sum c_i\, \Gamma(\cdot, m_i) \qquad (m_i \in H(R))$$

let

(6.10) $$V_g f = \sum c_i\, \Gamma(\cdot, U_g m_i).$$

If f' is another element of the same form (6.9) then

(6.11) $$[V_g f, V_g f'] = [f, f']$$

and we obtain the extension to all of $H(\Gamma)$. The verification that $\{V_g\}$ $(g \in G)$

is a group of unitary operators on $H(\Gamma)$ is done as before. We also have

$$(6.12) \qquad (V_g f)(m) = f(U_{g^{-1}} m)$$

for every f in $H(\Gamma)$ and g in G.

It turns out that, in effect, $\{V_g\}$ yields the desired extension of $\{U_g\}$ provided we pay due regard to the congruence relations stated in Theorem 2.2. Observe that $f \in H(\Gamma)$ is of the form (2.12). Under the congruence ψ^* between $H(\Gamma)$ and $\sum_p \oplus \sigma[\overset{p}{\otimes} H(R)]$ the element f in (2.12) goes into the element $\tilde{f}$ of the second Hilbert space given by

$$(6.13) \qquad \tilde{f} = \sum_p \sum_{n_1+\dots+n_r=p} \sum_{\lambda_1\dots\lambda_r} a^{n_1\dots n_r}_{\lambda_1\dots\lambda_r} e^{n_1\dots n_r}_{\lambda_1\dots\lambda_r}$$

If we now set

$$(6.14) \qquad \tilde{U}_g = \psi^* V_g \psi^{*-1}$$

it follows from (6.12) that $\{\tilde{U}_g\}$ is a unitary group on $\sum_p \oplus \sigma[\overset{p}{\otimes} H(R)]$ and $\tilde{U}_g\tilde{f}$ is given by the same expression (6.13) except that on the right hand side, $e^{n_1\dots n_r}_{\lambda_1\dots\lambda_r}$ is replaced by $(\tilde{U}_g e)^{n_1\dots n_r}_{\lambda_1\dots\lambda_r}$ where $(\tilde{U}_g e)^{n_1\dots n_r}_{\lambda_1\dots\lambda_r}$ is defined exactly as $e^{n_1\dots n_r}_{\lambda_1\dots\lambda_r}$, but we replace $e_{j_1} \otimes\dots\otimes e_{j_p}$ by $(U_g e_{j_i}) \otimes\dots\otimes (U_g e_{j_p})$. It is easy to see that $\{\tilde{U}_g\}$ is an extension of $\{U_g\}$ and furthermore that $\tilde{U}_g$ maps each Hilbert space $\sigma[\overset{p}{\otimes} H(R)]$ onto itself.

Let us assume that $\{x_t\}$ is a sample continuous process and that the homogeneous space T is compact. We may then take Ω to be the space of real continuous functions $C(T)$ and the random variables of the process to be the co-ordinate variables $x(t)$. For $g \in G$ define the transformation τ_g from $C(T)$ to itself by

$$(6.15) \qquad (\tau_g x\ (t) = x(gt).$$

In $C(T)$ we take $\underline{A}$ to be $\underline{B}(C)$, the σ-field of Borel sets. Let $\underline{B}(G)$ be the topological Borel field in G. Let us further assume that the map (g, x) $(g, x) \to \tau_g x$ is $\underline{B}(G) X \underline{B}(C)$ measurable. From the G-stationarity and (6.15) it follows that the transformations $\{\tau_g\}$ $(g \in G)$ are P_0-measure-preserving. Hence if u is any random variable (functional) on $L^2(C(T), P_0)$ then

$$(6.16) \qquad (S_g u)(x) = u(\tau_g x)$$

defines a group $\{S_g\}$ of unitary operators on $L^2(C(T), P_0)$. It can be shown that S_g corresponds under the congruence to $\tilde{U}_g$. Without going into further details the discussion of this section yields the following result.

Theorem 6.1. Let the above assumptions on $\{x_t\}$, T and $\{\tau_g\}$ hold and let $u \in L^2[C(T), P_0]$. Then u has the expansion (2.8) of Theorem 2.1. Furthermore we have the following. For each $p \geq 1$,

$$(6.17) \qquad S_g(\sigma[\overset{p}{\otimes} L_1(X)]) = \sigma[\overset{p}{\otimes} L_1(X)]$$

and

$$(6.18) \qquad \tilde{U}_g(\sigma[\overset{p}{\otimes} H(R)]) = \sigma[\overset{p}{\otimes} H(R)].$$

In principle the expansion yields a spectral decomposition of the covariance $B_u(g_1, g_2)$ of the strictly stationary process $\{S_g u\}$ $(g \in G)$.

References

[1]. Cameron, R. H. and Martin, W. T., The orthogonal development of nonlinear functionals in series of Fourier-Hermite functionals, Ann. Math. 48(1947).

[2]. Ciesielski, Z. Lectures on Brownian motion, heat conduction and potential theory, Math. Institute, Aarhus Univ. Denmark (1966).

[3]. Dudley, R. M., Gaussian processes on several parameters, Ann. Math. Statist., 36(1965), 771-788.

[4]. Gelfand, I.M. and Vilenkin, N. Ya. Some applications of Harmonic Analysis, (Generalized Functions, vol. 4), Moscow, (1961).

[5]. Hida, T. and Ikeda, N., Analysis on Hilbert space with reproducing kernel arising from multiple Wiener integral, Proc. 5th Berkeley Symp. Vol. II, Part 1(1965), 117-143.

[6]. Imamura, T., Meecham, W. C., and Siegel, A., Symbolic calculus of the Wiener process and Wiener-Hermite functionals, J. Math. Phy. 6, (1965), 695-706.

[7]. Ito, K., Multiple Wiener integral, J. Math. Soc. Japan, 3(1951), 157-169.

[8]. __________, Isotropic random current, Proc. 3rd Berkeley Symp. Vol. II, (1956), 125-132.

[9]. Kakutani, S., Spectral analysis of stationary Gaussian processes, Proc. 4th Berkeley Symp. Vol. 2(1961), 239-247.

[10]. Kallianpur, G., The role of reproducing kernel Hilbert spaces in the study of Gaussian processes, Advances in Probability Vol. 2, (ed. P. Ney) M. Dekker (1970), 51-83.

[11]. Kallianpur, G. and Nadkarni, M. G., Supports of Gaussian measures, Proc. 6th Berkeley Symp. (1970) (To appear).

[12]. Krein, M. G., Hermitian-positive kernels on homogeneous spaces, I. Amer. Math. Soc. Trans., Ser. 2, Vol. 34(1963), 69-108.

[13]. LePage, R., An isometry related to unbiased estimation, Chapter 5, Thesis, Univ. of Minnesota (1964).

[14]. Lévy, P., Processus Stochastiques et Mouvement Brownien. (Monographies des Probabilités, Fasc. 6) Gauthier-Villars, Paris (1948).

[15]. Molchan, G. M., On some problems concerning Brownian motion in Lévy's sense, Theory of Prob. and its Appl. (English Trans.) 12, (1967), 682-690.

References (cont.)

[16]. Neveu, J., Processus aleatoires Gaussiens, Séminaire de mathematiques superieures, U. of Montreal (1968).

[17]. Schwartz, L. Théorie des Distributions, Hermann, Paris, (1957 and 1959).

[18]. Treves, F., Topological vector spaces, distributions and kernels, Academic Press, New York and London (1967).

[19]. Wiener, N., Nonlinear problems in random theory, The M.I.T. Press, Cambridge, (1958).

[20]. Yaglom, A. M., Some classes of random fields in n-dimensional space related to random processes, Theory of Prob. and its Appl. (English Trans.), 2, (1957), 273-319.

[21]. Yosida, K., Functional Analysis, Springer-Verlag, (1965).

[22]. Zimmerman, G. J., Some sample function properties of the two-parameter Gaussian process, (to appear) (1971).

This work was supported by NSF Grant GP-20429

NON-ANALYTIC CHARACTER OF THE SHEAR-TENSOR DISTRIBUTION FUNCTION IN INCOMPRESSIBLE TURBULENCE

W. J. COCKE

Steward Observatory, University of Arizona

Tucson, Arizona 85721

I. INTRODUCTION

In this paper we examine the analytic properties of the probability density function of the shear tensor $T^s{}_{ab} = \frac{1}{2}(\partial_a U_b + \partial_b U_a)$, for incompressible, statistically isotropic turbulence.

The following basic probabilistic approach is used: It was shown in a previous paper (Cocke, 1969) that the probability density function for a real, symmetric tensor is completely specified if the tensor is assumed to be statistically isotropic and if the density function for its eigenvalues is also given. We argue here that for "almost any" eigenvalue density function, the derived marginal probability density f_d for any one of the diagonal components (say $T^s{}_{33}$) of T^s has a discontinuous first derivative at $T^s{}_{33} = T = 0$. (Hereafter, we abbreviate $T = T^s{}_{33}$ for the typical diagonal component.)

In Section II, we discuss the details of the probability transformations used, and present some general properties of the probability density $f_d(T)$. In Section III we use simple forms for the eigenvalue probability density function to derive examples which elucidate the nature of this singular behavior of $f_d(T)$, and then we give a few special counterexamples in which the discontinuity in the first derivative disappears. In particular, if the incompressibility assumption is relaxed, the discontinuity does not seem to occur.

Section IV deals with the general theory of this non-analytic property, and we show that it is plausible that this non-analyticity be present in real, physical turbulence. Indeed, it seems possible that $f_d(T)$ has a cusp at $T = 0$ or that $f_d(T)$

itself be discontinuous there. The limited amount of experimental data available indicates that something like this might in fact be true, but the data does not have sufficiently great resolution to permit firm conclusions at this time. (But see Note 1 at end of paper.)

II. GENERAL FORMALISM

The shear tensor T^s, being real and symmetric, has six independent components. Thus the basic probability space has six dimensions, and we have shown previously (Cocke, 1969) that these six quantities may be chosen to be the three real eigenvalues $(t_1, t_2, t_3) = \vec{t}$ and the three Euler angles (θ,ψ,ϕ) which specify the orthogonal rotation matrix $A = A(\theta,\psi,\phi)$ which diagonalizes the tensor T^s.

Actually, the transformation $T^s \rightarrow (\vec{t},\theta,\psi,\phi)$ is not unique, but is triple-valued, since the eigenvalue sets (t,u,v), (v,t,u), and (u,v,t) may clearly be obtained from the same T^s by different proper rotations. However, the inverse transformation $(\vec{t},\theta,\psi,\phi) \rightarrow T^s$ is singlevalued, and this is the transformation (with an important modification) which we employ in this paper.

The equations for this transformation may easily be obtained from the equation $T^s = \tilde{A}\bar{T}^s A$, where $\bar{T}^s$ is the diagonalized form of T^s. We have

$$T^s{}_{ab} = A_{1a}A_{1b}t_1 + A_{2a}A_{2b}t_2 + A_{3a}A_{3b}t_3 \qquad (1),$$

in which A_{ab} is given as a function of the Euler angles in many standard texts. We use here the conventions of Goldstein (1957).

The general task of this paper, as mentioned before, is to derive the probability density function for a typical diagonal component $T^s{}_{33} = T$, assuming the existence of a probability density function $F(\vec{t})$ for the eigenvalues. The complete joint probability density for the set $(\vec{t},\theta,\psi,\phi)$ was shown from the statistical isotropy postulate (Cocke, 1969) to be expressible in the form

$$\bar{f}(\vec{t},\theta,\psi,\phi) = F(\vec{t})\sin\theta/8\pi^2 \qquad (2),$$

where $F(\vec{t})$ is a symmetric function of $\vec{t}$.

We also assume that the fluid motion is incompressible, and hence Trace (T^s) $= \mathrm{Tr}T^s = t_1+t_2+t_3 = 0$. Therefore, we may write for the joint probability density for the eigenvalues $\vec{t}$

$$F(\vec{t}) = f(\vec{t})\,\delta(t_1+t_2+t_3) \qquad (3),$$

where $f(\vec{t})$ is also a symmetric function. If the turbulence is not statistically stationary, $f(\vec{t})$ also depends on the time, but that is of no concern to us here.

Since $f(\vec{t})$ is symmetric, its dependence on $\vec{t}$ may be conveniently expressed in terms of the rotational invariants $Tr(T^s)^n$, $n = 1,2,3,\ldots$. However, we now demonstrate that the restriction $TrT^s = 0$ allows us to limit the dependence to the invariants $Tr(T^s)^2 = \vec{t}^2$ and $Tr(T^s)^3$. In effect, setting $TrT = 0$, $\vec{t}^2 = c_1$ and $Tr(T^s)^3 = c_2$ immediately fixes any symmetric function of $\vec{t}$; i.e., it determines $\vec{t}$, except for order. We show this as follows:

Set $t_3 = -t_1-t_2$ and substitute into

$$\vec{t}^2 = t_1^2+t_2^2+t_3^2 = 2(t_1^2+t_1t_2+t_2^2) = c_1$$

This equation is readily solved for t_2 as

$$t_2 = -\tfrac{1}{2}t_1 \pm \tfrac{1}{2}(2c_1-3t_1^2)^{\frac{1}{2}} \qquad (4).$$

Further, we may write

$$Tr(T^s)^3 = t_1^3+t_2^3-(t_1+t_2)^3 = -3t_1t_2(t_1+t_2) \qquad (5),$$

and then substitute from Eq.(4), obtaining

$$Tr(T^s)^3 = 3t_1(t_1^2-\tfrac{1}{2}c_1) = c_2 .$$

This equation has three real roots, provided that $c_1^3 \geqslant 6c_2^2$. This latter relation is satisfied , since obviously we cannot have $Tr(T^s)^3$ proportionally very much larger than t^2, whereas the converse could be the case. These three roots must be the three values t_1,t_2,t_3 , since the problem is symmetric in these three variables.

Consequently, in the rest of the paper, we consider that $f(\vec{t})$ depends only on $\vec{t}^2$ and $Tr(T^s)^3$. In Section III we illustrate the discontinuity properties of $f_d(T)$ with simple functions for which $f(\vec{t})$ depends only on $\vec{t}^2$, whereas in Section IV we present a more general discussion in which dependence on $Tr(T^s)^3$ is permitted.

Since we are interested here only in the properties of a single typical diagonal component $T^s{}_{33}$, with all other components of T^s integrated out, it is convenient not to use the full transformation of Eq.(1), but rather a hybrid transformation in which the only changed variable is $t_3 \rightarrow T^s{}_{33} = T$. The formal change of variables is then written

$$t_1' = t_1\ ,\ t_2' = t_2,\ \ T = \sin^2\theta\,(t_1\sin^2\psi + t_2\cos^2\psi) + t_3\cos^2\theta\ ,$$

$$\theta' = \theta\ ,\ \ \psi' = \psi\ ,\ \ \phi' = \phi \qquad (6),$$

where the expression for T may easily be obtained from the relations in Goldstein (1957), as applied to Eq.(1). The Jacobian for this transformation is

$$J = |\partial(t_1', t_2', T, \theta', \psi', \phi') / \partial(\vec{t}, \theta, \psi, \phi)| = \cos^2\theta\ ,$$

and therefore the probability density f' for the transformed variables is

$$f'(t_1', t_2', T, \theta', \psi', \phi') = \bar{f}(\vec{t}, \theta, \psi, \phi) / \cos^2\theta \qquad (7).$$

One may then solve for t_3 as, with primes introduced,

$$t_3 = [T - \sin^2\theta'(t_1'\sin^2\psi' + t_2'\cos^2\psi')]\sec^2\theta' \qquad (8).$$

This expression may then be substituted into $\bar{f}$, and then $t_1', t_2', \theta', \psi', \phi'$ integrated out. Thus the probability density for T appears as

$$f_d(T) = \int \frac{dt_1\,dt_2\,\sin\theta\,d\theta\,d\psi}{4\pi\cos^2\theta}\, F\left(t_1, t_2, [T - \sin^2\theta(t_1\sin^2\psi + t_2\cos^2\psi)]\sec^2\theta\right)\ ,$$

where we have removed the primes, used Eqs.(2) and (7), and performed the integration with respect to ϕ.

The presence of the δ function in Eq.(3) allows us to integrate out another of the eigenvalue variables, say t_2, by using the expression $\delta(ax+b) = \delta(x+b/a)/|a|$. Substituting for t_3 in the expression $\delta(t_1+t_2+t_3)$ by means of Eq.(8) and then integrating with respect to t_2, one may obtain finally, after some manipulation,

$$f_d(T) = \int \frac{dt_1 \sin\theta\, d\theta\, d\psi}{4\pi |h_2|} f\left(t_1, -\frac{h_1}{h_2}t_1 - \frac{T}{h_2}, \frac{T}{h_2} + \left(\frac{h_1}{h_2} - 1\right)t_1\right) \quad (9),$$

where we have abbreviated

$$h_1 = \cos^2\theta - \sin^2\theta \sin^2\psi \,, \quad h_2 = \cos^2\theta - \sin^2\theta\cos^2\psi \quad (10).$$

We may now comment briefly on the origin of the non-analytic behavior of $f_d(T)$. The functions h_1 and h_2 have a common zero at $\cos\theta_o = \pm 3^{-\frac{1}{2}}$, $\psi_o = \pm\pi/4$, and when $f_d(T)$ is differentiated, successive powers of h_2 are brought into the integrand. We will see in the next section that if $f(\vec{t})$ depends only on $\vec{t}^{\,2}$, it is only the divergence at the point (θ_o, ψ_o) that causes $f_d'(T)$ to be discontinuous at $T = 0$. (From now on, a prime indicates d/dT or $d/d\vec{t}^{\,2}$.)

(See Note 2 at end of paper.)

III. SIMPLE EXAMPLES: DEPENDENCE ONLY ON $\vec{t}^{\,2}$

In this section we present some actual examples of $f_d(T)$, assuming that $f(\vec{t})$ depends only on $\vec{t}^{\,2}$. A discussion of the most general case, in which dependence on $Tr(T^s)^3$ is also permitted, is postponed to Section IV. We study the general discontinuity problem for $\vec{t}^{\,2}$ dependence, and then exhibit the discontinuity graphically for two special forms of $f(\vec{t})$, namely, the normal distribution and the uniform distribution.

Also, we discuss two counterexamples to the singularity, one a special type of functional dependence on $\vec{t}^{\,2}$, and another a case where the incompressibility restriction is relaxed.

We now assume the functional form $f(\vec{t}) = G(\vec{t}^{\,2})$, and evaluate the density $f_d(T)$ by means of Eq.(9). One may obtain, first,

$$\vec{t}^{\,2} = t_1^2 + (h_1 h_2^{-1} t_1 + T h_2^{-1})^2 + [T h_2^{-1} + (h_1 h_2^{-1} - 1)t_1]^2$$

$$= P t_1^2 + Q t_1 T + 2T^2 h_2^{-2} \quad (11),$$

where

$$P = 2(1 - h_2^{-1}h_1 + h_2^{-2}h_1^2) \,, \quad Q = (4h_1 - 2h_2)h_2^{-2} \quad (12).$$

One may change variables in Eq.(9) to $z = P^{\frac{1}{2}}t_1 + QT/(2P^{\frac{1}{2}})$, and after some manipulation find, with $f(\vec{t}) = G(\vec{t}^2)$,

$$f_d(T) = \int \frac{dz \sin\theta\, d\theta\, d\psi}{4\pi(2\Gamma)^{\frac{1}{2}}} G\left(z^2 + \frac{3T^2}{2\Gamma}\right) \quad (13),$$

where

$$\Gamma(\theta,\psi) = \tfrac{1}{2}h_2^2 P = 1 - 3\sin^2\theta\cos^2\theta - 3\sin^4\theta\sin^2\psi\cos^2\psi \quad (14).$$

Let us now examine more closely the possible non-analyticity, noting that $\Gamma(\theta_o,\psi_o) = 0$, where, again, $\cos\theta_o = \pm 3^{-\frac{1}{2}}$ and $\psi_o = \pm\pi/4$, but noting further that everywhere else, $\Gamma(\theta,\psi) > 0$. We see that $f_d(0)$ has a divergent integrand, but one can show that the integral exists. As for $f_d'(0)$, we write

$$f_d'(0) = \lim_{T\to 0} 3T \int \frac{dz \sin\theta\, d\theta\, d\psi}{4\pi(2\Gamma^3)^{\frac{1}{2}}} G'\left(z^2 + \frac{3T^2}{2\Gamma}\right) \quad (15).$$

Now, this limit might be double-valued, depending on whether $T \to 0$ through negative or positive values, but in general it is not possible to show this. However, if $f_d'(0)$ is non-zero, then it certainly is double-valued (or possibly infinite) since the integrand is symmetric in T. One may show that the integral $\int \sin\theta\,\Gamma^{-3/2} d\theta d\psi$ diverges, which fact suggests indeed a discontinuity in $f_d'(T)$.

However, it is necessary to resort to special cases in order to exhibit this property. The first case which we treat is the normal distribution, with $G(\vec{t}^2) = 3^{\frac{1}{2}}2(\pi\gamma^2)^{-1}\exp(-2\vec{t}^2/\gamma^2)$, where γ is a constant. Substitution into Eq.(13) and integration with respect to z yields the result

$$f_d(T) = \left(\frac{3}{\pi}\right)^{\frac{1}{2}} \int \frac{\sin\theta\, d\theta\, d\psi}{4\pi\gamma\,\Gamma^{\frac{1}{2}}} \exp\left(-\frac{3T^2}{\gamma^2\Gamma}\right) \quad (16).$$

The second special case is the uniform distribution, where $G(\vec{t}^2) = 3^{\frac{1}{2}}/(\pi\gamma^2)$ for $\vec{t}^2 < \gamma^2$, and $G(\vec{t}^2) = 0$, otherwise. We obtain

$$f_d(T) = \frac{3}{4\pi^2\gamma}\int \frac{\sin\theta\, d\theta\, d\psi}{\Gamma^{\frac{1}{2}}}\left(\frac{2}{3} - \frac{T^2}{\gamma^2\Gamma}\right)^{\frac{1}{2}} \qquad (17),$$

where the integral is restricted to a domain of integration such that the argument of the square root is positive.

In figure 1, we have plotted the functions given by Eqs.(16) and (17), with $\gamma^2 = 6$ for Eq.(16), and $\gamma^2 = 3/2$ for Eq.(17). These functions were evaluated by numerical integration on the University of Arizona CDC 6400 computer. The integrations were repeated with different mesh sizes and positions in order to reveal any instabilities in the numerical schemes. No instabilities were observed, and one must conclude that the discontinuities in $f_d'(T)$ at the origin shown in Figure 1 are real.

We now show a special form of $G(\vec{t}^2)$ for which the discontinuity in f_d' disappears. Let us take any function G which can be written in the form $G(\vec{t}^2,\gamma) = \gamma^{-2}g(\vec{t}^2/\gamma^2)$, such as the above normal and uniform densities. Further, suppose that G is a non-increasing function of t^2, so that the combined function

$$G^c(t^2) = [g(\vec{t}^2/\gamma_2^2) - g(\vec{t}^2/\gamma_1^2)]/(\gamma_2^2-\gamma_1^2)$$

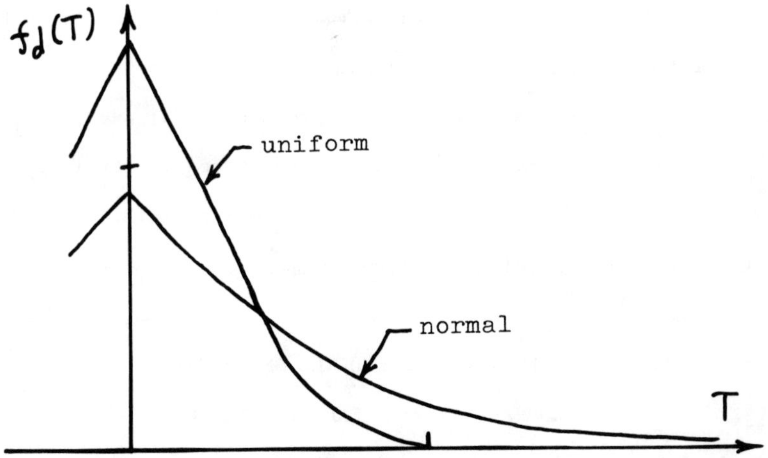

Figure 1. The densities $f_d(T)$ given by Eq.(16) (normal distribution, $\gamma^2 = 6$), and Eq.(17) (uniform distribution, $\gamma^2 = 3/2$).

is properly normalized, non-negative, and vanishes at $\vec{t}^2 = 0$. Then we show that the f_d^c derived from this density has $f_d^{c'}(0) = 0$.

Let $f_d(T,\gamma)$ be the density derived from $G(\vec{t}^2,\gamma)$. Then we have

$$f_d^c(T) = [\gamma_2^2 f_d(T,\gamma_2) - \gamma_1^2 f_d(T,\gamma_1)]/(\gamma_2^2-\gamma_1^2)$$

$$= [\gamma_2 \bar{g}(T^2/\gamma_2^2) - \gamma_1 \bar{g}(T^2/\gamma_1^2)]/(\gamma_2^2-\gamma_1^2),$$

where

$$\bar{g}(T^2/\gamma^2) = \int dz' \frac{\sin\theta\, d\theta\, d\psi}{4\pi(2\Gamma)^{\frac{1}{2}}}\, g\left(z'^2 + \frac{3T^2}{2\gamma^2\Gamma}\right),$$

with $z' = z/\gamma$.

Differentiating the above equation yields

$$\lim_{T\to 0} f_d^{c'}(T) = 2\lim_{T\to 0}\left[\frac{T}{\gamma_2}\bar{g}'\left(\frac{T^2}{\gamma_2^2}\right) - \frac{T}{\gamma_1}\bar{g}'\left(\frac{T^2}{\gamma_1^2}\right)\right]/(\gamma_2^2-\gamma_1^2)$$

$$= 2\lim_{\epsilon\to 0}\left[\epsilon\bar{g}'(\epsilon) - \epsilon\bar{g}'(\epsilon)\right] = 0,$$

independent of the sign of ε, even if $f_d'(T,\gamma)$ is double-valued at $T = 0$.

We also remark that one may relax the incompressibility assumption by dropping the δ function from Eq.(3). The author has evaluated $f_d(T)$ using the normal density for $G(\vec{t}^2)$ and replacing the δ function by the expression $(a/\pi)^{\frac{1}{2}}\exp[-a(t_1+t_2+t_3)^2]$, where a is a stiffness parameter. In this case, the discontinuity in f_d' disappears, but is recovered again in the limit $a \to \infty$. Thus, incompressibility would seem to be an essential ingredient for the discontinuity.

IV. GENERAL RESULTS: DEPENDENCE ON $Tr(T^S)^3$ AND COMPARISON WITH EXPERIMENT

Figure 2 shows some experimental results for $f_d(T)$ found by A. A. Townsend (Batchelor, 1967). We note that this experimental $f_d(T)$ does not appear to be symmetric in T, and thus it is not realistic to assume that G depends only on $\vec{t}^2$.

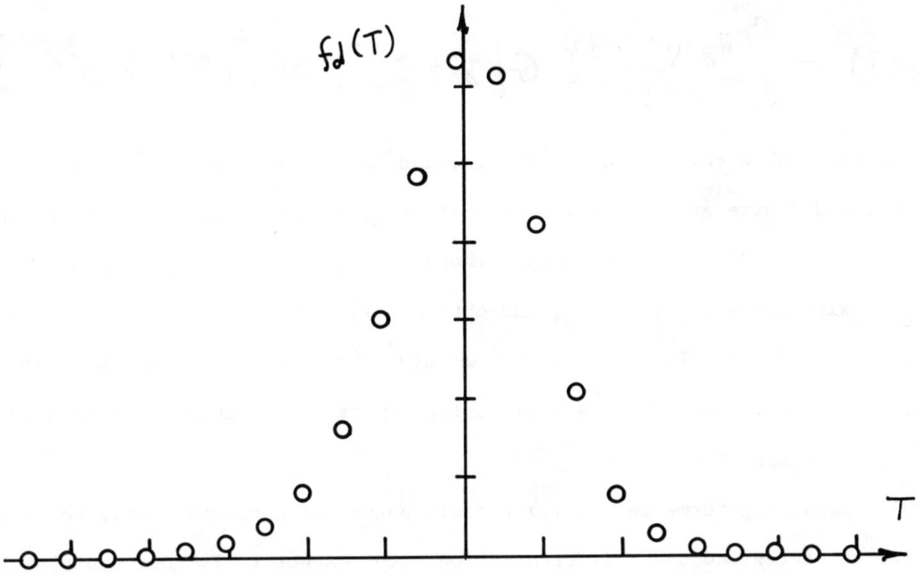

Figure 2. Experimental values of $f_d(T)$ by Townsend. (Adapted from Batchelor, 1967.)

Unfortunately, the data does not possess sufficient resolution for us to be able to say whether or not a discontinuity in f_d', or perhaps even a cusp or discontinuity in f_d itself, is present. However, the experimental data is more highly peaked than a normal density of the same dispersion (Batchelor, 1967).

To investigate the dependence on $Tr(T^s)^3$, it is expedient to evaluate this invariant by the same method used for $\vec{t}^2$ in the previous section. The same substitutions which lead to Eqs.(9) and (11) may be employed, with the assistance of Eq.(5) and the change of variable to $z = P^{\frac{1}{2}}t_1 + QT/(2P)^{\frac{1}{2}}$, to obtain

$$Tr(T^s)^3 = -3t_1t_2(t_1+t_2)$$

$$= 3P^{-\frac{1}{2}}(P^{-1}-\tfrac{1}{2})z^3 + \frac{3Q}{4P^2}(P-6)z^2T + \frac{3}{P^{\frac{1}{2}}}\left(\frac{3Q^2}{4P^2}+\frac{Q^2}{8P}-\frac{1}{h_2^2}\right)zT^2$$

$$+ \frac{3Q}{16P}\left(\frac{8}{h_2^2}-\frac{2Q^2}{P^2}-\frac{Q^2}{P}\right)T^3 \qquad (18).$$

Thus, Eq.(13) is generalized to read

$$f_d(T) = \int \frac{dz \sin\theta d\theta d\psi}{4\pi (2\Gamma)^{\frac{1}{2}}} G\left[z^2 + \frac{3T^2}{2\Gamma}, 3P^{-\frac{1}{2}}(P^{-1} - \tfrac{1}{2})z^3 + \cdots\right].$$

We first show that $P > 0$ everywhere, and that therefore $Tr(T^s)^3$ cannot diverge for $T = 0$ and finite z: Since $\Gamma = 0$ only at (θ_o, ψ_o) and h_2 is non-infinite, we need only to show that $P > 0$ in some neighborhood of (θ_o, ψ_o). We expand h_1 and h_2 about (θ_o, ψ_o), using $x = \theta - \theta_o$, $y = \psi - \psi_o$, and obtain $\Gamma(x,y) \approx 4(x^2+y^2)$, $h_2(x,y) \approx -2^{\frac{1}{2}}x + 4/3y$. Then, $P = 2\Gamma/h_2{}^2 \approx 8(x^2+y^2)/(2^{\frac{1}{2}}x - 4/3y)^2$, which is always greater than zero. Therefore, any discontinuity at $T = 0$ cannot be damped by $Tr(T^s)^3 \to \infty$ causing $G \to 0$.

The remaining terms in Eq.(18) contain powers of Q/P, which will be brought out into the integrand upon differentiation with respect to T. One can show that $Q/P \approx (-2^{-\frac{1}{2}}x - 4/3y)/(x^2+y^2)$, which can certainly be made to diverge at (0,0). Since Q/P occurs in the coefficient of z^2T in Eq.(18), differentiation of f_d would result in a term in the integrand proportional to $Q/(\Gamma^{\frac{1}{2}}P)$, which would cause a divergence in $f_d'(0)$. We would have

$$f_d'(0) = \int \frac{dz \sin\theta d\theta d\psi}{4\pi (2\Gamma)^{\frac{1}{2}}} \cdot \frac{3Q}{4P^2}(P-6)z^2 \dot{G}[z^2, 3P^{-\frac{1}{2}}(P^{-1} - \tfrac{1}{2})z^3]$$

+ contribution from Eq.(15),

where the dot indicates the derivative with respect to $Tr(T^s)^3$. It can be shown that this integral diverges logarithmically about (θ_o, ψ_o), and thus we might be faced with an $f_d(T)$ with a cusp at $T = 0$, or possibly $f_d(T)$ might itself be discontinuous there.

However, there seems to be no way to tell a priori what the case would be for real physical turbulence, and thus these conclusions must stand or fall with further experimental work. In any case, it might be possible to make some further theoretical progress which would enable us to understand more clearly what the presence or absence of such a discontinuity or singularity would mean.

REFERENCES

G. K. Batchelor, Theory of Homogeneous Turbulence, Cambridge University Press, p. 172 (1967).

W. J. Cocke, Phys. Fluids, 12, 2488 (1969).

H. Goldstein, Classical Mechanics, Addison-Wesley Publishing Company, p. 107 (1957).

NOTE 1.

The paper presented by C. Van Atta contains high-resolution data for which $f_d(T)$ has a relatively sharp peak (compared with a Gaussian) at the origin. However, the quantity actually investigated by Van Atta is $\langle u_3 \rangle^{-1} \partial U_3/\partial t$, which is similar to T^s_{33}, but is not quite the same.

NOTE 2.

In the previous paper (Cocke, 1969) it was demonstrated that if $f^s(T^s)$ is the joint density for T^s, then

$$f^s(T^s) = \bar{f}(\vec{t}, \theta, \psi, \phi)/|(t_1 - t_2)(t_2 - t_3)(t_3 - t_1)|$$

Since $f_d(T)$ can be derived from $f^s(T^s)$ by simply integrating out all the other components, one might suspect that any singularity in $f_d(T)$ would be due to singularities in the above equation. However, this is not the case, because the vanishing of Γ at θ_o, ψ_o has nothing to do with the zeros of the above denominator, which is a function of t only.

Actually, the cause of the non-analyticity of $f_d(T)$ is easily seen by noting that from Eq. (6) we have $T(\theta_o, \psi_o) = t_1 + t_2 + t_3 = 0$ from incompressibility. Thus for any given $T \neq 0$ in Eq. (9), it happens that when $\theta, \psi \to \theta_o, \psi_o$ in the integrand we must have $\vec{t} \to \infty$ to keep $T \neq 0$. However, if $T = 0$ it is not necessary that $\vec{t} \to \infty$. Thus the behavior of the integrand of $f_d(T)$ changes radically as $T \to 0$.

DYNAMO INSTABILITY AND FEEDBACK IN A STOCHASTICALLY DRIVEN SYSTEM

By H.K. Moffatt
Dept. of Applied Mathematics and Theoretical Physics
University of Cambridge, England

ABSTRACT

In §1, a method of treatment of the equation

$$\frac{\partial y_i}{\partial t} = \frac{\partial}{\partial x_j}\left(a_{ijk}(\underset{\sim}{x},t)\, y_k\right) + \lambda \nabla^2 y_i ,$$

where $a_{ijk}(\underset{\sim}{x},t)$ is a random tensor field of known statistical properties is reviewed, with particular reference (i) to the magnetohydrodynamic turbulent dynamo problem, and (ii) to the problem of diffusion of a passive scalar field by turbulent motion. The result of Steenbeck, Krause and Radler (1966), that in the former context dynamo action can occur (i.e. the ensemble average of $\underset{\sim}{y}$ can grow without limit) provided the statistics of the turbulent field lack reflexional symmetry, is discussed within the framework of the above general equation.

In §§ 2 and 3, the feedback mechanism in the magnetohydrodynamic context is considered. It is supposed that a velocity field lacking reflexional symmetry is generated in an electrically conducting fluid by a random body force of known statistical properties. Conditions are then conducive to the growth of large scale magnetic field perturbations. The growth is limited by the fact that the growing Lorentz force progressively modifies the statistical structure of the velocity field, until ultimately a statistical equilibrium is achieved. It is shown that in this equilibrium the magnetic energy density may exceed the kinetic energy density by a factor $O(L/\ell) \gg 1$, where L is the scale of the magnetic field, and ℓ the scale of the turbulence.

1. INSTABILITIES OF THE LINEAR DIFFUSION EQUATION WITH A RANDOM CONVECTIVE TERM

Let $a_{ijk}(\underset{\sim}{x},t)$ be a random tensor field, statistically homogeneous in $\underset{\sim}{x}$ and stationary in t, with $\langle a_{ijk}\rangle = 0$, the angular brackets representing an ensemble average. We shall consider the evolution of the random vector field $y_i(\underset{\sim}{x},t)$ satisfying the equation

$$\frac{\partial y_i}{\partial t} = \frac{\partial}{\partial x_j}\left(a_{ijk}\, y_k\right) + \lambda \nabla^2 y_i \tag{1.1}$$

and the initial condition $\underset{\sim}{y}(\underset{\sim}{x},0) = \underset{\sim}{y}_0(\underset{\sim}{x})$ where $\int \underset{\sim}{y}_0^2\, d\underset{\sim}{x} < \infty$.

Two particular choices of a_{ijk} correspond to well-known problems in the turbulence context. First, if

$$a_{ijk} = \epsilon_{ijl}\,\epsilon_{lmk}\, u_m(\underset{\sim}{x},t), \tag{1.2}$$

where $\underset{\sim}{u}$ is a (turbulent) velocity field, then with $\underset{\sim}{y} = \underset{\sim}{B}$, (1.1) becomes the induction equation of magnetohydrodynamics,

$$\frac{\partial \underset{\sim}{B}}{\partial t} = \nabla \wedge (\underset{\sim}{u} \wedge \underset{\sim}{B}) + \lambda \nabla^2 \underset{\sim}{B} \,. \qquad (1.3)$$

In this context, it has been established by Steenbeck, Krause and Radler (1966) (hereafter SKR) that a sufficient condition for unlimited growth of $\langle \underset{\sim}{B}^2 \rangle$ (i.e. for turbulent dynamo action) is that the statistical properties of the $\underset{\sim}{u}$-field should lack reflexional symmetry, i.e. should not be invariant under a change from a left-handed to a right-handed frame of reference. This situation is likely to occur whenever the external conditions lack reflexional symmetry, e.g. in a rotating fluid with a mean density gradient or a mean energy flux parallel to the rotation vector. The theory of SKR has crucial importance in geomagnetism and in cosmical electrodynamics, and a range of applications and elaborations have already been worked out. Two recent reviews of dynamo theory (Roberts 1971, Weiss 1971) between them give a comprehensive list of references, particularly to developments since 1966, which are largely the outcome of the SKR breakthrough.

Secondly, if

$$a_{ijk} = -\delta_{ij}\, u_k(\underset{\sim}{x}, t), \qquad (1.4)$$

then, with $\underset{\sim}{y} = \underset{\sim}{G} = \nabla\theta$, (1.1) becomes the gradient of the equation

$$\frac{\partial \theta}{\partial t} = -\underset{\sim}{u}\cdot\nabla\theta + \lambda \nabla^2 \theta \,, \qquad (1.5)$$

describing the convection and diffusion of a passive scalar contaminant. In this context, the effect of the turbulence on a localised θ-field is equivalent to that of an eddy diffusivity; the expression for this diffusivity was obtained by Taylor (1921) for $\lambda = 0$, and the first correction (for small λ) by Saffman (1962). These results may be recovered by the method described below, and in addition a simple expression for the limiting form of the eddy diffusivity for large λ may be obtained.

The method of treatment of (1.1) is as follows:

Let $\underset{\sim}{Y}(\underset{\sim}{x},t) = \langle \underset{\sim}{y}(\underset{\sim}{x},t)\rangle$ and let $\underset{\sim}{y}' = \underset{\sim}{y} - \underset{\sim}{Y}$; then from (1.1),

$$\frac{\partial Y_i}{\partial t} = \frac{\partial}{\partial x_j} E_{ij} + \lambda \nabla^2 Y_i \,, \tag{1.6}$$

where $E_{ij} = \langle a_{ijk} y_k' \rangle$. Subtracting this from (1.1),

$$\frac{\partial y_i'}{\partial t} = \frac{\partial}{\partial x_j}(a_{ijk} Y_k) + \frac{\partial}{\partial x_j}\left(a_{ijk} y_k' - \langle a_{ijk} y_k' \rangle\right) + \lambda \nabla^2 y_i' \,. \tag{1.7}$$

Equation (1.7), with the initial condition $y_i' = 0$, establishes a linear relation between y_i' and Y_k, say

$$y_i'(\underset{\sim}{x}, t) = \int K_{ij}(\underset{\sim}{x}, t; \underset{\sim}{x}', t')\, Y_j(\underset{\sim}{x}', t')\, d^3\underset{\sim}{x}'\, dt', \tag{1.8}$$

where K_{ij} is a functional of a_{ijk} and of the parameter λ. The range of integration is over all $\underset{\sim}{x}'$ and over $0 \leqslant t' \leqslant t$.

We can now construct the tensor E_{ij} needed in (1.6), viz.,

$$E_{ij}(\underset{\sim}{x}, t) = \int V_{ijk}(\underset{\sim}{\xi}, \tau)\, Y_k(\underset{\sim}{x}', t')\, d^3\underset{\sim}{\xi}\, d\tau \tag{1.9}$$

where $\underset{\sim}{\xi} = \underset{\sim}{x}' - \underset{\sim}{x}$, $\tau = t' - t$, and

$$V_{ijk}(\underset{\sim}{\xi}, \tau) = \langle a_{ijm}(\underset{\sim}{x}, t)\, K_{mk}(\underset{\sim}{x}, t; \underset{\sim}{x}' t') \rangle \,. \tag{1.10}$$

V_{ijk} depends only on $\underset{\sim}{\xi}$ and τ because of the assumed homogeneity and stationarity of a_{ijk}; it also depends implicitly on λ. We may certainly suppose that

$$V_{ijk}(\underset{\sim}{\xi}, \tau) \to 0 \quad \text{as} \quad |\underset{\sim}{\xi}| \to \infty \quad \text{and as} \quad \tau \to -\infty, \tag{1.11}$$

but the scales ℓ, t_o characteristic of this tensor (which could for example be defined in terms of its moments) may depend on λ as well as on the statistical properties of a_{ijk}.

Let us now consider solutions $\underset{\sim}{Y}(\underset{\sim}{x}, t)$ of (1.6) having scales L, T satisfying

$$L \gg \ell \,, \quad T \gg t_o \,.$$

Then in (1.9) we may expand $Y_k(\underset{\sim}{x}', t')$ in Taylor series about $(\underset{\sim}{x}, t)$, and integrate term by term, obtaining

$$E_{ij} = A^{(0)}_{ijk} Y_k + A^{(1)}_{ijkl} \frac{\partial Y_k}{\partial x_l} + \dots + B^{(1)}_{ijk} \frac{\partial Y_k}{\partial t} + \dots, \tag{1.12}$$

where

$$A^{(0)}_{ijk} = \int V_{ijk}\, d^3\xi\, d\tau, \quad A^{(1)}_{ijkl} = \int \xi_l V_{ijk}\, d^3\xi\, d\tau, \quad B^{(1)}_{ijk} = \int \tau V_{ijk}\, d^3\xi\, d\tau, \text{ etc.} \tag{1.13}$$

The range of integration in these moments is over all ξ, and over $-t \leqslant \tau \leqslant 0$; for $t \gg t_0$, the time range is effectively $-\infty < \tau \leqslant 0$, and the tensors defined by (1.13) are then simply constant tensors, determined solely by the statistical properties of a_{ijk} and by λ.

The expansion (1.12) may be rearranged, using (1.6) repeatedly, so that only space-derivatives appear; in this way we obtain

$$E_{ij} = A^{(0)}_{ijk} Y_k + A^{(1)}_{ijkl} \frac{\partial Y_k}{\partial x_l} + \dots, \tag{1.14}$$

where, in fact, $A^{(1)}_{ijkl} = \tilde{A}^{(1)}_{ijkl} + B^{(1)}_{ijm} A^{(0)}_{mlk}$, and higher coefficients may be similarly obtained. Equation (1.6) then becomes an equation with constant coefficients

$$\frac{\partial Y_i}{\partial t} = A^{(0)}_{ijk} \frac{\partial Y_k}{\partial x_j} + A^{(1)}_{ijkl} \frac{\partial^2 Y_k}{\partial x_j \partial x_l} + \dots + \lambda \nabla^2 Y_i. \tag{1.15}$$

The term involving $A^{(1)}_{ijkl}$ may be recognized as describing an 'eddy diffusion' effect which certainly appears in both the particular cases given by (1.2) and (1.4). The term involving $A^{(0)}_{ijk}$ is less familiar. It appears in the magnetohydrodynamic case (1.2), and is described in that context (for notational reasons) as the 'α-effect' by Steenbeck et al (1966)*; (would 'SKR-effect' or 'helicity effect' - see Moffatt 1970a - not be more apt?). It does not appear in the second situation described by (1.4), the reason being essentially that the kernel function V_{ijk} in (1.10) is in this case expressible as a gradient and so the

*Footnote: E.N. Parker anticipated the SKR-effect through inspired physical reasoning (Parker 1955); he used the symbol Γ where SKR use α !

expression for $A^{(0)}_{ijk}$ given by (1.13) vanishes.

The case when $a_{ijk}(\underset{\sim}{x},t)$ is statistically isotropic

Suppose now that $a_{ijk}(\underset{\sim}{x},t)$ is statistically isotropic, i.e. that its statistical properties are invariant under rotations of the frame of reference. Then, in particular, the tensors $A^{(m)}_{ijk\ldots}$ are isotropic, i.e.

$$A^{(0)}_{ijk} = \alpha \epsilon_{ijk} , \tag{1.16}$$

$$A^{(1)}_{ijkl} = \beta_1 \delta_{ij} \delta_{kl} + \beta_2 \delta_{ik} \delta_{jl} + \beta_3 \delta_{il} \delta_{jk} , \tag{1.17}$$

etc.

If the statistical properties of $a_{ijk}(\underset{\sim}{x},t)$ are invariant under reflexions of the frame of reference as well as rotations, then $A^{(m)}_{ijk\ldots} = 0$ when m is even, and in particular $\alpha = 0$. Only if the random field lacks reflexional symmetry can we have $\alpha \neq 0$. Note that α is a pseudo-scalar*, while β_1 , β_2 and β_3 are pure scalars.

When $\alpha \neq 0$, the first term on the right of (1.15) clearly dominates the development of $\underset{\sim}{Y}(\underset{\sim}{x},t)$ provided the scale L is sufficiently large, and we then have simply

$$\frac{\partial \underset{\sim}{Y}}{\partial t} = \alpha \nabla \wedge \underset{\sim}{Y} \qquad + \text{ negligible terms.} \tag{1.18}$$

Any mode for which

$$\nabla \wedge \underset{\sim}{Y} = (\operatorname{sgn} \alpha) K \underset{\sim}{Y} \tag{1.19}$$

then grows exponentially according to

$$\underset{\sim}{Y}(\underset{\sim}{x},t) = \underset{\sim}{Y}_0(\underset{\sim}{x}) e^{|\alpha| K t} , \tag{1.20}$$

if K is sufficiently small. For example, the Fourier components

$$\underset{\sim}{Y}_0(\underset{\sim}{x}) = \tilde{Y}_0(K)(\cos Kz, \mp \sin Kz, 0) \tag{1.21}$$

satisfy $\nabla \wedge \underset{\sim}{Y}_0 = \pm K \underset{\sim}{Y}_0$, and one of these Fourier components will grow exponentially if it is represented in the Fourier transform of the initial field $\underset{\sim}{y}_0(\underset{\sim}{x})$.

---

* Cf the mean helicity $\langle \underset{\sim}{u} \cdot (\nabla \wedge \underset{\sim}{u}) \rangle$.

In the reflexionally symmetric case, $\alpha = 0$, and (1.15) becomes

$$\frac{\partial \underset{\sim}{Y}}{\partial t} = (\beta_1 + \beta_3)\nabla\nabla\cdot\underset{\sim}{Y} + (\beta_2 + \lambda)\nabla^2\underset{\sim}{Y} \quad \text{+ negligible terms.} \tag{1.22}$$

In the first example, with $\underset{\sim}{Y} = \langle\underset{\sim}{B}\rangle$, $\nabla\cdot\underset{\sim}{Y} = 0$, and we simply have an 'eddy diffusivity'

$$\lambda_t^{(1)} = \beta_2 . \tag{1.23}$$

In the second example, with $\underset{\sim}{Y} = \langle\nabla\theta\rangle$, $\nabla\wedge(\nabla\wedge\underset{\sim}{Y}) = 0$ so that $\nabla\nabla\cdot\underset{\sim}{Y} = \nabla^2\underset{\sim}{Y}$, and we have an eddy diffusivity

$$\lambda_t^{(2)} = \beta_1 + \beta_2 + \beta_3 . \tag{1.24}$$

One would of course expect these diffusivities to turn out to be positive in real physical situations, and they do turn out to be positive in every limiting case amenable to analysis. However no general proof that they must invariably be positive appears to be available, and there remains the intriguing possibility that in certain circumstances the (net) diffusivity may be negative with interesting inverse diffusion consequencies.

<u>The case when $a_{ijk}(\underset{\sim}{x},t)$ is statistically anisotropic but reflexionally symmetric</u>

In these circumstances again, $A_{ijk}^{(0)}$ need not vanish. We might for example have a situation where

$$A_{ijk}^{(0)} = \lambda_i\,\delta_{jk} + \mu_j\,\delta_{ki} + \nu_k\,\delta_{ij} , \tag{1.25}$$

where $\underset{\sim}{\lambda}$, $\underset{\sim}{\mu}$ and $\underset{\sim}{\nu}$ are any three vectors. Putting

$$\underset{\sim}{Y}(\underset{\sim}{x},t) = \tilde{\underset{\sim}{Y}}(\underset{\sim}{K})\,e^{i\underset{\sim}{K}\cdot\underset{\sim}{x} + mt} , \tag{1.26}$$

in (1.15), and retaining only the O(K) term on the right-hand side, we then have

$$m\tilde{Y}_i = i\left(\lambda_i K_k + (\underset{\sim}{\mu}\cdot\underset{\sim}{K})\,\delta_{ki} + \nu_k K_i\right)\tilde{Y}_k . \tag{1.27}$$

The determinantal condition for a non-trivial solution gives a cubic equation for m with roots

$$m_1 = i\underset{\sim}{K}\cdot\underset{\sim}{\mu} \; , \quad m_{2,3} = -i\underset{\sim}{K}\cdot\underset{\sim}{\mu} \pm \sqrt{(\underset{\sim}{K}\cdot\underset{\sim}{\lambda})(\underset{\sim}{K}\cdot\underset{\sim}{\nu}) - K^2\underset{\sim}{\lambda}\cdot\underset{\sim}{\nu}} \, . \tag{1.28}$$

Hence we have an instability when K is in a direction satisfying

$$(\underset{\sim}{K}\cdot\underset{\sim}{\lambda})(\underset{\sim}{K}\cdot\underset{\sim}{\nu}) > K^2\underset{\sim}{\lambda}\cdot\underset{\sim}{\nu} \, . \tag{1.29}$$

[This situation cannot arise in the magnetohydrodynamic context, in which, because of the particular structure (1.2) of a_{ijk}, A_{ijk} is antisymmetric in i and j; the particular form (1.25) would then require $\underset{\sim}{\nu} = 0$ and $\underset{\sim}{\mu} = -\underset{\sim}{\lambda}$; the three roots $m_{1,2,3}$ are imaginary and there can be no question of instability.]

2. THE ESTABLISHMENT OF A NON-LINEAR EQUILIBRIUM

It has been shown in §1 that when the statistical properties of the a_{ijk}-field lack reflexional symmetry, equation (1.1) leads to exponential growth of Fourier components of the $\underset{\sim}{Y}$-field of sufficiently small wave-number. In a real physical situation, the growth must ultimately be limited by some back-reaction which modifies the statistical properties of the a_{ijk}-field. In the magnetohydrodynamic context, this back reaction is provided by the Lorentz force; the effect has been analysed when the $\underset{\sim}{u}$-field consists of a random field of decaying inertial waves in a rotating fluid (Moffatt 1970b). It has been further discussed in general terms by Roberts (1971). We shall here examine the magnetohydrodynamic situation, on the assumption that a body force $\underset{\sim}{f}(\underset{\sim}{x},t)$ of known statistical properties is applied to the fluid in order to generate the velocity field from which the magnetic field may be nourished and sustained. We assume that $\langle \underset{\sim}{f} \rangle = 0$. The equations governing the velocity field $\underset{\sim}{u}(\underset{\sim}{x},t)$ and the Alfvén velocity field $\underset{\sim}{v}(\underset{\sim}{x},t) = (\mu\rho)^{\frac{1}{2}} \underset{\sim}{B}(\underset{\sim}{x},t)$ are

$$\frac{\partial \underset{\sim}{u}}{\partial t} + \underset{\sim}{u}\cdot\nabla\underset{\sim}{u} = -\nabla\chi + \underset{\sim}{v}\cdot\nabla\underset{\sim}{v} + \nu\nabla^2\underset{\sim}{u} + \underset{\sim}{f} \, , \tag{2.1}$$

$$\frac{\partial \boldsymbol{v}}{\partial t} + \boldsymbol{u}.\nabla \boldsymbol{v} = \boldsymbol{v}.\nabla \boldsymbol{u} + \lambda \nabla^2 \boldsymbol{v} , \tag{2.2}$$

and

$$\nabla . \boldsymbol{u} = \nabla . \boldsymbol{v} = 0 . \tag{2.3}$$

As before, let $\boldsymbol{V}(\boldsymbol{x},t) = \langle \boldsymbol{v} \rangle$ and suppose that the scale L of $\boldsymbol{V}$ is large compared with the scale ℓ of any relevant statistical property of the $\boldsymbol{u}$-field generated. The equation satisfied by $\boldsymbol{V}$ is

$$\frac{\partial \boldsymbol{V}}{\partial t} = \nabla \wedge \langle \boldsymbol{u} \wedge \boldsymbol{h} \rangle + \lambda \nabla^2 \boldsymbol{V} , \tag{2.4}$$

where $\boldsymbol{h} = \boldsymbol{v} - \boldsymbol{V}$, and the main objective is to obtain an expression for $\langle \boldsymbol{u} \wedge \boldsymbol{h} \rangle$ in terms of $\boldsymbol{V}$ so that (2.4) may be integrated (see the expressions (2.6) and (2.13) below).

Assuming that $\langle \boldsymbol{u} \rangle = 0$ (and this is something that will require further consideration in retrospect), the equations for $\boldsymbol{u}$ and $\boldsymbol{h}$ are

$$\left.\begin{aligned} \frac{\partial \boldsymbol{u}}{\partial t} &= -\nabla \chi + \boldsymbol{V}.\nabla \boldsymbol{h} + \boldsymbol{h}.\nabla \boldsymbol{V} + \nu \nabla^2 \boldsymbol{u} + \boldsymbol{f} + N.L. \\ \frac{\partial \boldsymbol{h}}{\partial t} &= \boldsymbol{V}.\nabla \boldsymbol{u} - \boldsymbol{u}.\nabla \boldsymbol{V} + \lambda \nabla^2 \boldsymbol{h} + N.L. \end{aligned}\right\} \tag{2.5}$$

where N.L. indicates terms quadratic in the fluctuating quantities $\boldsymbol{u}$ and $\boldsymbol{h}$. We shall suppose that the forcing field $\boldsymbol{f}$ is sufficiently weak and the field $\boldsymbol{V}$ (which grows exponentially on the linear analysis of §1) sufficiently strong, for these non-linear terms to be negligible. We may also neglect the terms $\boldsymbol{h}.\nabla \boldsymbol{V}$ and $\boldsymbol{u}.\nabla \boldsymbol{V}$ on account of the slow variation of the field $\boldsymbol{V}$. We are then left with two coupled linear equations describing forced Alfvén waves in an effectively uniform field $\boldsymbol{V}$.

We can now Fourier transform these equations and construct the convolution $\langle \boldsymbol{u} \wedge \boldsymbol{h} \rangle$ in terms of the spectrum tensor $F_{ij}(\boldsymbol{k},\omega)$ of $\boldsymbol{f}$; omitting the details, which are straightforward, we obtain

$$\langle \boldsymbol{u} \wedge \boldsymbol{h} \rangle_i = - \int \frac{\lambda k^2 (\boldsymbol{k}.\boldsymbol{V}) [i \epsilon_{ijk} F_{jk}(\boldsymbol{k},\omega)}{(\lambda^2 k^4 + \omega^2) |\sigma|^2} \, d^3\boldsymbol{k} \, d\omega , \tag{2.6}$$

where

$$\sigma = -(\omega + i\nu k^2) + (\underset{\sim}{k}.\underset{\sim}{V})^2(\omega + i\lambda k^2)^{-1}. \tag{2.7}$$

The quantity in square brackets under the integral sign is real in view of the Hermitian symmetry condition $F^*_{jk}(\underset{\sim}{k},\omega) = F_{kj}(\underset{\sim}{k},\omega)$ for the spectrum tensor of a real vector field. It is no surprise that the expression is non zero only if $F_{jk}(\underset{\sim}{k},\omega)$ lacks reflexional symmetry. The expression (2.6) is evidently non-linear in $\underset{\sim}{V}$ due to the dependence of σ on $\underset{\sim}{V}$ in the denominator. As $\underset{\sim}{V}$ grows in strength, $|\langle \underset{\sim}{u} \wedge \underset{\sim}{h}\rangle|$ ultimately decreases, and the mechanism responsible for the growth of $\underset{\sim}{V}$ is correspondingly controlled.

Evaluation of the integral (2.6) for given $F_{jk}(\underset{\sim}{k},\omega)$ is in general a very difficult (if not impossible) matter. It may be sufficient for this lecture to consider a choice of $F_{jk}(\underset{\sim}{k},\omega)$ which is very special, but which nevertheless serves to illustrate the type of effect that may be expected in general. Suppose that, locally,

$$\underset{\sim}{V} = (V, 0, 0), \tag{2.8}$$

and suppose that

$$\underset{\sim}{f}(\underset{\sim}{x},t) = \text{Re}\, \underset{\sim}{\hat{f}}\, e^{ik_0(x-Vt)}, \quad \underset{\sim}{\hat{f}} = \frac{1}{\sqrt{2}} f_0(0,1,-i). \tag{2.9}$$

This is a forcing wave at the 'resonance' frequency $\omega = k_oV$, with a helical structure. Note that

$$\underset{\sim}{\tilde{f}} \wedge \underset{\sim}{\tilde{f}}^* = f_0^2(i,0,0), \tag{2.10}$$

and correspondingly,

$$i\epsilon_{ijk}F_{jk}(\underset{\sim}{k},\omega) = -\langle f_0^2\rangle \delta_{i1}\,\delta(\underset{\sim}{k}-\underset{\sim}{k}_0)\,\delta(\omega - k_0V), \tag{2.11}$$

where $\underset{\sim}{k}_o = (k_o, 0, 0)$. We put angular brackets round f_o^2 to indicate that this amplitude may be a random variable with an ensemble average. Suppose further that the dissipative effects are weak, <u>viz</u>. that

$$\nu k_0 \ll V \ , \quad \lambda k_0 \ll V, \tag{2.12}$$

so that, to lowest order, $\sigma = -i(\nu+\lambda)k_o^2$; then, from (2.6),

$$\langle \underset{\sim}{u} \wedge \underset{\sim}{h} \rangle = \frac{\lambda \langle f_0^2 \rangle \underset{\sim}{V}}{k_0^3 (\lambda+\nu)^2 V^2} \ . \tag{2.13}$$

<u>The implications for dynamo theory</u>

Let us now suppose that $\underset{\sim}{V}$ varies slowly, and that the statistical properties of the $\underset{\sim}{f}$-field are such that the formula (2.13) holds everywhere. This is admittedly unrealistic, in that it requires a forcing helicity wave whose wave-vector $\underset{\sim}{k}_o$ is everywhere locally aligned with the mean field $\underset{\sim}{V}$; but it is the only assumption that leads to a reasonably simple treatment of the feedback phenomenon. Equation (2.4) then becomes

$$\frac{\partial \underset{\sim}{V}}{\partial t} = A \nabla \wedge \left(\frac{\underset{\sim}{V}}{V^2} \right) + \lambda \nabla^2 \underset{\sim}{V} \ , \tag{2.14}$$

where $A = \lambda(\lambda+\nu)^{-2} k_0^{-3} \langle f_0^2 \rangle$. This equation admits solutions of the form

$$\underset{\sim}{V} = V_0(t) \left(\cos Kz \, , \, -\sin Kz \, , \, 0 \right) , \tag{2.15}$$

provided $V_o(t)$ satisfies

$$\frac{dV_0}{dt} = \frac{AK}{V_0} - \lambda K^2 V_0 \ . \tag{2.16}$$

For consistency, we require $K \ll k_o$. The particular interest of solutions of this type is that they satisfy the 'force-free' condition (1.19), and are therefore preferentially amplified on the basis of linear theory. According to the non-linear theory of this section, the structure of such modes remains unaltered when the back-reaction is included, but the growth rate is modified.

The solution of (2.16) is given by

$$\frac{1}{2}V_o^2 = \frac{A}{2\lambda} + Ce^{-2\lambda K^2 t}, \tag{2.17}$$

where C is a constant determined by the value of V_o at some time $t = t_1$ beyond which the approximations (2.12) are satisfied. In a time of order $(\lambda K^2)^{-1} = L^2/\lambda$, the magnetic energy density $M(t) = \frac{1}{2}V_o^2$ (there is in this situation negligible energy in the fluctuating magnetic field) asymptotically attains the value

$$M_{ult} = \frac{A}{2\lambda} = \frac{\langle f_o^2\rangle}{2(\lambda+\nu)^2 k_o^3 K} . \tag{2.18}$$

The kinetic energy density is given by

$$E_{ult} = \frac{1}{2}\langle \underset{\sim}{u}^2\rangle = \frac{\langle f_o^2\rangle}{2|\sigma|^2} = \frac{\langle f_o^2\rangle}{4(\lambda+\nu)^2 k_o^4} . \tag{2.19}$$

Hence

$$\frac{M_{ult}}{E_{ult}} = \frac{2k_o}{K} = \frac{2L}{\ell} \gg 1 . \tag{2.20}$$

The magnetic energy in the ultimate steady state is therefore necessarily an order of magnitude greater than the kinetic energy of the background velocity field.

3. DISCUSSION

There are three questions that arise in the course of the above analysis that require special comment: (i) Can we be sure that a mean velocity field $\underset{\sim}{u}_o = \langle \underset{\sim}{u}\rangle$ does not develop as a result of the non-uniform Reynolds stress distribution implied by the non-uniform large-scale magnetic field? (ii) Is the particular assumption (2.11) concerning the spectrum tensor of the forcing field too special for the result to have any significance? (iii) Is it at all realistic to consider a forcing field whose statistical properties lack reflexional symmetry? Let us take these questions in turn.

(i) Generation of a mean velocity field

The mean momentum equation, from (2.1) is

$$\frac{\partial \mathbf{u}_o}{\partial t} + \mathbf{u}_o . \nabla \mathbf{u}_o = -\nabla \chi_o + \mathbf{V} . \nabla \mathbf{V} - \nabla . \langle \mathbf{u}\mathbf{u} - \mathbf{h}\mathbf{h} \rangle + \nu \nabla^2 \mathbf{u}_o , \tag{3.1}$$

so that a mean velocity field will develop if either $\mathbf{V}.\nabla\mathbf{V}$ or $\nabla.\langle \mathbf{uu} - \mathbf{hh} \rangle$ is rotational; if these terms are irrotational, they can be accommodated by a mean (effective) pressure field $\chi_o(\mathbf{x},t) = \langle \chi \rangle$.

For the particular field (2.15) considered above $\mathbf{V}.\nabla\mathbf{V}$ is certainly irrotational; indeed $\mathbf{V}\wedge(\nabla\wedge\mathbf{V}) = 0$ since the growing field is force-free, so that $\mathbf{V}.\nabla\mathbf{V} = \frac{1}{2}\nabla \mathbf{V}^2$. Furthermore $P_{ij} = \langle u_i u_j - h_i h_j \rangle$ is a symmetric tensor axisymmetric about the direction of $\mathbf{V}$, and depending only on z and t, since $\mathbf{V}$ depends only on z and t. Clearly $P_{13} = P_{23} = 0$, since Oz is a possible principal axis of P_{ij} for all z ; hence

$$\frac{\partial}{\partial x_j} P_{ij}(z,t) = \frac{\partial}{\partial z} P_{i3}(z,t) = \frac{\partial}{\partial z} P_{33}(z,t)\,\delta_{i3} = \frac{\partial}{\partial x_i} P_{33}(z,t) ,$$

and so this term is also irrotational. Hence no mean flow will develop, but a mean pressure field periodic in z will be established.

For a more general field than (2.15) consisting of a superposition of such Fourier components, but having different wave-vectors $\mathbf{K}$, it seems certain that a mean velocity field $\mathbf{u}_o(\mathbf{x},t)$ <u>will</u> develop, but what its structure will be and how it will modify equation (2.16) remains an open question.

(ii) The effect of a forcing field with a continuous spectrum

The assumption (2.11) is indeed too restrictive, and not only because the result (2.13) would then appear to have only local and not global significance. There is some justification for choosing the resonant frequency $\omega = k_o V$, since the response to a random $\mathbf{f}$-field containing a continuous spectrum of wave-numbers $\mathbf{k}$ and frequencies ω will (for weak dissipative effects) peak around the resonant frequencies of the undamped system. However, in singling out a particular resonant frequency, we overestimate the response, which for a continuous spectrum, is controlled by a narrow band of frequencies around the resonant frequency. Moreover, we should take account of all the resonant frequencies given by the Alfvén dispersion relation $\omega = \pm \underset{\sim}{k}.\underset{\sim}{V}$. These effects can be taken into account in an asymptotic evaluation of the integral (2.6) for particular choices of continuous spectrum tensors; but it is better first to take some account of the third question (iii) raised above.

(iii) What physical mechanism can generate a lack of reflexional symmetry?

It was recognized by Steenbeck et al (1966) that, as mentioned in the introductory paragraphs, a lack of reflexional symmetry is likely to arise only in a rotating fluid in which a definite direction relative to the rotation vector $\underset{\sim}{\Omega}$ can be distinguished, i.e. only if the 'external conditions' themselves lack reflexional symmetry. A random force field $\underset{\sim}{f}$ is unlikely in itself to lack reflexional symmetry. If however a reflexionally symmetric $\underset{\sim}{f}$-field acts upon a rotating fluid, the resulting $\underset{\sim}{u}$-field will lack reflexional symmetry if the statistical properties of the $\underset{\sim}{f}$-field lack symmetry about planes perpendicular to $\underset{\sim}{\Omega}$, e.g. if (with the convention $\omega > 0$) only waves for which $\underset{\sim}{k}.\underset{\sim}{\Omega} > 0$ are present.

An analysis similar to that described in §2, but with the addition of a Coriolis force $2\underset{\sim}{\Omega} \wedge \underset{\sim}{u}$ in equation (2.1) has been carried out (Moffatt 1971), and full account is taken of the considerations under heading (ii) above. In this calculation, it is supposed that the spectrum of $\underset{\sim}{f}$ is isotropic over the half-space $\underset{\sim}{k}.\underset{\sim}{\Omega} > 0$ and reflexionally symmetric, and characterised by wave-numbers $O(\ell^{-1})$ and frequencies $O(\omega_0)$. Under the assumptions

$$\nu \ll \lambda \ll \Omega \ell^2 , \quad V \gg \lambda/\ell , \quad V \gg \ell\omega_0 , \quad \omega_0 \ll \Omega ,$$

a result analogous to (2.20) is obtained, viz.,

$$\frac{M_{ult}}{E_{ult}} = C\left(\frac{\Omega}{\omega_0}\right)^{\frac{1}{2}}\left(\frac{\nu}{\lambda}\right)^{\frac{1}{2}} \frac{L}{\ell} , \tag{3.3}$$

where C is a dimensionless constant of order unity. Again M_{ult} can be much greater than E_{ult} if the length scale L available for the growth of a mean magnetic field is sufficiently large compared with the length-scale ℓ of the background forcing field.

REFERENCES

Moffatt, H.K. 1970a,b. J. Fluid Mech. 41, 435 and 44, 705.

Moffatt, H.K. 1971. (submitted for publication).

Parker, E.N. 1955. Astrophys. J. 122, 293.

Roberts, P.H. 1971. Lectures in applied mathematics (ed. W.H. Reid). Providence: American Mathematical Society.

Saffman, P.G. 1960. J. Fluid Mech. 8, 273.

Steenbeck, M., Krause, F. and Radler, K.-H., 1966. Z. Naturforsch. 21a, 369.

Taylor, G.I. 1921. Proc. Lond. Math. Soc. 20, 196.

Weiss, N.O. 1971. Quart. J. Roy. Astr. Soc. (to appear).

THE STATISTICAL MECHANICS OF THE GUIDING CENTRE PLASMA*

J. B. Taylor
Culham Laboratory, Abingdon, Berkshire, England

W. B. Thompson
University of California at San Diego, La Jolla, California

1. INTRODUCTION

While studying possible origins of Bohm diffusion in a magnetically confined plasma we have been led to investigate the statistical properties of the two dimensional guiding centre plasma, an intrinsically non-linear mechanical system which exhibits some properties in common with the two dimensional turbulence of an ideal fluid.

To construct a diffusion coefficient, one normally considers particles as executing a random walk, with an average step length ℓ , and a step frequency ν , whereupon dimensional arguments yield a diffusion coefficient $D \sim \nu \ell^2$. The path of a charged particle in the plane perpendicular to a strong magnetic field forms circles of radius $r_L \sim \frac{m v_\perp}{eB}$, v being the particle velocity and B the magnetic field strength. Collisions move the centres of these circles at random, and for a gas of charged particles, the collision frequency $\nu \sim n v^{-3}$ (e.g. Chandrasekhar) n being the particle density, hence one expects a cross field diffusion $\sim nv^{-3} . v^2 B^{-2} \sim nB^{-2} T^{-1/2}$. It is upon the anticipation of such a slow diffusion rate that the hopes were based of magnetically confining a gas at temperatures high enough for a significant release of thermonuclear energy. It has been somewhat disappointing, therefore, that in many confinement experiments the diffusion scaled as T/B, the Bohm scaling.

Attempts to explain this unfortunate process have usually invoked dynamical instabilities leading to charge fluctuations and to electric fields $\underline{E}$. An electric field then produces a motion of the charge $V_D = \frac{E \times B}{B^2} c$ the electric drift, which is superposed on the gyromotion, and it is this random drift that produces diffusion.

---

*Supported in part by the U. K. Atomic Energy Authority and in part by the U. S. Atomic Energy Commission, Contract AT(04-3)-34 P.A. 85-13

We have explored a model which represents the limit of this process--a model in which the plasma is represented as a system of randomly distributed charged filaments aligned parallel to a strong uniform magnetic field. The charges give rise to electric fields, hence to drift motions and to diffusion. In the present paper, we derive from the equations of motion, and a plausible statistical assumption a kinetic equation for the spectrum of charge fluctuations. Despite its unusual form, this equation preserves the constants of the motion, has the thermal equilibrium (Gibbsian) spectrum as its unique stationary solution, and at least a suggestion of a relaxation process. The time dependent correlation function is shown to oscillate, but nonetheless in a coarse grained sense to produce Bohm diffusion. (Taylor and MacNamara)

Statistical Mechanics

One's attitude toward the Gibbsian statistical mechanics of a system of this kind may go through a cycle. It should apply, and if so, should yield for the charge correlation function $q(k) = \int \langle \rho(\underline{x})\,\rho(\underline{x}')\,e^{i\underline{k}.(\underline{x}-\underline{x}')} \rangle \, d^2\underline{x}\, d^2\underline{x}'$ the value for a non magnetized gas of charged particles

$$q(k) = \frac{n\lambda^2 k^2}{1+\lambda^2 k^2}$$

(Landau and Lifschitz).

On the other hand, the system is mechanically somewhat peculiar. There are no forces, only workless constraints, and no kinetic energy, only potential. One thus feels doubtful about invoking Gibbs. However, the equations of motion can be written in Hamiltonian form, the canonical conjugate coordinates being the pair of cartesian coordinates (x_i, y_i) for each particle, and the Hamiltonian being $\sim$ the total potential $\mathcal{H} = \frac{c\Phi}{B} = \frac{c}{B}\sum_i\sum_j \left(\frac{e_i}{\ell}\right)\left(\frac{e_j}{\ell}\right) \log|\underline{r}_i - \underline{r}_j|$ which is a constant. The potential of any single particle, however, i.e. $\sum_j \left(\frac{e_i}{\ell}\right)\left(\frac{e_j}{\ell}\right) \log|\underline{r}_i - \underline{r}_j|$ is no constant and again the arguments of Gibbs may be expected to apply.

If Gibbs arguments are applicable, then the distribution function should be given by $K e^{-\beta\mathcal{H}} \sim e^{-\Phi/kT}$. Since Φ depends only on the pair correlation of particles, there will always be some solution for which particles are distributed completely at random although it may correspond to an infinite temperature. If $\Phi(\underline{r}_i - \underline{r}_j)$ falls off sufficiently rapidly for large separations, and is not too singular at small separations $e^{-\Phi/kT}$ may be finite for a random distribution of particles at a finite temperature. This is the behaviour toward which computer solutions to this problem will be driven.

2. THE KINETIC EQUATION

a) Equations of motion

We will now derive an equation for the two particle correlation function, most conveniently in Fourier component form. To do so, we start from the equations of motion for the particle density

$$\frac{\partial n}{\partial t} + \underline{\nabla} \cdot n \underline{v} = 0 \tag{1}$$

and combine this with the prescription for $\underline{v}$

$$\underline{v} = \frac{\underline{E} \times \underline{B}}{B^2} c = \frac{\underline{E} \times \underline{b}}{B} c = - \frac{\underline{\nabla} \Phi \times \underline{b}}{B} c \tag{2}$$

and with Poisson's equation determining Φ

$$\nabla^2 \Phi = -4\pi \left(\frac{e}{\ell}\right) n(\underline{x}) \tag{3}$$

where we have assumed for each filament a charge per unit length of (e/ℓ).

If now, we introduce the Fourier components, writing

$$n(\underline{x}) = \sum_{\underline{k}} \rho(\underline{k}) e^{i \underline{k} \cdot \underline{x}} \qquad \text{etc.}$$

Then

$$\Phi(\underline{k}) = \frac{4\pi \left(\frac{e}{\ell}\right)}{k^2} \rho(\underline{k})$$

$$\underline{v}(\underline{k}) = \left(\frac{4\pi e c}{B \ell}\right) i \frac{\underline{b} \times \underline{k}}{k^2} \rho(\underline{k})$$

and

$$\frac{\partial \rho(\underline{k})}{\partial t} - \underline{k} \cdot \sum_{\underline{k}'} \left(\frac{4\pi e c}{B \ell}\right) \frac{\underline{b} \times \underline{k}'}{k'^2} \rho(\underline{k}') \rho(\underline{k} - \underline{k}') = 0 \tag{4}$$

This is the fundamental dynamical equation of the problem, and at this point contact is made with the two dimensional turbulence of an ideal inviscid fluid. In such a system, the vorticity $\psi(\underline{x})$ necessarily directed along the ignorable coordinate acts as a conserved scalar.

$$\frac{\partial \psi}{\partial t} + \underline{\nabla} \cdot \underline{v} \psi = 0 \tag{1'}$$

Moreover, the velocity $\underline{v}$ for the incompressible case may be derived from a stream function

$$\underline{v} = -\underline{u}_z \times \underline{\nabla} \phi \tag{2'}$$

and on constructing the vorticity in terms of ϕ

$$\nabla^2 \phi = -\psi \tag{3'}$$

hence, again introducing Fourier componants

$$\frac{\partial \psi(\underline{k})}{\partial t} - \sum_{\underline{k}'} \underline{u}_z \cdot \frac{\underline{k}' \times \underline{k}}{k'^2} \psi(\underline{k}') \psi(\underline{k} - \underline{k}') = 0 \tag{4'}$$

which aside from the constant $\frac{4\pi ec}{B\ell}$ is (4).

b) <u>The kinetic equation</u>

To develop an equation for $q_k = \langle \rho_k \rho_{-k} \rangle$ we differentiate twice with respect to time and consider

$$\ddot{q}_k = \langle \ddot{\rho}_k \rho_{-k} + \dot{\rho}_k \dot{\rho}_{-k} \rangle + \text{c. c.}$$

Using (4) for the time derivatives of ρ_k and abbreviating $\frac{4\pi ec}{B\ell}$ to α

$$\ddot{q}_k = 2 \mathcal{R} \alpha^2 \Big\langle \sum_{\underline{k}', \underline{k}''} \underline{b} \cdot \frac{\underline{k}' \times \underline{k}}{k'^2} \Big\{ \underline{b} \cdot \frac{\underline{k}'' \times \underline{k}'}{k''^2} \rho_{k''} \rho_{k'-k''} \rho_{k-k'} \rho_{-k}$$
$$+ \underline{b} \cdot \frac{\underline{k}'' \times (\underline{k} - \underline{k}')}{k''^2} \rho_{k''} \rho_{k-k'-k''} \rho_{k'} \rho_{-k} - \frac{\underline{b} \cdot \underline{k}'' \times \underline{k}}{k''^2} \rho_{k''} \rho_{k-k'} \rho_{k'} \rho_{-k-k''} \Big\} \Big\rangle \tag{5}$$

At this point we introduce the plausible statistical hypothesis that only terms of the form $\langle \rho_k \rho_{-k} \rangle$ survive on averaging and that products of four ρ_k's may be factored as products of pairs, thus $\langle \rho_k \rho_{-k} \rho_{k'} \rho_{-k'} \rangle = \langle \rho_k \rho_{-k} \rangle \langle \rho_{k'} \rho_{-k'} \rangle = q_k q_{k'}$
Note that for any spatially homogeneous system

$$\langle \rho_k \rho_{k'} \rangle = q_k \, \delta(\underline{k} + \underline{k}')$$

from translational invariance, and for a random distribution

$$\langle \rho_k \rho_{k'} \rangle = N\,\delta(\underline{k} + \underline{k}')$$

and

$$\langle \rho_{k_1}\rho_{k_2}\rho_{k_3}\rho_{k_4} \rangle = N^2\,\delta(\underline{k}_1 + \underline{k}_2)\,\delta(\underline{k}_3 + \underline{k}_4) + \text{perms} + N\,\delta(\underline{k}_1 + \underline{k}_2 + \underline{k}_3 + \underline{k}_4)$$

hence the hypotheses becomes valid for large N

With this statistical assumption (5) becomes

$$\ddot{q}_k = 2\alpha^2 \sum_{\underline{\ell},\, \underline{m}} \delta(\underline{k} + \underline{\ell} + \underline{j})\, \frac{(\underline{b}.\,\underline{\ell} \times \underline{k})^2}{\ell^2} \,. \left\{ \left(\frac{1}{k^2} - \frac{1}{\ell^2}\right) q_k q_\ell + \left(\frac{1}{\ell^2} - \frac{1}{j^2}\right) q_\ell q_j + \left(\frac{1}{j^2} - \frac{1}{k^2}\right) q_j q_k \right\} \tag{6}$$

where we have eliminated dashes by employing more letters for the $\underline{k}$. Note first that the quantity $\delta(\underline{k} + \underline{\ell} + \underline{j})\,(\underline{b}\cdot\underline{\ell} \times \underline{k})^2$ is symmetric under any permutation of the $\underline{k}$, $\underline{\ell}$, $\underline{j}$, while the term in the brace is antisymmetric under such an interchange. Employing these symmetries and using the notation

$$A_{k\ell j} = 2\alpha^2\,\delta(\underline{k} + \underline{\ell} + \underline{j})\,(\underline{b}\cdot\underline{k} \times \underline{\ell})^2$$

$$B_{kj} = \frac{1}{k^2} - \frac{1}{j^2}$$

$$\ddot{q}_k = \sum_{\underline{\ell},\,\underline{j}} \frac{1}{2} A_{k\ell j} B_{\ell j} \left[B_{\ell j}\, q_\ell q_j + B_{jk}\, q_j q_k + B_{k\ell}\, q_k q_\ell \right] \tag{7}$$

$$= \sum_{\underline{\ell},\,\underline{j}} \frac{1}{2} A_{k\ell j} B_{\ell j} C_{k\ell j}$$

c) Constants of motion

Note that if in (7) we sum over $\underline{k}$, then we can write, interchanging the dummy indices $\underline{k}$, $\underline{\ell}$, $\underline{j}$, and adding

$$\sum_{k} \ddot{q}_k = \sum_{k,l,j} \frac{1}{6} A_{klj} \left[B_{lj} + B_{jk} + B_{kl} \right] C_{klj} = 0 \tag{8}$$

as the first bracket on the right vanishes.

Moreover

$$\sum_{k} \frac{\ddot{q}_k}{k^2} = \sum_{k,l,j} \frac{1}{6} A_{klj} \left[\frac{B_{lj}}{k^2} + \frac{B_{jk}}{l^2} + \frac{B_{kl}}{j^2} \right] C_{klj} = 0 \tag{9}$$

as again the first bracket is easily shown to vanish. From the conservation equation for ρ_k (1) and the divergenceless nature of $\underline{v}$: $\int \rho_k^2 \, d\tau = \text{const.}$ and this is the constant represented by (8).

Since $\langle E^2 \rangle = \left(4\pi \frac{e}{l}\right)^2 q^2/k^2$, and since the energy

$$\mathcal{E} = \frac{1}{4\pi} \int E^2 \, d\tau = \text{const}$$

(9) represents the conservation of energy.

In the fluid case, (8) represents the constancy of the total of the squared vorticity, and (9) again the conservation of energy. The basic equation of motion is, of course, exactly the equation for local conservation of vorticity.

d) Thermal equilibrium

We now investigate the nature of the stationary solutions to (7), first establishing an entropy-like theorem. (7) can be written

$$\frac{\ddot{q}_k}{q_k} = \sum_{l,j} \frac{1}{2} A_{klj} \frac{B_{lj}}{q_k} C_{klj} \tag{10}$$

Now, if we sum this over $\underline{k}$ and use the trick employed in deriving (8)

$$\begin{aligned} \sum_{k} \frac{\ddot{q}_k}{q_k} &= \sum_{k,l,j} \frac{1}{2} A_{klj} \left[\frac{B_{lj}}{q_k} + \frac{B_{jk}}{q_l} + \frac{B_{kl}}{q_j} \right] C_{klj} \\ &= \sum_{k,l,j} \frac{1}{2} A_{klj} \left[\frac{B_{lj}}{q_k} + \frac{B_{jk}}{q_l} + \frac{B_{kl}}{q_j} \right]^2 q_k q_l q_j \end{aligned} \tag{11}$$

substituting the $C_{k l j}$ from (7).

Since the q_k must be positive, it follows that $\sum_{k} \ddot{q}_k / q_k \geq 0$

Moreover, it can only vanish if the bracketted quantity vanishes.

It is clear that q_k = constant satisfied this, which corresponds to completely uncorrelated particles. To get a general solution, write $q_k = \frac{k^2}{\psi(k^2)}$ whereupon

$$(j^2 - \ell^2)\psi(k^2) + (k^2 - j^2)\psi(\ell^2) + (\ell^2 - k^2)\psi(j^2) = 0$$

or

$$(k^2 - 2k\ell\cos\theta)\psi(k^2) - (\ell^2 - 2k\ell\cos\theta)\psi(\ell^2) + (\ell^2 - k^2)\psi(\ell^2 + k^2 - 2k\ell\cos\theta) = 0$$

which requires that ψ be linear in k^2. Hence $q_k \sim \frac{k^2}{A + Bk^2}$. This may be identified with the result of statistical mechanics

$$q_k = Q(\underline{k}) = \frac{n\lambda^2 k^2}{1 + \lambda^2 k^2} . \tag{12}$$

Such solutions, while usually of great importance in kinetic theory, are usually of less significance in the study of turbulence where dissipative processes play a central role. Nonetheless, even in that study, the existence of an equipartition solution, and an equation which exhibits that as its unique stationary solution, may not be without interest.

e) Relaxation

To discuss the dynamics of small departures from equilibrium we may study the linearized form of (7).

Write $q_k = Q_k(1 + f_k(t))$ whereupon

$$\ddot{f}_k = -\sum_{\ell, j} \tfrac{1}{2} A_{k\ell j} \frac{B_{\ell j}}{Q_k} Q_k Q_\ell Q_j \left[\frac{B_{\ell j}}{Q_k} f_k + \frac{B_{jk}}{Q_\ell} f_\ell + \frac{B_{k\ell}}{Q_j} f_j \right] \tag{13}$$

From this equation the linearized form of the conservation laws (8) and (9) immediately follow. Moreover, (13) is a linear integral equation for the f_k. Thus, it admits eigensolutions of the form $f_k(s)e^{ist}$. For such an eigensolution we can prove that all the s must be real and that the eigensolution for s = 0 corresponds to a perturbed thermal equilibrium. Multiply (13) by f_k^* and sum over $\underline{k}$, then interchange the $\underline{k}$, $\underline{\ell}$, $\underline{j}$ to obtain

$$s^2 \sum_{\underline{k}} f_k(s) f_k^*(s) = \sum_{\underline{k},\ell,j} \frac{1}{6} A_{k\ell j} Q_k Q_\ell Q_j \left| \frac{B_{\ell j} f_k}{Q_k} + \frac{B_{jk} f_\ell}{Q_\ell} + \frac{B_{k\ell} f_j}{Q_j} \right|^2$$

and the R.H.S. is clearly positive definite, thus all the $f_k(s)$ oscillate and the system appears reversible. An argument due to Landau and Lifschitz suggests that this need not hold. Any initial disturbance can be expanded in the eigenfunctions of (13), and at later times may be written

$$f_k(t) = \sum_s a(s) f_k(s)\, e^{ist}$$

After a long time the contributions due to $a(s \neq 0)$ will phase mix away, and in a time-averaged sense

$$\langle f_k(t) \rangle = \frac{1}{\Delta t} \int_t^{t+\Delta t} f_k(t')\, dt' \longrightarrow a(0) f_k(0)$$

just perturbed thermal equilibrium. This, of course, will not be true for those special disturbances that can initially be represented by a single, or a few eigenfunctions.

3. DIFFUSION

a) The diffusion model

The motivation for this kinetic theory study has been the cross field diffusion problem. For the two dimensional model we may write a Fokker-Planck equation for the coarse grained density

$$\frac{\partial n}{\partial t} + \frac{\partial}{\partial \underline{x}} \frac{\partial}{\partial \underline{x}} \frac{1}{2} \langle \Delta \underline{x}\, \Delta \dot{\underline{x}} \rangle n = 0$$

and for $\langle \Delta \underline{x}\ \Delta \dot{\underline{x}} \rangle$ using the guiding centre motion

$$\underline{\underline{D}} = \frac{1}{2} \langle \Delta \underline{x}\, \Delta \dot{\underline{x}} \rangle = \frac{1}{2} \frac{c^2}{B^2} \left\langle \int_{-\infty}^{t} \underline{b} \times \underline{E}(\underline{x}, t)\ \underline{b} \times \underline{E}(\underline{x}(t')\, t')\, dt' \right\rangle$$

The fluctuating electric fields $\underline{E}$ may be expressed in terms of the ρ_k thus

$$\underline{E}(\underline{x}, t) = -4\pi i \left(\frac{e}{\ell}\right) \sum_{\underline{k}} \underline{k}\, \rho_k(t)\, e^{i \underline{k} . \underline{x}}$$

$$\underline{\underline{D}} = \frac{1}{2} \frac{c^2}{B^2} \left(4\pi \frac{e}{\ell}\right)^2 \left\langle \sum_{\underline{k}} (\underline{b} \times \underline{k})(\underline{b} \times \underline{k}) \int_{-\infty}^{t} \rho_k(t) \rho_{-k}(t')\, e^{i \underline{k} . [\underline{x}(t') - \underline{x}(t)]}\, dt' \right\rangle$$

In forming the averages here, we think of the fluctuating fields as being given, and of the particles whose diffusion is studied as being dropped into the existing, equilibrium fields--the test particle model. In this model the $\Delta \underline{x}$ is considered as statistically independent of the $\underline{E}$, hence

$$e^{i\underline{k}.\Delta \underline{x}(t')} = e^{-k^2 D t}$$

and the diffusion coefficient becomes

$$\underline{\underline{D}} = \frac{1}{2}\left(\frac{4\pi ec}{lB}\right)^2 \frac{1}{(2\pi)^2}\int_{-\infty}^{\infty} d^2\underline{k} \int_{-\infty}^{t} \frac{(\underline{b}\times\underline{k})(\underline{b}\times\underline{k})}{k^4} \langle \rho_{-k}(t)\,\rho_k(t')\rangle e^{-k^2 D t}\, dt' \tag{14}$$

To evaluate this we need the function $\langle \rho_{-k}(t)\,\rho_k(t')\rangle = S_k(\tau)$, the autocorrelation function, which in steady state depends only on τ .

b) <u>The Autocorrelation function</u>

To determine $S_k(\tau)$, we again use the equations of motion and construct a second derivative

$$\ddot{S}_k(\tau) = \alpha^2 \Big\langle \rho_{-k} \sum_{\underline{k}',\underline{k}''} \frac{\underline{b}.\underline{k}'\times\underline{k}}{k'^2}\Big[\frac{\underline{b}.\underline{k}''\times\underline{k}'}{k''^2}\rho_{k''}\rho_{k'-k''}\rho_{k-k'} + \frac{\underline{b}.\underline{k}''\times(\underline{k}-\underline{k}')}{k''^2}\rho_{k''}\rho_{k-k'-k''}\rho_{k'}\Big]\Big\rangle$$

where the ρ_{-k} is evaluated at t and the remaining ρ at $t-\tau$. Again invoking the random phase approximation, which should be valid at least for short times.

$$\frac{d^2 S_k}{d\tau^2} = \alpha^2 \sum_{\underline{k}'} \frac{[\underline{b}.\underline{k}'\times\underline{k}]^2}{k'^2}\Big\{\Big[\frac{1}{k^2}-\frac{1}{k'^2}\Big]q_{k'} + \Big[\frac{1}{(\underline{k}-\underline{k}')^2}-\frac{1}{k^2}\Big]q_{k-k'}\Big\} S_k$$

$$= \sum_{j,l} \frac{1}{4} A_{kjl} B_{jl}\left[B_{kj} q_j + B_{lk} q_k\right] S_k(\tau) \tag{15}$$

In thermal equilibrium we may use

$$B_{kj} Q_k Q_j + B_{j\ell} Q_j Q_\ell + B_{\ell k} Q_\ell Q_k = 0$$

to write

$$\frac{d^2 S_k}{d\tau^2} = -\frac{1}{4}\left[\sum_{j,\ell} A_{kj\ell} B_{j\ell}^2 \frac{Q_j Q_\ell}{Q_k}\right] S_k \tag{16}$$

Thus the coefficient on the right is negative definite, and in thermal equilibrium, the correlation function $S_k(\tau)$ oscillates with τ, and since at $\tau = 0$, $S_k = Q_k$, it is given by

$$S_k(\tau) = Q_k \cos \Omega_k \tau \tag{17}$$

Using (15) and the equilibrium values for Q_k and replacing the sum by an integral

$$\begin{aligned} \Omega_k^2 &= \frac{n\alpha^2\lambda^2}{(2\pi)^2}\int_V \left\{\frac{\ell^2 - 2\ell k\cos\theta}{1+\lambda^2(k^2+\ell^2-2\ell k\cos\theta)} - \frac{\ell^2 - k^2}{1+\lambda^2\ell^2}\right\}\sin^2\theta\,\ell\,d\ell\,d\theta \\ &= \frac{n\alpha^2}{\lambda^2} F(\lambda^2 k^2) \end{aligned} \tag{18}$$

where $F(u)$ is readily shown to be

$$F(u) = -\frac{1}{8\pi}\frac{(1+u)^2}{u}\left[\frac{u}{1+u} - \log(1+u)\right]$$

For small k

$$\Omega_k^2 \longrightarrow \frac{\pi n e^2 c^2 k^2}{B^2 \ell^2} \tag{19}$$

We may now use this value for $S_k(\tau)$ to write an expression for the diffusion coefficient.

c) The Diffusion Ceefficient

We now substitute our calculated value for the autocorrelation function (17) in (14), and carry out the integration over the directions of k obtaining

$$D = 4\pi\left(\frac{e}{\ell}\right)^2 \frac{c^2}{B^2}\int\frac{dk}{k}\int Q_k \cos\Omega_k\tau\, e^{-k^2 D\tau}\,d\tau$$

$$D = 4\pi\left(\frac{e}{\ell}\right)^2\frac{c^2}{B^2}\int\frac{dk}{k}\,Q_k\,\frac{k^2D}{\Omega_k^2+k^4D^2}$$

$$= 4\pi n\left(\frac{e}{\ell}\right)^2\frac{c^2}{B^2}\int\frac{dk}{k}\cdot\frac{\lambda^2k^2}{1+\lambda^2k^2}\cdot\frac{k^2D}{\Omega_k^2+k^4D^2}$$

(20)

Since this integral is singular at small k in the absence of Ω_k, and since $\Omega_k^2 \longrightarrow k^2\log k^2$ for large k^2, we may use a small k approximation, $\Omega_k^2 = \pi n\left(\frac{e}{\ell}\right)^2\frac{c^2}{B^2}k^2$

Expressing D^2 as

$$D^2 = \left(\pi n\left(\frac{e}{\ell}\right)^2\left(\frac{c}{B}\right)^2\lambda^2\right)x^2$$

$$= \pi\,(n\lambda^2)\left(\frac{ec}{\ell B}\right)^2x^2$$

$$= \frac{1}{16\pi}\left[\frac{kT}{eB}\right]^2\frac{x^2}{n\lambda^2}$$

(20) becomes $$1 = 2\int\frac{dy}{(1+y)}\,\frac{1}{(1+x^2y)} = \frac{2}{(x^2-1)}\log x^2$$

which is satisfied by $x^2 \simeq 3.5$. Hence, in thermal equilibrium the diffusion coefficient may be written

$$D \simeq \frac{1}{4}\,\frac{ckT}{n_D^{1/2}\,eB}$$

where $n_D = n\lambda^2$ is the number of particles in a Debye circle.

BIBLIOGRAPHY

S. Chandrasekhar, "Plasma Physics," Chicago, p. 180 (1960).

L. D. Landau and L. M. Lifschitz, "Statistical Physics," Pergamon, (1959).

L. D. Landau and L. M. Lifschitz, "Fluid Mechanics," Pergamon, p. 107 (1959).

L. Onsager, Nuovo Cimento Sup. VI, p. 279 (1949).

J. B. Taylor and B. McNamara, Phys. Fluids 14, 1492 (1971).

G. Vahala and D. Montgomery, "Kinetic Theory of a Two-Dimensional Plasma," J. of Plasma Physics (to be published), (1971).

STRANGE ATTRACTORS AS A MATHEMATICAL EXPLANATION OF TURBULENCE

David Ruelle*

Institute for Advanced Study, Princeton, New Jersey 08540

ABSTRACT

We discuss a mechanism for the generation of turbulence in viscous fluids. According to this mechanism, first proposed by D. Ruelle and F. Takens [9], the solutions of the equations of motion describing a turbulent flow are asymptotic to "strange attractors."

1. INTRODUCTION

Various mathematical interpretations have been proposed for the physical phenomenon of turbulent fluid motion. J. Leray [8], followed by some Russian authors, remarks that a solution of the Navier-Stokes equation with good initial data may after a certain time have large velocity gradients. In that case the conditions under which the Navier-Stokes equation is physically acceptable are no longer satisfied. This may lead mathematically to breakdown of the uniqueness of the solution and physically to turbulence. It is however not proved that turbulent motion always involves large velocity gradients. Another idea was put forward by E. Hopf [5], and defended by L. D. Landau and E. M. Lifshitz [6]. They simply propose that turbulent motion is quasi-periodic, i.e., the physical parameters x describing the fluid have a time dependence of the form

$$x(t) = f(\omega_1 t, \ldots, \omega_m t)$$

where f has period 2π in each of its arguments separately and the frequencies $\omega_1, \ldots, \omega_m$ are not rationally related. A time dependence of this sort at least superficially reproduces the complicated and irregular behavior characteristic of turbulent motion. In physical language one would say that turbulence results from the superposition of modes with different frequencies. One has however, to ask how non-linear effects may affect such a superposition. This problem can be investigated with the methods of modern differential analysis and it is found that quasiperiodicity does generally not survive the non-linear perturbations. It is however also found that non-linear effects may lead to a complicated and apparently erratic motion, with very sensitive dependence on initial conditions. This happens when the solution of the equations of motion is asymptotic to a "strange attractor." It is proposed [9] that turbulence is due to this phenomenon. Interestingly, the

*On leave of absence from Institut des Hautes Etudes Scientifiques, 91 Bures-sur-Yvette, France.

great sensitivity to initial conditions found here is not shared by the quasi-periodic motions discussed earlier.

2. QUALITATIVE THEORY OF DIFFERENTIAL EQUATIONS

Let $B = \{x: |x| \leq R\}$ be the closed ball with radius R in n-dimensional Euclidean space E^n. We let $\mathfrak{X}^k$ be the space of vector fields on B which have continuous derivatives up to the order $k \geq 1$ (C^k vector fields). The space $\mathfrak{X}^k$ is a Banach space with respect to the norm

$$\|X\| = \sup_{1 \leq i \leq n} \ \sup_{(\rho) \leq k} \ \sup_{x \in B} \left| \frac{\partial^{(\rho)}}{\partial x^{\rho}} X^i(x) \right|$$

where
$$\frac{\partial^{(\rho)}}{\partial x^{\rho}} = \left(\frac{\partial}{\partial x^1}\right)^{\rho_1} \cdots \left(\frac{\partial}{\partial x^n}\right)^{\rho_n}$$

and $|\rho| = \rho_1 + \ldots + \rho_n$. A subset of $\mathfrak{X}^k$ is called residual if it contains a countable intersection of dense open subsets of $\mathfrak{X}^k$. Baire's theorem implies that a residual set is again dense in $\mathfrak{X}^k$; therefore a residual set S may be considered in some sense to be a "large" subset of $\mathfrak{X}^k$. A property of a vector field $X \in \mathfrak{X}^k$ which holds on a residual subset of $\mathfrak{X}^k$ is called generic.*

It is easy to extend the definitions of residual set and genericity to a compact manifold M: cover M with a finite number of balls B_i and let $\mathfrak{X}^k(M)$ be the Banach space of C^k vector fields on M with the norm $\sup_i \|\cdot\|_i$. We note at this point a consequence of a theorem of Peixoto†: the vector fields on the 2-torus T^2 are generically not quasi-periodic. We say that a vector field is quasi-periodic if one can choose coordinates such that T^2 appears as a square with opposite sides identified, and the vector field is constant in these coordinates.

Given $X \in \mathfrak{X}^k$, the integral curve $x(\cdot)$ through x_0 satisfies $x(0) = x_0$ and $dx(t)/dt = X(x(t))$; it is defined at least for sufficiently small $|t|$. If $x(t) \equiv x_0$, i.e. $X(x_0) = 0$ we have a critical point of the vector field X. If $x(\tau) = x_0$ and $x(t) \neq x_0$ for $0 < t < \tau$ we have a closed orbit of (smallest) period τ. It may happen that the integral curves through all points near a critical point (or a closed

†See for instance R. Abraham and J. Marsden [1].

*The notion of genericity based on residual sets is not very satisfactory when physical applications are considered. It is indeed known that a residual subset of the interval [0,1] may have Lebesgue measure 0. In fact the important known generic properties of vector fields are usually generic in some stronger sense but we shall not elaborate on this point.

orbit) tend to this critical point (or closed orbit) as $t \to \infty$. This is the case for attracting critical points (and closed orbits) to be introduced later. We will now show that there exist more complicated sets which can attract the integral curves through the nearby points. We construct such a strange attractor below, for a 4-dimensional vector field.

First we introduce the idea of Poincaré map. Imagine that the integral curves of a n-dimensional vector field intersect transversally a piece of n-1-dimensional surface Σ. If $x \in \Sigma$ and $P(x)$ is the first intersection of the integral curve through x with Σ, P is a Poincaré map. Take now $n - 1 = 3$, and assume that P maps a solid ring U (bounded by a 2-torus) into itself as shown in Fig. 1., i. e. winding twice around the central hole. The set $\Omega = \bigcap_{n>0} P^n U$ is difficult to describe, it is locally the product of a Cantor set and a line interval. If $x_0 \in U$, $x_n = P^n x_0$ tends to Ω in an apparently erratic manner depending sensitively on the original position x_0.

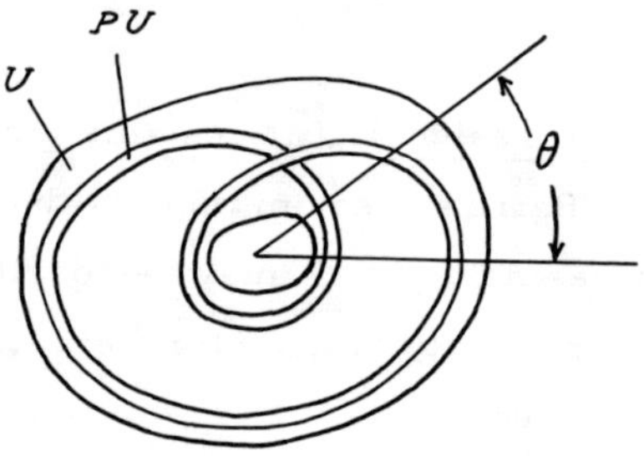

To demonstrate this sensitivity, let θ be an angular coordinate in U (see Fig. 1). If the origin is properly chosen, P roughly multiplies θ by 2, and therefore $\theta(x_n) \simeq 2^n \theta(x_0)$ (mod. 2π): any angular deviation is rapidly magnified. One can now construct a 4-dimensional vector field X which has P as Poincaré map; corresponding to Ω, X has a "strange attractor" Ω^* which is locally the product of a Cantor set and a piece of 2-dimensional manifold. The integral curves of X tend asymptotically to Ω^*, in a complicated and apparently erratic manner depending sensitively on the initial conditions.

The possible generic types of asymptotic behavior of integral curves of vector fields have not been completely classified. It seems however that, apart from attracting critical points and attracting closed orbits, the behavior described above is typical: complicated and apparently erratic with sensitive dependence on initial conditions. (See S. Smale [11] and R. F. Williams [12].)

3. THE PROBLEM OF ONSET OF TURBULENCE AS A BIFURCATION PROBLEM

To study the onset of turbulence, we consider a viscous fluid system on which a steady external action is exerted, described by a parameter μ (typically a Reynolds number). When $\mu = 0$, there is no external action; we assume that the time evolution leads asymptotically to a state of equilibrium independent of initial condition. When μ is increased the following succession of events is typically observed in experiments. First, the equilibrium state is replaced by a time independent (steady) non-equilibrium state. For larger values of μ a time periodic motion is observed. Finally, as μ is further increased, the flow becomes turbulent.

It is necessary to digress here briefly on other changes in flow type which are often observed, leading from one steady motion to another steady motion. This is, for instance, the case when the Couette flow is replaced by Taylor cells in the flow between concentric cylinders. These transitions occur in experimental situations which have a certain symmetry (vertical translations and reflections in the case of flow between rotating cylinders). The transitions from steady motion to steady motion then change the symmetry pattern of the flow (symmetry breakdown). We shall henceforth restrict our attention to systems without symmetry, the above phenomena will then not occur.

We shall consider the time evolution of a viscous fluid system as determined by an evolution equation

$$\frac{dx}{dt} = X_\mu(x) \tag{1}$$

Since x represents a velocity field, (1) is a differential equation in an infinite dimensional space. This infinite dimensional character leads to mathematical difficulties: if (1) is the Navier-Stokes equation, no general existence and uniqueness theorem for the solution has been proved. As we shall see later, turbulence has probably nothing to do with these difficulties. We shall thus think of (1) as a differential equation in a finite dimensional space; what we shall say extends in fact to the infinite dimensional case for well-behaved vector fields X_μ.

We are thus led to studying the changes in the qualitative behavior of solutions of (1) as the parameter μ is varied. Such a study is the object of <u>bifurcation theory</u>. Our problem is, however, very undetermined because we know very little about the nature of the vector field X_μ. Actually, we know that for $\mu = 0$, this vector field is attracting to one point: the equilibrium state. For $\mu > 0$ it is reasonable to study <u>generic</u> deformations of the situation at $\mu = 0$ (genericity in the sense of Section 2). In other words we shall ignore possibilities of deformation which are in some sense exceptional. This point of view could lead to serious error if, by some law of nature which we have overlooked, X_μ happens to be in a special class with exceptional properties. This would, for instance, happen if X_μ described <u>conservative</u> (Hamiltonian) rather than <u>dissipative</u> systems, or if we had systems with a non-trivial symmetry group. It appears, however, that a three dimensional viscous fluid without symmetry conforms to the pattern of generic behavior which we discuss below. Our discussion should in fact apply to very general dissipative systems.

4. A MATHEMATICAL MECHANISM FOR TURBULENCE

Let X_μ be a vector field on a finite dimensional space, depending on the real parameters μ. In accordance with what was said above we shall describe a pattern of changes of X_μ when μ increases, which is physically acceptable and leads to something like turbulence.

For $\mu = 0$, the equation

$$\frac{dx}{dt} = X_\mu(x) \tag{1}$$

has the solution $x = 0$ (corresponding to equilibrium). We assume that the eigenvalues of the Jacobian matrix $A^j{}_k$ defined by

$$A^j{}_k = \frac{\partial X_0^j}{\partial x^k}(0)$$

have all strictly negative real parts; this corresponds to the fact that the critical point 0 is attracting. The Jacobian determinant does not vanish, hence by the implicit function theorem there exists a differentiable function $\xi(\cdot)$ such that

$$\xi(0) = 0, \quad X_\mu(\xi(\mu)) = 0 .$$

In the hydrodynamical picture, $\xi(\mu)$ describes a steady state.

We follow now $\xi(\mu)$ as μ increases. For sufficiently small μ the Jacobian matrix $A^j{}_k(\mu)$ defined by

$$A^j{}_k(\mu) = \frac{\partial Xj_\mu}{\partial x^k}(\xi(\mu))$$

has only eigenvalues with strictly negative real parts (by continuity). This means that the critical point $\xi(\mu)$ is attracting, the corresponding steady state is therefore stable.

We assume that, as μ increases further, successive pairs of complex conjugate eigenvalues of the Jacobian cross the imaginary axis, for $\mu = \mu_1, \mu_2, \mu_3, \ldots$*. For $\mu > \mu_1$, the critical point $\xi(\mu)$ is no longer attracting. It has been shown by E. Hopf [4] that there is a one-parameter family of closed orbits of X_μ near ξ_μ for μ close to μ_1. Depending on the sign of some coefficient these closed orbits appear for $\mu > \mu_1$ or $\mu < \mu_1$. Let us assume that they appear for $\mu > \mu_1$, they are then attracting (limit cycles). Such a closed orbit is physically interpreted as a

*Another less interesting possibility is that a real eigenvalue vanishes. When this happens the attracting fixed point $\xi(\mu)$ generically coalesces with another fixed point (non-attracting) and disappears. If a symmetry group is present this generic behavior is changed and one may observe bifurcation of critical points with a new symmetry type (that is what happens when Taylor cells replace the Couette flow).

periodic motion, its amplitude increases with μ (in fact like $(\mu-\mu_1)^{1/2}$). The Hopf bifurcation, just described for a finite dimensional vector field, can also be shown to occur when (1) is the Navier-Stokes equation (N. N. Brushlinskaja [2] and D. H. Sattinger [10]).

The situation where several pairs $\lambda_1, \bar{\lambda}_1, \ldots, \lambda_m, \bar{\lambda}_m$ of complex eigenvalues have crossed the imaginary axis is more complicated. To understand it we follow R. Jost and E. Zehnder [7] , and make a change of coordinates to put the vector field X in normal form. Generically it is possible to replace the r real coordinates of x by n/2 complex coordinates $z_1, \ldots, z_n$ so that the vector field X takes the form Z:

$$\dot{z}_j = Z_j(z) = (\lambda_j + \sum_{k=1}^{n} a_{jk}|z_k|^2)z_j + O_4 .$$

Ignoring the correction O_4 for the moment we find

$$\frac{d}{dt}|z_j|^2 = [\,2(\mathrm{Re}\,\lambda_j) + \sum_{k=1}^{n} (\mathrm{Re}\,a_{jk})|z_k|^2\,]\,|z_j|^2 .$$

It follows that, for suitable $\mathrm{Re}\,a_{jk}$, the vector field Z_j has, apart from higher order terms, an invariant attracting m-torus $|z_1| = c_1 > 0, \ldots, |z_m| = c_m > 0$. If the perturbation O_4 is sufficiently small $\dot{Z}_j$ itself has an invariant m-torus T^m (see,for instance, M. Hirsch, C. C. Pugh, M. Shub [3] ; see also the remarks in [2]).

It remains to study the time evolution on T^m. If $O_4 = 0$, we have $z_j(t) = z_j(0)\, e^{i\omega_j t}$ with $\omega_j = \mathrm{Im}\,(\lambda_j + \sum a_{jk} c_k^2)$. Therefore the original coordinates satisfy

$$x(t) = f(\omega_1 t, \ldots, \omega_m t)$$

where f has period 2π in each argument separately: we have a quasi-periodic motion. We know, however, that the perturbation O_4 will destroy this quasi-periodicity. (See the consequence of Peixoto' s theorem mentioned in Section 2.) It is shown in [9] that if $m \geq 4$, a small perturbation of the quasi-periodic motion on T^m leads to strange attractors of the type described in Section 2*. It is proposed that turbulence arises when the integral curve describing the time evolution of a viscous fluid system is asymptotic to a strange attractor (in the general sense of an attracting set different from a critical point or a closed orbit, and having a certain character of genericity).

5. DISCUSSION

The explanation of turbulence proposed above avoids the non-genericity problem encountered with the quasi-periodic explanation. A distinctive feature

*It is not known if this is also the case for $m = 3$. For $m = 2$, strange attractors do not appear.

is the sensitive dependence on initial condition associated with a strange attractor but not quasi-periodic motion. This is also reflected in the fact that the time correlation functions for the motion on a strange attractor tend to zero (as observed experimentally)--this is not so for quasi-periodic motions.

It would be very interesting to associate a specific strange attractor to a specific turbulent flow. This is a difficult problem, one could attempt to solve it only for very simple flows which could be analyzed in detail. Perhaps a turbulent slug in the flow down a pipe would provide a reasonable example.

BIBLIOGRAPHY

[1] R. Abraham and J. Marsden. Foundations of Mechanics. Benjamin, New York, 1967.

[2] N. N. Brushlinskaja. The behavior of solutions of the equations of hydrodynamics when the Reynolds number passes through a critical value. Doklady A. N. S. S. S.R. 162, No. 4 (1965).

[3] M. Hirsch, C. C. Pugh, and M. Shub. Invariant manifolds. Bull. A.M.S. 76, 1015-1019 (1970).

[4] E. Hopf. Abzweigung einer periodischen Lösung von einer stationären Lösung eines Differentialsystems. Ber. Math.-Phys. Kl. Sächs. Akad. Wiss. Leipzig 94, 1-22 (1942).

[5] E. Hopf. A mathematical example displaying features of turbulence. Commun. Pure Appl. Math. 1, 303-322 (1948).

[6] L. D. Landau and E. M. Lifshitz. Fluid Mechanics. Pergammon, New York, 1959.

[7] R. Jost and E. Zehnder. To be published.

[8] J. Leray. Sur le mouvement d'un fluide visqueux emplissant l'espace. Acta Math. 63, 193-248 (1934).

[9] D. Ruelle and F. Takens. On the nature of turbulence. Commun. Math. Phys. 20, 167-192 (1971).

[10] D. H. Sattinger. Bifurcation of periodic solutions of the Navier-Stokes equations. Arch. Rat. Mech. Anal. 41, 66-80 (1971).

[11] S. Smale. Differentiable Dynamical Systems. Bull. A.M.S. 73, 747-817, (1967).

[12] R. F. Williams. One-dimensional non-wandering sets. Topology 6, 473-487 (1967).

RANDOM GEOMETRIC PROBLEMS SUGGESTED BY TURBULENCE[o]

by

S. Corrsin
Mechanics Department, The Johns Hopkins University

A three-dimensional, unsteady, random fluid velocity field, even subject to constraints such as mass and momentum balances, is a rich system. In trying to understand turbulent motion, we inevitably run into interesting statistical questions. My intention is to outline a few geometric and kinematic problems in turbulent flows and to mention some simpler, non-physical, statistical processes and questions inspired by the turbulent experience.

The term "geometry" in the title is meant to include kinematics, which is a kind of geometric evolution -- or evolutionary geometry.

An outline of turbulent geometry is aided by categorizing according to dimensionality: points, lines, surfaces and volumes (limited to three spatial dimensions in physical cases).* For each we can ask three kinds of questions: (1) what types are "naturally" identifiable in turbulent flows? (2) what roles do they play, or what properties do they have? (3) what stochastic games can we invent which share some of the difficulties of the turbulent case, but are more tractable?

The heuristic outline in this paper is far from exhaustive; it is intended only to suggest problems, some of which have been solved in part.

Some of the mathematical questions raised may already have been worked out in publications unknown to me. I can only apologize for my lack of professional familiarity with the literature of mathematical statistics, signal processing, pattern recognition, etc. It is likely that relevant turbulence references have been missed as well.

---

[o] Supported in part by U. S. Office of Naval Research

* We should note the non-integer dimensional concepts introduced by Mandelbrot (e.g. 1967) into the mathematical representation of real processes.

I. SOME POINT QUESTIONS IN TURBULENT FLOW

A. Points of Possible Physical Interest.

(1) fluid material points

(2) positions in space

(3) "initial" positions of material points

(4) field extrema, e.g. temperature or pressure maxima or minima, zero velocity ("stagnation") points

(5) field "saddle points"

(6) line/surface intersections

B. Specific Turbulence Point Questions

(1) Dispersion of fluid points from a fixed location. In the simplest cases the mean displacement is zero, so we seek the mean square displacement as a function of time. To identify the "particle" conveniently, we express displacement in "material coordinates", often called "Lagrangian coordinates." Taylor's (1921) analysis showed the simple dependence of mean square displacement on velocity autocovariance function in this frame. But the statistical data and theoretical results most commonly available are the low order moments of the velocity field in "spatial coordinates", often called "Eulerian." The kinematic challenge, then, is to estimate the Lagrangian frame velocity moments as functions of Eulerian moments. Lumley (1962a, 1962b) has shown that an exact connection exists only at the functional level. Reviews have been given by Batchelor and Townsend (1956), Hinze (1959), Corrsin (1962), Tennekes and Lumley (1971), and others.

(2) The inverse problem: "dispersion" of starting positions for material points passing a fixed spatial location at time t. In devising a theory for gradient heat transfer in isotropic turbulence, Corrsin (1952) assumed that this mean square (Eulerian) "backward displacement" had the same statistical properties as Taylor's (Lagrangian) forward displacement. Simple analysis (Corrsin, 1971a) shows that this is formally correct. Also the mean square initial position "displacement" can be related to an Eulerian velocity autocorrelation function just as the ordinary mean square displacement

is to the Lagrangian one.

(3) Relative dispersion. Two point joint dispersion from two fixed starting positions has been studied by Richardson (1926), Obukhov (1941), Brier (1950), Batchelor (1950, 1952a), Kistler (1956), Lin (1960), Roberts (1961), Corrsin (1962), Kraichnan (1966), and others. Reviews have been presented by Batchelor and Townsend (1956) and by Corrsin (1962).

Apparently no one has yet studied the problem of "relative backward dispersion", i.e. the relative starting positions of two material points which pass two Eulerian coordinate positions at any specified time.

(4) Probability of return to starting point. In the simplest random walks on lattices, Polya (1921) showed that the probability of return to the starting point is 1.0 in one- and two-dimensional walks, but is smaller than 1.0 for three or more dimensions, decreasing monotonically with increasing dimensionality (e.g., Feller, 1968). No one seems to have explored the question in turbulent material point dispersion. As remarked by Lumley (1962b), a similar phenomenon may occur in appropriate continuous velocity fields.

The return probabilities for fields which are inhomogeneous or non-stationary, or have superimposed mean velocity gradients may also prove interesting.

(5) Statistical properties of "signals" detected by a moving probe. In turbulent fluid having a mean speed relative to a "fixed" observer (the common experimental case), the data recorded are effectively by a probe moving relative to the simplest Eulerian frame, the frame moving with the mean flow velocity. It is important to be able to transform the measured moments into those which would be measured by (hypothetical) probes moving with the mean flow. Lacking a general solution to that problem, experimenters, led by Favre, Gaviglio and Dumas (1951), have turned to multiple probes and appropriate time-delay devices.

For mean flow speeds much larger than the root-mean-square turbulent velocities

Taylor (1938b) suggested the approximation of interpreting time derivatives of the signal as space derivatives in the mean flow direction (Lumley, 1965; Comte-Bellot and Corrsin, 1971). When the "Taylor approximation" is not fully justified, or when the probe is mounted on a generally moving vehicle, such as an airplane or submarine, we meet general questions on the interpretation of moments of random fields as "sampled" along deterministic paths in space-time.

The Eulerian $\longrightarrow$ Lagrangian moment problem is that of interpreting moments of random fields as sampled along a special class of random paths in space-time: this random path is deterministically related to the random field being sampled.

(6) The displacement of a (hypothetical) point which "tracks" along a turbulent streamline at any instant. For any turbulence field, there exists at each instant a continuous, random field of "streamlines", defined in fluid mechanics as lines everywhere parallel to the velocity (vector) field. Kraichnan (1970) has computed some statistical properties (in simulated turbulence) of the field as sampled along streamlines. He has also raised questions analogous to the Polya problem for streamlines used as trajectories.

The "backward" tracking statistics along the streamline which ends at a stagnation point on the nose of a small probe inserted into turbulent flow may be important for understanding the dynamic readings of "pressure tubes" in turbulence.

II. SIMPLER STOCHASTIC PROBLEMS SUGGESTED BY I.

A. The "Walk on a Random Field."

Classical random walks are, in the language of fluid mechanics, Lagrangian processes. They have no Eulerian field representation, where the random variables are expressed as field functions of spatial position and time. Therefore a different class of random walks was invented, with both Eulerian and Lagrangian properties (Lumley and Corrsin, 1959; Lumley, 1960). Since the Eulerian description is a random velocity field in space-time, these might be called "walks on random fields". A major goal is to compute the material point displacement statistics as a function of Eulerian field moments. A simple (Markoffian) class of fields was treated in the first paper cited above. A more general class was pursued on a computer by Patterson and Corrsin (1966). This work might well be extended to more than one space dimension and to other properties which have been studied in classical random walks, including relative dispersion problems.

B. The Polya Problem in Random Walks Which are Inhomogeneous or Unstationary, or Have Superimposed Mean Velocity Fields or Probability Biases.

As a simple example, consider the probability of return to the starting point for a walk with step length (or step "speed") which decreases or increases monotonically with step number (or "time"). Another direct generalization is the walk whose steps change monotonically with distance from the starting point (analogous to Brownian motion in an inhomogeneous medium?).

C. Statistical Properties of a Random Function (or Sampling) of a Random Function.

Suppose we are given the joint statistical properties of the random functions $f(x)$ and $\xi(x)$. What are those of $g(x) \equiv f[x + K\xi(x)]$? K is constant. When f and ξ are statistically independent there is, of course, no difficulty (Corrsin, 1962). When we seek the covariance of g, we can get some explicit results for small enough K by power series.

There is a growing literature on such "delay" function problems in the control theory journals and elsewhere.

D. The "Self-Dependent Function."

One of the peculiarities of the Eulerian $\longrightarrow$ Lagrangian moment transformation is that the random sampling trajectory in each realization of the random Eulerian velocity field is deterministically dependent on the form of the individual realization function (see, for example, Corrsin, 1961. This suggests a class of functional analysis problems which involve what might be called "self-dependent functions". A non-stochastic example: suppose $f[f(x)] = \phi(x)$, with $\phi(x)$ given. What is $f(x)$? A particular solution is $f(x) = x^{\left(\frac{\log \phi}{\log x}\right)}$.

In a random case we may, for example, ask for the covariance of

$b(x) \equiv a[x + F\{a(x)\}]$, given the form of $F(n)$ and covariance of $a(x)$. In turbulence, F is a functional.

III. SOME LINE QUESTIONS IN TURBULENT FLOW

A. Lines of Possible Physical Interest

(1) particle paths (the trajectories followed by fluid material points)

(2) streamlines

(3) fluid material lines, including "streak lines" (the line of material points which have passed a fixed or moving "tagging" position)

(4) vortex lines

(5) strain-rate principal axis lines

(6) scalar field gradient lines

(7) (in 2-dimensions) scalar field constant-value lines, such as isotherms, isobars, etc.

B. Some Line Shape and Evolution Questions in Turbulence.*

(1) Average line length per unit volume and/or per unit of projected area or length along spatial surfaces or lines. The search for connections among various statistical measures in random geometry dates at least as far back as Buffon (1777). Some of the work has been summarized by Kendall and Moran (1963). Turbulence-inspired relations on the topic of this paragraph were deduced by Corrsin (1955), and generalized by Corrsin and Phillips (1961), Pawula (1963; 1968) (motivated by statistical communica-

---

* simple excursion statistics of path lines and streamlines were included in "points"

tion theory) and Lumley (1964).

(2) Treatment of random, multiple-valued functions. The study of turbulent flows sometimes forces us to cope with functions which are continuous and smooth but have "overhang", i.e. multiple-values. Apparently mathematicians and statisticians have not concerned themselves with such entities. Corrsin and Phillips (1961) showed a simple relation between contour length and numbers of intersections with sampling lines or planes. Lumley (1964) went farther, generalizing the concept of probability density function for these and even "worse" functions. His generalized density integrates to the "average valuedness" of the multiple-valued function (instead of to 1.0), and reduces to the classical density for traditional (single-valued) ones.

(3) Mean growth of fluid material lines in stationary, isotropic, constant density turbulence. Interest in fluid material line growth in turbulent flow originated with Taylor's (1938a) inference of the mean growth of vortex lines from experimental determination of the corresponding (vorticity amplification) term in the dynamic equation for mean square vorticity. At large enough Reynolds numbers, vortex lines are approximately fluid lines.

Of course vortex line behavior is not typical of fluid line behavior, even in inviscid flow, where all vortex lines are fluid lines. Fluid line growth was first examined by Batchelor (1952b), who offered the plausible conjectures that (a) fluid lines do in fact grow on the average, and that (b) after large enough time they grow exponentially. Reid (1955) confirmed Batchelor's conjectures under the assumption that fourth order cumulants could be neglected. Corrsin and Karweit (1969) presented some small-time data on actual line growth.

Cocke (1969) in a landmark paper gave the first formal proof of growth, and his proof has been simplified by Orszag (1970). It appears that growth can be demonstrated even more simply as a consequence of the tendency of two fluid points to separate (Corrsin, 1971b).

(4) Other line properties. It may be interesting to explore such line properties as mean square curvature and torsion, average densities of inflection points, etc.

(5) Knottedness of vortex lines. Moffatt (1969) has demonstrated the importance of the degree of "knottedness" of isolated vortex tubes ("filaments") in otherwise irrotational flow fields. It is unknown whether vortex field lines in turbulence are knotted.

(6) Statistical properties of streamlines in "two-dimensional turbulence." Although the physical existence of two-dimensional turbulence is controversial (some atmospheric motions behave in some ways like two-dimensional flows), it is worth noting that the existence of the scalar "stream function" simplifies the writing of moment equations related to streamline behavior.

IV. SIMPLER STOCHASTIC PROBLEMS RELATED TO III

A. Random Geometry Involving Lines.

It is clear that the questions discussed under III. B. (1) have intrinsic mathematical interest apart from their appearance in turbulence problems. Furthermore they occur in many other areas of scientific research (Kendall and Moran, 1963). Lumley's generalization of the probability density function also has interest outside of turbulent flows.

B. Line Growth Processes Associated with Random Walks.

Random "material line growth" processes can be based on the random walk. In fact, the increase in mean distance between two points walking on the same lattice may be interpreted as line growth, if we imagine the two points always connected by a straight line. It is virtually obvious from elementary random walk theory that this "simple line growth" is at a power law rate. For independent walks it generalizes trivially to an infinite row of points. The correlated walk case will be more interesting.

We can also devise a random line walk process showing growth analogous to Batchelor's exponential conjecture for turbulence. We need only set rules which increase the number of randomly walking points in proportion to the line length. One such "compound line growth" process is defined by creating a new walking point wherever the growing line crosses a lattice line (Karweit and Corrsin, 1971).

It may be interesting to extend this work to non-stationary walks and to the "walks on random fields" (Section II. A).

V. SOME SURFACE QUESTIONS IN TURBULENT FLOW

A. Surfaces of Possible Physical Interest.

(1) fluid material surfaces (including "streak surfaces")

(2) trajectory surfaces swept out by fluid lines

(3) (in 3 dimensions) constant-property surfaces, such as isotherms and isobars

(4) interfaces between fluid regions of sharply differing statistical properties in the turbulence itself. A familiar example is the boundary between turbulent and non-turbulent fluid (see below).

B. Some Surface Shape and Evolution Questions in Turbulence

(1) Statistical properties of the turbulent-flow/potential-flow interface at a "free-stream" boundary. This is the physical entity observed by Corrsin (1943) and pursued quantitatively in the experiments of Townsend (1948), Corrsin and Kistler (1955), Fiedler and Head (1966), Kovasznay, Kibens and Blackwelder (1970), and others. This irregular boundary surface is often multiple-valued, so Lumley's (1964) ideas will be useful.

(2) Statistical properties of the turbulent-flow/laminar-flow interface. A stage sometimes observed in the transition from a fully laminar shear flow to a fully turbulent one features isolated turbulent patches of fluid growing within the laminar region [Charters (1943), Emmons (1951), Rotta (1956), Lindgren (1960, etc.), Klebanoff, Tidstrom and Sargent (1962), Kovasznay, Komoda and Vasudeva (1962), and many

others]. Presumably the turbulent patch grows both by direct shear at the interface and by destabilizing the laminar fluid. No satisfactory dynamic theory exists [the most serious effort seems to have been by Mitchner (1952)], but the interface geometry may be relevant.

(3) Mean growth of fluid material surfaces in stationary, isotropic, constant density turbulence. As pointed out by Batchelor (1952b), this topic relates to "indelible" scalar field gradient amplification in a way much like that in which material line growth relates to "indelible" solenoidal vector field amplification. He conjectured that (a) these surfaces grow on the average, and that (b) asymptotically this growth is exponential. The line growth proofs of Cocke and Orszag presumably can be adapted to prove surface growth.

(4) Free-surface geometry on random surface waves. Although random ocean waves, for example, need not involve turbulent motion, it is possible that many of them do, especially at large amplitudes (see, for example, Phillips, 1966). Longuet-Higgins (1957; 1960) has explored many of the geometrical properties of single-valued surfaces, especially those suitable for describing waves.

(5) Evolution of constant-property surfaces in the presence of molecular diffusion. With molecular diffusion (e.g. of heat in a temperature field) the problem is enlivened by a local "propagation velocity" down the scalar gradient. Gibson (1968) has investigated some of the consequences during isotropic turbulent mixing.

VI. SOME SIMPLER STOCHASTIC PROBLEMS RELATED TO V

A. Random Geometry Involving Surfaces.

In turbulence studies and other areas people have had to look for measures of average surface area per unit volume and/or per unit of projected area on a spatially defined coordinate surface. Such matters have been discussed by Corrsin (1955), Corrsin and Phillips (1961), Lumley (1964) and others (see, for example Kendall and Moran 1963).

In addition to area questions we may ask about such detailed characteristics as curvature distributions, extremum and saddle point densities, etc.

B. Surface Growth Processes Associated with Random Walks.

It is obvious that the simple and compound line growth processes mentioned in IV. B (Karweit & Corrsin, 1971) can be extended to surfaces as well.

VII. SOME VOLUME QUESTIONS IN TURBULENT FLOW

A. Volumes of Possible Physical Interest.

(1) fluid material volumes

(2) regions enclosed by constant-property surfaces

(3) regions containing sharply contrasting statistical properties of the turbulence — especially regions containing the relatively localized "fine-structure" of the velocity field (Batchelor and Townsend, 1949; see also Batchelor, 1953; Kuo and Corrsin, 1971a, 1971b).

B. Some Volume Shape and Evolution Questions in Turbulence.

(1) Turbulent distortion of an initially spherical, "very small" fluid material blob. The obvious questions are (a) what do we mean by "very small?" (b) what is a complete catalog of possible shape classes? (c) is there a preferred class on the average? (d) if so, which one is it? The work of Townsend (1951a, 1951b), Batchelor (1952a, 1952b), Reid (1955) and Betchov (1956) has contributed to some partial answers to these questions.

"Very small" means small enough to experience essentially uniform strain during a significant time interval. This requires an initial diameter much smaller than the "Kolmogorov microscale." Then the distorted shape will be a general ellipsoid, so the classes are characterized simply by principal axis ratios. The statistically preferred axis ratios are still an open question. Townsend's (1951b) hot spot data are inconclusive on this point, but led him to speculate that his (rather large) "spot" was distorting into a crinkled line or ribbon. His high wave number spectra, compared with vortex-sheet and vortex-line theoretical models of turbulent fine structure (1951a), favor the former. The models have been questioned by Batchelor (1959), and perhaps a more important result is Townsend's observation that the mean product of the principal strain rates is measured to be negative, thus precluding the possibility that the mean local deformation is an axi-symmetric stretching.

Betchov (1956) invoked formal mathematical inequalities to infer a tendency to distortion into highly oblate ellipsoids, with two axes much larger than the third.

(2) Statistical shapes of the velocity fine-structure regions in fully turbulent fluid. These regions are not as small as the Kolmogorov microscale, so their shapes are irregular. A first crude attempt to classify and measure these regions has been made by Kuo and Corrsin (1971b). Restricting the interpretation to only three shape categories, which might be described as "sphere-like," "rod-like" and "slab-like," we find a tendency to the rod-like class.

(3) Tendencies toward isotropy in the isotropic turbulent mixing of initially oriented scalar fields. Nothing definitive seems to have been done on this problem. Related experiments, not designed for this purpose (Mills, Kistler, O'Brien and Corrsin, 1958), showed that when a typical isotropic turbulence generating grid is heated, the concomitant temperature fluctuation field becomes essentially isotropic at the same downstream distance as the velocity field. Batchelor's (1952a) 2-particle dispersion study is relevant, as is an unpublished report by O'Brien (1963). O'Brien finds no rigorous approach to isotropy, but he makes the conjecture that the very small structure will approach isotropy with an inverse time scale of the order of the r.m.s. turbulent vorticity. He speculates that the very large structure will approach isotropy as the mean square fluid point displacement becomes much greater than the square of the (Eulerian) integral length scale of the turbulence.

A related topic is the angular dispersion of fluid material lines (Corrsin & Karweit, 1968, 1971b) and surfaces.

VIII. SIMPLER STOCHASTIC PROBLEMS RELATED TO VII

A. Taxonomy of Random Shapes, and the Corresponding "Pattern Recognition" Problem.

There is a large and burgeoning published literature on questions of "pattern recognition," much of it in the electrical engineering and communication theory journals.

B. Approaches Toward or Away From Isotropy in Random Walk Problems.

In a special sense the 4-direction symmetric walk on a square, 2-dimensional lattice may be viewed as isotropic motion. A useful exercise would be the computation of the approach toward isotropy of the two particle walk, first when they move independently, then when the walks are correlated. This is an exercise in the angular dispersion of the line connecting them.

More challenging problems would be (a) the case with several particles alined at the start, (b) angular dispersion in the compound line growth problem (a computer experiment on this is included in Corrsin & Karweit (1971b) , (c) a two particle "walk on a random field."

As counterpoint problems we might study the degrees of anisotropy imposed on initially isotropic fields of points by assorted species of non-isotropic walks.

C. Convective Approaches Toward or Away From Isotropy in Simple Random Flow Ensembles.

Because of the "parametrically driven" character of the heat and mass flow equations (i.e. the convective terms $u_k \frac{\partial T}{\partial x_k}$, where u_k is a velocity component, x_k is a spatial coordinate), the two temperature (or concentration) fields resulting from convection by two simple flows cannot be merely added to yield the field which would result from convection by the combined flows. The equation is "quasi-non-linear." We may ask, however, for the average behavior of a random ensemble of simple flows. A first attempt at this exercise (Corrsin and Karweit, 1971a) suggests that an isotropic ensemble of single Fourier velocity modes does tend to make an oriented scalar field more nearly isotropic in the ensemble average.

D. Statistical Properties of Intermittent Functions.

In probing and measuring flows which have turbulent/non-turbulent interfaces and other statistical two-state behaviors, we cannot avoid "intermittent" random functions. The simplest mathematical cases, such as a Gaussian process multiplied by an independent random (on-off) square wave, are text book problems. Cases in which the discontinuous state changes are statistically or deterministically related to the details of the signal pose more interesting questions.

REFERENCES

Batchelor, G. K. Quart. J. Roy. Met. Soc. 76, 133 (1950).

Batchelor, G. K. Proc. Cambr. Phil. Soc. 48, 345 (1952a).

Batchelor, G. K. Proc. Roy. Soc. A 213, 349 (1952b).

Batchelor, G. K. "The Theory of Homogeneous Turbulence," Cambr. Univ. Press (1953).

Batchelor, G. K. J. Fluid Mech. 5, 113 (1959).

Batchelor, G. K. & Townsend, A. A. Proc. Roy. Soc. A 199, 238 (1949).

Batchelor, G. K. & Townsend, A. A. in "Surveys in Mechanics, G. I. Taylor 70th Anniversary Volume" ed. by G. K. Batchelor & R. M. Davies, Cambr. Univ. Press (1956).

Betchov, R. J. Fluid Mech. 1, 497 (1956).

Brier, G. W. J. Meteorol. 7, 283 (1950).

Buffon, G. "Essai d'Arithmétique Morale," Suppl. à l'Histoire Nat. 4 (1777).

Charters, A. C., N.A.C.A. Tech. Note 891 (1943).

Cocke, W. J. Phys. of Fluids 12, 2488 (1969).

Comte-Bellot, G. & Corrsin, S. J. Fluid Mech. 48, 273 (1971).

Corrsin, S. N.A.C.A. Adv. Confid. Rept. 3L23 (1943) [reissued as N.A.C.A. Wartime Rept. W-90].

Corrsin, S. J. Appl. Phys. 23, 113 (1952).

Corrsin, S. Quart. Appl. Math. 12, 404 (1955).

Corrsin, S. in "Mécanique de la Turbulence," ed. du C.N.R.S., Gordon & Breach, 27 (1962).

Corrsin, S. (submitted to Phys. of Fluids, 1971a).

Corrsin, S. (submitted to Phys. of Fluids, 1971b).

Corrsin, S. & Karweit, M.J. Bull. Am. Phys. Soc. 13, 805 (1968).

Corrsin, S. & Karweit, M. J. J. Fluid Mech. 39, 87 (1969).

Corrsin, S. & Karweit, M. J. (in these proceedings 1971a).

Corrsin, S. & Karweit, M. J. (submitted to J. Fluid Mech. 1971b).

Corrsin, S. & Kistler, A. L., N.A.C.A. Rep. 1244 (1955).

Corrsin, S. & Phillips, O. M. J. Soc. Ind. & Appl. Math. 9, 395 (1961).

Favre, A., Gaviglio, J. & Dumas, R. Publ. Sci. et Tech. de Minist. de l'Air, No. 251, 293 (1951).

Feller, W. "An Introduction to Probability Theory and Its Applications", (Vol. I, 3rd ed., p. 360), J. Wiley & Sons (1968).

Fiedler, H. & Head, H. R. J. Fluid Mech. 25, 719 (1966).

Gibson, C. H. Phys. of Fluids 11, 2305 (1968).

Hinze, J. O. "Turbulence", McGraw-Hill (1959).

Karweit, M. J. & Corrsin, S. (in these proceedings, 1971).

Kendall, M. G. & Moran, P. A. P. "Geometrical Probability", C. Griffin (1963).

Kistler, A. L. "Measurement of Joint Probability in Turbulent Dispersion of Heat from Two Line Sources," Part II of Ph.D. diss., The Johns Hopkins Univ. (1956).

Klebanoff, P. S., Tidstrom, K. D. & Sargent, L. M. J. Fluid Mech. 12, 1 (1962).

Kovasznay, L. S. G., Kibens, V. & Blackwelder, R. F. J. Fluid Mech. 41, 283 (1970).

Kovasznay, L. S. G., Komoda, H. & Vasudeva, B. R. Proc. 1962 Ht. Trans. & Fluid Mech. Inst. 1, Stanford Univ. Press (1962).

Kraichnan, R. H. Phys. of Fluids 9, 1937 (1966).

Kraichnan, R. H. Phys. of Fluids 13, 22 (1970).

Kuo, A. Y.-S. & Corrsin, S. J. Fluid Mech. (in press, 1971a).

Kuo, A. Y.-S. & Corrsin, S. (submitted to J. Fluid Mech., 1971b).

Lin, C.C. Proc. Nat. Acad. Sci. 46, 1147 (1960).

Longuet-Higgins, M. S. Phil. Trans. Roy. Soc. A 249, 321 and 250, 157 (1957).

Longuet-Higgins, M. S. J. Opt. Soc. Am. 50, 838 (in 3 parts) (1960).

Lumley, J. L. Appl. Sci. Res. A 10, 153 (1960).

Lumley, J. L. J. Math. Phys. 3, 309 (1962a).

Lumley, J. L. in "Mecanique de la Turbulence", Ed. du C.N.R.S., Gordon & Breach, 17 (1962b).

Lumley, J. L. J. Math. Phys. 5, 1198 (1964).

Lumley, J. L. Phys. of Fluids 8, 1056 (1965).

Lumley, J. L. & Corrsin, S. in "Advances in Geophysics" 6,179, Acad. Press (1959).

Mandelbrot, B. Science 156, 636 (1967).

Mills, R. R., Kistler, A. L., O'Brien, V. & Corrsin, S. N.A.C.A. Tech. Note 4288 (1958).

Mitchner, M. "The Propagation of Turbulence into a Laminar Boundary Layer", Ph.D. diss., Harvard Univ. (1962).

Moffatt, H. K. J. Fluid Mech. 35, 117 (1969).

O'Brien, E. E. "On the Behavior of Passive Scalars in a Turbulent Fluid", unpub. rep., Coll. of Eng'g. State Univ. of N.Y. at Stony Brook (1963).

Obukhov, A. M. Izv. Akad. Nauk., Ser. Georgr. i. Geofiz. 5, 453 (1941).

Orszag, S. A. Phys. of Fluids 13, 2203 (1970).

Patterson, G.S. & Corrsin, S. in "Dynamics of Fluids & Plasmas" (ed. by S. I. Pai et al.) Acad. Press (1966).

Pawula, R. F. IEEE Trans. on Info. Theo. IT-9, 208 (1963).

Pawula, R. F. IEEE Trans. on Info. Theo. IT-14, 770 (1968).

Phillips, O. M. "The Dynamics of the Upper Ocean" Cambr. Univ. Press (1966).

Polya, G. _Math. Annalen 84_, 149 (1921).

Reid, W. H. _Proc. Cambr. Phil. Soc. 51_, 350 (1955).

Rotta, J. Ing.-Archiv 24, (1956).

Richardson, L. F. _Proc. Roy. Soc. A 110_, 709 (1926).

Roberts, P. H. _J. Fluid Mech. 11_, 257 (1961).

Taylor, G. I. _Proc. London Math. Soc. A 20_, 196 (1921).

Taylor, G. I. _Proc. Roy. Soc. A 164_, 15 (1938a).

Taylor, G. I. _Proc. Roy. Soc. A 164_, 476 (1938b).

Tennekes, H. & Lumley, J. L. "_A First Course in Turbulence_" M.I.T. Press (in press, 1971)

Townsend, A. A. _Proc. Roy. Soc. A 208_, 534 (1951a)

Townsend, A. A. _Proc. Roy. Soc. A 209_, 418 (1951b)

Townsend, A. A. _Austral. J. Sci. Res. A 1_, 161 (1948)

SIMPLE AND COMPOUND LINE GROWTH IN RANDOM WALKS [#]

by

M. J. Karweit [*]
and
S. Corrsin
Mechanics Department, The Johns Hopkins University

Cocke (1969) has proved Batchelor's (1952) conjecture that fluid material lines grow on the average in stationary, isotropic, constant density turbulence. No firm experiment or theoretical analysis has yet supported or contradicted Batchelor's second plausible conjecture, i.e. that this growth would be asymptotically exponential. Reid (1955) showed, however, that the cumulant discard hypothesis [now known to be risky (O'Brien &Francis, 1962; Ogura, 1963)] yields exponential growth of the mean square length of infinitesimal lines.

Our purpose here is to introduce line growth problems associated with discrete random walks. The "line" is drawn as straight segments connecting "particles" which were initially adjacent. We propose two classes: (I) "simple line growth," in which the number of walking points is fixed, and (II) "compound line growth," in which additional walking points may be created at each step, so that the mean number of walking points per unit of unit of line length may approach a constant.

"Simple line growth" is merely the relative dispersion between sequential particles, thus is expected to give a power-law dependence of length on time (i.e. on number of steps). "Compound line growth," however, should be qualitatively more rapid. It is in fact de-

---

Supported by U.S. Office of Naval Research and The Johns Hopkins University. We should also like to acknowledge the encouragement of The Society for Statistical Geometry.

* Also Chesapeake Bay Institute

signed to behave like Batchelor's interpretation of the turbulent case.

Figures 1a and 1b illustrate two members of the contrasting ensembles, after 3 steps. The numbered points are at the same positions in the two realizations. Since the points walk independently in these examples, two points

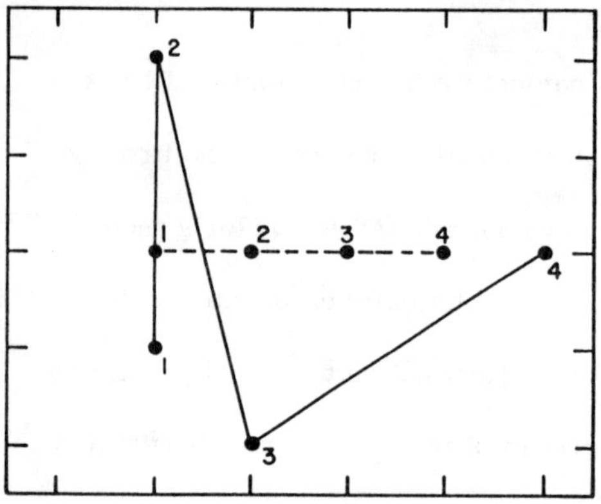

(a) "simple"

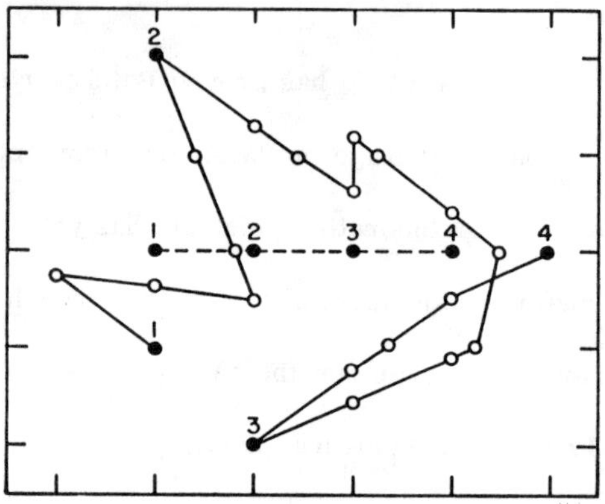

(b) "compound"

Figure 1. Line growth generated by random walks after 3 steps of the primary points.

suffice for computation, although more have been included in the figure.

I. Simple Line Growth

On a plane square lattice we begin with a row of points (or "particles") along the x -axis. The line connecting them is the "material line" in its initial position. Each particle has probability 1/4 of moving in each of the 4 Cartesian directions. We restrict to independently walking points, hence need consider only the distance $\Delta(n)$ between two which are initially adjacent, as a function of step number, n , the equivalent of time in the simplest case.

$$\Delta(n) = \left\{ \left[X'(n) - X(n) \right]^2 + \left[Y'(n) - Y(n) \right]^2 \right\}^{\frac{1}{2}} \qquad , (1)$$

where X, Y and X', Y' are the Cartesian coordinates of the two "particles," and are four statistically independent variables. Because of the independence, the joint probability density is simply the product of four binomial densities, but summing the product of (1) and this density seems formidable, so we confine the analysis to a direct asymptotic estimate for $n >> 1$.

We write $X' - X \equiv \xi$ and $Y' - Y \equiv \eta$. Then the "conditional" average of $\Delta(n+1)$ [for any $\Delta(n)$] can be written as the appropriately weighted sum of the outcomes of the 16 possible combinations of changes in (ξ, η):

$$\xi(n+1) - \xi(n) = \pm 1 \text{ or } \pm 2 \quad ; \quad \eta(n+1) - \eta(n) = \pm 1 \text{ or } \pm 2 \qquad .(2)$$

The step length is 1.0 in each direction. The conditional average change in line length is, writing $\xi(n)$ as ξ and $\eta(n)$ as η,

$$\bar{\Delta}(n+1) - \Delta(n) = \frac{1}{16}\Big\{4\sqrt{\xi^2+\eta^2} + \sqrt{\xi^2+(\eta-2)^2} + \sqrt{\xi^2+(\eta+2)^2}$$

$$+ \sqrt{(\xi-2)^2+\eta^2} + \sqrt{(\xi+2)^2+\eta^2} + 2\sqrt{(\xi-1)^2+(\eta-1)^2} + 2\sqrt{(\xi-1)^2+(\eta+1)^2}$$

$$+ 2\sqrt{(\xi+1)^2+(\eta-1)^2} + 2\sqrt{(\xi+1)^2+(\eta+1)^2}\Big\} - \sqrt{\xi^2+\eta^2} \qquad .(3)$$

To find an asymptotic $(n >> 1)$ form, we assume that this will imply $\Delta >> 1$. We then factor Δ out of each square root in (3) and use the binomial expansion on each, as far as terms quadratic in Δ^{-1}. The resulting estimate is

$$\bar{\Delta}(n+1) - \Delta(n) \xrightarrow{\sim} \frac{1}{2}\Delta^{-1} \qquad .(4)$$

This is independent of the individual values of $\xi(n)$ and $\eta(n)$. $\bar{\Delta}(n+1)$ is still a conditional average. To get the complete result, we would have to average (4) in terms of the proper joint densities of $\xi(n)$ and $\eta(n)$. But (4) is sufficient to tell us that the <u>continuous limit</u> will be a <u>power law</u> of form

$$\langle \bar{\Delta} \rangle \sim t^{1/2} \quad ,(5)$$

the same behavior as the more accessible r. m. s. value, $\langle \Delta^2 \rangle^{1/2}$.

We have run a 1000-realization computer experiment as a check, and find

$$\langle \Delta_c \rangle \approx 1.24 \; t^{1/2} \quad .(6)$$

II. An Example of Compound Line Growth

As before, two points (X, Y) and (X', Y') perform simultaneous, independent random walks on a 2-dimensional square lattice (unity on a side). At each step each moves either ± 1 in the x-direction or ± 1 in the y-direction. As before the "material line" is initially on the x-axis, with the initial collection of "particles" occupying all lattice points on this axis, although we need treat only two.

The generalization is that a new independently walking point is created on the "material line" wherever a segment of it crosses a lattice line, after each set of single steps. As these new interior points move independently (with unity step length) thereafter, the line bends into more and more connected segments, whose sum defines its length. Note that this modification of the game imposes an upper bound length scale on the structure of the growing line: no segment (distance between neighboring particles) can be greater than $\sqrt{2}$.

In the realization of Figure 1b, several new points (joints?) have been created in the first 3 steps. The step length is maintained at 1.0 for the new points as well. The rate at which a line grows is evidently monotonic with the rate of generation of new points per step, per segment. Estimation of this rate turns out to be a problem in counting.

To begin, we recognize 5 categories of segments, classified according to how their end points lie on the lattice lines (Figure 2):

Class	Description
A	on perpendicular grid lines, excluding intersections
B	on parallel grid lines, excluding intersections
C	on the same grid line, with one on an intersection

D	on the same grid line, at adjacent intersections
E	on different grid lines, with one on an intersection

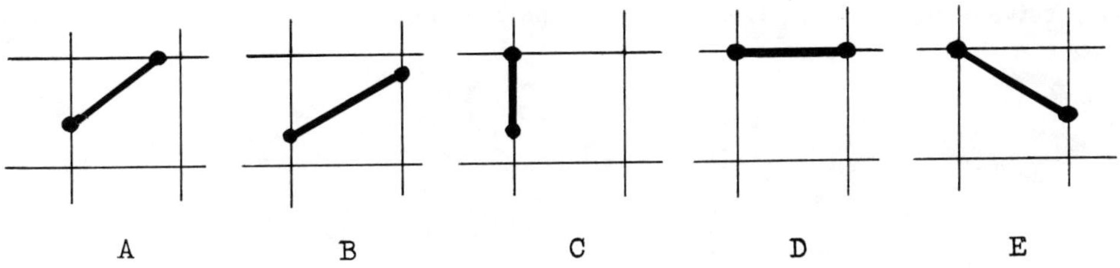

Figure 2: The 5 classes of line segments in compound line growth

After the process has advanced a few steps from the start we ordinarily find some members of all 5 categories on the "material line." The transition from the configuration after n steps to that after $n+1$ steps can be computed by considering the outcomes of all 16 possible changes for a member of each class. For one member of each class (before the step) we can then count the population of each class after the step. This gives us a "transition matrix" $\underline{\underline{M}}$ which can be said to operate on the nth "population vector" $\underline{P}_n$ to give the $(n+1)^{st}$ "population vector" $\underline{P}_{n+1}$:

$$\frac{1}{16}\begin{pmatrix} 20 & 11 & 4 & 0 & 0 \\ 10 & 15 & 2 & 0 & 0 \\ 0 & 8 & 6 & 0 & 10 \\ 0 & 0 & 3 & 12 & 12 \\ 0 & 0 & 10 & 12 & 12 \end{pmatrix} \begin{pmatrix} A \\ B \\ C \\ D \\ E \end{pmatrix}_n = \begin{pmatrix} A \\ B \\ C \\ D \\ E \end{pmatrix}_{n+1} \quad (7)$$

or

$$\underline{\underline{M}}\,\underline{P}_n = \underline{P}_{n+1} \quad .(7a)$$

The transition from configuration $\underline{P}_n$ to $\underline{P}_{n+k}$ simply recycles eqn. (7):

$$\underline{\underline{M}}^k\,\underline{P}_n = \underline{P}_{n+k} \quad .(8)$$

We are especially interested in the possible existence of stationary asymptotic behavior, and in the corresponding growth rate if stationarity can be found. Such behavior can be characterized by simple proportionality of the population vectors in any two successive configurations, i.e. we try out the condition

$$\underline{P}_{n+1} = G \, \underline{P}_n \tag{9}$$

or

$$\underline{P}_{n+k} = G^k \, \underline{P}_n \tag{10}$$

for large n . G is the average growth per step, the growth rate.

Combining (8) and (9),

$$\left\{ \underline{\underline{M}} - G \, \underline{\underline{1}} \right\} \underline{P}_n = 0 \quad , \tag{11}$$

which specifies our eigenvalue problem. The relevant solution turns out to be

$$\left.\begin{array}{l} G = 1.903 \\ \text{with } \underline{P}_n \text{ components} \\ \begin{array}{ll} A = 0.2491 & D = 0.1303 \\ B = 0.2602 & E = 0.2004 \\ C = 0.1599 & \end{array} \end{array}\right\} . \tag{12}$$

This asymptotic result is independent of the starting configuration of the material line.

Having found a constant positive G , we can in principle compute the average growth rate in line length, if we were first to obtain a constant asymptotic average segment length, $\langle \ell \rangle$. The product would give a geometric progression for line growth. In the continuous limit $(n \to \infty)$, this would imply exponential growth.

A theoretical estimate of $\langle \ell \rangle$ has not yet been made, and a computer experiment on this problem is beyond our resources. Therefore we turn to a modified process which is simpler to compute.

III. Compound Line Growth - A Different Case and a Computer Experiment

In order to get a rough check on the above analysis, and to obtain other statistical properties easily, we turn to a closely related process which can be done not only theoretically but also as a short computer experiment. This consisted of randomly walking a pair of points as above, recording the actual components of the "population vector" after each step, then randomly (with uniform probability) selecting a single one of these segments for use in the next step. The distances between the pairs of points were also recorded. This gave an estimate of average line segment length. The process was carried out for 50 steps $(n = 50)$, easily far enough to reach the asymptotic state*, with results independent of starting conditions. The "ensemble average" was taken over 1000 realizations.

This computer experiment is clearly different from the process analyzed in Section II, because it keeps only a single segment after each step. Differences are discussed in the Appendix.

It is not difficult to work out the theoretical probability transition matrix for this modified process. Instead of equation (7) we have

$$\frac{1}{16}\begin{pmatrix} 11\frac{2}{3} & 3\frac{2}{3} & 2 & 0 & 2\frac{1}{6} \\ 4\frac{1}{3} & 8\frac{1}{3} & 1 & 0 & 2 \\ 0 & 4 & 4\frac{5}{6} & 0 & 3 \\ 0 & 0 & 1\frac{1}{6} & 10 & 1 \\ 0 & 0 & 7 & 6 & 7\frac{5}{6} \end{pmatrix} \begin{pmatrix} A' \\ B' \\ C' \\ D' \\ E' \end{pmatrix}_n = \begin{pmatrix} A' \\ B' \\ C' \\ D' \\ E' \end{pmatrix}_{n+1} . \quad (13)$$

Note that the elements of this transition matrix are (properly normalized) conditional probabilities. $\underline{\underline{M}}$ in (7) is not literally a probability matrix.

A stationary state eigenvalue problem was worked out for this case in just the same way as for the previous one. The computer experiment for this case also gave asymptotic average population vector components. The theory and experiment compared as follows:

---

* a close approximation to the asymptotic state was reached by $n = 20$

Modified Compound Line Growth Models
Population Vector Components

Class:	A'	B'	C'	D'	E'
Theory:	0.374	0.272	0.140	0.054	0.160
Experiment:	0.385	0.274	0.145	0.042	0.154

We note that these are roughly the same as in the previous process.

The theoretical and experimental number "growth rates" for this process turns out to be

$$G' = 1.851 \quad ; \quad G'_{Exp.} = 1.872 \qquad , (14)$$

which are surprisingly close to the previous case, considering the differences in the population vectors.

The asymptotic average segment lengths for the 5 categories were computed to be

$\bar{\ell}_{A'}$	$\bar{\ell}_{B'}$	$\bar{\ell}_{C'}$	$\bar{\ell}_{D'}$	$\bar{\ell}_{E'}$	
0.640	1.091	0.486	1.000	1.143	(15)

By forming the average of these, weighted with the population vector components, we arrive at an asymptotic average segment length

$$\langle \ell' \rangle = 0.835$$

If we assume that the category average segment lengths in (15) are also a fair approximation for the previous process (Section II), then its asymptotic average segment length is

$$\langle \ell \rangle \approx 0.880$$

The density of average segment lengths tabulated in the computer experiment, Figure 3, shows a considerable peak at the lattice scale, 1.0.

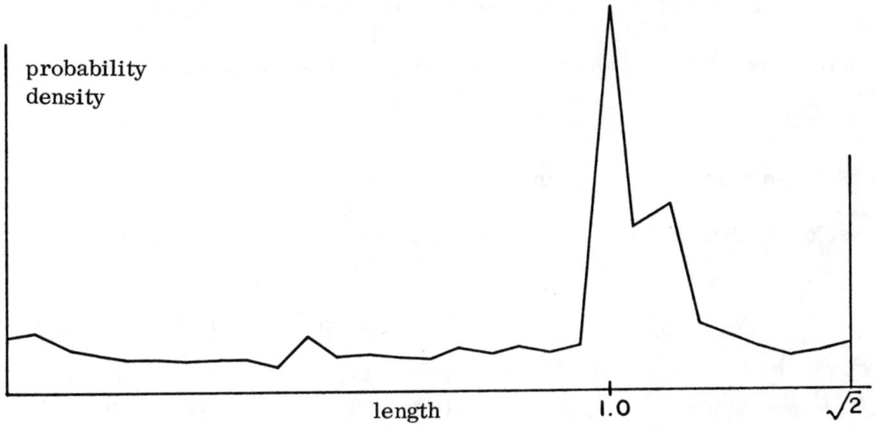

Figure 3. Probability density (actually "frequency function") of segment lengths in compound line growth computer experiment.

The crucial point is not the particular numerical value of $\langle \ell' \rangle$ or $\langle \ell \rangle$; it is the fact that both of these compound line growth processes have asymptotic stationary growing states in which the increase in average length per step is proportional to the average length. In the continuous limit, replacing n by time, t, we get the exponential number and length growth,

$$\langle N \rangle = G^t \quad ; \quad \langle L \rangle = \langle \ell \rangle G^t \qquad . (16)$$

References

Batchelor, G. K. Proc. Roy. Soc. A 213, 349 (1952).

Cocke, W. J. Phys. of Fluids 12, 2488 (1969) [see also Orszag, S. A. Phys. of Fluids, 13, 2203 (1970)].

O'Brien, E. E. & Francis, G. C. J. Fluid Mech. 13, 369 (1962).

Ogura, Y. J. Fluid Mech. 16, 33 (1963).

Reid, W. H. Proc. Cambr. Phil. Soc. 51, 350 (1955).

Appendix

An illustration of why the transition matrix for the modified compound process (Section III) is not the same as that for the more "natural" compound process (Section II) is as follows:

Suppose a line segment is in class A, and all 16 possible moves of its end points are considered. The following elements can be generated for the next step:

①	A	⑤	A	⑨	A, B	⑬	B, A
②	A, A, A	⑥	A	⑩	B, A	⑭	B, B, A
③	A, B, B	⑦	A, B	⑪	A	⑮	A, A, A,
④	A, B	⑧	B, A	⑫	A	⑯	A

The "natural" compound growth random walk makes use of <u>all</u> elements within each of the 16 possible moves, and we see that there are 20 opportunities to generate an A and 10 opportunities to generate a B. Hence the expected relative density of (A, B) is $(\frac{2}{3}, \frac{1}{3})$.

The modified process, on the other hand, uses only one element from each of the possible moves and the probability of obtaining a given type must be conditioned by how many other types occur in that particular move. The relative density of (A, B) is then calculable as

$$A = \frac{1}{16}\left(1 + 1 + \frac{1}{3} + \frac{1}{2} + 1 + 1 + \frac{1}{2} + \frac{1}{2} + \frac{1}{2} + \frac{1}{2} + 1 + 1 + \frac{1}{2} + \frac{1}{3} + 1 + 1\right) = 0.729$$

$$B = \frac{1}{16}\left(\frac{2}{3} + \frac{1}{2} + \frac{1}{2} + \frac{1}{2} + \frac{1}{2} + \frac{1}{2} + \frac{1}{2} + \frac{2}{3}\right) = 0.271$$

It is interesting that the mean growth rates and average segment lengths are so nearly equal for the two compound processes, when the transition probabilities and population vectors are appreciably different.

THE MIXING OF SCALAR STRIPES

BY AN ISOTROPIC ENSEMBLE OF SINGLE VELOCITY MODES#

by

S. Corrsin and M. J. Karweit*
Mechanics Department, The Johns Hopkins University

It seems plausible to expect that the mixing of a homogeneous, initially oriented scalar field (without mean gradients) by stationary isotropic turbulence would tend to make the scalar field increasingly isotropic. Nothing definitive seems to have been done on this problem. Related experiments, not designed for this purpose (Mills, Kistler, O'Brien and Corrsin, 1958), showed that when a typical isotropic turbulence generating grid is heated, the concomitant temperature fluctuation field becomes essentially isotropic at the same downstream distance as the velocity field. When this combined isotropic field was distorted by passage through a contraction, both were observed to return slowly toward isotropy in a straight duct following the contraction (Mills and Corrsin, 1959).

On the theoretical side, Batchelor's (1952) 2-particle dispersion study is relevant, but the only direct attack we have seen on the problem is an unpublished report by O'Brien (1963). He finds no rigorous approach to isotropy, but makes the conjecture that the very small structure will approach isotropy with an inverse time scale of the order of the r.m.s. turbulent vorticity. He speculates that the very large structure will approach isotropy as the mean square fluid point displacement becomes much greater than the square of the (Eulerian) length scale of the turbulence.

A related topic is the angular dispersion of fluid material lines (Corrsin & Karweit 1968, 1971) and surfaces.

Supported by U. S. Office of Naval Research and The Johns Hopkins University

* Also Chesapeake Bay Institute

With adequate modern computer facilities, it would not be difficult to carry out detailed turbulent mixing calculations using the partial differential equation for heat (or mass) transfer.

Here we go to the opposite extreme, and consider one of the simplest convection problems carrying any flavor of the turbulent case: the non-diffusive, plane convection of a single Fourier mode of passive scalar field by an isotropic ensemble of single Fourier modes of velocity.

The transport equation for $\vartheta(\underline{x}, t)$ in an Eulerian frame, without molecular diffusion, is

$$\frac{\partial \vartheta}{\partial t} + \underline{u} \cdot \nabla \vartheta = 0 \quad , (1)$$

where $\underline{u}$ is the velocity field. We assign the initial condition

$$\vartheta(x, 0) = \vartheta_0 \cos(\ell x + \beta) \quad . (2)$$

Any single realization of the ensemble of velocity fields is

$$u(\underline{x}) = U \cos\phi \cos(\underline{k} \cdot \underline{x} + \alpha) \; ; \; v(\underline{x}) = U \sin\phi \cos(\underline{k} \cdot \underline{x} + \alpha) \quad , (3)$$

with the added constraint of the constant density mass balance equation,

$$\underline{u} \cdot \underline{k} = 0 \quad . (4)$$

Equation (1) can be easily integrated by the "method of characteristics"; in this non-diffusive case the characteristics are merely the (straight line) particle paths.

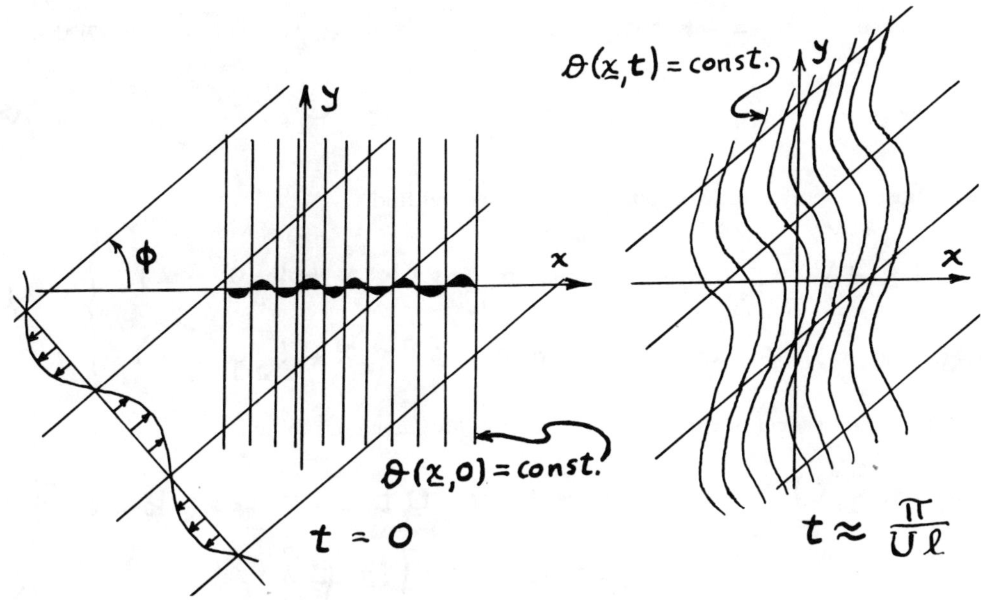

Figure 1: Convective distortion of a sinusoidal scalar field by a sinusoidal velocity field

We used the equivalent strategy of solving in a Lagrangian frame, then transforming the solution back to the Eulerian frame. In (Lagrangian) material coordinates $\underline{X}_o$, equation (1) is

$$\frac{\partial}{\partial t}\,\Theta(\underline{X}_o, t) = 0 \qquad ,(5)$$

where $\underline{X}(\underline{X}_o, t)$ is the material position field and $\underline{X}_o$ is defined by $\underline{X}_o \equiv \underline{X}(\underline{X}_o, 0)$.

The solution of (5) is simply

$$\Theta(\underline{X}_o, t) = \Theta(\underline{X}_o, 0) \qquad .(6)$$

To transform back to $(\underline{x}, t)$ we need the inversion $\underline{x}_o(\underline{x}, t)$ of the position field $\underline{X}(\underline{X}_o, t)$, given $\underline{u}(\underline{x})$. The answer is obvious because the points merely move in straight lines at constant speeds, but in principle we must solve the simultaneous equations

$$\left.\begin{aligned} \frac{\partial}{\partial t} X(\underline{X}_o, t) &= U\cos\phi\,\cos(k_x X + k_y Y + \alpha) \\ \frac{\partial}{\partial t} Y(\underline{X}_o, t) &= U\sin\phi\,\cos(k_x X + k_y Y + \alpha) \end{aligned}\right\} \qquad .(7)$$

Since $k_x = -k \sin\phi$ and $k_y = k \cos\phi$, we deduce from (7) that

$$\frac{\partial}{\partial t}\left[k_x X + k_y Y\right] = 0 \qquad .(8)$$

We integrate, apply "initial conditions", invert and find

$$\left.\begin{aligned} x_0(\underline{x},t) &= x - Ut\cos\phi\cos(k_x x + k_y y + \alpha) \\ y_0(\underline{x},t) &= y - Ut\sin\phi\cos(k_x x + k_y y + \alpha) \end{aligned}\right\} \qquad ,(9)$$

and the Eulerian form of the solution, equation (6), turns out to be

$$\vartheta(x,y,t) = \vartheta_0 \cos\left\{\ell\left[x - Ut\cos\phi\cos(-kx\sin\phi + ky\cos\phi + \alpha)\right] + \beta\right\} \qquad .(10)$$

We should like to compute the history of the ϑ-field correlation function for an isotropic ensemble of velocity fields, corresponding to averaging over ϕ. To give the necessary homogeneity, we average also with respect to the position phase angles, α and β.

The correlation function expressions have been partially integrated for the two extreme cases, $\rho(0,\eta,t)$, along the original "stripe" direction, and $\rho(\xi,0,t)$, perpendicular to that. After some analysis, we find

$$\rho(0,\eta,t) = \frac{1}{\pi}\int_0^{\pi} J_0(F)\,d\phi \qquad ,(11)$$

where $F = 2\ell Ut\cos\phi\sin\left(\frac{k\eta}{2}\cos\phi\right)$,

and

$$\rho(\xi,0,t) = \frac{\cos(\ell\xi)}{\pi}\int_0^{\pi} J_0(G)\,d\phi \qquad ,(12)$$

where $G = 2\ell Ut\cos\phi\sin\left(\frac{k\xi}{2}\sin\phi\right)$.

J_0 is the zero order Bessel Function of first kind. Evidently these forms behave properly for $t = 0$: the first is just constant at 1.0 for all η ; the second is a simple cosine function.

For $t > 0$ both begin to look more like traditional correlation functions. Figure 2 shows isometric plots of equations (12) and (13) from $lUt = 1$ to $lUt = 40$ for the special case in which the velocity fields have the same wave number as that of the initial scalar field, $k = l$.

Figure 3 is a direct comparison of the two functions at $lUt = 25$.

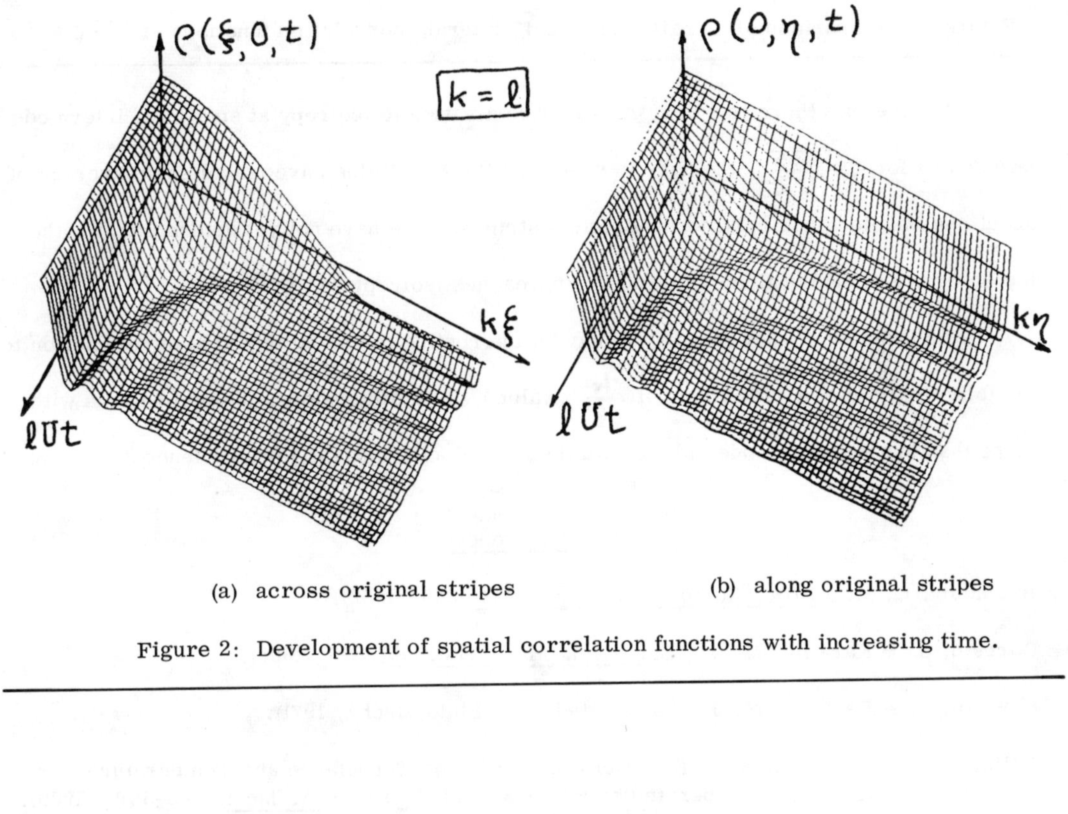

(a) across original stripes (b) along original stripes

Figure 2: Development of spatial correlation functions with increasing time.

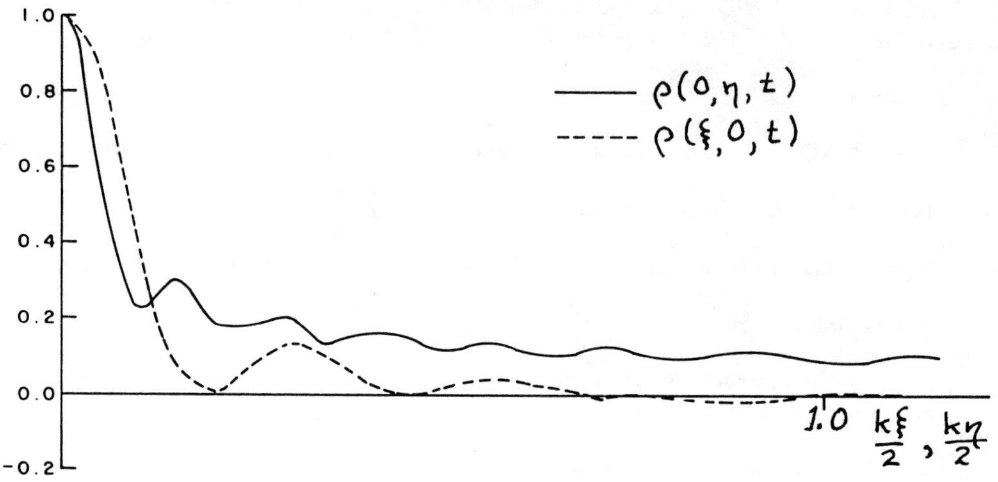

Figure 3 : Comparison of "longitudinal" and "lateral" correlation functions at $U\ell t = 25$

We see that there is a very rough approximation to isotropy at small and intermediate scales for $t = \frac{25}{U\ell}$, the time in which the fluid particles have travelled an average of about one wavelength normal to the original stripes. We have not yet determined whether the ensemble of scalar fields rigorously approaches isotropy as $t \to \infty$.

As time and money permit, it may be interesting to extend this kind of calculation to the following: (a) a wide range of $\frac{k}{\ell}$ values, (b) 3 dimensions, (c) realizations with more than one velocity mode, (d) non-stationary velocity modes, (e) turbulence.

References

Batchelor, G. K. Proc. Cambr. Phil. Soc. 48, 345 (1952).

Corrsin, S. & Karweit, M. J. Bull. Am. Phys. Soc. 13, 805 (1968).

Corrsin, S. & Karweit, M. J. (submitted to J. Fluid Mech., 1971).

Mills, R. R. & Corrsin, S. " Effect of Contraction on Turbulence and Temperature Fluctuations Generated by a Warm Grid," N.A.S.A. Memo 5-5-59W (1959).

Mills, R. R., Kistler, A. L., O'Brien, V. & Corrsin, S. " Turbulence and Temperature Fluctuations behind a Heated Grid," N.A.C.A. Tech. Note 4288 (1958).

O'Brien, E. E. " On the Behavior of Passive Scalars in a Turbulent Fluid," unpub. rep. Coll. of Eng'g, State Univ. of N. Y. at Stony Brook (1963).

POSSIBLE REFINEMENT OF THE LOGNORMAL HYPOTHESIS CONCERNING THE DISTRIBUTION OF ENERGY DISSIPATION IN INTERMITTENT TURBULENCE

by

Benoit B. Mandelbrot

Mathematical Science Department
IBM Thomas J. Watson Research Center
Yorktown Heights, New York 10598

ABSTRACT: Obukhov, Kolmogoroff and Yaglom, and in effect (independently) the geologist deWijs, have argued that energy dissipation in intermittent turbulence is lognormally distributed. However, this hypothesis will be shown to be probably untenable: depending upon the precise formulation chosen, it is either unverifiable or inconsistent. The present paper proposes a variant of the generating model leading to the lognormal. This variant is consistent, appears tractable, and for sufficiently small values of the parameter μ (in Yaglom's notation) it yields the lognormal hypothesis as a good approximation. As μ increases, the approximation worsens, and for high enough values of μ, turbulence ends by concentrating in very few huge "blobs". Still other consistent alternative models of intermittency yield distributions that diverge from the lognormal in the opposite direction; these various models in combination suggest several empirical tests.

INTRODUCTION

A striking feature of the distributions of turbulent dissipation in the oceans and the high atmosphere is that both are extremely "spotty" or "intermittent" in a hierarchical fashion. In particular, both are very far from being homogeneous in the sense of the 1941 Kolmogoroff-Obukhov theory. Nevertheless, many predictions of this classic theory have proved strikingly accurate. Self-similarity and the $k^{-5/3}$ spectrum not only have been observed but are found to hold beyond their assumed domain of applicability. An unexpected embarrassment of riches, and a puzzle!

For many workers, studying turbulence is synonymous with attempting to derive its properties, including those listed above, from the Navier Stokes equations of fluid mechanics. But one can also follow a different tack and view intermittency and self-similar statistical hierarchies as autonomous phenomena. Earlier examples of this approach in the literature are few in numbers, but they go very far back in time and have involved several disciplines. In the field of cosmology, intermittency had already been faced during the eighteenth century, and its study underwent bursts of activity in 1900-1920 and today. The study of intermittency also arose, independently and near simultaneously, in the fields of turbulence (including work by Obukhov, and then Gurvich, Kolmogoroff, Novikov, Stewart and Yaglom) in the study of geomorphy and especially of the distribution of rare minerals (including work by deWijs, and then Matheron), and finally in my own work concerning the many non-thermal noises which go under such names as "burst noise", "impulse noise", "flicker noise", and "1: f noise", and may be considered as forms of electromagnetic turbulence. As it happens, despite obvious specific differences, all the specific workers in these fields have followed the same few generic paths. What has brought these various applications together is not yet clear: It may be either common underdevelopment or genuine kinship.

In this Note, we shall be concerned with one of these generic paths, which may be designated as the "method of self similar random multiplicative perturbations". The first of its two widely distinct sources was a footnote remark in Landau and Lifshitz (5) concerning the 1941 Kolmogoroff theory of self similar homogeneous turbulence, a remark taken up by Obukhov (12), and discussed and developed by Kolmogoroff himself (4), Yaglom (14) and Gurvich and Yaglom (3). The second source lied in works by deWijs (1), and then Matheron (8) and his school, on the distribution of rare minerals. Using the vocabulary of turbulence, let η and L designate the Kolmogoroff micro- and macro-scales, and ϵ ($\underline{x}$, r, η, L) be the average energy dissipation over the cube of side r and center $\underline{x}$. Obukhov and Kolmogoroff hypothesize, and deWijs and Yaglom attempt to derive, the property that log ϵ ($\underline{x}$, r, η, L) is a normal random variable of variance equal to $A(\underline{x}, t) + \mu \log (L/r)$, where the term $A(\underline{x}, t)$ depends on the characteristics of the large scale motion and μ is a parameter, possibly a univeral constant. The above assertion is usually called "Kolmogoroff's third hypothesis". In addition, the expectation of $\log \epsilon$ is ordinarily assumed equal to $-(\mu/2)\log(L/r) - A(\underline{x}, t)/2$. Finally, the averages $\epsilon(\underline{x}, \eta, \eta, L)$ corresponding to the micro-scale are assumed to have a certain correlation function of the form required by self similarity; for large values of Δx, one should have $E[\epsilon(\underline{x} + \Delta\underline{x}, \eta, \eta, L)\, \epsilon(\underline{x}, \eta, \eta, L)]$ proportional to $(L/|\Delta\underline{x}|)^{\mu}$. This last expression will be called "Gurvich-Zubkovskii correlation". Observe that neither this correlation nor Kolmogoroff's "third" hypothesis involve η explicitly, which expresses that they obey Kolmogoroff's "second" hypothesis of 1941 (unchanged in 1962).

The purposes of the present paper are, first to show that the above "third hypothesis" raises serious conceptual difficulties, and in fact appears untenable; secondly, to propose an improved alternative. The practical relevance of my criticism has not so far been established. It depends upon the value of the parameter μ and in each field of application it will have to be investigated specifically.

CRITIQUES OF VARIOUS FORMS OF LOGNORMALITY

To make the historical background more precise: Obukhov introduces lognormality as an "approximate hypothesis" on the ground that the lognormal "represents any essentially positive characteristic", Kolmogoroff treats lognormality as a "third hypothesis" to be derived from other assumptions, and deWijs and Yaglom derive lognormality from a "cascade" argument. Each approach requires a separate reexamination.

Obukhov's approximate hypothesis -- precisely because it is approximate-- can only be examined on pragmatic grounds. Its weakness is that it cannot support the elaborate calculations of moments which have been built on it, because the population moments of the random variable $\exp(Y)$ are overly sensitive to small deviations of Y from normality. "Random variable" will be shortened to "r.v.". For example, consider a normal r.v. (G), a Poisson r.v. (P), and a Bernouilli r.v. (B) obtained as sum of a large number H of binomial r.v.'s B_h. When their respective means and variances are equal and large, those three r.v.'s are indeed considered by probabilists as "nearly identical", but this concept of "near identity" tells little about higher moments of the same order of G, P, and B; a fortiori, the moments of e^G, e^P and e^B of all orders are so influenced by the tails of the various distributions that their values may be very different. For example, suppose they all have the same mean δ and variance δ, the possible values and probabilities of the binomials B_h contributing to B being designated by B', B'', π' and π'', with $\pi' B' H < \delta$, $\pi'' B'' B < \delta$, and $\pi' B' > \pi'' B''$. We have then

$$E(e^G)^p = \exp(p\delta + \delta p^2/2) = \exp[\delta(p+p^2/2)]$$
$$E(e^P)^p = \exp(-\delta + \delta e^p) = \exp[\delta(e^p - 1)]$$
$$E(e^B)^p = \exp E(e^{\Sigma B_h})^p = (Ee^{pB_h})^H = (e^{p\pi' B'} + e^{p\pi'' B''})^H .$$

For large δ, $E(e^B)^p$ increases like $e^{pH\pi' B'}$, that is, less rapidly than $e^{p\delta}$; $E(e^G)^p$ increases more rapidly, and $E(e^P)^p$ even more rapidly. Even for k=1, the values, respectively $E(e^G) = \exp(1.5\delta)$ and $E(e^P) = \exp(1.7\delta)$, are very different. The coefficients of variation,

$$\frac{E[(e^G)^2]}{[E(e^G)]^2} = e^{\delta}, \text{ and } \frac{E[(e^P)^2]}{[E(e^P)]^2} = e^{(e-1)^2\delta} \sim e^{3\delta}$$

differ even more, and higher order moments differ strikingly. In short, B and P are not good approximations to the normal, and the significance of moment calculations under Obukhov's approximate hypothesis of lognormality is entirely unclear.

This last finding must be reviewed from the viewpoint of the observation, due to Orszag (13), that the moments of the lognormal increase too fast to satisfy the so-called Carleman criterion. Consequently, lognormal y intermittent turbulence is not determined by its moments. The moments of Poisson intermittent turbulence increase even more rapidly, while those of the binomial do satisfy the criterion. However, we have noted that this property is sensitive to minor deviations from normality, so that I hesitate to consider this question of determinacy as solved.

As to Kolmogoroff's suggestion that the lognormal hypothesis be considered as strictly valid, it encounters a different kind of difficulty; let us show indeed that the joint assumptions that the r.v.'s $\log \epsilon(\underline{x}, r, \eta, L)$ are normal, for every $\underline{x}$ and every r, are incompatible with the assumption that the correlation of ϵ follows the self similar (Gurvich Zubkovskii) form. Indeed, it $A(\underline{x}, t)$ is replaced by the constant term $\exp \bar{\epsilon}$, the lognormal hypothesis yields

$$E[r^3 \epsilon(\underline{x}, r, \eta, L)]^p = \bar{\epsilon}^p r^{3p - p(p-1)\mu/2} L^{p(p-1)\mu/2}$$

When r reaches its maximum value, which is r=L, all these moments reduce to $\bar{\epsilon}^p r^{3p}$, as they should. But we must examine more closely how do they tend to this limit. Suppose that $\mu>3$ and focus on the second moment (p=2). The fact that the exponent of r in the above expression takes the value $6-\mu<3$ expresses that when r is doubled, $E[r^3\epsilon]^2$ is multiplied by a factor <u>less than</u> 8. On the other hand, the fact that Gurvich Zubkovskii correlation is positive implies that the factor in question must be <u>greater than</u> 8. This is the contradiction that has been announced. When $\mu<3$, the contradiction moves up to higher moments, namely to moments such that p satisfies $3p-p(p-1)\mu/2<3$, i.e. $p/3>2/\mu$. In the sequel, this last criterion will be encountered repeatedly.

Additional internal contradiction: when the variables log ϵ corresponding to 8 neighboring small cubes obtained by subdividing a bigger cube are lognormal, consistency would require the variables log ϵ corresponding to the big cube to be also lognormal. However, sums(and hence averages) of independent lognormal variables are themselves <u>not</u> lognormal, which suggests that if and when the eight small cubes variables are nearly independent statistically, the above requirement is violated. In particular, when μ is very large, the correlation between variables over neighboring small cubes is very small, which suggests that the dependence is small and the said requirement is violated.

To sum up, for moderately large values of μ, the lognormal hypothesis could only be consistent with some special rule of dependence for which the correlation function is not positive. I can't imagine any such rule, and circumstantial evidence to be described below makes me doubt such rule exists. This suggests that <u>Kolmogoroff's strict hypothesis is probably untenable</u>.

As a result, one may expect the third form of lognormality, namely the form obtained as the conclusion of the deWijs-Yaglom(WY) cascade arguments, to be flawed. Let us review WY step by step. First step: one paves space by a regular grid of eddies: the elementary eddies are cubes of side η, eddies of the next stage are cubes of 2η (to be specific) so each contains 8 elementary eddies, etc... Second step: assuming that $r=\eta 2^n$ for some integer n while $L=\eta 2^N$ for some integer N, one rewrites $\epsilon(\underline{x}, r, \eta, L)$ as the product

$$\frac{\epsilon(\underline{x}, r, \eta, L)}{\epsilon(\underline{x}, 2r, \eta, L)} \cdot \frac{\epsilon(\underline{x}, 2r, \eta, L)}{\epsilon(\underline{x}, 4r, \eta, L)} \cdots \frac{\epsilon(\underline{x}, 2^{N-1}r, \eta, L)}{\epsilon(\underline{x}, 2^N r, \eta, L)} \cdot \epsilon(\underline{x}, L, \eta, L)$$

Third step: one identifies the last term as $\bar{\epsilon}$, and one assumes the ratios in the above expression to be independent identically distributed r.v.'s. Fourth step: one applies the central limit theorem to the sum of the logarithms of the above rations. Conclusion: when the cube of center $\underline{x}$ and side r is one of the above eddies, the distribution of the corresponding ϵ ($\underline{x}$, r, η, L) is lognormal.

To this WY generative model, the preceding criticisms of Obukhov's and Kolmogoroff's approaches apply unchanged. In addition, WY predictions concern the eddies themselves, so that direct verification is impossible. On the other hand, when our cube of center $\underline{x}$ and radius r is not an eddy, ϵ ($\underline{x}$, r, η, L) is not lognormal. For example, when r is large and one cube overlaps over several big eddies, ϵ ($\underline{x}$, r, η, L) is the average of several near independent lognormal variables, which we have seen implies it is not lognormal. To establish the distribution of ϵ over an arbitrary cube, one would have to average the distribution corresponding to cubes having the same r and overlapping various numbers of eddies.

AN ALTERNATIVE GENERATOR BASED UPON LIMIT LOGNORMAL PROCESSES

The basic difficulty with the WY cascade argument is, I think, due to the fact that it imposes local conservation of dissipation. This is expressed by the fact the various random ratios of the form ϵ ($\underline{x}$, r/2, η, L)/ϵ ($\underline{x}$, r, η, L) corresponding to different parts of an eddy are required to average to one. Especially when μ is large, this requirement implies that such ratios are strongly negatively correlated, which feature is foreign to the Gurvich Zubkovskii correlation, but (as we saw) is needed in the Kolmogoroff argument. By way of contrast, the variant of the model of multiplicative perturbations proposed in the present paper can be characterized by the feature that conservation of dissipation is assumed not on the local but only on the global level. That is, this model visualizes the cascade process as combined with powerful mixing motion and with exchanges of energy that disperse dissipation and free the above ratios from having to average to one. Moreover, in order to satisfy self similarity more closely, the hierarchy of eddy breakdown is taken as continuous rather than discrete. Under these conditions, one can relate ϵ to a sequence of random functions (r.f.'s) F'($\underline{x}$, λ, L) such that the r.f.'s log F'($\underline{x}$, λ, L) are Gaussian, with the variance $\mu\log(L/\lambda)$, the expectation $-(\mu/2)\log(L/\lambda)$ and a spectral density equal to $\mu/2k$ for $1/L<k<1/\lambda$, and to 0 elsewhere. Consequently, the covariance C(s, λ) of log F'($\underline{x}$, λ, L) will be assumed to satisfy $C(s)=\lim_{\lambda\to\infty} C(s,\lambda) = -\mu\log(2\pi e^{\gamma} s/L)$ (γ is the Euler constant, whose value is about .577). For fixed $\underline{x}$ and L, F'($\underline{x}$, λ, L) is clearly a sequence of lognormal r.v.'s whose expectation is identically 1, while their variance, and hence their skewness and their kurtosis, all increase without bound as $\lambda\to\infty$. F($\underline{x}$, r, λ, L) will then be defined as the integral of F' over a cube of center $\underline{x}$ and side r (which need not be any specific cube designated as "eddy"), and will be viewed as r^3 ϵ ($\underline{x}$, r, λ, L), namely as the approximate total dissipation that only takes account of perturbations whose wavelength lies between λ and L.

Note that, in contrast to the deWY model, there is no specific grid of eddies in the present model. A resemblance with WY is that when $\eta<\lambda\ll r_1<r_2<r_3\ll L$ and $r_2/r_1=r_3/r_2$, $F(\underline{x}, r_2, \lambda, L)/F(\underline{x}, r_1, \lambda, L)$ and $F(\underline{x}, r_3, \lambda, L)/F(\underline{x}, r_2, \lambda, L)$ have idential distributions. A difference is that those ratios need not be independent. Another difference is that in WY the randomness in $F(\underline{x}, L, \lambda, L)$ lies entirely beyond the model, being given by the term $A(\underline{x}, t)$, while in the present variant part of "$A(\underline{x}, t)$" is derived as due to eddy action.

Our task is to derive the distribution of $F(\underline{x}, r, \lambda, L)$. In particular, the smallest value of λ is η, and we must check whether or not the distribution of $F(\underline{x}, r, \eta, L)$ for $r>\eta$ is independent of η. If it is, then Kolmogoroff's second hypothesis, unchanged from 1941 to 1962, is satisfied.

Our procedure will be to keep $\underline{x}$ and r fixed and view $F(\underline{x}, r, \lambda, L)$ as a r.f. of λ. From the mathematical viewpoint, this r.f. happens to be a "martingale" and of course $F\geq 0$, so its behavior is covered by Doob's classical "convergence theorem for positive martingales" (2, p319): it shows that $\lim_{\lambda\to 0} F(\underline{x}, r, \lambda, L)=F(\underline{x}, r, 0, L)$ exists, and it suggests it may be legitimate to view $F(\underline{x}, r, \lambda, L)$ for small but positive λ, say for $\lambda=\eta$, as differing from the limit $F(\underline{x}, r, 0, L)$ by a "perturbation term". However, the convergence theorem leaves two possibilities open: $F(\underline{x}, r, 0, L)$ could either be nondegenerate, that is, have a positive probability of being finite and positive, or be degenerate, that is, almost surely reduced to 0. When $F(\underline{x}, r, 0, L)$ is nondegenerate, then for small but positive values of λ, such as $\lambda=\eta$, a perturbation term dependent on λ is required, but its value is negligible so $F(\underline{x}, r, 0, L)$ is a reasonable approximation and one may consider that Kolmogoroff's second hypothesis holds. But when $F(\underline{x}, r, 0, L)$ is degenerate, then for small λ, $F(\underline{x}, r, \lambda, L)$ either nearly vanishes, with probability nearly 1, or is extraordinarily large, with a very small probability. This last probability tends to zero with λ, but it is finite if $\lambda>0$, which is why the normalizing constraint $EF(\underline{x}, r, \lambda, L)$ could be imposed without contradiction. Nevertheless, the perturbation term is non-negligible, so $F(\underline{x}, r, 0, L)$ is a bad approximation and Kolmogoroff's second hypothesis fails. This shows the importance of determining which of the above alternatives holds for given μ. More precisely, we shall seek when and to which extent the lognormal approximation to $F(\underline{x}, r, L)$ is reasonable.

The main results of the investigation below are as follows. $F(\underline{x}, r, L)$ is degenerate when $\mu>6$ and non-degenerate when $\mu<6$. In the latter case, the moment $EF^p(\underline{x}, r, 0, L)$ is finite when $\mu < 6/p$ and is infinite when $\mu > 6/p$, which suggests that for large value of u,

$$\Pr \{ F(\underline{x}, r, 0, L) > u) \sim C(r, L)\, u^{-6/\mu}.$$

Thus, as $\mu \to o$, $F(\underline{x}, r, 0, L)$ acquires an increasing number of finite moments, which moreover will be shown to converge towards those of the lognormal. This establishes constructively that for small μ, the cascade scheme of deWijs and Obukhov can be modified to avoid the difficulties listed but without significant change in the prediction; for large μ, the required changes are significant. The transition criterion $\mu = 6/p$ has already been encountered when discussing the inconsistency of Kolmogoroff's strict hypothesis: the same high moments that used to behave inconsistently no longer do so, because they are infinite throughout. Proofs of the above assertions will be given in the following two sections, differing by the mathematical tools used.

THE LIMIT LOGNORMAL MODEL FOR $\mu > 6$

First step: A feature of very skew lognormal distributions is that their expectation is overwhelmingly due to occasional large values. Thus, let $\log F'$ be Gaussian with variance $\mu \log(L/\lambda)$ and expectation $-(\mu/2)\log(L/\lambda)$, implying $F' \equiv 1$, and let $N(\lambda)$ be a function such that $\lim_{\lambda \to o} N(\lambda) = \infty$ while $\lim_{\lambda \to o} N(\lambda)/\sqrt{\log (L/\lambda)} = 0$. Also define the functions Threshold (λ, L), F'_+ and F'_- as follows:

$$\text{Threshold } (\lambda, L) = (L/\lambda)^{\mu/2} \exp [-N(\lambda) \sqrt{\mu \log (L/\lambda)}\,]$$

$$F'_+ = F' \text{ and } F'_- = 0 \text{ when } F'(\underline{x}, \lambda, L) \geq \text{Threshold } (\lambda, L)$$

$$F'_+ = 0 \text{ and } F'_- = F' \text{ when } F'(\underline{x}, \lambda, L) < \text{Threshold } (\lambda, L)$$

Finally, let F_+ and F_- be the integrals of F'_+ and F'_-.

The motivation of the above definitions lies in the value of the expectation

$$EF_+(\underline{x}, r, \lambda, L) = \frac{r^3}{\sqrt{2\pi\mu \log(L/\lambda)}} \int \exp\left\{ x - \frac{[x + (\mu/2)\log(L/\lambda)]^2}{4\mu \log(L/\lambda)} \right\} dx$$

with integration from $\log[\text{Threshold}(\lambda, L)]$ to infinity. Transforming the integrand into

$$\exp \left\{ - \frac{[x - (\mu/2) \log(L/\lambda)]^2}{4\mu \log(L/\lambda)} \right\},$$

and then changing the variable of integration, we obtain

$$EF_+(\underline{x}, r, \lambda, F) = \frac{r3}{\sqrt{2\pi}} \int_{-N(\lambda)}^{\infty} \exp(-z^2/2)dz$$

This shows that the contribution of F_- to F is asymptotically negligible, and that the above $N(\lambda)$ has been appropriately chosen to make F arbitrarily closely approximated by F_+. Moreover, for $\lambda>0$, the function F' is a. s. continuous, so the variation of F is a. s. concentrated on those intervals where $F'=F'_+$.

Second step: Over any cube of side λ, F', and therefore F'_+, is near constant. Hence $F(\underline{x}, r, \lambda, L) \sim \lambda^3 \Sigma F'_+(\underline{x}, \lambda, L)$ with summation carried out over those points of a regular lattice of side λ for which F' lies above Threshold(λ, L).

Third step. We suspect that there exist circumstances under which $\lim_{\lambda \to 0} F(\underline{x}, r, \lambda, L)=0$. At least some among such circumstances may also fulfill the stronger sufficient condition that in a cube of side L the random number of lattice sites for which $F'_+>0$ tends to 0 almost surely. In turn, a sufficient condition for the latter property is that the number in question should tend to 0 on the average. This expected number equals

$$(L/\lambda)^{-3} \Pr\{ F'(\underline{x}, \lambda, L) > \text{Threshold}(\lambda, L)\}$$

In terms of the r. v. $[\log F' + (\mu/2)\log(L/\lambda)] / \sqrt{\mu \log(L/\lambda)}$, which is a reduced Gaussian G, the probability in question becomes

$$\Pr\{ G > \sqrt{\mu \log(L/\lambda)} - N(\lambda)\}$$

Using the well known tail approximation of G, the expected number in question is approximately equal to

$$(L/\lambda)^3 \exp[-(\mu/2)\log(L/\lambda)] \,/\, \sqrt{2\pi}\sqrt{\mu \log(L/\lambda)}$$

$$= \frac{(L/\lambda)^{3-(\mu/2)}}{\sqrt{2\pi\mu \log(L/\lambda)}}$$

Note that this last approximation is independent of $N(\lambda)$ for $\lambda \to 0$. For it to tend to 0 with λ, a sufficient condition is $\mu>6$. (It is also necessary, but this is besides the point; see below). It follows that, when $\mu>6$, $\lim_{\lambda \to 0} F(\underline{x}, r, \lambda, L)=0$ almost surely. Obviously, the limit is far from being distributed lognormally.

(The preceding argument is somewhat heuristic; but in a one dimensional version of the limit lognormal process, it could be readily made rigorous by using the Rice formula. This is one more reason why it would be desirable to generalize the Rice formula to a higher dimension.)

Digression: In the case $\mu<6$, the preceding argument would suggest that the bulk of the variation of F is concentrated over about $(L/\lambda)^{3-\mu/2}$ sites of side λ. As $\lambda \to 0$, each site either is eliminated or becomes subdivided into numerous subsites. However, from the mathematical viewpoint, this last suggestion is very incomplete because the fact that the r. f. $X(\lambda)$ satisfies $\lim_{\lambda \to 0} X(\lambda)<\infty$ does not exclude the possibility that $\lim_{\lambda \to 0} X(\lambda)=0$ almost surely. Worries of such nature are usually disregarded in applications, but in this instance the misbehavior of F for $\mu>6$ suggests that extreme care is necessary and that different tools are needed.

THE LIMIT LOGNORMAL MODEL FOR $\mu<6$

In this Section, the moments $EF^p(\underline{x}, r, 0, L)$ will be evaluated for integer p, and then compared with the moments $E[rF'(\underline{x}, r, L)]$. This last log normal r. v. is interesting because it provides some kind of link with the WY model. Indeed, it is tempting to argue as follows: $F'(\underline{x}, r, L)$ varies little over a cube of side r, while the ratio $F'(\underline{x}, \lambda, L)/F'(\underline{x}, r, L)$, which equals $F'(\underline{x}, \lambda, r)$, varies rapidly. So it seems reasonable to hope that, over the cube of side r, this last ratio averages to its expectation equal to one. This means one may hope that $F'(\underline{x}, r, \lambda, L)$ is approximated by $rF'(\underline{x}, r, L)$ reasonably. The question is, is such really the case?

The case p=2. Integrating over the domain where all coordinates of $\underline{u}$ and $\underline{v}$ lie between 0 and r, we have

$$EF^2(\underline{x}, r, \lambda, L) = E\iiiint\!\!\iint \exp\left[\log F'(\underline{u}, \lambda, L) + \log F'(\underline{v}, \lambda, L)\right] d\underline{u}\, d\underline{v}$$

$$= \iiiint\!\!\iint E\exp\left[\log F'(\underline{u}, \lambda, L) + \log F'(\underline{v}, \lambda, L)\right] d\underline{u}\, d\underline{v}$$

The exponand is a Gaussian r.v. of expectation $-\mu\log(L/\lambda)$ and variance $2\mu\log(L/\lambda)+2C(|\underline{u}-\underline{v}|,\lambda)$. As a result

$$EF^2(\underline{x},r,d,L) = \iiiiiint \exp[C(|\underline{u}-\underline{v}|,\lambda)]\,d\underline{u}\,d\underline{v}$$

Suppose that r is fixed, with $r \ll L$, and let $\lambda \to 0$. The preceding integral continues to converge if and only if (iff) $\mu/2<3/2=3/p$, in which case its limit for $d \to 0$ equals

$$(2\pi e^{\gamma}/L)^{-\mu} \iiiiiint |\underline{u}-\underline{v}|^{-\mu}\,d\underline{u}\,d\underline{v}$$

Alternatively, the integration being now carried over the variables $\underline{u}'=\underline{u}/r$ and $\underline{v}'=\underline{v}/r$, whose values vary from 0 to 1, the limit of the above second moment is equal to

$$r^{6-\mu}L^{\mu}\left[(2\pi e^{\gamma})^{-\mu} \iiiiiint |\underline{u}'-\underline{v}'|^{-\mu}\,d\underline{u}'\,d\underline{v}'\right]$$

By way of contrast, for the would be approximating lognormal $rF'(\underline{x},r,L)$ has a second moment equal to $r^{6-\mu}L^{\mu}$. The ratio of approximating, namely the ratio between the limit and the approximate moment, is the quantity in brackets. As $\mu \to 0$, its integrand and $2\pi e^{\gamma}$ tend to 1, and so does the ratio itself.

Suppose that it is true that $\lim_{\lambda \to 0} EF^p(\underline{x},r,\lambda,L)=EF^p(\underline{x},r,0,L)$. (Unfortunately, as has been mentioned, the preceding calculation does not suffice to establish this last equality.) It would follow that as $\mu \to 0$, $rF'(\underline{x},r,L)$ becomes a good second order approximation to $F(\underline{x},r,0,L)$.

One establishes that similarly for $p \geq 3$, $\lim_{\lambda \to 0} EF^p(\underline{x},r,\lambda,L)<\infty$ iff $\mu/2<3/p$, suggesting that $rF'(\underline{x},r,L)$ is a good approximation to $F(\underline{x},r,0,L)$ up to the order $6/\mu$. When μ is very small, F^p has very many finite moments and its low order moments lie near those of the lognormal, in which sense F itself is near lognormal.

Now we must tackle the mathematical difficulty concerning the identity, or lack of it, between $\lim_{\lambda\to 0} EF^p(\underline{x}, r, \lambda, L)$ and $EF^p(\underline{x}, r, 0, L)$. I am able to answer this question only in part. Let $P(\mu)$ be the largest integer satisfying $\mu/2<3/P$. When $P\geq 3$, which implies $\mu<2$, a standard theorem on martingales due to Doob (2, p. 319, Theorem VII. 4.1, clause iii) suffices to establish that $\mu<2$ is a sufficient condition for $F(\underline{x}, r, 0, L)$ to be nondegenerate, meaning that $\Pr\{F(\underline{x}, r, 0, L)>0\}>0$. In addition, it establishes that $EF^p(\underline{x}, r, 0, L)=\lim_{\lambda\to 0} EF^p(\underline{x}, r, \lambda, L)$. In particular, since $2<P$, the above obtained $\lim_{\lambda\to 0} EF^2(\underline{x}, r, \lambda, L)$ is indeed the second moment of $F(\underline{x}, r, 0, L)$. A bit of additional manipulation establishes that for all p $EF^p(\underline{x}, r, 0, L)=(r/L)^{p(p-1)\mu/2}E(\mu, p)$, where $0< E(\mu, p) <\infty$ if $r\ll L$ and $p<P$, and $E(\mu, 1)\equiv 1$.

MISCELLANEOUS REMARKS

Different forms of correlation. The preceding theory of epsilon and also the Gurvich-Zubkovskii correlation, concern cubes, of side either r or η. Experimental measurements, on the contrary, seem rather to concern averages of epsilon along thin cylinders of fixed uniform cross section and varying length r. Appropriate changes must be made.

The problem of the predicted probability distribution of $F(\underline{x}, r, 0, L)$ and of experimental verification. One question must be touched: are the above results special to the specific lognormal model, or are they of greater generality? It has been noted that in the scheme of multiplicative perturbations the set on which the bulk of variation of $X(t, f)$ occurs is influenced greatly by the tails of the distribution of $\log X'(t, f)$, about which the central limit theorem gives no control. More generally, it happens that different models of multiplicative perturbations, which a priori seem to differ by inconsequential details, may lead to different predictions for the distribution of Kolmogoroff's "epsilon". In addition, the alternative models of intermittency belonging to the second broad class mentioned in the introduction, namely the models of Novikov and Stewart (11) and Mandelbrot (6), lead to still different concentration sets, and to probability distributions less scattered than the lognormal. In other words, the model is non robust to the extreme, and appropriately selected variants could account both for distributions that are more scattered and less scattered than the lognormal. In truth, the theory in its present stage offers little that the experimentalist may want to verify.

Generative models of the law of Pareto. The interplay we have observed between multiplicative perturbations and the lognormal and Pareto distributions has incidental applications in other fields of science where very skew probability distributions are encountered, notably in economics. Having mentioned the fact, I shall leave its elaboration to another more appropriate occasion.

Figure 1. Computer simulated approximations to one dimensional self-similar limit lognormal r. f. 's $F(x, r, \lambda, L, \mu)$ for successive values of x, all multiple of r.

Method of construction: First, computer simulation, using IBM System 360/Model 91, has generated $\log F'(x, \lambda, L=10^7, \mu=2)$, defined as a Gaussian r. f. 's for discrete x with $1 \leq x \leq 560,000$, whose spectral density is approximately equal to $1/k$ for $1/L<k<1/\lambda$. For selected values of μ and λ, $F(x, r, \lambda, L, \mu)$ was computed for x multiple of 1000, using the formula $\Sigma_{u=1}^{1000} [F'(u, \lambda, 10^7, 2)]^{\mu/2}$ and then one has computed the ratio $R(x, \lambda, \mu)$ between F and the median of the values of F along the sample. For each given value of μ, the output of the program is a set of Calcomp tracings, drawn across a broad strip of paper. All the programs were written by Hirsh Lewitan using the fast fractional Gaussian noise algorithm in Mandelbrot (7). On the next three pages, portions of these graphs are shown for $\mu=0.5$, $\mu=1$, $\mu=4$ respectively, with λ decreasing down the page.

Analysis of the results. The theory predicts that when μ is small (graphs A, B, C) the ratio $R(x, \lambda, \mu)$ converges to a limit, that is, soon ceases to vary. This is clearly confirmed by simulation. The ostensible limit is clearly non Gaussian, but not extremely so. As μ increases (D, E. F) the point of ostensible convergence moves towards decreasing values of λ, and the non Gaussian character of the ostensible limit of R become increasingly apparent. In particular, an increasing proportion of the cumulated F becomes due to a decreasing number of sharp peaks and blobs. (The peaks are drawn on **reduced** scale for the sake of legibility.) When μ exceeds 2 (G, H, I) (2 is the critical value in one dimension as $\mu=6$ was in three dimension) the ostensible convergence of R to a limit ceases.

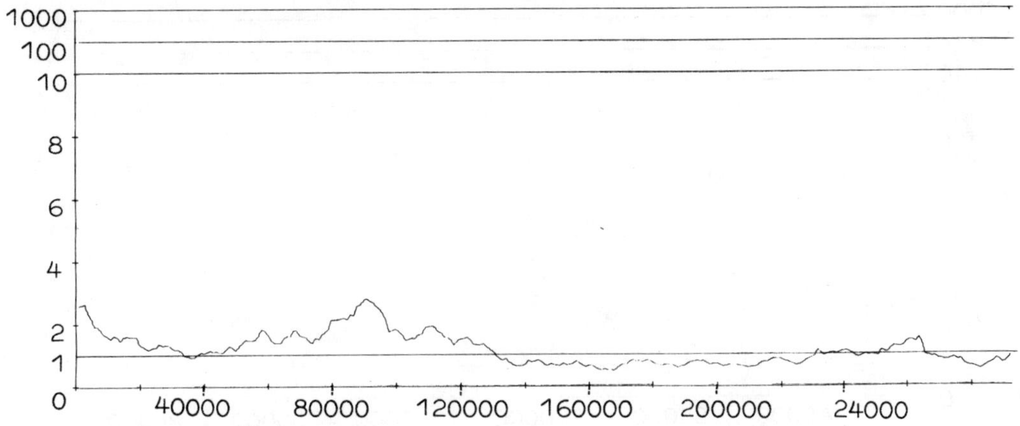

A: μ =0.5; λ large

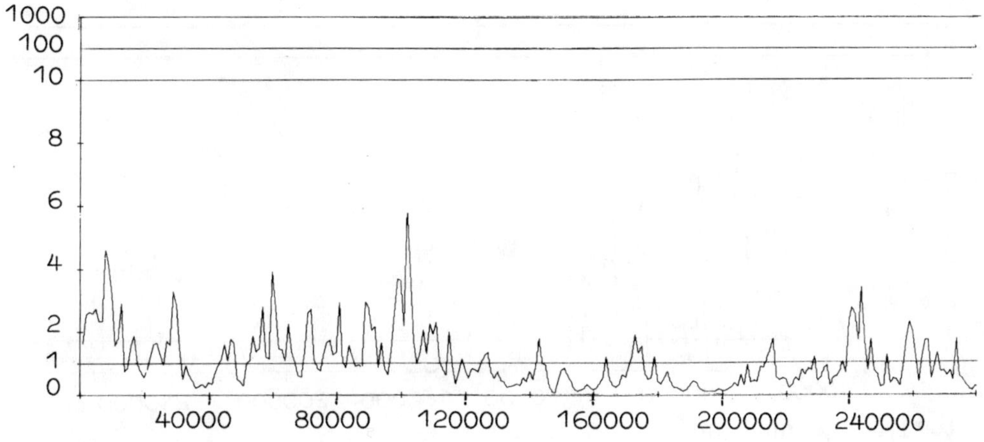

B: μ =0.5; λ medium

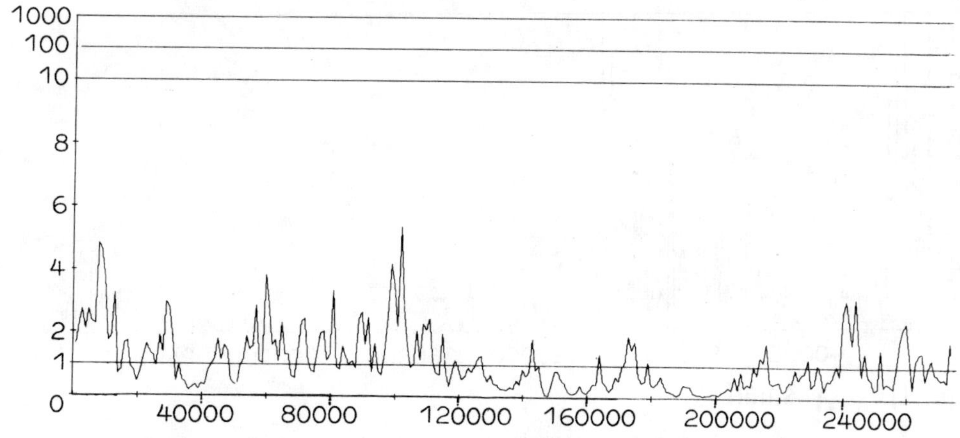

C: μ =0.5; λ small

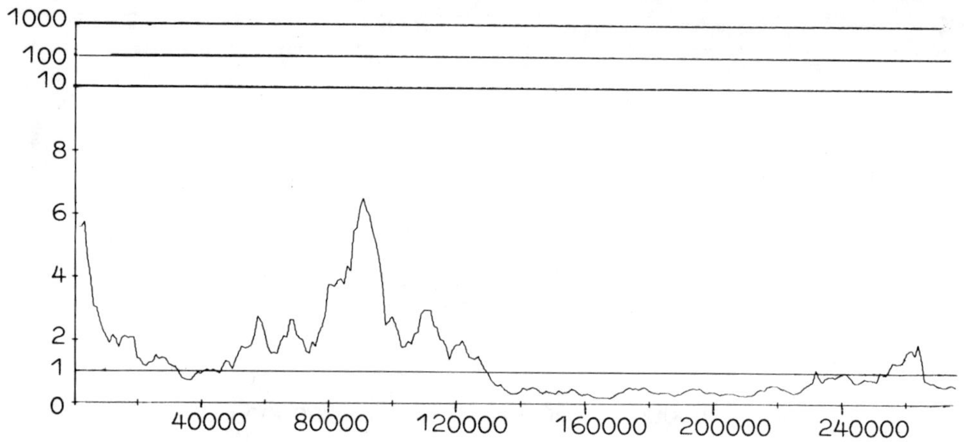

D: $\mu = 1$; λ large

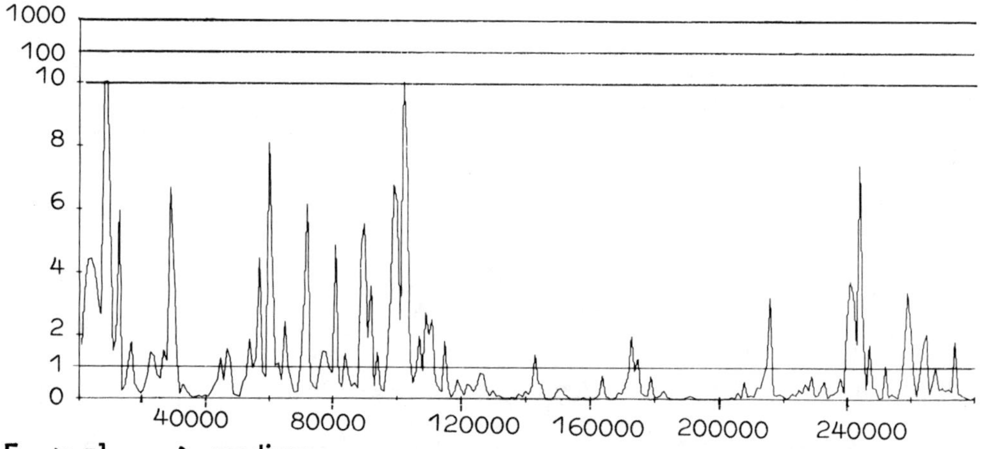

E: $\mu = 1$; λ medium

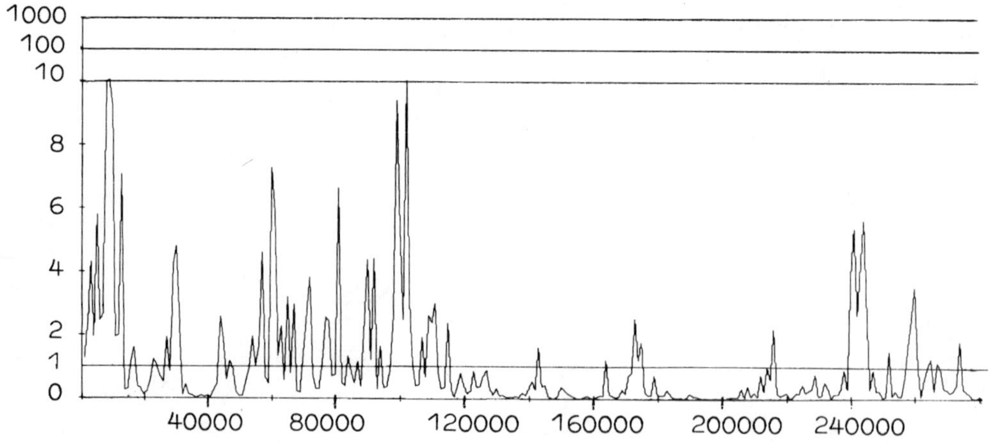

F: $\mu = 1$; λ small

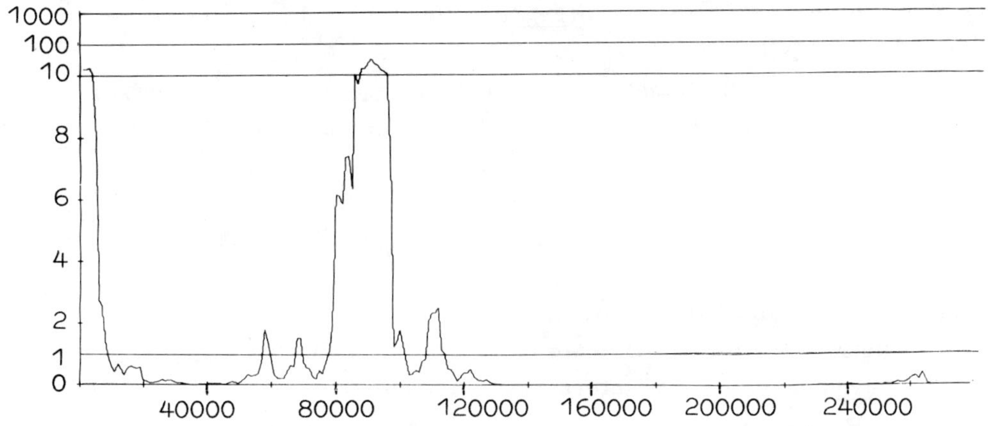

G: μ=4; λ large

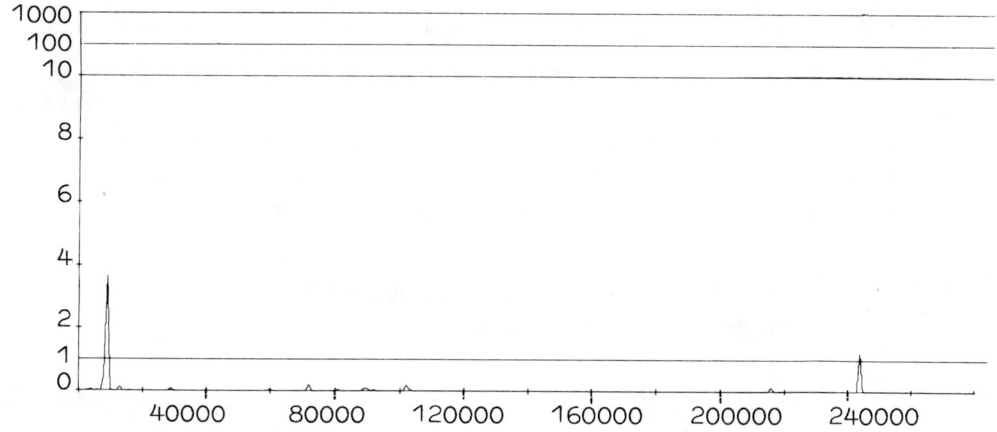

H: μ=4; λ medium

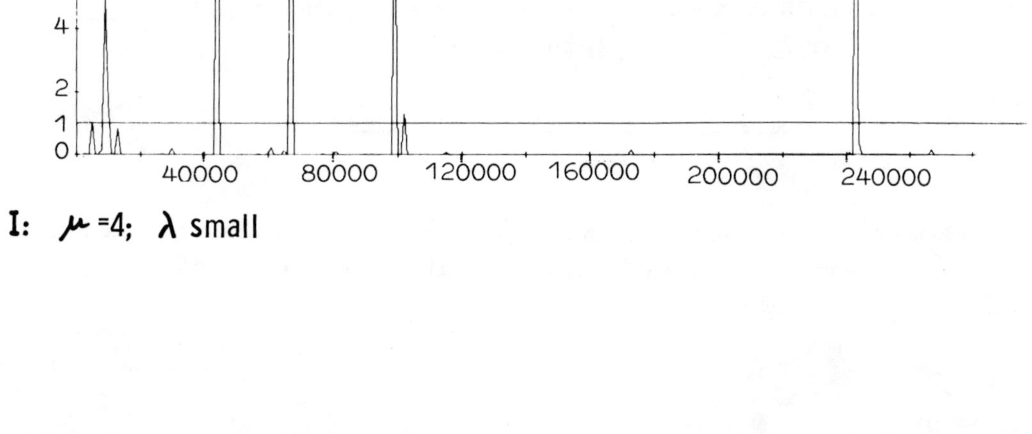

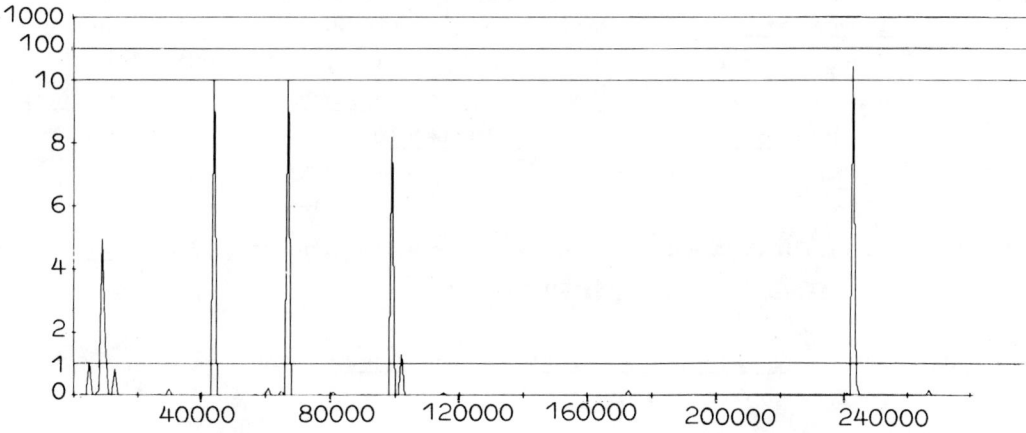

I: μ=4; λ small

REFERENCES

(1) deWijs, H. J. "Statistics of ore distribution", Geologie en Mijnbouw, Part 1, Vol. 13 (1951), pp. 365-75; Part 2, Vol. 15 (1953) pp. 12-24.

(2) Doob, J. L. Stochastic processes. New York, J. Wiley, 1953.

(3) Gurvich, A. S. and Yaglom, A. M. "Breakdown of eddies and probability distribution for small scale turbulence", in Boundary Layers and Turbulence. (Kyoto International Symposium, 1966), a supplement to Physics of Fluids, Vol. 10, 1967, pp. S-59-65.

(4) Kolmogoroff, A. N. "A refinement of previous hypotheses concerning the local structure of turbulence in a viscous incompressible fluid at high Reynolds number." J. Fluid Mech. Vol. 13, No. 1 (1962), 82-85. See also "Clarification of ideas on the local structure of turbulence in an incompressible viscous fluid for large Reynolds numbers" (in French and Russian). Mécanique de la Turbulence (Colloques Internationaux du CNRS, Marseille, 1961), Paris, 1962.

(5) Landau, L. D. and Lifshitz, E. M., Fluid Mechanics, London, Pergamon Press; Reading, Addison Wesley.

(6) Mandelbrot, B. "Self-Similar Error Clusters in Communications Systems, and the Concept of Conditional Probability", Institute of Electrical and Electronics Engineers, IEEE Transactions on Communications Technology, Vol. COM-13 (1965), pp. 71-90. See also his "Sporadic random functions and conditional spectral analysis: self-similar examples and limits. Proceedings of the Fifth Berkeley Symposium on Mathematical Statistics and Probability, held in 1965, University of California Press, 1967 Vol. III, pp. 155-179.

(7) Mandelbrot, B. A fast fractional Gaussian noise generator, Water Resources Research, Vol. 7, 1971, pp. 543-553.

(8) Matheron, G. Traité de Géostatistique Appliquée, Tome 1. Paris, Technip, 1962.

(9) Monin, A. S. and Yaglom, A. M. "On the laws of small scale turbulent flow of liquids and gases". (translated from the Russian) Russian Math-

ematical Surveys, Vol. 18 (1963) pp. 89-109.

(10) Monin, A. S. and Yaglom, A. M. Statistical fluid mechanics (translated from the Russian) Cambridge, Mass., MIT Press 1971

(11) Novikov, E. A. and Stewart, R. W. "Intermittency of turbulence and the spectrum of fluctuations of energy dissipation". Isvestia Akademii Nauk SSSR, Seria geofizika. Vol. 3 (1969) pp. 408.

(12) Obukhov, A. M. "Some specific features of atmospheric turbulence", J. Fluid Mech. Vol. 13, No. 1 (1962), pp 77-81. See also Proc. Sympos. on Turbulence in Geophysics, J. Geophys. Res. Vol. 67, No. 8 (1962) pp. 3011-3014.

(13) Orszag, S. A. "Indeterminacy of the Moment Problem for Intermittent Turbulence". Physics of Fluids, Vol. 13 (1970) pp. 2211-2212.

(14) Yaglom, A. M. "The influence of the fluctuation in energy dissipation on the shape of turbulent characteristics in the inertical interval". In Russian: Doklady Akademii Nauk SSSR, Vol. 166, 1966, pp. 49-52. In English: Soviet Physics-Doklady, Vol. II (1966) pp. 26-29

SOME OBSERVED PROPERTIES OF ATMOSPHERIC TURBULENCE

John A. Dutton and Dennis G. Deaven

Department of Meteorology
The Pennsylvania State University
University Park, Pennsylvania 16802

1.0 INTRODUCTION

Turbulence is the phenomenon with which the atmosphere closes its kinetic energy cycle. It arises at the end of a complex chain of events and is an important part of the atmosphere's response to the thermal forcing that sets it in motion.

Atmospheric turbulence has been studied for many years because of its obvious importance in diffusion, aeronautics, wind-loading of structures, and all boundary layer processes. But it is only recently that the theoretical and practical implications of the presence of turbulence for large-scale meteorology have been perceived.

The essential realization has been that as the time scale of interest expands, so does the wave-number range of the phenomena that must be considered. Meteorology has long been divided into specialties by wave-number boundaries -- the continental scales of the weather producing systems being treated independently of the smaller scales associated with boundary layers or clear air turbulence. But now that weather predictions a week or more in advance are being considered, it becomes essential to consider all components of the atmosphere's energy cycle. For when we consider longer range predictions, or to what extent solutions are unique over extended periods, we allow enough time for the small-scale random components to materially affect the evolution of the large-scale components. And thus questions about the properties and role of atmospheric turbulence in the atmosphere's general circulation are crucial problems in current theoretical and numerical meteorology. Turbulence is now recognized as part of the essential physics of atmospheric motion that must be understood for more important reasons than mathematical curiosity.

The Origin of Atmospheric Turbulence. The atmosphere is a fluid in a continual motion caused by temperature gradients created on the largest scale by the sphericity of the Earth and on the smallest by surface inhomogenieties. On the global scale, the radiation patterns that have evolved result in heating at high temperature and cooling at low temperature. This process provides the decrease in entropy that motivates the motion.

The energy cycle describes the processes with which the atmosphere reacts to the destruction of entropy. Kinetic energy is created by the interaction of the motion fields with those of the thermodynamic variables, and is transformed by the non-linear processes into a relatively smooth spectral distribution of energy.

As the flow proceeds, air parcels with quite different velocities and temperatures are brought together by the convective process and the variations in wind and temperature become concentrated in the thin layers -- the most pronounced being in the boundary layer, in the frontal zones associated with weather systems, and in the vicinity of the jet stream near the tropopause.

As these gradients become sharp enough, the unstable modes of fluid flow are stimulated and turbulence is created in the transfer of energy from large to small scales. But then the turbulence leads to mixing that reduces the gradients and results in the transport of heat and momentum from one region to another. And as with all turbulence, the dissipation mechanism converts the kinetic energy to thermal energy increasing the entropy, providing heat to be radiated to space, and completing the cycle.

Problems Posed by Atmospheric Turbulence. The various problems posed by turbulence have acquired a new urgency because of the attempts to develop improved long-range numerical weather prediction models. If models are to correctly simulate the motions of the atmosphere a week or more in advance, then surely they will have to simulate the energy budget or energy cycle of the atmosphere correctly. But turbulence occurs on a scale much smaller than those used in the computations and it cannot be resolved by the standard data acquisition techniques either.

Meteorologists have been surprised, perhaps, by the importance of clear air turbulence in the free atmosphere as a component of the atmospheric energy budget. The total dissipation rate in the atmosphere is still a controversial subject, but it appears to lie in the range 5 to 8 watts/m^2, with 2 or 3 watts/m^2 being contributed in the boundary layer. The next most important contribution occurs in the altitude range that includes the polar-front jet stream. Thus Kung (1966, 1967,

1969) using standard meteorological data and Trout and Panofsky (1969) using aircraft data have both estimated a dissipation rate of 1.3 watts/m^2 for the range of 25,000 to 40,000 ft. Some of the components of this estimate are given in Table 1, which shows that the majority of the dissipation is occurring in air reported turbulent by aircraft pilots.

Table 1. Dissipation in the layer 25,000-40,000 ft.

	Turbulence Category			
	Smooth	Light	Moderate	Severe
Dissipation Rate cm^2/sec^3	1.5	30	85	675
Percentage Rate of Occurrence	88.1	7.9	4.0	0.05
Dissipation in the Layer (watts/m^2)	0.25	0.45	0.48	0.12
Percentage of Dissipation	19	35	37	9

The problem posed for numerical weather prediction is that not enough is known yet about atmospheric turbulence to be able to parameterize these dissipation rates, or the turbulent fluxes of heat, momentum, and energy, as functions of the large-scale variables. The difficulty is illustrated in Figure 1. Most of the time the spectral density follows the curve marked "smooth air", with some of the energy available in the shear of large-scale wind patterns being drained off by the work against the stabilizing effects of stratification. But in other cases when the outbreaks of moderate or severe turbulence that have plagued aviation occur, there is a distinct change in the spectrum and an energy bulge appears at the wavelength at which the Reynolds stresses are presumably concentrating the kinetic energy that is being realized at the expense of the large-scale flow.

The details of this process are not understood, although it certainly represents some form of instability involving a rapid transfer of kinetic energy from large to smaller wavelengths. The evidence suggesting that clear air turbulence is a manifestation of the Kelvin-Helmholtz instability has been reviewed recently by Dutton and Panofsky (1970).

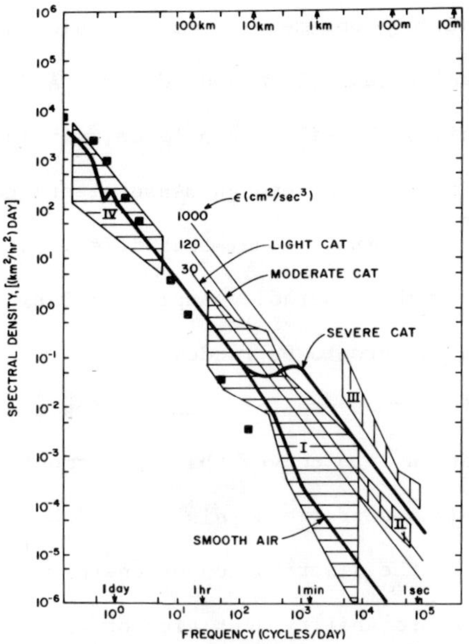

Figure 1.1 Illustration of the difference between smooth air and clear air turbulence of various intensities in the free atmosphere, adapted from Vinnichenko and Dutton (1969). The shaded boxes represent I, aircraft measurements taken in the U.S.S.R.; II, aircraft measurements at jet stream level in Colorado; III, aircraft measurements in thunderstorms; and IV, radiosonde winds in the U.S.S.R. The solid squares are Lagrangian balloon measurements of turbulence. For references to the original data, see the paper cited above.

The presence of turbulence has important consequences in the study of the predictability of larger-scale atmospheric flow, for it produces a random component of the flow that cannot be resolved by synoptic measuring techniques nor treated by the numerical integration methods now in use. In the atmosphere, the turbulence provides a component of the total flow that evolves stochastically, and thus eventually even the gross features are no longer predictable by deterministic means.

Ways of Looking at Turbulence. The view that the turbulence problem is in large measure one of scale provides some understanding of the difficulties we face. The point is that turbulence is not stochastic in the sense that the evolution is truly random, but rather that it appears random because it is not possible to comprehend or even manipulate the vast number of measurements that would be necessary to resolve the turbulent field to the point that we could actually follow the dynamic processes that are working deterministically at that scale. For example, a modern numerical model of the general circulation might involve 2,000 grid points at each of 10 levels. This number of points is equivalent to a

27 x 27 x 27 grid, so that in a one-meter cube we would have a mesh of 3.7 cm and thus we could resolve wavelengths 7.4 cm and longer. A direct integration of the equations of motion on this scale will not produce much information about turbulence.

Thus we are in the dilemma that enough measurements to permit us to resolve the dynamics would overwhelm our computational facilities, but the number of measurements we can handle gives such a sporadic sampling relative to the dynamical processes that the motion appears to be random.

Two techniques are in common use for transforming the inscrutable time or space histories of turbulent motion into curves that vary smoothly and thus can be comprehended. The first is spectral analysis, in which we disregard the information on phase and concentrate on the distribution of energy with respect to wavelength. The second is reduction to probability densities or distributions in which we disregard all information about the temporal or spatial sequencing of events and consider only the relative frequency of occurrence of velocities or other variables.

In this paper we shall concentrate on the observed probabilistic characteristics of atmospheric turbulence, taking the view that despite the difficulties and uncertainties present in observations of atmospheric turbulence, this collection of results does reveal some features that a successful theory of atmospheric turbulence will have to explain.

2.0 DESCRIPTION OF THE DATA

Samples of turbulence obtained at four different altitudes with instrumented aircraft are analyzed in this paper to determine whether certain statistical properties may be considered to be characteristic of atmospheric turbulence. Three of the data sets used here are part of those collected in measurement programs sponsored by the U.S. Air Force, one was obtained by the National Aeronautical Establishment (NAE) of Canada.

These measurement programs were intended primarily to produce information of value for aircraft design. Some of the characteristics of the data are given in Table 2, and the references cited there give detailed documentation of the methods and instrumentation used for the measurements as well as the summaries of the observational results. A discussion of the general approach and some of the difficulties of measuring turbulence with aircraft is given in Dutton (1971).

Data sets from these collections have been analyzed in other papers. Among attempts to utilize the data in studies of the physical properties of turbulence are analyses of the energy budget and environment of stratospheric turbulence given by Dutton (1969) and Delay and Dutton (1971). Some of the statistical properties of the data are discussed in the references cited in Table 2 and by Dutton, Thompson and Deaven (1969), Dutton and Deaven (1969), and Dutton (1970).

Table 2. Description of the Data Sample.

HICAT		
Severe CAT	Δx = 17 meters	N = 10,452 per component
Grand Junction, Colorado	Δt = 0.08 secs	$\bar{\sigma}$ = 1.65 m/sec
60,000 ft	Run lengths 16-56 km	$\bar{\varepsilon}$ = 73 cm^2/sec^3
	Low pass filtered	$\bar{\mu}_4$ = 3.63
Reference: Crooks, Hoblit and Mitchell, 1968.		
CAT (NAE)		
Mountain CAT	Δx = 9.55 meters	N = 12,288 per component
Sierra Nevada Mountains	Δt = 0.05 secs	$\bar{\sigma}$ = 2.09
33,000 ft	Run lengths 39 km	$\bar{\varepsilon}$ = 374
	Band pass filtered (wavelengths greater than 3.6 km removed)	$\bar{\mu}_4$ = 4.09
Reference: Mather, 1967.		
LO-LOCAT 750		
Boundary Layer Turbulence	Δx = 1.87 meters	N = 49,152 per component
Kansas Plains (unstable conditions)	Δt = 0.01 secs	$\bar{\sigma}$ = 1.17
	Run lengths 15.3 km	$\bar{\varepsilon}$ = 99
750 ft	Band pass filtered (wavelengths greater than 4.6 km removed)	$\bar{\mu}_4$ = 3.54
Reference: Jones, Mielke, and Jones, 1970.		
LO-LOCAT 250		
Boundary Layer Turbulence	Δx = 1.87 meters	N = 147,456 per component
Kansas Plains (stable conditions	Δt = 0.01 secs	$\bar{\sigma}$ = 0.84
	Run lengths 15.3 km	$\bar{\varepsilon}$ = 76
250 ft	Band pass filtered (wavelengths greater than 4.6 km removed)	$\bar{\mu}_4$ = 3.1
Reference: Jones, Mielke, and Jones, 1970.		

A particular problem of data quality inherent in aircraft measurements merits mention. All of the data used here have been low-pass filtered to remove noise, but problems also occur at low frequencies. In aircraft data, both gyro drift and the resolution of accelerometers at the small accelerations encountered in passing through long wavelength eddies or waves lead to errors. For these reasons, the data is often high-pass filtered as well in order to remove energy at those wavelengths at which errors may occur. Thus we are forced to perform statistical analyses on data in which the low-frequency components are either not those of the actual turbulence or are missing completely. The filtering applied to these data sets is specified in Table 2.

The measurements used here are resolved into components that are longitudinal (u), lateral (v), and vertical (w) with respect to the flight path of the airplane. The linear trends in each component were removed and the time history was normalized with its variance. All of the analyses were performed on these standardized data samples, and the graphs in Section 3, 4, and 5 show average quantities for each of the four data sets.

Values of the dissipation rate, ε, were computed from the energy spectrum, with the usual assumption that the -5/3 power law behavior observed is an extension of the inertial subrange so that the expression

$$\Phi(k) = \alpha\, \varepsilon^{2/3}\, k^{-5/3} \tag{2.1}$$

can be solved for ε when the constant α is known (see Pasquill and Panofsky, 1963). Here the spectral estimates in the middle third of the high-frequency half of the spectrum were used in order to compute estimates of ε automatically for each component. Figure 2.1 and 2.2 show the relationships between the dissipation rate and the original standard deviation of the data samples. The ε plotted is the average of the estimates obtained for each data run.

The kurtosis of the individual data sets is plotted against the original standard deviation in Figure 2.3. It is clear that the average kurtosis for this collection of data exceeds the value of 3 associated with Gaussian distributions.

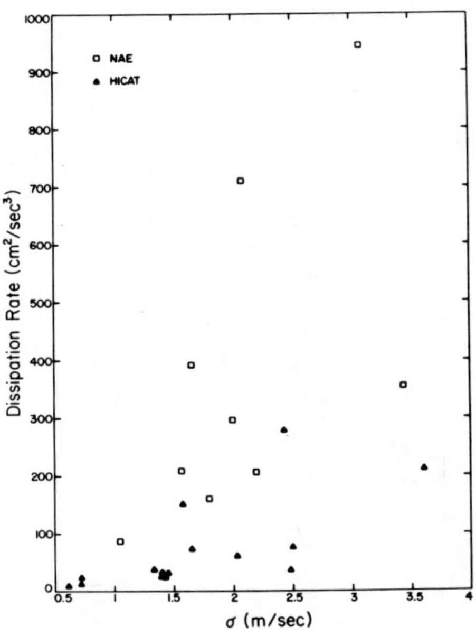

Figure 2.1 The relation between dissipation rate (ε) and standard deviation (σ) for the HICAT and NAE data.

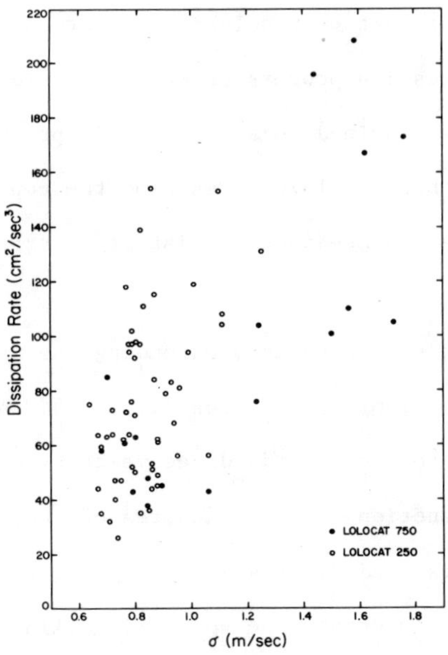

Figure 2.2 The relation between dissipation rate (ε) and standard deviation (σ) for the LO-LOCAT data sets at 250 and 750 ft.

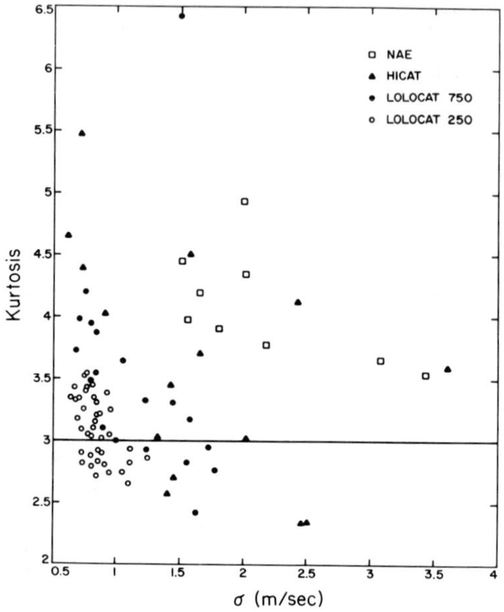

Figure 2.3 The relationship between the kurtosis and standard deviation (σ) for the four data sets.

3.0 PROBABILITY DENSITIES AND DISTRIBUTIONS

The first question that must be resolved about atmospheric turbulence is whether it is a realization of a Gaussian process or may be considered to be so in an approximate sense. We have examined this question in previous papers (Dutton, Thompson and Deaven, 1969; Dutton, 1970), reaching the conclusion that both the basic frequency functions and exceedance statistics departed from those of a Gaussian process.

We begin here with the same question, extending the use of probability functions to powers of the velocity variables. Figures 3.1 and 3.2 show the probability density and distribution functions for the standardized velocity components, Figures 3.3 and 3.4 the same probability functions for the squares of the data, Figures 3.5 and 3.6 show the third power results, and Figures 3.7 and 3.8 give the probability functions for the fourth power. In each graph, the solid line illustrates the Gaussian behavior with the curves for higher orders derived from the standard transformation law

(3.1) $$p_y(y) \left| \frac{dy}{dx} \right| = p_x(x)$$

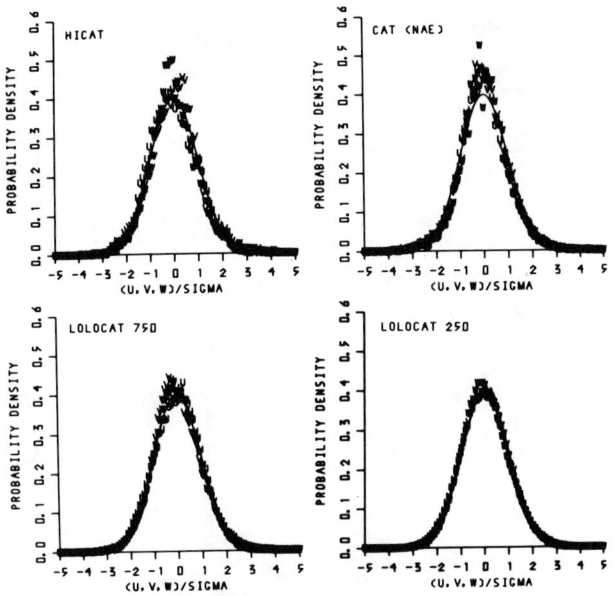

Figure 3.1 Probability density functions for the standardized data. The Gaussian case is illustrated by a solid line.

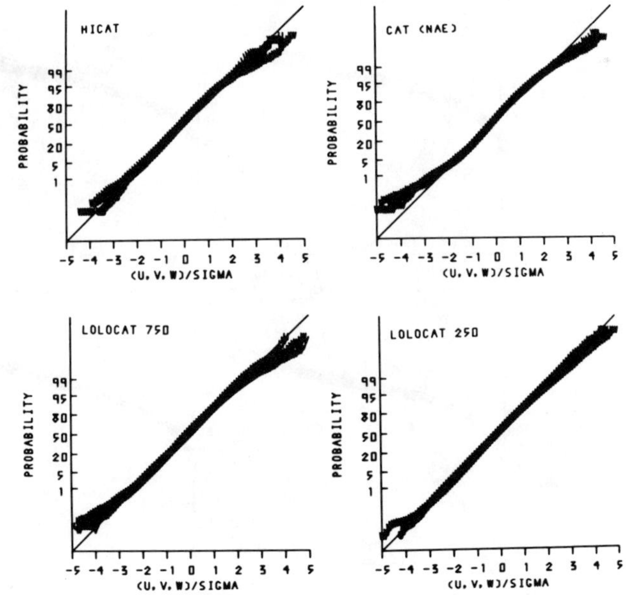

Figure 3.2 Probability distribution functions for the standardized data. The Gaussian case is illustrated by a solid line.

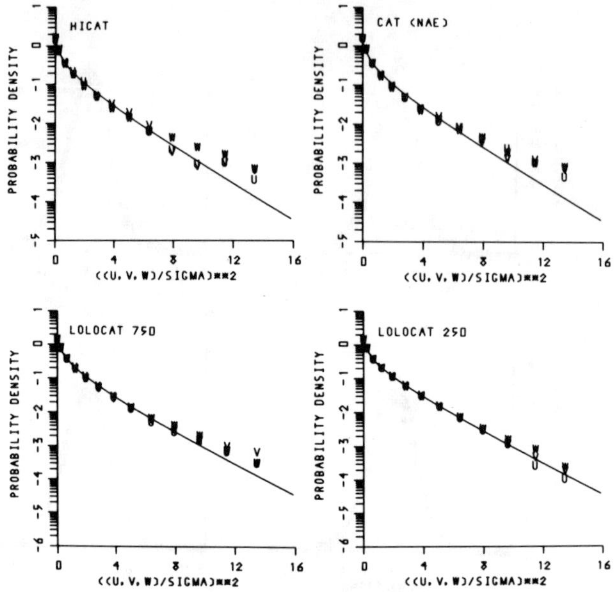

Figure 3.3 Probability density functions for the square of the standardized data. The vertical axis is scaled logarithmically, the integers denoting powers of 10. The Gaussian case is illustrated by a solid line.

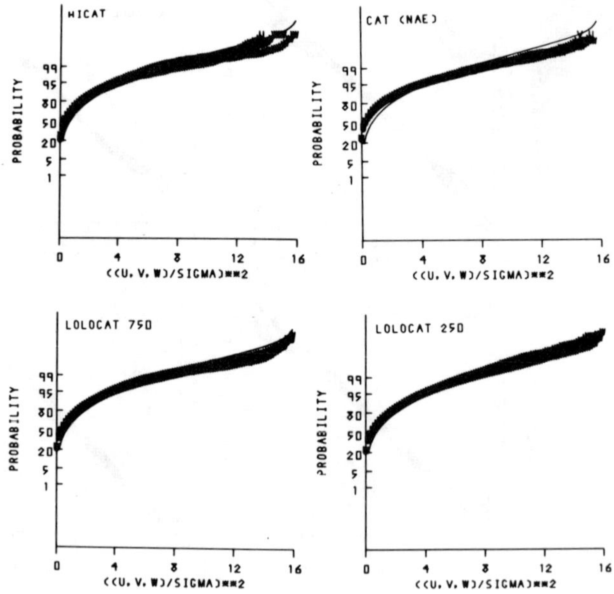

Figure 3.4 Probability distribution functions for the square of the standardized data. The vertical axis is scaled logarithmically, the integers denoting powers of 10. The Gaussian case is illustrated by a solid line.

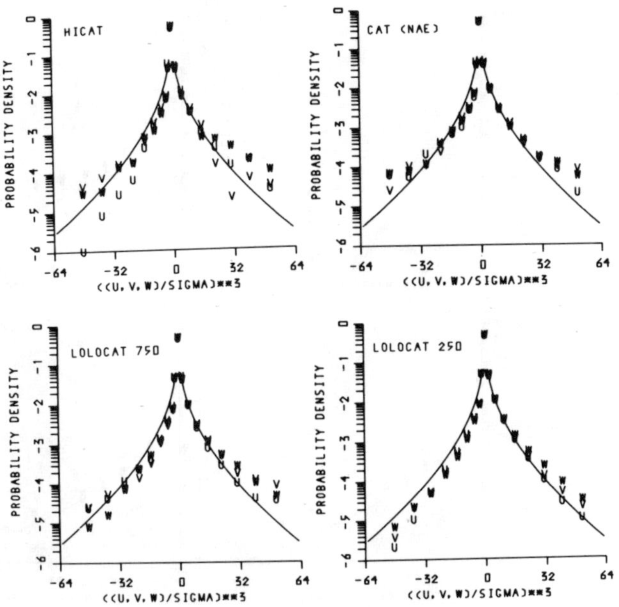

Figure 3.5 Probability density functions for the third power of the standardized data. The vertical axis is scaled logarithmically, the integers denoting powers of 10. The Gaussian case is illustrated by a solid line.

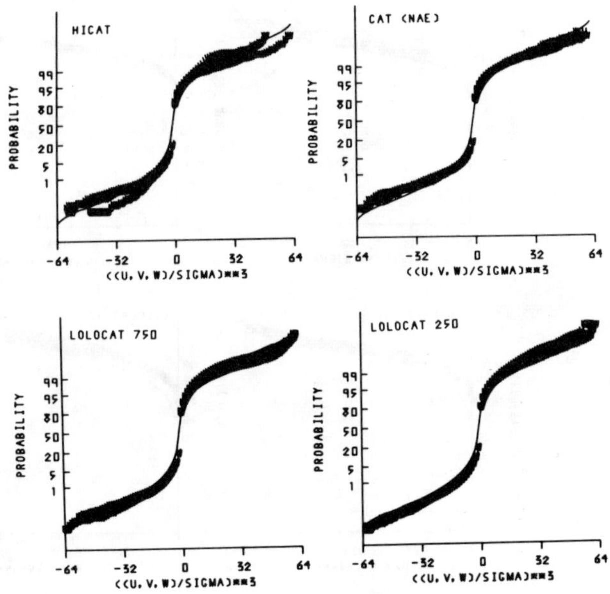

Figure 3.6 Probability distribution functions for the third power of the standardized data. The Gaussian case is illustrated by a solid line.

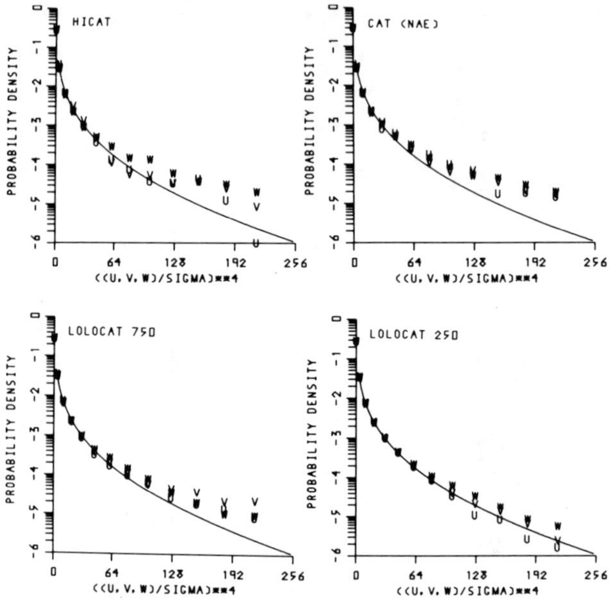

Figure 3.7 Probability density functions for the fourth power of the standardized data. The vertical axis is scaled logarithmically, the integers denoting powers of 10. The Gaussian case is illustrated by a solid line.

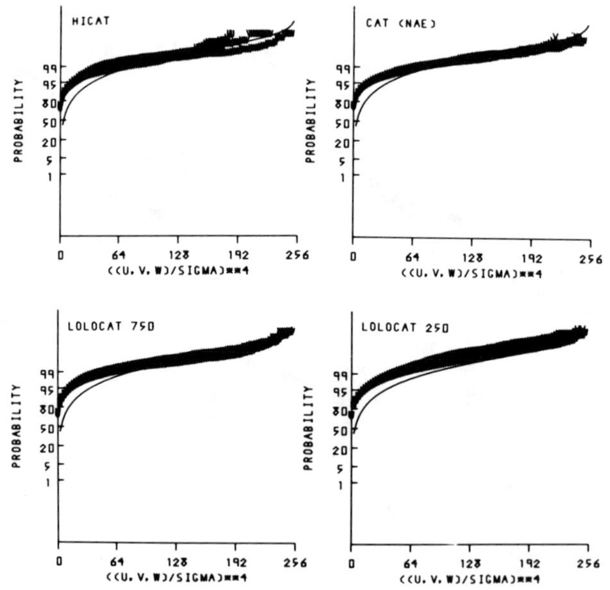

Figure 3.8 Probability distribution functions for the fourth power of the standardized data. The Gaussian case is illustrated by a solid line.

that connects the densities p_y and p_x in a range in which y(x) is a monotonic function of x. The densities for higher order powers have singularities at the origin that lead to slight errors in the numerical integration used to obtain the Gaussian curves shown on the graphs of the higher order distributions.

This series of graphs makes it clear that atmospheric turbulence has both more large gusts and more zeros than would be expected if the distribution were Gaussian. This property of the time series is reflected in the increasing deviation from Gaussian behavior seen in the higher order probability functions.

It is worth noting that the data sets have enough values that smoothness of the probability functions is maintained to the fourth order. Any function chosen to approximate the first order densities should be tested with (3.1) against the higher order densities as well.

The study of differences of the velocity components at various lags reveals aspects of the structure of turbulence that cannot be analyzed quantitatively from the statistical properties of the original records. Here we consider the increments defined by

$$(3.2) \qquad U(x,L) = u(x) - u(x + L)$$

and again shall use the standardized form U/σ_U. The structure function D(L) that occurs in turbulence theory is defined with these increments as

$$(3.3) \qquad D(L) = \overline{U^2(x,L)}$$

in which the overbar denotes an average over the data sample. The theoretical importance of the structure function derives from the fact that it should be dependent only on internal conditions of the turbulent flow. The statistical significance of the increments arises from the fact that they characterize the spatial variations in the turbulence so that the distribution of increments provides insight into the uniformity of spatial structure of the velocity fields. Finally, if the series u(x) were Gaussian, then the distribution of increments would also be Gaussian. Hence the distribution of increments provides specific comparison of the spatial structure of turbulence to a Gaussian series. The increments are also useful in the study of whether the data is self-similar in the sense of Mandelbrot (see, for example, Dutton and Deaven, 1969).

The probability functions for the increments of the four data sets are shown in Figures 3.9-3.16. The curves demonstrate that the increments with small lags deviate more sharply from Gaussian behavior than the increments at large lags. Moreover, the increments at short lags show a pronounced excess of both zeros and large values. Thus in comparison to a Gaussian series, the turbulent velocities tend to either vary less or more at a given lag. This suggests an intuitive picture in which the record is relatively constant at a given value and then shifts suddenly to a new value, a phenomenon that may be described by saying that the velocity gradients tend to be concentrated.

The impression gained from the probability functions presented in this section and from Fig. 2.3 is that the HICAT and NAE data deviate most from Gaussian behavior and that the LO-LOCAT 250 data is the most nearly Gaussian. The obvious question is whether this reflects a true dependence of statistical structure on altitude and thus on generating mechanisms, a dependence on instrumentation, or is a manifestation of the number of samples in the data sets.

We argue that the results demonstrate dependence on generating mechanism. The data of Figure 2.3 show that although the total number of samples in the LO-LOCAT data is indeed greater, the individual runs each possess a smaller kurtosis than those typical of the other data sets. Moreover the LO-LOCAT 750 and LO-LOCAT 250 data were both obtained with the same airplane and processed with the same computer routines. Thus the difference lies in proximity to the Earth's surface and in the difference in stratification. The 250 ft data, obtained under stable conditions, thus appears to illustrate the effect of both stability and nearness of the boundary in giving the turbulence a less sporadic structure than that encountered at higher altitudes.

The HICAT and NAE data were obtained under conditions of large-scale stable thermal stratification and the turbulence is maintained by the interaction of the Reynolds stresses with larger-scale shear patterns. This interaction produces local variations in the energy source and gives the velocity field of the turbulence a

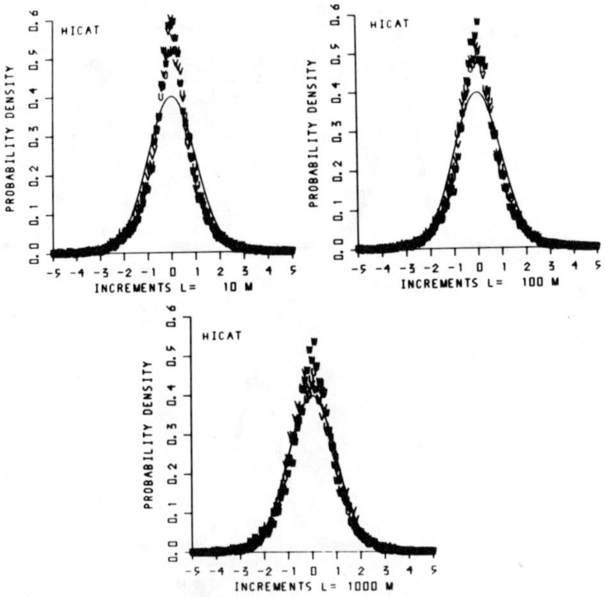

Figure 3.9 Probability density functions for the standardized increments of the standardized HICAT data. The intervals 10, 100 and 1000 m indicated on the graphs signify actual lags of 17, 102 and 996 m. The Gaussian case is illustrated by a solid line.

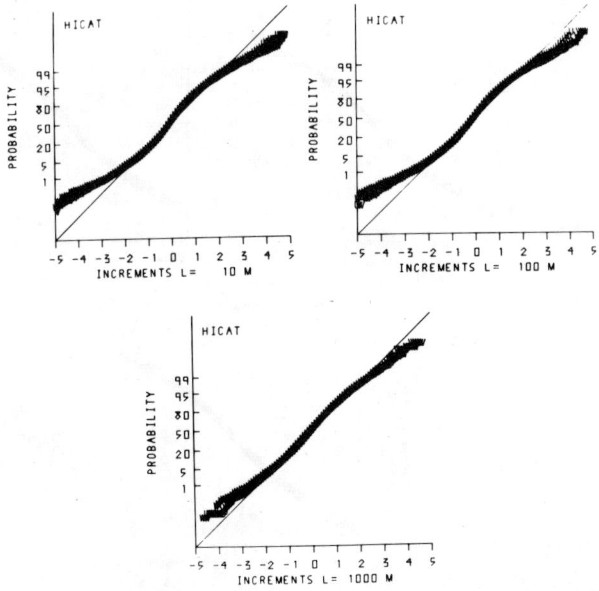

Figure 3.10 Probability distribution functions for the standardized increments of the standardized HICAT data. The intervals 10, 100 and 1000 m indicated on the graphs signify actual lags of 17, 102 and 996 m. The Gaussian case is illustrated by a solid line.

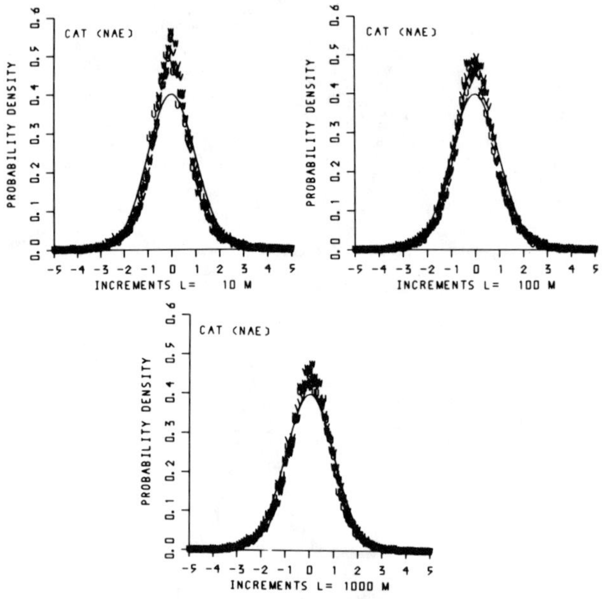

Figure 3.11 Probability density functions for the standardized increments of the standardized NAE data. The intervals 10, 100 and 1000 m denote actual intervals of 9.55, 95.5 and 955 m. The Gaussian case is illustrated by a solid line.

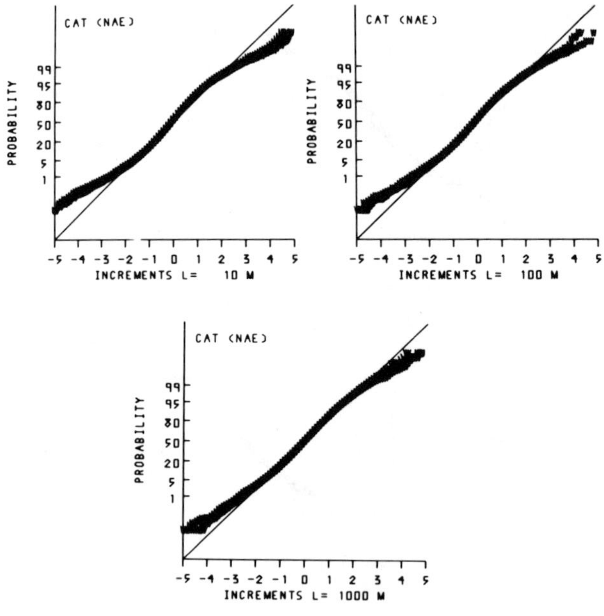

Figure 3.12 Probability distribution functions for the standardized increments of the standardized NAE data. The intervals of 10, 100 and 1000 m denote actual intervals of 9.55, 95.5 and 955 m. The Gaussian case is illustrated by a solid line.

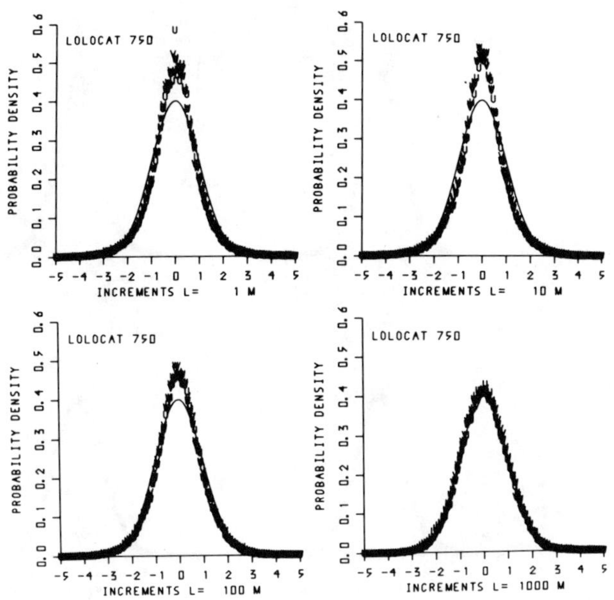

Figure 3.13 Probability density functions for the standardized increments of the standardized LO-LOCAT 750 ft data. The intervals of 1, 10, 100 and 1000 m denote actual intervals of 1.87, 9.35, 99.1 and 1000 m.

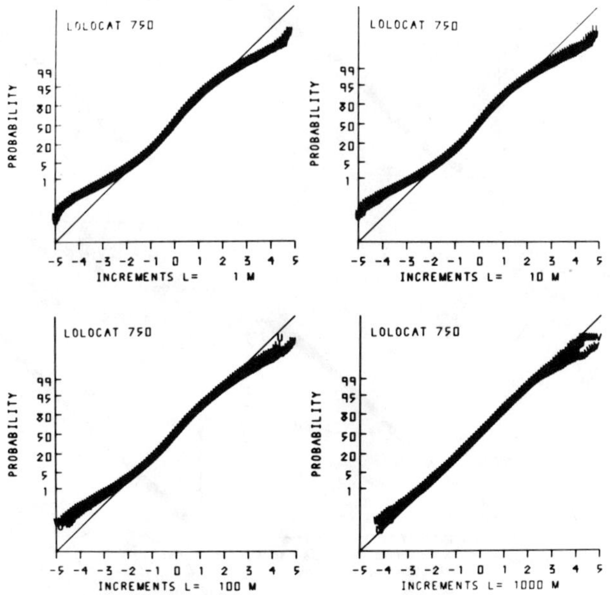

Figure 3.14 Probability distribution functions for the standardized increments of the standardized LO-LOCAT 750 ft data. The intervals of 1, 10, 100 and 1000 m denote actual intervals of 1.87, 9.35, 99.1 and 1000 m.

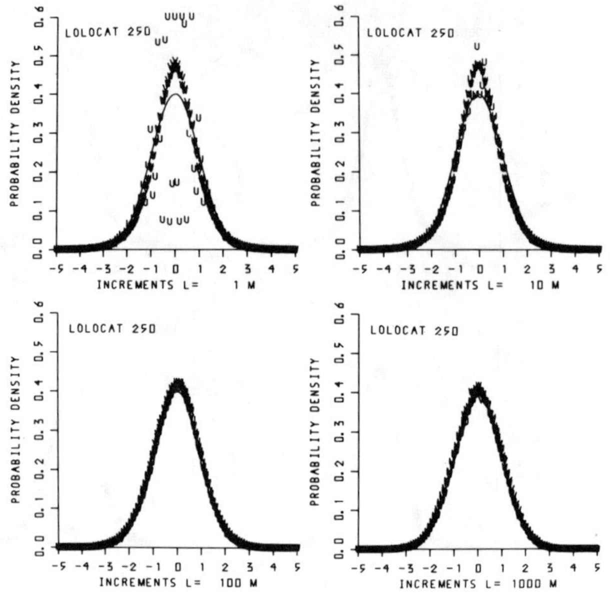

Figure 3.15 Probability density functions for the standardized increments of the standardized LO-LOCAT 250 ft data. The intervals of 1, 10, 100 and 1000 m denote actual intervals of 1.87, 9.35, 99.1 and 1000 m.

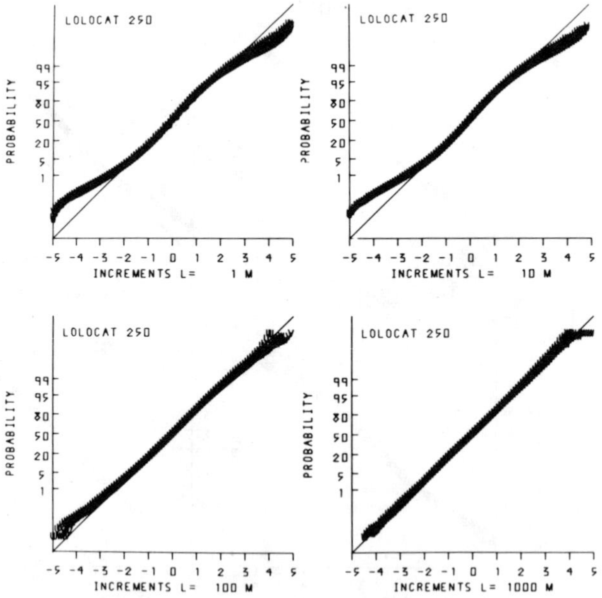

Figure 3.16 Probability distribution functions for the standardized increments of the standardized LO-LOCAT 250 ft data. The intervals of 1, 10, 100 and 1000 m denote actual intervals of 1.87, 9.35, 99.1 and 1000 m.

sporadic structure. Some of the physical aspects of this process are discussed by Lester (1970), Dutton (1970), and Delay and Dutton (1971).

Thus the departure from Gaussian behavior appears to be intimately related with the sporadic structure or intermittent in intensity -- the lack of homogeniety - of atmospheric turbulence. We shall consider intermittency in further detail in the next section.

4.0 MEASURES OF INTERMITTENCY

Both pilots and observant airline passengers know that atmospheric turbulence possesses a distinct element of surprise -- apparently homogeneous turbulence is punctuated by large bumps that occur without warning. Interest in this intermittent property has increased recently, motivated in part by the realization that it represents an aspect of the physics of atmospheric turbulence that is not now understood.

In the previous section it was shown that atmospheric turbulence has larger gusts than would a realization of a normal process -- a fact that is reflected in higher kurtosis factors for turbulence than for a Gaussian process. In other words, a relatively large fraction of the variance is contributed by a relatively small fraction of the total record. This concept has been used as a description of intermittency by Dutton, Lane et al. (1969).

In this section, then, we shall consider how moments such as the variance, skewness, and kurtosis accumulate as a function of the fraction of the total record length. These measures of intermittency represent transformations of the probability functions. It is easy to see how they are derived numerically: the record is rearranged so that the observations are arranged by size; the desired curves are then obtained by summing the appropriate power of these observations and plotting the result against the fraction of observations used in the sum.

The same curves may be obtained from the probability density functions. For example, the fraction of the total variance contributed by observations with absolute value greater than $|u|$ is

$$(4.1) \qquad F_2(u) = \frac{\int_{-\infty}^{-|u|} x^2 \, p_u(x) \, dx + \int_{|u|}^{\infty} x^2 \, p_u(x) \, dx}{\int_{-\infty}^{\infty} x^2 \, p_u(x) \, dx}$$

Similarly, the fraction of record occupied by observations with absolute value greater than $|u|$ is

$$(4.2) \qquad R_2(u) = \int_{-\infty}^{-|u|} p_u(x) \, dx + \int_{-\infty}^{-|u|} p_u(x) \, dx$$

Thus if the density function is known, both F_2 and R_2 can be determined as functions of $|u|$ and so F_2 is known as a function of R_2. This technique provides a method for determining curves appropriate for a Gaussian process that may be used for comparison to observed data.

Figures 4.1, 4.2, and 4.3 show the accumulation of variance, skewness, and kurtosis as a function of percentage of record for the four data sets. Although the departure from Gaussian for the variance is not pronounced, the deviations of the accumulation of skewness and kurtosis are quite distinct.

It is noteworthy that the kurtosis factor graphs in Figure 4.3 show that the three most non-Gaussian data sets exhibit a very rapid accumulation of the fourth moment in less than 10 percent of the record, and that the more rapid accumulation than Gaussian continues throughout the record.

These curves, as shown analytically by (4.1) and (4.2), are transformations of the probability density functions that reveal non-Gaussian structure quite dramatically. They too provide what appear to be sensitive criteria for testing density functions proposed as approximations to those of turbulence.

Various forms of exceedance statistics have been used for studying statistical structures and the frequency of occurrence of large gusts. A discussion of some of them and references are given in Dutton (1970). Figure 4.4 shows an exceedance statistic that is widely used in aeronautical engineering. The quantity $N(y)$ is defined to be the number of crossings with positive slope per unit time of level y. Hence $N(y)/N(0)$ is the ratio of the number of crossings to y to the number of crossings of zero. If y and its derivative are independent, then as shown in

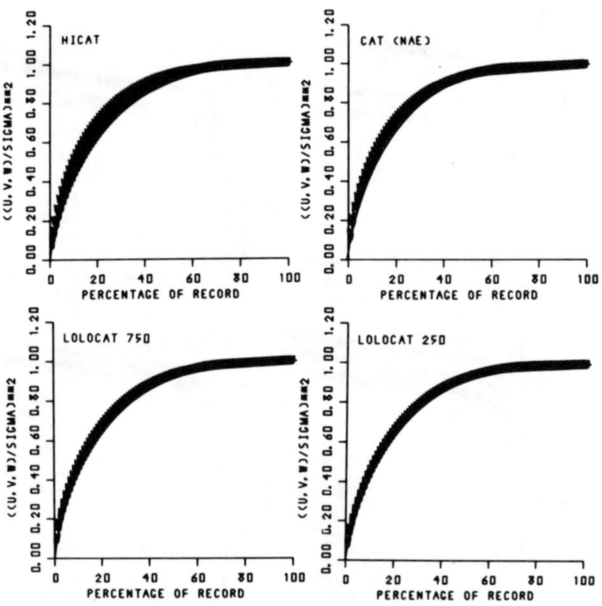

Figure 4.1 Accumulation of variance (on the vertical axis) of the standardized data as a function of the percentage of record. The Gaussian case is illustrated by a solid line. Note that the final symbol on the right gives the average skewness for the data set.

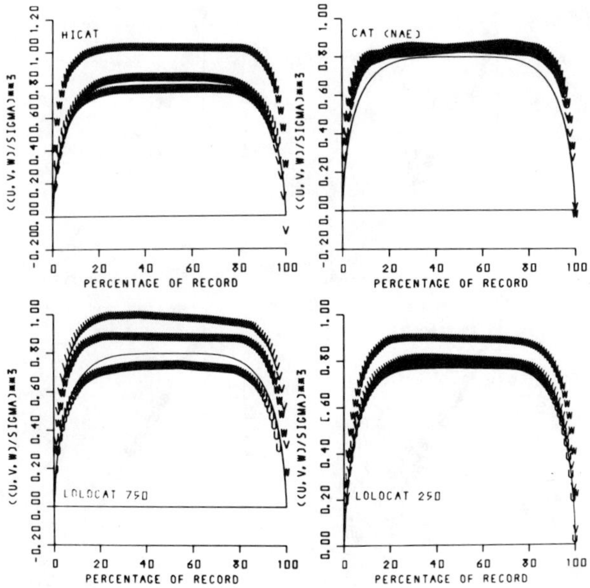

Figure 4.2 Accumulation of skewness (on the vertical axis) of the standardized data as a function of percentage of record. The Gaussian case is illustrated by a solid line. Note that the final symbol on the right gives the average skewness for the data set.

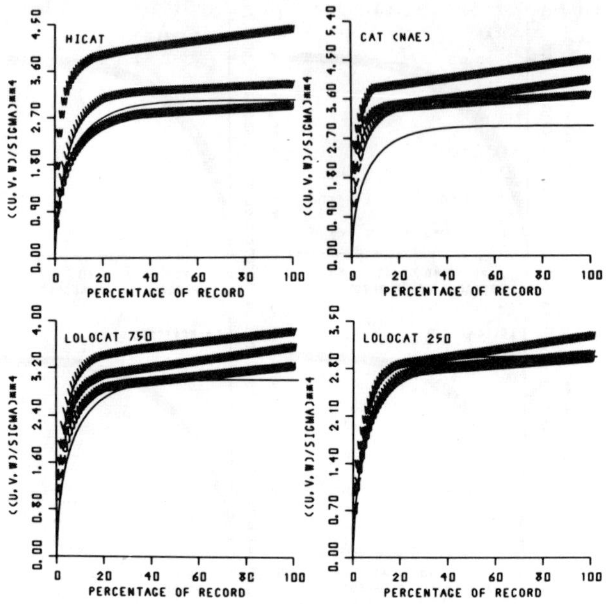

Figure 4.3 Accumulation of kurtosis (on the vertical axis) as a function of percentage of record for the standardized data. The Gaussian case is illustrated by a solid line. Note that the final symbol on the right give the average kurtosis for the data set.

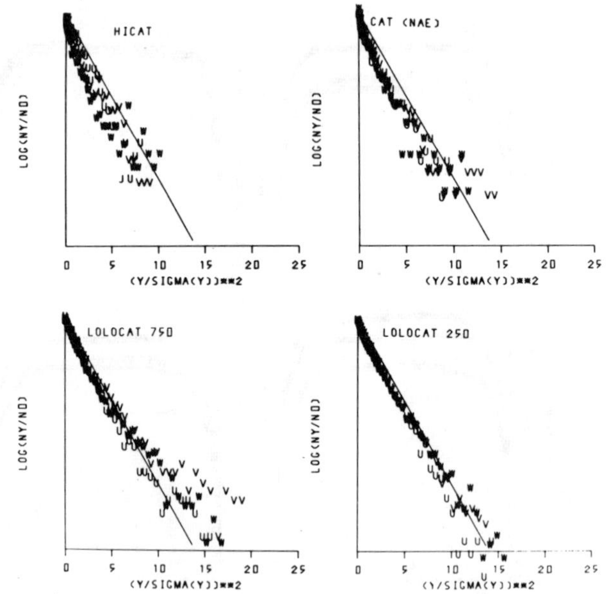

Figure 4.4 The ratio for the standardized data of the number, N(y), of crossings of level y with positive slope to the number, N(0), of crossings of zero with positive slope as a function of y/σ_y.

Dutton (1970), N(y)/N(0) is identical to p(y)/p(0) where p denotes the probability density. The values of N(y)/N(0) are compared to the Gaussian case in Figure 4.4.

5.0 SPECTRAL PROPERTIES

So far in this paper we have been concerned with statistical quantities that do not take account of the sequence of events throughout the record. Investigators of turbulence have found the sequencing revealed directly by the record to be generally unfathomable, and have turned to various techniques to produce smooth functions that summarize this information.

We consider first the correlation function, defined for a record of length L by

$$\phi_{fg}(\xi) = \frac{1}{L}\int_{o}^{L} f(x)\; g(x+\xi)\; dx \tag{5.1}$$

The autocorrelation functions ϕ_{uu}, ϕ_{vv}, ϕ_{ww} computed for the data sets by the discrete analog of (5.1) are shown in Figure 5.1 on a logarithmic lag scale. In general, the correlation of the horizontal components u and v is higher in midrange than that of the vertical component w. The variation in the functions at lags greater than about 1000 m may be due to the filtering applied to all of the data sets except HICAT.

The cross correlation functions ϕ_{uv}, ϕ_{uw}, and ϕ_{vw} are shown in Figure 5.2. The hints of a wavy structure with length about 1000 km that appear in ϕ_{wv} for the HICAT and NAE data sets seem to be reinforced by the behavior of ϕ_{uw} and ϕ_{vw} at the same point.

Both sets of correlation functions are quite smooth up to lags in the range 100 to 1000 m and qualitatively the same for all the data sets. If we take into account the variations expected between turbulence in the boundary layer and the free atmosphere, then the change in slope and scattering at lags greater than about 100 m probably may be attributed to the presence of the larger scale structures associated with transfer of energy to the turbulent motion.

As an alternate to correlation functions, Fourier transform techniques provide a decomposition of the velocity record into a sum or integral of harmonic functions. The important property is that the square modulus of the transform integrates to the variance or energy of the velocity and is usually much smoother and has more apparent

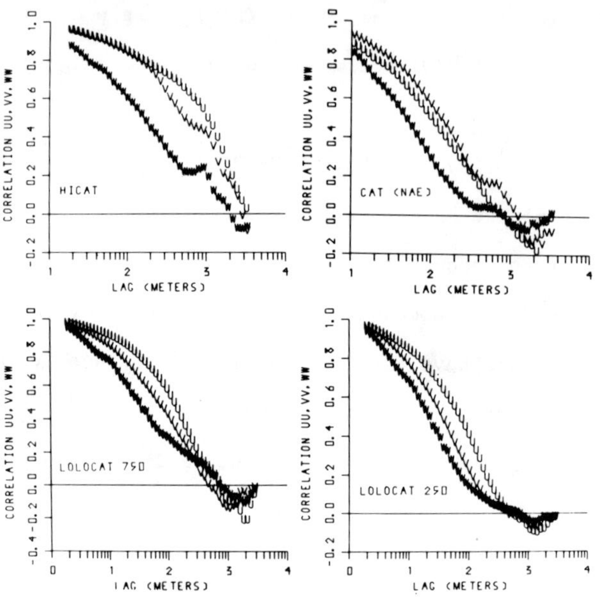

Figure 5.1 The autocorrelation functions for the standardized data plotted as a function of the lag on a logarithmic scale. The integers on the horizontal axis denote powers of 10.

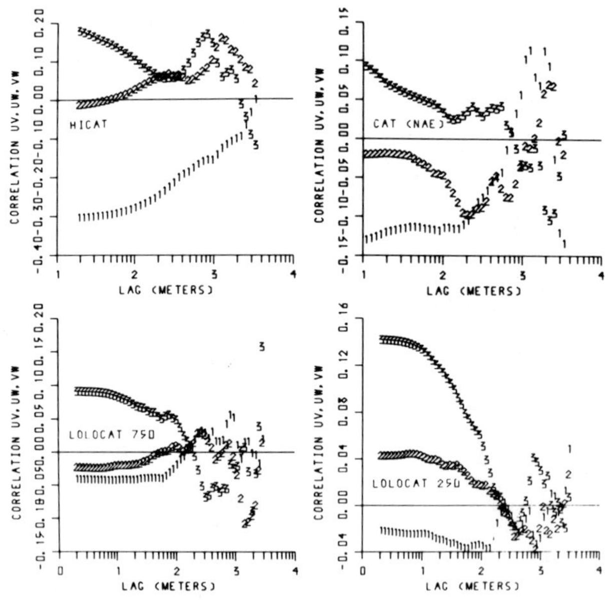

Figure 5.2 Cross correlation functions for the standardized data. The plotted 1 refers to the u-v correlation functions, the 2 to u-w, the 3 to v-w. For further details see the legend for Fig. 5.1.

structure than the energy (or squared velocity) record itself. Thus if $\hat{u}(k)$ is the Fourier transform (suitably defined) of the record, then the spectral density $\Phi(k)$ is defined (in practice, in a smoothed version) as $|\hat{u}(k)|^2$ and has the property that

(5.2) $$\sigma^2 = \int_o^\infty \Phi(k)\, dk$$

Thus the quantity $\Phi(k)dk$ shows how much of the variance σ^2 is contributed by harmonic functions having wavelengths in the interval $(k, k + dk)$. Similarly if $\Phi_n(k)$ is defined as $|\hat{u^n}(k)|^2$, then

(5.3) $$\overline{u^{2n}} = \int_o^\infty \Phi_n(k)\, dk$$

where the overbar denotes an average over the record. Thus the nth order spectral density $\Phi_n(k)$ shows how the contributions to the 2nth moment are arranged by harmonic. Here we shall consider the spectra Φ, Φ_2, Φ_3 corresponding to the second, fourth and sixth moments about the mean.

The spectra Φ, Φ_2, and Φ_3 for each component are shown in Figures 5.3-5.6. The spectra plotted here are in the form $k\Phi(k)$, which derives from the relation

(5.4) $$\int_o^\infty \Phi(k)\, dk = \int_{-\infty}^\infty k\Phi(k)\, d\ln(k)$$

and allows a logarithmic wave number scale to be used. The quantity $k\Phi(k)$ is also plotted logarithmically to allow power-law behavior to be inferred. Note that slopes on these graphs will be one more than the slope on the usual log-log plot.

If we assume the existence of an inertial subrange in which all spectral properties depend only on ε and k, we can infer the form of the spectra. Taking [ ] to denote dimensions, we have

(5.5) $$[\Phi_n] = [u^{2n}]\cdot L = L^{2n+1}\, T^{-2n}$$

The simplest form in which Φ_n depends only on ε and k is

(5.6) $$\Phi_n(k) = \alpha_n\, \varepsilon^\gamma\, k^\gamma$$

where the α_n are constants. Hence

(5.7) $$L^{2n+1}\, T^{-2n} = L^{2\gamma-\lambda}\, T^{-3\gamma}$$

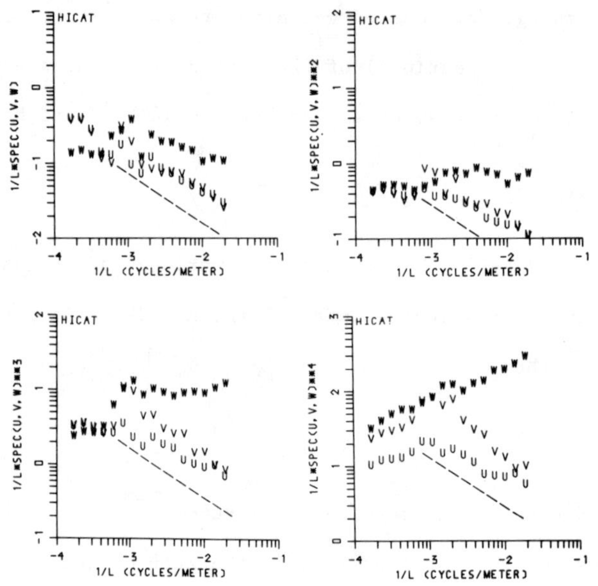

Figure 5.3 The spectra of the first, second, third and fourth powers of the HICAT data. Note that the function $k\Phi(k)$ is scaled logarithmically on the vertical axis and that k is scaled logarithmically on the horizontal axis, with the integers denoting powers of 10. The dashed line gives the -2/3 slope that illustrates a -5/3 power law on a plot with these coordinates.

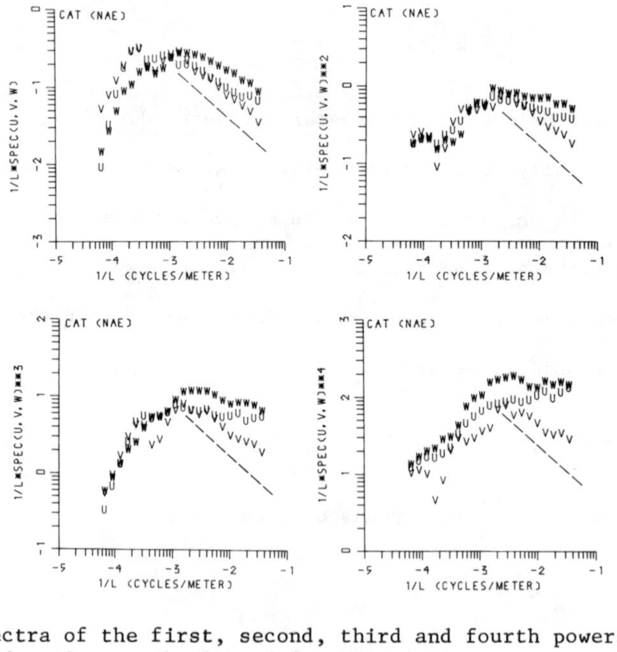

Figure 5.4 The spectra of the first, second, third and fourth powers of the NAE data. For further details see the legend for Fig. 5.3.

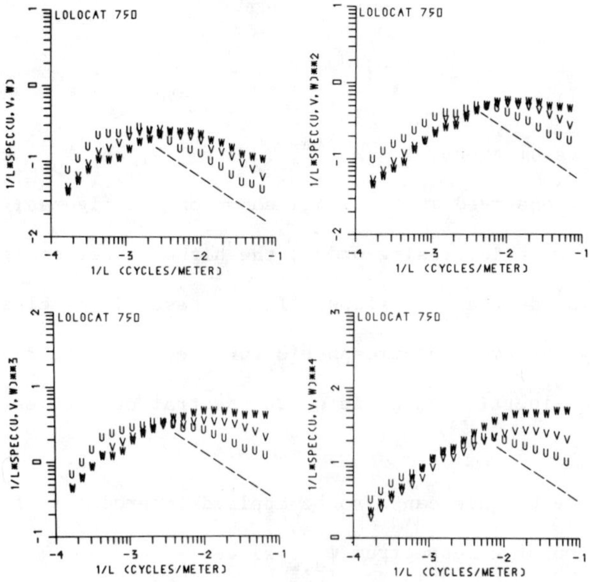

Figure 5.5 The spectra of the first, second, third and fourth powers of the LO-LOCAT 750 ft data. For further details see the legend for Fig. 5.3.

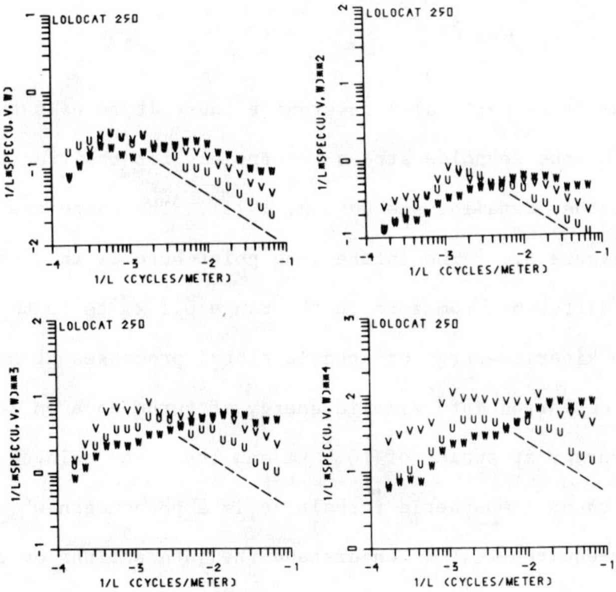

Figure 5.6 The spectra of the first, second, third and fourth powers of the LO-LOCAT 250 ft data. For further details see the legend for Fig. 5.3.

and so

$$(5.8) \qquad \gamma = \frac{2n}{3} \qquad\qquad \lambda = -\frac{2n+3}{3}$$

Therefore on these assumptions, $\Phi \sim k^{-5/3}$, $\Phi_2 \sim k^{-7/3}$, and $\Phi_3 \sim k^{-3}$. But clearly this behavior is not observed in the range shown on the figures, and rather than increasing in slope with increasing order, the higher order spectra either retain the -5/3 power law or decrease in slope. In any case, it is clear that the -5/3 power law generally observed in atmospheric turbulence in the range 100 to 1000 m does not indicate an inertial range in which spectral properties depend only on ε and k.

The transform techniques can also be applied to products of variables, so that a suitable definition of a cospectrum $\psi_{u,w}(k)$ as

$$(5.9) \qquad \psi_{u,w} = \mathrm{Re}\ \{\hat{u}\ \bar{\hat{w}}\}$$

gives

$$(5.10) \qquad \overline{uw} = \int_0^\infty \psi_{u,w}(k)\ dk$$

This cospectrum is of particular interest because it reveals the wavenumber decomposition of the Reynolds stress responsible for the transformation of mechanical energy (for further details, see Dutton, 1971). The cospectra for the four data sets are shown in Figure 5.7. The interesting point here is that the cospectra are significantly different from zero in the range 0.1 km to 10 km. Thus it is in this range that the kinetic energy created in global processes at scales of thousands of kilometers is converted into kinetic energy of turbulence on its way to dissipation into thermal energy at scales of 10^{-5} km and less. As pointed out in the introduction, then, atmospheric turbulence is a phenomenon whose explanation and comprehension requires that we understand the interactions over a vast range of scales.

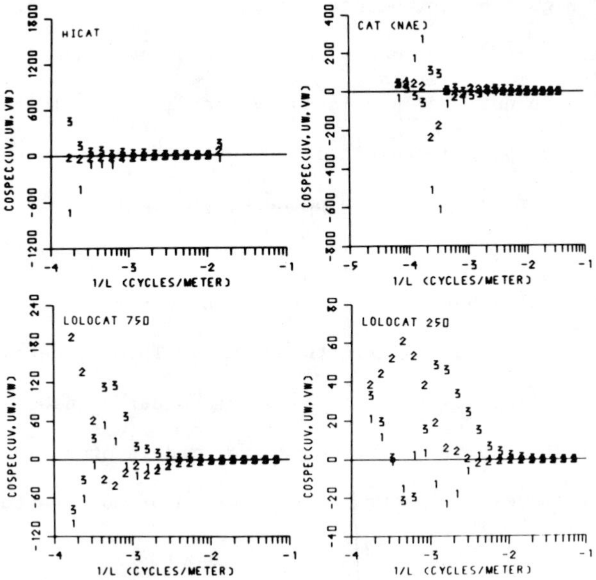

Figure 5.7 The cospectrum (or spectral form of the Reynolds stresses) for the standardized data. The cospectrum is scaled on the vertical axis against wave number on a logarithmic scale on the horizontal axis. The plotted 1 denotes the cospectra for u and v, the 2 for u and w, the 3 for v and w.

5.0 CONCLUSIONS

The observational results given present a formidable theoretical challenge. There are two possibilities: Either a model must be developed, from first principles, that yields probability distributions and associated statistical functions as well as the spectral distributions of the variance and higher order, or, as suggested by Prof. Parzen in this symposium, a transformation of the data must be found so that an existing model applies.

Specification of a model may be expected to be difficult because of the complexity of the structure. First, we must approximate or predict the non-Gaussian densities with sufficient accuracy that the transformations to densities of higher order powers, to accumulation of moments as a function of fraction of record, and to exceedance statistics agree with the empirical results. Joint distributions must be found or predicted that will give the correct behavior for the increments. And finally, sequencing of events must be arranged so that the spectral distributions of the variance and higher order moments are correctly predicted. As shown in the previous section, there is a broad band of wavenumbers in which the slope of the spectral densities is not that predicted by the dimensional reasoning of inertial theory. Still these spectra exhibit an organized behavior with increasing order and we shall have to find a set of hypotheses that will imply the observed structure.

We view our results here as a contribution to the effort to determine the actual statistical structure of atmospheric turbulence. This is certainly a necessary part of the development of statistical models of atmospheric turbulence, and these results suggest that a first step may be to search for models of series that allow probabilistic structure and spectral distribution of moments to interact.

6.0 REFERENCES

Crooks, W. M., F. M. Hoblit and F. A. Mitchell, 1968: Project HICAT -- High altitude measurements and meteorological correlations. Technical Report AFFDL-TR-68-12, Air Force Flight Dynamics Laboratory, Wright-Patterson Air Force Base, Ohio.

Dutton, J. A. and D. G. Deaven, 1969: A self-similar view of atmospheric turbulence. Radio Science, 4, 1341-1349.

Dutton, J. A. and R. D. Delay, 1971: An analysis of conditions associated with an occurrence of stratospheric CAT. In press, Journal of the Atmospheric Sciences, October.

Dutton, J. A. and J. A. Lane, 1969: Intermittency of small-scale structures. Radio Science, 4, 1357-1359.

Dutton, J. A. and H. A. Panofsky, 1970: Clear air turbulence: a mystery may be unfolding. Science, 167, 937-944.

Dutton, J. A., G. J. Thompson and D. G. Deaven, 1969: The probabilistic structure of clear air turbulence - some observational results and implications. Clear Air Turbulence and Its Detection, Y. H. Pao and A. Goldburg, Eds., Plenum Press, New York.

Dutton, J. A., 1969: An energy budget for a layer of stratospheric CAT. Radio Science, 4, 1137-1142.

Dutton, J. A., 1970: Effects of turbulence on aeronautical systems. Progress in Aerospace Sciences, 11, 67-109, D. Kucheman, Ed., Pergamon Press, New York.

Dutton, J. A., 1971: Clear air turbulence, aviation, and atmospheric science. In press, Reviews of Geophysics and Space Sciences, August.

Fichtl, G. H. and J. A. Dutton, 1971: Higher order spectra of atmospheric turbulence. In preparation.

Jones, J. W., R. H. Mielke and G. W. Jones, 1970: Low altitude atmospheric turbulence LO-LOCAT Phase III. Technical Report AFFDL-TR-70-10, Air Force Flight Dynamics Laboratory, Wright-Patterson Air Force Base, Ohio.

Lester, P. F., 1970: Some physical and statistical aspects of clear air turbulence. Atmospheric Science Paper No. 165, Colorado State University, Fort Collins, Colorado.

Kung, E. C., 1966: Large-scale balance of kinetic energy in the atmosphere. Monthly Weather Review, 94, 627-640.

Kung, E. C., 1967: Diurnal and long-term variations of the kinetic energy generation and dissipation for a five-year period. Monthly Weather Review, 95, 593-606.

Kung, E. C., 1969: Further study on the kinetic energy balance. Monthly Weather Review, 97, 573-581.

Mather, G. K., 1967: Some measurements of mountain waves and mountain wave turbulence made using the NAE T-33 turbulence research aircraft. DME/NAE Quarterly Bulletin No. 1967(2).

Pasquill, F. and H. A. Panofsky, 1963: The constant of the Kolmogorov Law. Quart. J. Roy. Met. Soc., 89, 550-551.

Trout, D. A. and H. A. Panofsky, 1969: Energy dissipation near the tropopause. Tellus, 21, 355-358.

Vinnichenko, N. K. and J. A. Dutton, 1969: Empirical studies of atmospheric structure and spectra in the free atmosphere. Radio Science, 4, 1115-1126.

SOME MEASUREMENTS OF THE FINE STRUCTURE

OF LARGE REYNOLDS NUMBER TURBULENCE

by

J. C. WYNGAARD

Air Force Cambridge Research Laboratories

Bedford, Mass. 01730

Y. H. PAO

Flow Research, Inc., Kent, Wash. 98031

ABSTRACT

Spectra and moments of the streamwise velocity derivative measured in the atmospheric surface layer are presented. Many of the properties are found to depend on turbulence Reynolds number in ways consistent with Kolmogoroff's refined theory.

INTRODUCTION

Kolmogoroff's original (1941) hypothesis held that in turbulence of sufficiently large Reynolds number (Re), the fine structure is isotropic, determined only by the mean dissipation rate $\langle\epsilon\rangle$ and the kinematic viscosity ν, and independent of the large-scale flow. It was recognized early that large Re values were difficult to achieve in the laboratory, and in fact Batchelor (1956) felt that laboratory flows could not provide complete confirmation of the theory.

Experimenters have turned to geophysical flows in quest of large Re, and have found values of R_λ (based on the Taylor microscale λ, a turbulent velocity scale and ν) upwards of a few thousand in the lower atmosphere. By contrast, early laboratory flows (summarized by Batchelor, 1956) covered the 10 - 100 range in R_λ, and the largest values to date seem to be 700 in grid turbulence (Kistler and Vrebalovich, 1966) and 780 in a jet (Gibson, 1963).

When Kolmogoroff (1962) refined his hypothesis, replacing $\langle\epsilon\rangle$ with its locally averaged value, he gave new significance to the Re value. As shown later, the refinement implies that fine structure is Re-dependent. Documentation of these dependencies requires data from a wide Re range, and there now seems to be renewed interest in laboratory flows. Their Re values have very little overlap with those in geophysics, and the combination shows promise of giving a wide enough Re range to reveal even weak trends.

The results introduced here are used to investigate some of these trends. They were obtained from a modest hot-wire experiment carried out during the much more extensive atmospheric surface-layer measurement program described by Haugen et al (1971). Single hot wires, 5μ in dia and 1.2mm long, were operated in the linearized constant temperature mode (DISA 55D05-15 units) at 5.66, 11.3 and 22.6m above a horizontally homogeneous Kansas prairie. Streamwise velocity derivatives $\partial u_1/\partial t$ were obtained from four-pole Butterworth filters (Wyngaard and Lumley, 1967) which differentiate and low-pass filter (24 db/octave) to eliminate high-frequency noise. The filtered $\partial u_1/\partial t$ signals were then recorded on FM tape. The spectral data in this paper were obtained at Boeing Scientific Research Laboratories by digital processing of the velocity derivatives, using the fast Fourier transform.

INTERPRETATION OF FINE-STRUCTURE DATA

It is seldom possible to measure the desired turbulence properties directly. For example, while theoretical discussions concern the behavior of wave number spectra, we measure frequency spectra and rely on a "frozen field" assumption to relate the two. While predictions are made for the structure of the smallest eddies, our observations come from sensors which smooth out and distort their detail in sometimes strange ways. In this section we will review briefly the impact of some of these problems, which are particularly important in the study of Re trends.

The "frozen field", or Taylor's (1938) hypothesis, takes temporal fluctuations as convected streamwise spatial ones. In fact the flow velocity fluctuates, and there are a number of other reasons why the hypothesis can introduce errors. Lumley (1965) has shown that if various criteria are satisfied all but the fluctuating convection velocity effect can be ignored at high frequencies. In his model of this process, the measured and true one-dimensional spectra of the streamwise velocity fluctuations are found to be related by

$$\Phi^m = \Phi + \frac{\langle v_1^2 \rangle}{2U^2}\left(\kappa_1^2 \Phi'' + 2\Phi + 4\kappa_1 \Phi'\right) - \left\{\frac{\langle v_2^2 \rangle + \langle v_3^2 \rangle}{U^2}\right\}\left(\kappa_1 \Phi' + \Phi\right) \tag{1}$$

where Φ is the streamwise wave number (κ_1) spectrum of u_1, Φ^m is its value inferred from the frequency (ω) spectrum through $\kappa_1 = \omega/U$, and primes denote differentiation. U is the mean flow velocity, and v_1, v_2, v_3 are its fluctuation components. (1) is useful in assessing the accuracy of the frozen-field approximation

$$\left\langle \left(\frac{\partial^n u_1}{\partial x_1^n}\right)^2 \right\rangle \simeq \frac{1}{U^{2n}} \left\langle \left(\frac{\partial^n u_1}{\partial t^n}\right)^2 \right\rangle \tag{2}$$

To do this, we note that

$$\left\langle \left(\frac{\partial^n u_1}{\partial x_1^n}\right)^2 \right\rangle = \int_0^\infty \kappa_1^{2n} \Phi \, d\kappa_1 \tag{3}$$

and

$$\frac{1}{U^{2n}} \left\langle \left(\frac{\partial^n u_1}{\partial t^n}\right)^2 \right\rangle = \int_0^\infty \kappa_1^{2n} \Phi^m d\kappa_1 \tag{4}$$

Carrying out the integration in (4) and using (1) we find

$$\left\langle \left(\frac{\partial^n u_1}{\partial t^n}\right)^2 \right\rangle = U^{2n} \left\langle \left(\frac{\partial^n u_1}{\partial x_1^n}\right)^2 \right\rangle \left\{ 1 + (2n^2 - n)\frac{\langle v_1^2 \rangle}{U^2} + \frac{2n}{U^2}\left(\langle v_2^2 \rangle + \langle v_3^2 \rangle\right) \right\} \tag{5}$$

(5) indicates that significant overestimates of spatial derivative variances are made in high-intensity turbulence. For n = 1, we note that

$$\left\langle \left(\frac{\partial u_1}{\partial t}\right)^2 \right\rangle = U^2 \left\langle \left(\frac{\partial u_1}{\partial x_1}\right)^2 \right\rangle \left\{ 1 + \frac{\langle v_1^2 \rangle}{U^2} + 2\frac{\langle v_2^2 \rangle}{U^2} + 2\frac{\langle v_3^2 \rangle}{U^2} \right\} \tag{6}$$

which was derived by Heskestad (1965) in a different way.

(1) and (5) will be useful in interpreting the spectral data introduced later. Little is known of the validity of applying Taylor's hypothesis to higher-order statistics, and further theoretical work here would be valuable.

Another problem is accurately resolving small-scale motions with sensors having both temporal and spatial averaging properties (Corrsin, 1963). In geophysical flows the spatial averaging effects are usually more serious. Uberoi and Kovasznay (1953) have laid the basic groundwork here, and Wyngaard (1968, 1969, 1971) has calculated one-dimensional spectral transfer characteristics of hot-wire arrays used to measure velocity, vorticity, and temperature fluctuations. Fig. 1, which is based on these calculations, shows streamwise vorticity (ω_1) and u_1 spectra and the forms that would be measured with hot wire arrays of scale $\ell = 10\eta$. These one-dimensional (1-D) spectra are related to $E(\kappa)$, the three-dimensional (3-D) velocity spectrum (called the energy spectrum function by Batchelor, 1956) through

$$\begin{aligned} \phi_{u_1} &= \int_{\kappa_1}^{\infty} \frac{E(\kappa)}{\kappa}\left(1-\frac{\kappa_1^2}{\kappa^2}\right) d\kappa \\ \phi_{\omega_1} &= \int_{\kappa_1}^{\infty} \frac{\kappa^2 E(\kappa)}{\kappa}\left(1-\frac{\kappa_1^2}{\kappa^2}\right) d\kappa \end{aligned} \tag{7}$$

$E(\kappa)$ was assumed to have a $\kappa^{-5/3}$ inertial range followed by an exponential break at $\kappa \simeq 1/\eta$ (Pao, 1965). Two features of the results (which do not depend on the details of the break) are important here. First, while both 1-D and 3-D u_1 spectra have the same ($\kappa^{-5/3}$) inertial subrange form, this is not true for ω_1, whose 3-D spectrum goes as $\kappa^{1/3}$ while the 1-D form (Fig. 1) monontonically decreases. Experiments yield 1-D spectra, while theoretical predictions are usually for the 3-D kind, and if, as with vorticity, their shapes are different, comparisons can be hazardous.

A second feature of Fig. 1 is that measured u_1 spectra are affected by spatial averaging only for $\kappa_1 > 1/\ell$, while vorticity spectra are affected, both in amplitude and slope, at all κ_1. The explanation is that the 1-D spectra, at a given κ_1, receive their dominant contributions from different scales of motion in the two

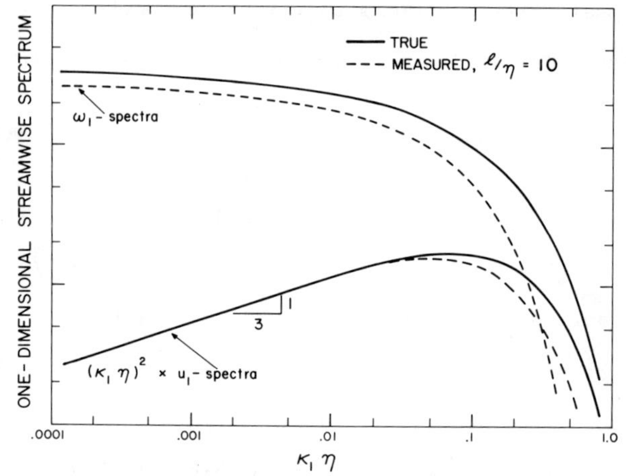

Fig. 1. Calculated effect of hot-wire length on measured velocity and vorticity spectra.

cases. For velocity, (7) shows that the 1-D spectral level is determined principally by motions with κ of the order of κ_1. For vorticity, the $\kappa^{1/3}$ inertial range increase of the 3-D spectrum insures that the 1-D spectrum at <u>any</u> κ_1 is dominated by contributions from κ near $1/\eta$.

These considerations also complicate the interpretation of measured spectra of velocity derivatives raised to a power n. If these are treated as isotropic scalars the spectra are related by

$$\phi_n = \int_{K_1}^{\infty} \frac{F_n(K)}{K}\, dK \tag{8}$$

Novikov (1965) predicts that the inertial range of F_n goes as κ^p where $p = -\frac{1}{2}$ for $n = 2$ but is positive for larger n. The situation for $p > 0$ is like that for vorticity; the 1-D spectrum is flat in the inertial range and its measured level at any κ_1 depends on the small-scale resolving ability of the instrumentation.

The final point is that in geophysical turbulence, velocity and temperature fluctuations are usually found together. If hot-wire sensors are used, the temperature fluctuations (θ) contaminate the u_1 signal:

$$u_1^m = u_1 + a\Theta \tag{9}$$

The temperature sensitivity a depends on operating conditions (Rose, 1962). It leads to overestimates of derivative variances by an amount (in "frozen" isotropic turbulence)

$$\left\langle \left(\frac{\partial^n u_1}{\partial x_1^n} \right)^2 \right\rangle^m = \left\langle \left(\frac{\partial^n u_1}{\partial x_1^n} \right)^2 \right\rangle + a^2 \left\langle \left(\frac{\partial^n \theta}{\partial x_1^n} \right)^2 \right\rangle \tag{10}$$

Contamination to the same order appears in a third moment:

$$\left\langle \left(\frac{\partial u_1}{\partial x_1} \right)^3 \right\rangle^m = \left\langle \left(\frac{\partial u_1}{\partial x_1} \right)^3 \right\rangle + 3a^2 \left\langle \left(\frac{\partial u_1}{\partial x_1} \frac{\partial \theta}{\partial x_1} \frac{\partial \theta}{\partial x_1} \right) \right\rangle \tag{11}$$

The parasitic term in (11) has an important role in temperature gradient dynamics (Wyngaard, 1971b) and, like that in (10), can be expressed as a moment of the temperature spectrum. Little is now known of the details of its high wavenumber end, but if it has a fuller tail than the velocity spectrum these effects could be troublesome in certain situations.

KOLMOGOROV'S REFINED THEORY AND THE ROLE OF INTERMITTENCY

Since 1941 a number of workers have noted, on theoretical and experimental grounds, a basic feature of turbulence: the intermittency of the dissipating structure. It has become clear that the energy associated with large wave numbers is very unevenly distributed in space. The original statement of the Kolmogoroff theory made no provision for this, and Obukhov (1962) suggested that it could be included by replacing the mean dissipation $\langle \epsilon \rangle$ with its locally averaged value $\tilde{\epsilon}$. Kolmogoroff (1962) did refine his theory, which as in 1941 was phrased in terms of two-point velocity differences, and suggested that they depend on ν and $\tilde{\epsilon}$, where the averaging is over a spherical volume whose poles are the points under consideration. $\tilde{\epsilon}$ was taken to have a log-normal probability distribution and new predictions were made for certain properties of the structure function.

This refinement was timely, because large Re fine structure data have become increasingly difficult to reconcile with the original hypothesis. Most striking, perhaps, is the trend of K, the kurtosis of $\partial u_1/\partial x_1$:

$$K = \left\langle \left(\frac{\partial u_1}{\partial x_1} \right)^4 \right\rangle \left\langle \left(\frac{\partial u_1}{\partial x_1} \right)^2 \right\rangle^{-2} \tag{12}$$

By the original theory K should be an absolute constant at sufficiently large Re, but Kuo (1970) shows that existing data monotonically increase with R_λ, with K ranging from about 4 at R_λ of 20 to about 10 at $R_\lambda = 1000$. This is about the limit for laboratory flows, but the next decade in R_λ can be covered in the atmospheric surface layer, where K values continue their increase (Wyngaard and Tennekes, 1970). There is no evidence, then, that the probability density of $\partial u_1/\partial x_1$ is approaching a universal form at large Re, as originally predicted. Fig. 2 shows probability densities (computed from $\partial u_1/\partial t$ signals) measured at R_λ values of 200 (in a mixing layer), and 2400 and 10,000 (in Kansas). The normalization is

$$\begin{aligned} \left(\frac{\partial u_1}{\partial t} \right)^* &= \frac{\partial u_1}{\partial t} \left\langle \left(\frac{\partial u_1}{\partial t} \right)^2 \right\rangle^{-1/2} \\ \beta^* &= \left\langle \left(\frac{\partial u_1}{\partial t} \right)^2 \right\rangle^{1/2} \beta \end{aligned} \tag{13}$$

so that

$$\begin{aligned} \int_{-\infty}^{\infty} \beta^* dy &= 1 \\ \int_{-\infty}^{\infty} \beta^* y^4 dy &= K \end{aligned} \tag{14}$$

The increasing probability of very large values of the velocity derivative with increasing R_λ is consistent, of course, with increasing K. The departure from Gaussian behavior is unmistakable: $R_\lambda = 10^4$, derivative values of 10 times the standard deviation are 10^{18} times more likely than in a Gaussian process.

This behavior is allowed by the refined hypothesis, in which the logarithm of $\tilde{\varepsilon}$ is assumed to be normally distributed with a variance increasing with Re. Wyngaard and Tennekes (1970) showed that this leads to predicted power law increases

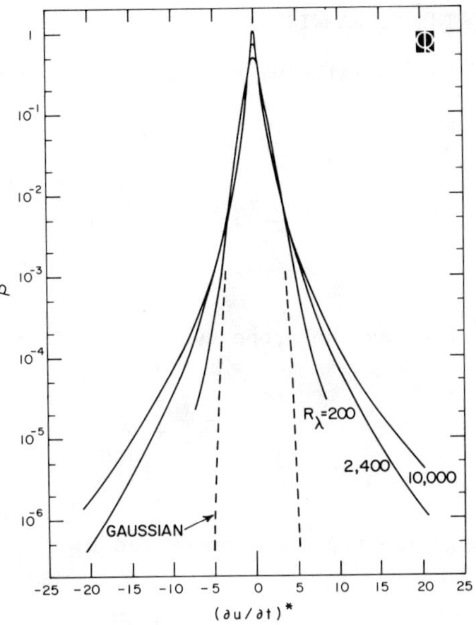

Fig. 2. Measured probability densities of a velocity derivative for three R_λ values.

of skewness (S) and kurtosis of $\partial u_1/\partial x_1$ with increasing Re, while $-S \propto K^{3/8}$. Fig. 3, compiled from a number of sources, supports this prediction.

It is clear that some predictions change markedly in the refined hypothesis, and that it is rather successful in predicting the trend of existing S and K data. In following sections other fine-structure data are considered in the new context.

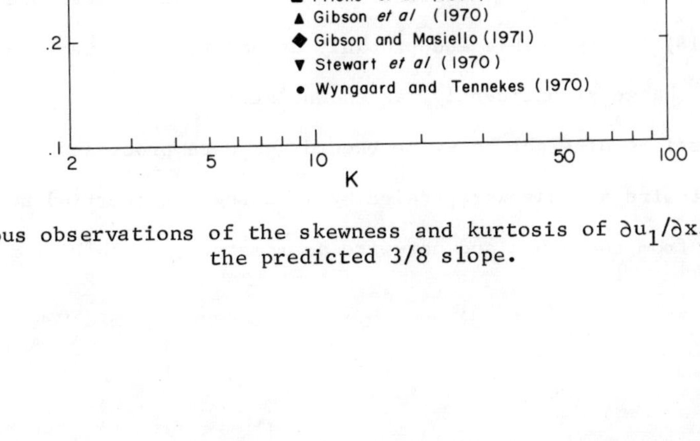

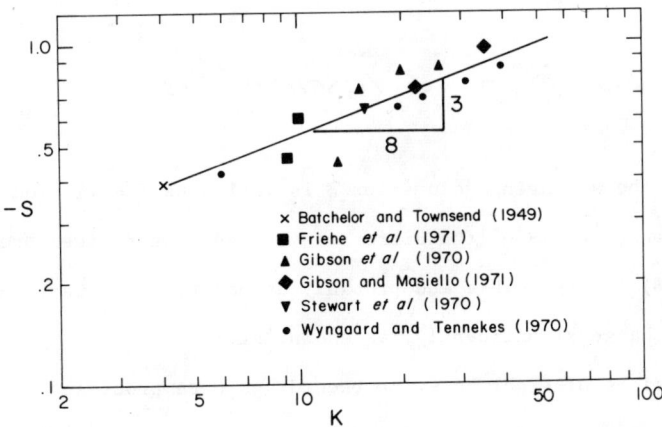

Fig. 3. Various observations of the skewness and kurtosis of $\partial u_1/\partial x_1$ compared with the predicted 3/8 slope.

THE STREAMWISE VELOCITY SPECTRUM

At the lower end of the equilibrium range, where ν is no longer important, Kolmogorov's form for E is

$$E = \alpha \langle \tilde{\epsilon}^{2/3} \rangle \kappa^{-5/3} \tag{15}$$

Our one-dimensional spectrum has the property

$$\langle u_1^2 \rangle = \int_0^\infty \Phi \, d\kappa_1 \tag{16}$$

and is related to E through isotropy by (Lumley and Panofsky, 1964)

$$2E = \kappa^3 \frac{\partial}{\partial \kappa} \left(\frac{1}{\kappa} \frac{\partial \Phi}{\partial \kappa} \right) \tag{17}$$

The inertial range form is then

$$\begin{aligned} \Phi &= \frac{18}{55} \alpha \langle \tilde{\epsilon}^{2/3} \rangle \kappa_1^{-5/3} \\ &= \alpha_1 \langle \tilde{\epsilon}^{2/3} \rangle \kappa_1^{-5/3} \end{aligned} \tag{18}$$

The older hypothesis predicts the inertial range form

$$\Phi = \alpha_1 \langle \epsilon \rangle^{2/3} \kappa_1^{-5/3} \tag{19}$$

If the scale of the averaging volume for $\tilde{\epsilon}$ is varied as $1/\kappa$ (Yaglom, 1966), or fixed (Wyngaard and Tennekes, 1970) the difference between the refined prediction (19) and the original (18) is negligible and probably unmeasureable. Certain smaller-scale predictions do change significantly, as shown later.

15-min portions of eight hot-wire recordings were processed to give α_1 estimates. The hot-wire signals were scaled by matching the inertial subrange levels of the u_1-spectra from the sonic and hot-wire anemometers, which were mounted at the same levels (5.66, 11.3 and 22.6m; see Haugen et al, 1971). Fig. 4 shows the matching.

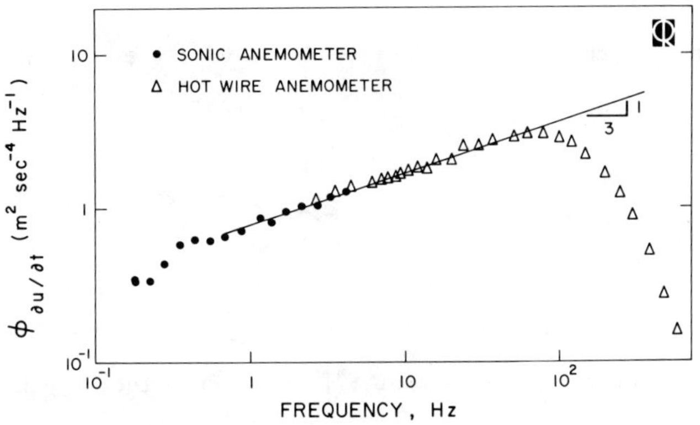

Fig. 4. The hot-wire signals were scaled by matching their spectra with those from sonic anemometers.

The resulting hot-wire u_1 spectra had well-defined inertial ranges covering about two decades in κ_1 and provided the α_1 estimates in Fig. 5. These are plotted against the stability index z/L, where z is height above ground and L is the Monin-Obukhov length (Lumley and Panofsky, 1964). The α_1 estimates show no stability dependence and very small scatter (standard deviation 0.02) about a mean of 0.53.

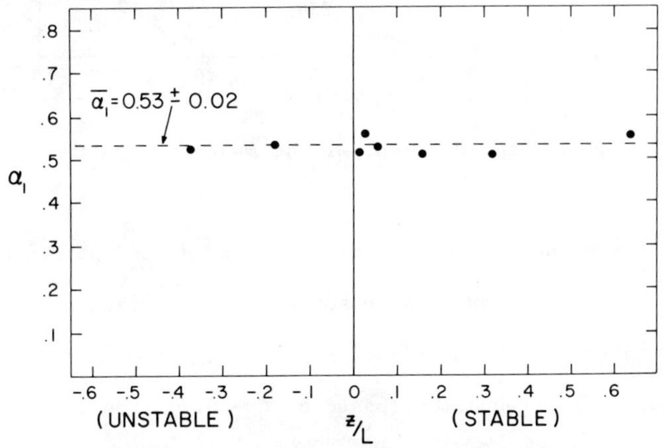

Fig. 5. Estimates of the one-dimensional spectral constant.

For the velocity fluctuation levels during these runs ($\langle v_1^2\rangle \simeq \langle v_2^2\rangle + \langle v_3^2\rangle \simeq 0.025\ U^2$), Lumley's model predicts overestimates of the inertial subrange level by 1.5% and $\langle\epsilon\rangle$ by 7.5%. Therefore α_1 was slightly underestimated, and our corrected value is $\alpha_1 = 0.55$, or $\alpha = 1.7$.

The spectrum at larger wave numbers is conveniently represented by the dimensionless form Φ^*, defined by

$$\Phi^* = \Phi(\langle\epsilon\rangle \nu^5)^{-1/4} \tag{20}$$

and a function of the dimensionless wave number $\kappa_1\eta$ where η is the Kolmogorov microscale $\nu^{3/4}\langle\epsilon\rangle^{-1/4}$. The second moment of Φ^*, which from isotropy has the property

$$\frac{1}{15} = \int_0^\infty (\kappa_1\eta)^2\, \Phi\, d(\kappa_1\eta) \tag{21}$$

is shown in Fig. 6. The fourth moment is shown in Fig. 7. All eight runs extended

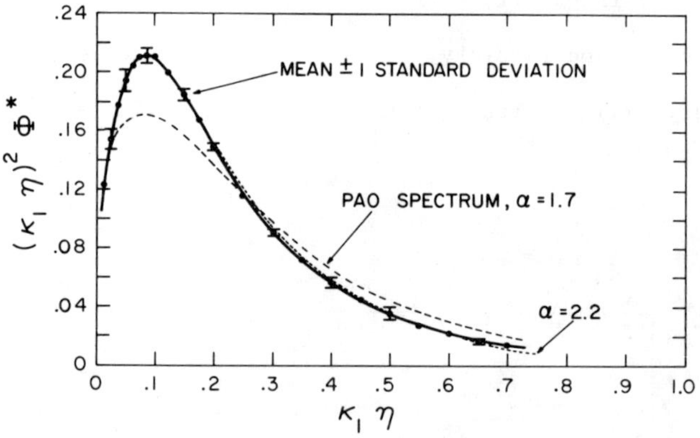

Fig. 6. Second moment of the measured spectrum.

to $\kappa_1\eta = 0.8$, but the number diminishes beyond that and the points near $\kappa_1\eta = 1.5$ are based on only two runs. For this reason standard deviations of the band-averaged spectral estimates are shown only to $\kappa_1\eta = 0.8$.

Calculated spectral attenuation (based on $\eta = 0.7$mm and 1.2mm wire length) was about 10% at $\kappa_1\eta = 1.5$, so the data need no correction on this account. Although

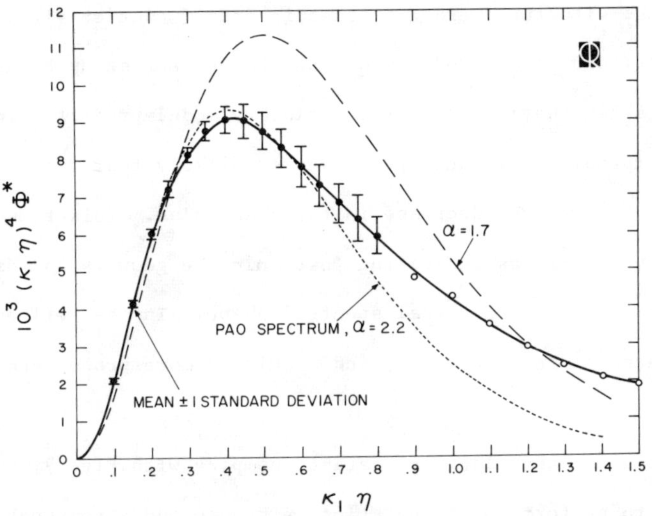

Fig. 7. Fourth moment of the measured spectrum.

the temperature measurements did not extend to such small scales, their contamination effects in velocity measurement can be estimated from the temperature variance budget results (Wyngaard and Coté, 1971). On this basis the dissipation overestimate (Eq. (10) for n = 1) was typically 2 - 3% in daytime (unstable) conditions and an order of magnitude less at night. Depending on the high wave number shape of the temperature spectrum, the third moment error (11) could be as much as six times this.

In isotropic turbulence Φ^* can be related, through the vorticity variance equation (Batchelor and Townsend, 1947), to S. At large Re this equation reduces to a balance between production by vortex stretching and destruction by viscous forces, and in dimensionless form is (Wyngaard and Tennekes, 1970)

$$S = -116 \int_0^\infty (\kappa_1\eta)^4 \Phi^* \, d(\kappa_1\eta) \tag{22}$$

In view of the constraints (21) and (22) and the evidence that -S increases with Re, we expect that as Re increases, Φ^* decreases in the lower portion of the dissipative range and increases at the upper end. Existing laboratory data on $(\kappa_1\eta)^2\Phi^*$, with R_λ in the 100 - 300 range, have peak values averaging about 0.25. Geophysical data, with R_λ values about an order of magnitude larger, tend to show lower peak values, with the average of oceanic data (Grant et al, 1963), earlier atmospheric data (Pond et al 1963, 1966) and the present study being about 0.20. Although Lumley's fluctuating convection velocity model predicts similar behavior with increasing turbulence intensity, it seems unlikely that this mechanism accounts for much of the observed 20% decrease in the peak value. This trend would have been interpreted ten years ago as indicating that only the geophysical data had sufficiently large Re to give a universal spectral shape. In the refined view the trend is expected because of intermittency, and should be more pronounced in even larger Re flows.

There are few, if any, large Re data to compare with Fig. 7, but we can check (22). In order to perform the integration we had to add a reasonable-looking tail to the spectrum, which stops at $\kappa_1\eta = 1.5$, and the result, corrected for the 10 - 15% overestimate implied by Lumley's model, is about -0.85. The directly-measured (from $\partial u_1/\partial t$) S for these runs was in the range -0.70 to -0.85, which gives fairly good agreement.

Another spectral shape parameter is

$$A = \frac{\left\langle\left(\frac{\partial u_1}{\partial x_1}\right)^2\right\rangle\left\langle\left(\frac{\partial^3 u_1}{\partial x_1^3}\right)^2\right\rangle}{\left\langle\left(\frac{\partial^2 u_1}{\partial x_1^2}\right)^2\right\rangle^2} = \frac{\int_0^\infty (k_1\eta)^2\,\Phi^*\,d(k_1\eta)\int_0^\infty (k_1\eta)^6\,\Phi^*\,d(k_1\eta)}{\left\{\int_0^\infty (k_1\eta)^4\,\Phi^*\,d(k_1\eta)\right\}^2} \tag{23}$$

Early measurements, reported by Batchelor and Townsend (1949) and Batchelor (1956), appeared to support the original prediction that A be a constant at sufficiently large Re.

Because of the uncertainties in the extended tail in our data, the A value, Fig. 8, is bracketed by maximum and minimum estimates. According to (4) the convection velocity fluctuations caused only a slight (about 5%) overestimate of A. The other points in Fig. 8 are from a mixing layer at $R_\lambda = 200$ (Wyngaard and Tennekes, 1970) and from the low Re data summarized by Batchelor (1956). The sparse data suggest that A increases with Re; to get an idea of the behavior implied by the refined theory, we write

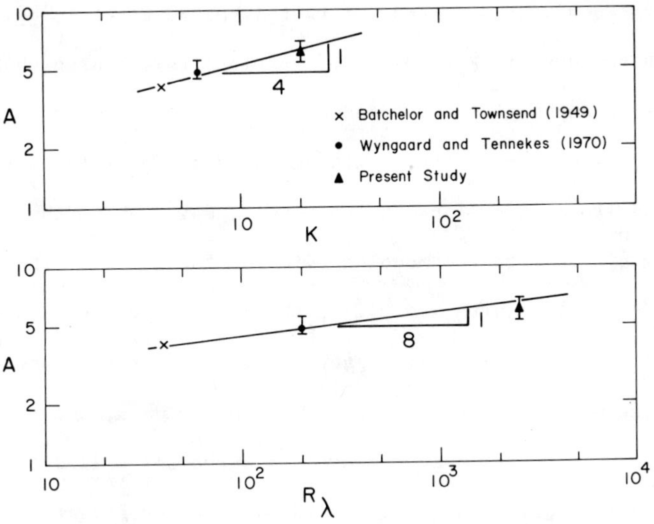

Fig. 8. Observations of A compared with the predicted trends.

$$\left\langle \left(\frac{\partial^n u_1}{\partial x_1^n} \right)^2 \right\rangle \propto \left\langle \tilde{\epsilon}^{\frac{n+1}{2}} \nu^{\frac{1-3n}{2}} \right\rangle \tag{24}$$

so that

$$A \propto \frac{\langle \tilde{\epsilon} \rangle \langle \tilde{\epsilon}^2 \rangle}{\langle \tilde{\epsilon}^{3/2} \rangle^2} \tag{25}$$

If $\tilde{\epsilon}$ is distributed log-normally, it follows that (Obukhov 1962)

$$A \propto \exp(\sigma^2/4) \tag{26}$$

where σ^2 is the variance of the logarithm of $\tilde{\epsilon}$. It can be shown in this way that K, the kurtosis of $\partial u_1/\partial x_1$, varies as $\exp(\sigma^2)$. If the same averaging volume for $\tilde{\epsilon}$ is appropriate for all these properties, we get

$$A \propto K^{1/4} \tag{27}$$

Existing K data show a $R_\lambda^{1/2}$ trend, so that A should go as $R_\lambda^{1/8}$. These predictions are shown with the data in Fig. 8. The agreement is encouraging but more data are needed before the details of the trend are certain.

We have compared our spectrum with the prediction of Pao's (1965) closure model, which is similar to one developed by Corrsin (1964). While the observed inertial subrange spectrum agrees with the model prediction for α = 1.7, the dissipation spectrum observations (Fig. 6) require α = 2.2 in the model, and even higher wave number behavior (Fig. 7) is not fit well by any α. It is possible that spectra predicted by other closure models might agree better with our data, but it seems that unless the model includes intermittency effects the agreement will deteriorate at sufficiently large Re. Keller and Yaglom (1970) have investigated these effects through

$$E(K) = \langle E(K;\tilde{\epsilon})\rangle = \int_0^\infty E(K;\tilde{\epsilon})\,\beta(\tilde{\epsilon})\,d\tilde{\epsilon} \tag{28}$$

where $\beta(\tilde{\epsilon})$, the probability density of locally-averaged dissipation, is such that $\log \tilde{\epsilon}$ is Gaussian. They find that at large wave numbers E depends on β but is practically independent of the spectral shape at fixed $\tilde{\epsilon}$. As intermittency (Re) increases, the averaged spectrum develops an increasingly fuller tail, which is consistent with our observations.

THE SPECTRUM OF $(\partial u_1/\partial t)^2$

The constant μ in Kolmogoroff's (1962) hypothesis for the variance of $\log \tilde{\epsilon}$, where the averaging is over a sphere of radius r,

$$\sigma^2_{\log \tilde{\epsilon}} = a(x,t) + \mu \log (L/r) \tag{29}$$

also appears in Yaglom's (1966) intermittency model prediction for the spectrum of dissipation fluctuations:

$$E_{\epsilon\epsilon} \propto K^{-1+\mu} \tag{30}$$

If $0 < \mu < 1$, then the 1-D spectrum has $\kappa_1^{-1+\mu}$ behavior as well. Present techniques do not permit measurement of dissipation fluctuations, and usually it is assumed that the $(\partial u_1/\partial x_1)^2$ spectrum, for example, also goes as $\kappa_1^{-1+\mu}$. Gibson <u>et al</u> (1970) have reviewed μ values found in this way.

Fig. 9 shows the measured $(\partial u_1/\partial t)^2$ spectrum as found from the average of seven 15-min runs. The scaling is

$$\int_0^\infty \Psi^* \, d(\kappa_1 \eta) = 1 \tag{31}$$

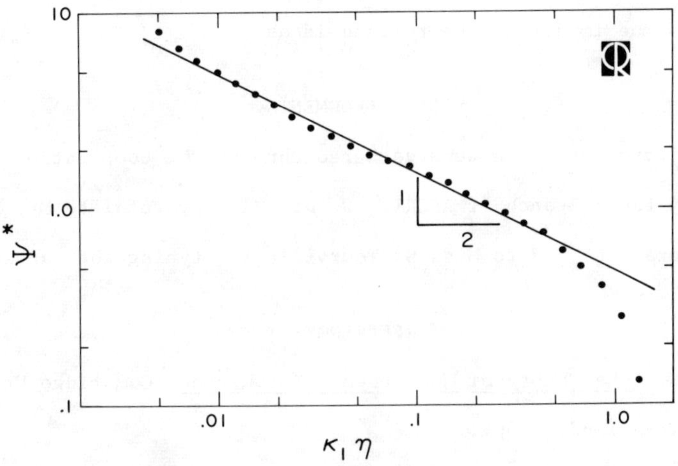

Fig. 9. The measured $(\partial u_1/\partial t)^2$ spectrum.

The inertial range behavior is found to be

$$\Psi^* = 0.5\,(\kappa_1 \eta)^{-0.5} \tag{32}$$

Recent laboratory jet (Friehe <u>et al</u>, 1971, $R_\lambda = 540$) and atmospheric (Gibson <u>et al</u>, 1970) data also show -0.5 slopes and values of about 0.5 for the spectral constant in (32).

DISCUSSION

The Reynolds number dependencies revealed by these fine-structure data agree well with the simple intermittency model predictions suggested by the refined Kolmogoroff hypothesis. Certainly many more experiments are needed to verify these trends, but there is no reason to doubt that geophysical measurements can provide the necessary high-quality data. With several groups now making both velocity and temperature fine-structure measurements we can anticipate much progress in the near future.

Possibly these trends are important in physical problems. Batchelor (1962) stresses the diverse applications of the original Kolmogoroff ideas, and as experiments reveal more about dissipation fluctuations and Re trends, it should be possible to assess the impact of the refined ideas.

ACKNOWLEDGMENTS

The data introduced here were gathered through the cooperation of all members of the Boundary Layer Branch of AFCRL. We are also grateful to Mr. R. Sizer for preparing the drawings and to Miss S. Tourville for typing the manuscript.

REFERENCES

Batchelor, G. K., The Theory of Homogeneous Turbulence (Cambridge University Press, Cambridge, England, 1956).

Batchelor, G. K., in Mecanique de la Turbulence (Centre National de la Recherche Scientifique, Paris, 1962), p. 94.

Batchelor, G. K., and Townsend, A. A, Proc. Roy. Soc. A 190 534-550 (1947).

Batchelor, G. K., and Townsend, A. A., Proc. Roy. Soc. A 199, 238-255 (1949).

Corrsin, S., Handb. Phys. 8, 524-590 (Springer-Verlag, Berlin, 1963).

Corrsin, S., Phys. Fluids 7, 1156-1159 (1964).

Friehe, C. A., Van Atta, C. W., and Gibson, C. H., to be presented at AGARD Specialist's Meeting on Turbulent Shear Flows, London, 1971.

Gibson, C. H., Stegen, G. R., and McConnell, S., Phys. Fluids 13, 2448-2451 (1970).

Gibson, C. H., and Masiello, Proceedings, Symposium on Statistical Models and Turbulence, San Diego, 1971.

Gibson, M. M., J. Fluid Mech. 15, 161-173 (1963).

Grant, H. L., Stewart, R. W., and Moilliet, A., J. Fluid Mech. 12, 241-263 (1962).

Haugen, D. A., Kaimal, J. C., and Bradley, E. F., Quart. J. Roy. Meteorol. Soc. 97, 168-180 (1971).

Heskestad, J. Appl. Mech. 87, 735-739 (1965).

Keller, B. S., and Yaglom, A. M., Izv. AN SSSR, Mek. jid. i gaza, No. 3, 70-79 (1970).

Kistler, A. L., and Vrebalovich, T., J. Fluid Mech. 26, 37-47 (1966).

Kolmogoroff, A. N., Compt. Rend. (Dokl.) Akad. Sci. USSR 30, 301-305 (1941).

Kolmogoroff, A. N., J. Fluid Mech. 13, 82-85 (1962).

Kuo, A. Y., Ph.D. Thesis, The Johns Hopkins University (1970).

Lumley, J. L., and H. A. Panofsky, The Structure of Atmospheric Turbulence (Interscience, New York, 1964).

Lumley, J. L., Phys. Fluids 8, 1056-1062 (1965).

Novikov, E. A., Izv. Akad. Nauk SSSR Ser. Geofiz. 1, 788-796 (1965).

Obukhov, A. M., J. Fluid Mech. 13, 77-81 (1962).

Pao, Y. H., Phys. Fluids 8, 1063-1075 (1965).

Pond, S., Stewart, R. W., and Burling, R. W., J. Atmos. Sci. 20, 319-324 (1963).

Pond, S., Smith, S. D., Hamblin, P. F., and Burling, R. W., J. Atmos. Sci. 23 376-386 (1966).

Rose, W. G., J. Appl. Mech. 29, 554-558 (1962).

Stewart, R. W., Wilson, J. R., and Burling, R. W., J. Fluid Mech. 41 141-152 (1970).

Taylor, G. I., Proc. Roy. Soc. A, 164, 476-490 (1938).

Uberoi, M. S., and Kovasznay, L. S. G., Quart. Appl. Math. 10, 375-393 (1953).

Wyngaard, J. C., J. Sci. Instr. Ser. 2 1, 1105-1108 (1968).

Wyngaard, J. C., J. Sci. Instr. Ser. 2 2, 983-987 (1969).

Wyngaard, J. C., Phys. Fluids 14 (to appear, 1971a).

Wyngaard, J. C., J. Fluid Mech. (to appear, 1971b).

Wyngaard, J. C.,and Lumley, J. L., J. Appl. Meteorol. 6, 952-955 (1967).

Wyngaard, J. C., and Tennekes, H., Phys. Fluids 13, 1962-1969 (1970).

Wyngaard, J. C., and Coté, O. R., J. Atmos. Sci. 28, 190-201 (1971).

Yaglom, A. M., Dokl. Akad. Nauk SSSR 166, 49-52 (1966) [Sov. Phys. Dokl. 11, 26-29].

STATISTICAL SELF-SIMILARITY AND INERTIAL SUBRANGE TURBULENCE

C. W. Van Atta†

and

J. Park

Institute for Pure and Applied Physical Sciences and
Department of Engineering Sciences
University of California, San Diego
La Jolla, California

INTRODUCTION

This paper discusses some experimental observations and comparisons with simple statistical hypotheses of the probability density and moments of increments of the fluctuating fluid velocity in the inertial subrange in high Reynolds number turbulent flow. The term increment as used here means the difference in the values of the fluid velocity measured at two points separated by a distance r. Conflicting theories for the behavior of the moments (structure functions) have been discussed by Kolmogorov (1941, 1962), Yaglom (1966) and others. No theoretical predictions for the probability density itself are available, although one interesting formulation of the problem has been given by Kuznetsov (1967). Since no theoretical results based on the Navier-Stokes equations presently exist, Dutton and Deaven (1969) have suggested that the hypothesis of statistical self-similarity of the increments, which resembles Kolmogorov's original theory in some respects and which furnishes some rather explicit predictions for the behavior of the probability density and its moments as a function r, is an attractive one for comparison with experimentally observed behavior in the inertial subrange. They made the first comparisons of the hypothesis with some experimental data obtained in the atmospheric boundary layer over land. Here

---

†Also Scripps Institution of Oceanography

we shall make a similar comparison with some data obtained in the atmospheric boundary layer over the ocean and discuss the relation with some measured moments, predictions of the conflicting theories, and the previous measurements.

STATISTICAL SELF-SIMILARITY HYPOTHESIS

The concept of a self-similar stochastic process was developed by Mandelbrot, and has been applied by him and his co-workers to a number of physical problems. The first application to fluid turbulence appears to be that of Dutton and Deaven (1969), who tested the degree of self-similarity of measured one-dimensional probability densities $p(u(t) - u(t+T)) = p(\Delta u; T)$ of the increments $\Delta u = u(t) - u(t+T)$ of the longitudinal fluctuating velocity in atmospheric turbulence.

From Mandelbrot and Van Ness (1960), the notation $\{X(t)\} \triangleq \{Y(t)\}$ means that the two random functions $X(t)$ and $Y(t)$ have the same distribution functions. Let $u(t)$ be the fluctuating velocity at a fixed point as a function of time t. The increments of the random function $u(t)$ are said to be self-similar with parameter $\alpha (\alpha \geq 0)$ if for any h

$$\{u(t+T) - u(t)\} \triangleq \{h^{-\alpha}[u(t+hT) - u(t)]\} .$$

Letting $u(t+T) - u(t) = y_T$, $u(t+T') - u(t) = y_{T'}$, $h = T'/T$, in terms of the distribution function P for each variable we have

$$P\{y_T \leq \zeta\} = P\{h^{-\alpha} y_{T'} \leq \zeta\} = P\{y_{T'} \leq h^{\alpha}\zeta\} . \tag{1}$$

Letting p_T and $p_{T'}$ be the probability densities of y_T and $y_{T'}$, respectively, (1) becomes

$$\int_{-\infty}^{\zeta} p_T(\xi)\, d\xi = \int_{-\infty}^{h^{\alpha}\zeta} p_{T'}(\eta)\, d\eta ,$$

and differentiating with respect to ζ we have

$$p_T(y) = h^{\alpha} p_{T'}(h^{\alpha} y) \tag{2}$$

which relates the probability densities of the increments of a self-similar process measured at different lags.

Dutton and Deaven have provided a subjective physical argument for the relevance of the self-similarity hypothesis to the inertial subrange of a turbulent flow. They note that when an inertial subrange with spectral slope of $-5/3$ on a doubly logarithmic plot exists, the amplitudes of the harmonics that make up the motion are proportional to the $1/3$ power of the wavelength, and turbulence time histories have a certain characteristic appearance. They argue that if one takes a section of a time history and expands its length considerably without changing the time scale the new time history will not look like turbulence. If the velocity scale were also expanded to some extent, the new plot might again resemble turbulence. For inertial subrange turbulence, to preserve the relation between amplitude and length, after doubling the record length one must increase the amplitude by $2^{1/3}$ in order to have a new function that is "similar" to the original. If the length scale of any random process is expanded by a factor h, the amplitude must be expanded by the factor h^{α} for self-similarity.

In general, if $P\{y_T \leq \zeta\}$ is the distribution function of the original process, we could have

$$P(\zeta) = P_h(\zeta_h)$$

where ζ_h is the variable obtained by magnification of time by a factor h and magnification of amplitude by some other factor. For a self-similar transformation we specify the amplitude magnification as h^{α}, and

$$P(y_T \leq \zeta) = P(y_{T'} \leq h^{\alpha}\zeta)$$

in agreement with equation (1), which followed from the formal definition of self-similarity given by Mandelbrot and Van Ness.

Since the measurements discussed here provide $u(t)$ at one spatial location only, we shall assume that a given time lag may be related to a spatial separation r between two distinct points according to $r = -UT$, where U is the mean velocity of the flow. This assumption is one form of what is commonly known as Taylor's hypothesis.

To test experimental data for self-similarity one may compare the behavior predicted by equation (2) with measured probability densities of velocity fluctuation increments obtained with different time lags T or spatial separations r. However, instead of computing transformed probability densities for different values of α from (2), we note that one may accurately and conveniently test for self-similarity without assuming a value for α by merely plotting $\sigma p(y)$ versus y/σ, a fairly standard way of presenting probability densities which normalizes the area under the probability curves to a value of one. Suppose we have a self-similar process, i.e.,

$$p_T(y) = h^\alpha p_{T'}(h^\alpha y) \ .$$

For the purposes of our comparison we define a new random variable

$$X = y/\sigma\,[y(t)] = y/\sigma_T$$

then for the distribution function we have

$$P\{X \le x\} = P\{y \le \sigma_T x\} = \int_{-\infty}^{\sigma_T x} p_T(y)\,dy \ .$$

If we plot the probability densities for two different values of T, then, since $\sigma_{T'} = h^\alpha \sigma_T$ the functions f_{x_1} and f_{x_2} plotted are

$$f_{x_1}(x) = \sigma_T p_T(\sigma_T x) = \sigma_T h^\alpha p_{T'}(h^\alpha \sigma_T x) = \sigma_{T'} p_{T'}(\sigma_{T'} x) = f_{x_2}(x) \ . \tag{3}$$

Thus, if a process is self-similar the probability densities for all values of T will be invariant when plotted in the normalized form, independent of the value of α.

KOLMOGOROV THEORY FOR MOMENTS OF INERTIAL SUBRANGE TURBULENCE

According to Kolmogorov's original theory, in the inertial subrange the moments of $p(\Delta u; r)$ or so called structure functions of nth-order are given by the expression

$$\langle (u - u')^n \rangle = C_n (\varepsilon r)^{n/3} , \tag{4}$$

where r is the separation distance of two points, u and u' are the components of the fluctuating velocity at the two points in the direction of the separation, ε is the mean rate of dissipation of kinetic energy, and the C_n are universal constants. From the Kármán-Howarth equation, Kolmogorov (1941b) found the value of C_3 to be $-\frac{4}{5}$. The values of the remaining C_n have not been determined theoretically, but simple linear relations between C_{2n+3} and C_{2n}, as well as lower bounds for some C_n with n even have been found by Kuznetsov (1967). The expression (4) applies only to separations in the inertial subrange $\eta \ll r \ll L$, where $\eta = (\nu^3/\varepsilon)^{1/4}$ is the Kolmogorov length scale, ν is the kinematic viscosity, and L is the characteristic length scale of the energy containing eddies. In the original theory the behavior in the inertial subrange is assumed to be independent of L. As noted by Dutton and Deaven, the variances of a self-similar process, whose squares define the second-order structure function, are related according to $\sigma_{T'} = \langle \Delta u_{T'}^2 \rangle^{1/2} = h^\alpha \sigma_T$, and the choice $\alpha = 1/3$ is a natural one for the classical Kolmogorov inertial subrange. In general, for arbitrary n we find from (2) that

$$\langle y_{T'}^n \rangle = h^{n\alpha} \langle y_T^n \rangle = (T'/T)^{n\alpha} \langle y_T^n \rangle$$

or, if r and T are related according to Taylor's hypothesis, $r = -Ut$,

$$\langle \Delta u^n \rangle = C_n r^{n\alpha} \tag{5}$$

Hence all the structure functions associated with a random variable whose increments are self-similar with $\alpha = 1/3$ have the same r dependence as that predicted by Kolmogorov's original theory for structure functions of velocity fluctuations in the inertial subrange of a turbulent viscous fluid.

According to Landau and Lifshitz (1959), the original Kolmogorov theory does not take into account the influence of the statistical distribution of fluctuations $\tilde{\varepsilon}(\bar{x}, t) - \varepsilon$ in the local dissipation rate. The severe intermittency of the smaller scales of high Reynolds number turbulence produces a large dispersion in $\tilde{\varepsilon}$, and according to the model of the cascade process described by Yaglom (1966), and Gurvich and Yaglom (1967), the variance of $\ln \tilde{\varepsilon}_r$ ($\tilde{\varepsilon}_r$ is the average of $\tilde{\varepsilon}$ over a region of size r) is given by the expression

$$\sigma^2_{\ln \tilde{\varepsilon}_r} = A(\bar{x}, t) + \mu \ln L/r \ ,$$

where μ $(0 < \mu < 1)$ is a universal constant and $A(\bar{x}, t)$ depends on the macrostructure of the flow. The probability distribution of $\tilde{\varepsilon}_r$ is found to be lognormal and the spectrum of fluctuations in local dissipation is given by

$$E_{\tilde{\varepsilon}\tilde{\varepsilon}}(k) \sim k^{-1+\mu}$$

Equation (4) is replaced by the expression given by Kolmogorov (1962),

$$\langle (u - u')^n \rangle = \tilde{C}_n(\bar{x}, t)(\varepsilon r)^{n/3} (L/r)^{\mu n(n-3)/18} \tag{6}$$

where $\tilde{C}_n(\bar{x}, t)$ are not absolute constants but may depend on the macrostructure of the flow.

We note that the modified theory, which gives n^{th} order moments which vary like $r^{\frac{n}{3} - \frac{\mu n}{18}(n-3)}$ rather than $r^{n/3}$, is not consistent with the self-similarity hypothesis.

The differences between the original ($\mu = 0$) theory and the modified theory for $n = 2$ are very small, and, as discussed by Yaglom (1966), measurements for the case $n = 2$ (second-order structure functions and corresponding power spectra) are currently not sufficiently precise to provide a basis for rejection of either theory in favor of the other. The predictions of the two theories are, of course, the same for $n = 3$, as the linear behavior in this case is an independent consequence of the Navier-Stokes equations deduced by Kolmogorov (1941). For increasing n, the predictions of the original and modified theories diverge sharply with the differences growing at an increasing rate as n increases. For each value of $\mu (\mu \neq 0)$, the theoretical value for the power of r reaches a maximum value of $\frac{1}{2}(1 + 1/\mu + \mu/4)$ at $n = \frac{3}{2}(1 + 2/\mu)$ and the variation with n is symmetrical about this value.

THEORETICAL RELATIONS FOR THE PROBABILITY DENSITY

Kuznetsov (1967) has derived differential equations for the characteristic function of the probability density of the difference in velocity $\Delta u_j = v_j = u_j(x^1) - u_j(x^2)$ as a function of the distance between two points $r_j = x_j^1 - x_j^2$. Defining $\varphi = \exp(i\lambda_j v_j)$, the mean value of φ is the characteristic function of $p(\bar{v})$, $f(\lambda) = \int \exp(i\lambda_j v_j)\, p(\bar{v})\, d^3 v$, where $p(\bar{v})$ is the probability distribution of v_j. The mean value of

$$\varepsilon_{kj}^{(s)} = \frac{\partial u_k(x^s)}{\partial x_\ell^{(s)}} \frac{\partial u_j(x^s)}{\partial x_\ell^{(s)}}$$

is proportional to the dissipation of kinetic energy of the flow. Kuznetsov assumes that φ and $\varepsilon_{kj}^{(s)}$ are statistically independent for large Reynolds numbers and $r_j \gg \eta = (\nu^3/\varepsilon)^{1/4}$, using a plausibility argument based on the statistical independence of Fourier coefficients for different wave numbers. He also assumes that

$$\varphi \text{ and } \psi_k = \frac{\partial p(x^{(1)})}{\partial x_k^{(1)}} - \frac{\partial p(x^{(2)})}{\partial x_k^{(2)}}$$

are uncorrelated, where p is here the fluid pressure and ψ_k is the difference in the pressure gradient terms of the Navier-Stokes equation written for two separate points. Under these assumptions, multiplying the Navier-Stokes equations by $\lambda_k \varphi$ and taking the mean value leads to the following equation for f:

$$\frac{\partial f}{\partial t} = i \frac{\partial^2 f}{\partial \lambda_k \, \partial r_k} + 2\nu \frac{\partial^2 f}{\partial r_k^2} + 2\nu \langle \varepsilon_{jk} \rangle \lambda_j \lambda_k f$$

For $r \gg \eta$, the viscous term is small, and using the isotropic form of $\langle \varepsilon_{jk} \rangle = \frac{1}{3} \varepsilon \delta_{jk}$ the equation for f becomes

$$\frac{\partial f}{\partial t} = i \frac{\partial^2 f}{\partial \lambda_k \, \partial r_k} + \frac{2}{3} \varepsilon \lambda_k^2 f$$

In the inertial subrange, from dimensional analysis and isotropy f can depend only on the two variables

$$z = (\lambda_k^2)^{1/2} (\varepsilon r)^{1/3}, \qquad y = \lambda_j r_j / r \, (\lambda_k^2)^{1/2}$$

This similarity form is analogous to the original Kolmogorov theory, as only ε (not ε_r) is assumed relevant and L does not appear in the scaling. Kuznetsov substitutes a series expansion for f(y, z) in the resulting equation for f(y, z), and the vanishing of appropriate coefficients of terms in the expansion then leads to relations between moments of order 2n and order 2n + 3 which still involve certain unknown constants. Thus, although Kuznetsov obtains relations between certain pairs of moments of $p(\Delta u; r)$ corresponding to the original Kolmogorov theory, and the variation with r of each of the moments is known, the absolute values of the individual moments are not known, and so the probability density cannot be completely determined via the characteristic function. The corresponding equation satisfied by the joint probability density $p(v_1, v_2, v_3; r_1, r_2, r_3)$ of the difference of velocities at two different points in homogeneous, isotropic, turbulence in the inertial subrange is

$$\frac{\partial p}{\partial t} = -v_k \frac{\partial p}{\partial r_k} - \frac{2}{3} \varepsilon \frac{\partial^2 p}{\partial v_k^2}$$

where v_1, v_2, v_3 and r_1, r_2, r_3, are the three components of the velocity difference and the separation distance, respective. If the original Kolmogorov scaling is assumed, the self-similar scaling implied by equation (3) is consistent with any solution for $p(\Delta u; r)$ which may be derived from the Navier-Stokes equations. In terms of the dimensionless variables

$$\omega = (v_k^2)^{1/2} / (\epsilon r)^{1/3} = v(\epsilon r)^{-1/3}$$

and

$$\alpha = v_i r_i / vr$$

which express a similarity scaling and dimensional dependence that is equivalent to the original Kolmogorov theory, the equation for the dimensionless probability density $\Pi = P \epsilon r$ in the inertial subrange becomes

$$\frac{\partial}{\partial \omega}\left(\omega^2 \frac{\partial \Pi}{\partial \omega}\right) + \frac{\partial}{\partial \alpha}\left[(1-\alpha^2)\frac{\partial \Pi}{\partial \alpha}\right] - \omega^4 \alpha/2 \frac{\partial \Pi}{\partial \omega} + \frac{3\omega^3}{2}(1-\alpha^2)\frac{\partial \Pi}{\partial \alpha} - \frac{3}{2}\omega^3 \alpha \Pi = 0$$

The one-dimensional probability density $p(v_1; r_1) = p(\Delta u; r)$ is then

$$p(v_1; r_1) = \int_{-\infty}^{\infty}\int_{-\infty}^{\infty} p(v_1, v_2, v_3; r_1, 0, 0)\, dv_2 dv_3$$

$$= (\epsilon r)^{-1} \iint_{-\infty}^{\infty} \Pi(\omega, \alpha)\, dv_2\, dv_3$$

Transforming to polar coordinates ρ and θ, where $\rho^2 = v^2 - v_1^2$, we have

$$p(v_1; r_1) = 2\pi/\epsilon r_1 \int_0^{\infty} \Pi \quad (v_1^2 + \rho^2)^{1/2} \quad (\epsilon r_1)^{1/3}, \; v_1/(v_1^2 + \rho^2)^{1/2} \quad \rho\, d\rho$$

Letting $x = v(\epsilon r)^{-1/3}$, this becomes

$$p(v_1; r_1) = 2\pi(\epsilon r_1)^{-1/3} \int_{v_1/(\epsilon r_1)^{1/3}}^{\infty} \Pi(x, v_1/(\epsilon r)^{1/3} x)\, x dx$$

and consequently

$$(\epsilon r_1)^{1/3}\, p(v_1; r_1) = g(v_1/(\epsilon r_1)^{1/3}) \tag{7}$$

i.e., $(\epsilon r_1)^{1/3}\, p(v_1; r_1)$ is a function only of the variable $v_1/(\epsilon r_1)^{1/3}$. Since

$\sigma_{r_1} \sim r_1^{\alpha}$ for a self-similar process, equation (7) expresses the same invariant scaling as the self-similarity hypothesis with $\alpha = 1/3$ as given by equation (3). Thus, if the original arguments of Kolmogorov apply, $p(\Delta u; r) = p(v_1; r_1)$ is necessarily self-similar within the inertial subrange. According to this view, self-similarity is not merely a hypothesis but may be regarded as a necessary consequence of the Navier-Stokes equations.

PRESENT EXPERIMENTS AND DATA ANALYSIS

The present data were obtained during steady trade wind conditions over the open ocean from FLIP, the stable floating instrument platform of the Scripps Institution of Oceanography, during the Barbados Oceanographic and Meteorological Experiment, in May 1969. The fluctuating component of the turbulent velocity field in the mean wind direction, u, was measured at several heights from 3 to 30 m above mean water level with a single vertically oriented hot-wire 5 microns in diameter and 1 mm long. Here, only the data for $z = 3$ m will be presented in detail. The hot-wire was operated in the constant resistance mode using a DISA 55D05 anemometer, and the anemometer output was linearized with a DISA linearizer. The linearized hot-wire signal was FM tape recorded and later played back and sampled in the laboratory with a twelve-bit analog to digital converter at a rate of 521.5 samples/sec. The signal was dc coupled throughout the entire recording-playback operation. Further details of the experimental conditions and apparatus, calibrations, and data sampling have been given by Van Atta and Chen (1970). The present digital data were in fact a part of the data available from this earlier study of structure functions and the present measurements therefore directly complement those results

From these earlier results we can operationally define the extent of the inertial subrange, the range of separations r over which we can expect to make a critical test of the self-similarity hypothesis. If the probability distribution were self-similar then all the moments would of course behave as predicted by equation (5). The converse is not necessarily true, i.e., apparent self-similar behavior of the moments alone does not imply that the probability density itself is self-similar.

The previous computations of structure functions (which extended up to fourth-order) showed that second and third-order moments varied according to equation (4) or equation (5) with $\alpha = 1/3$ over a considerable range in r. These moments are therefore consistent with possible self-similarity of the probability density. The variation of the fourth order moment with r was closer to that predicted by equation (6) (using the measured value of $\mu = 1/2$) but the power of r predicted by equation (6) is still only about 8% smaller than the self-similar value for $\alpha = 1/3$. As determined from either spectra or structure functions, the lower bound of the inertial subrange is at about $10\ \eta$, or about 0.8 cm for the 3 m data. To estimate the upper bound, it is necessary to use the appearance of deviations from inertial subrange form in the structure functions rather than in the energy spectra, since the latter follow the inertial subrange $k^{-5/3}$ law (where k is the wave number) for scales so large that the flow cannot possibly be locally isotropic. The most sensitive criterion appears to be the behavior of the third-order structure function, which for the 3 m data breaks sharply away from the linear theoretical inertial subrange relation at a length scale of about $10^4\ \eta$ or roughly 8 m. The deviations from simple power law behavior of structure functions of second and fourth order are much more gradual and do not afford as sharp a criterion as the third-order moment in the present case. For the present self-similarity comparison, we shall take the inertial subrange as extending from $r = 0.8$ cm to $r = 8$ m.

PROBABILITY DENSITY FUNCTIONS OF VELOCITY DIFFERENCES

The probability density functions of the velocity differences were computed for a number of intervals ranging from the smallest separation for consecutively adjacent data samples to a separation interval of 2047 data points, corresponding to spatial separations in the range from 1.4 cm to 28.3 m, respectively. A total of 615,000 velocity differences were used to compute each of the probability density functions shown in Figs. 1 through 4. As expected, for the smallest separation the measured distribution is practically identical to that for the time derivative of the velocity for the same data. For the smaller values of separation r, the probability densities are very far from Gaussian, having a shape characteristic of a strongly intermittent random variable; i.e., the probability of very small values of $\Delta u/\sigma_{\Delta u}$ and values of $|\Delta u|/\sigma_{\Delta u}$ greater than 3 is considerably larger than would be expected for a Gaussian distribution, while the probability of intermediate values is smaller. The probability densities exhibit several interesting nonsymmetrical features. Upon close inspection, the most probable value of Δu is found not to be zero, as the maximum value of the probability occurs not at $\Delta u/\sigma_{\Delta u} = 0$, but at a small positive value of $\Delta u/\sigma_{\Delta u}$. This value appears to increase from about 0.01 at the smallest separations to about 0.025 at the intermediate separations considered, and then may decrease again at larger separations. For $|\Delta u|/\sigma_{\Delta u} < 1.5$, the probability density appears to be roughly symmetrical about this maximum value, so that the inner cross-over points with the Gaussian distribution occur at a smaller value of $\Delta u/\sigma_{\Delta u}$ for negative $\Delta u/\sigma_{\Delta u}$ (at about -0.05 for the smaller separations) than for positive values of $\Delta u/\sigma_{\Delta u}$ (about +0.065). Thus in the range of $\Delta u/\sigma_{\Delta u}$ from zero to about 1.5, the probability of a given positive value of Δu is greater than for corresponding negative values. For larger values of Δu, symmetry around the positively displaced origin is replaced by a small but crucial asymmetry, in which the probability of large negative values is larger than that for large positive values. The negative contribution to the odd moments of this small surplus of large negative values dominates the positive contribution from the smaller values (say $|\Delta u|/\sigma_{\Delta u} \leq 1.5$) for which the probability density is seen to be locally positively skewed. Hence, all the odd moments, such as the skewness factor of the velocity derivative and the third- and fifth-order structure functions, are negative.

As the separation distance increases, the probability density of the velocity difference continuously approaches the Gaussian distribution, except for the extreme tails of the distribution. Here, both the absolute and relative changes are much smaller. The negative skewness is still apparent by inspection from the density curve for the largest separation considered here. The persistence of this characteristic is of course directly related to the fact that even for the largest separations we have considered the odd-order structure functions are either still increasing with separation distance or near their maxima (see Figs. 5 and 6). It appears likely that for even larger separations the odd-order structure functions will eventually decrease and approach zero as expected, the probability density of Δu will approach even closer to the Gaussian distribution, and the skewness associated with the tails of the density function will disappear.

COMPARISON WITH STATISTICAL SELF-SIMILARITY

From the data in Figs. 1 through 4 we have noted that there is a continuous change in the normalized probability density as the separation r is varied. The most rapid variation occurs in the vicinity of the maximum in the density. The changes for larger values of Δu, while continuous with changing r, are fairly small for values of the separation between one cm and one meter, where the second-order structure function is most closely proportional to $r^{2/3}$. It is clear that the probability densities exhibit systematic measurable departures from self-similarity within the inertial subrange, where the lower-order structure functions behave nearly as expected for self-similarity. However, the degree of self-similarity of the data in this range for all but the smallest values of $\Delta u/\sigma_{\Delta u}$ is remarkable. The smaller values of $\Delta u/\sigma_{\Delta u}$ in the non-self-similar range make a relatively small contribution to the higher-order structure functions, and the higher-order moments will depend much more strongly on the tails of the distribution as the order increases. As shown in Figs. 3 and 4, the tails also change systematically with changing r and are only very roughly self-similar even for small values of r. Small systematic differences for large values of $\Delta u/\sigma_{\Delta u}$ are of course much more important in determining the behavior of the higher moments than the behavior for small values of $\Delta u/\sigma_{\Delta u}$. This suggests that moments of increasingly higher order will deviate further and further from the predictions of self-similarity as the order increases, a conjecture tentatively verified by some preliminary results discussed later. Dutton and Deavon compared the behavior predicted by equation (2) with measured probability distributions of velocity fluctuation increments obtained at a height of 18 m in the atmospheric boundary layer from a meteorological tower at Cape Kennedy, and from airborne measurements in the troposphere and stratosphere obtained in operation HICAT. None of the sets of data they used exhibits an extensive range in which $\langle \Delta u^2 \rangle \sim r^{2/3}$, even for separations one would normally expect to lie within the inertial subrange, so for their data there is strictly no change of rigorous self-similarity with $\alpha = 1/3$. They compared their measured probability densities of $u(t+T) - u(t)$ for different lags with those computed from equation (2), based on the measured density at one particular fixed lag for each set of data. They made the comparison for both $\alpha = 1/3$ and for one particular measured value of α, α_m, which was determined from the values of the measured probability densities at the origin according to

$$\alpha_m = \ln[p_T(0)/p_{T'}(0)]/\ln(T'/T).$$

Picking a reference lag of 7 m for their boundary layer data, they concluded that the density function for a lag of 14 m is predicted fairly well using either $\alpha = 1/3$ or the measured α_m of 0.26 - 0.28, although systematic differences as large as 10% are apparent for both their horizontal and vertical velocity component data. For lags of 70 and 700 m the predictions using $\alpha = 1/3$ show very large disagreement with the measured data. Comparisons with the calculations using α_m, which is found to decrease with lag to 0.21 - 0.22 show closer agreement with the measurements, but there are still large differences, especially when one realizes that agreement at the origin is automatically built into the comparison when using α_m. Agreement of the self-similarity hypothesis and the airborne measurements is generally even poorer, except for one HICAT case at the smallest difference in lags (a factor of two) for which the differences are about the same as for the boundary layer data. The values of α_m range from 0.05 to 0.31.

Dutton and Deaven felt that for short and moderate values of the lag the predictions with $\alpha = 1/3$ and α_m were both good, and remark that while the prediction with $\alpha = 1/3$ fails at large values of the lag, the predictions with α_m are still remarkably good. Noting also that, because of the observed non-self-similar behavior of their structure functions, one should not expect $\alpha = 1/3$ to work for their data anyway, they conjectured that the behavior of the distribution of increments seems to be well represented by the predictions of the self-similar hypothesis when α_m is used.

In the present measurements we have made the same type of comparison as Dutton and Deaven, but our comparison is made with considerably more accuracy over a range in which the lower-order structure functions and spectra are consistent with self-similarity, removing the additional uncertainty arising in the previous work and allowing a less ambiguous and more exact test of the self-similarity hypothesis for one case of atmospheric turbulence. Comparing our data with the results of Dutton and Deaven, we note that the smallest lag which they considered in their comparisons was 7 meters, which corresponds almost exactly to the second largest separation in the plots of the present data, and for which the probability density is given in Figs. 2 and 4. Although our density function for $r = 15$ m (not shown) is clearly somewhat closer to a Gaussian than the one for $r = 7$ m, to the accuracy of the comparison made by Dutton and Deaven the two densities are essentially identical. Their data and the present measurements are thus consistent in this range, perhaps because this is the only range in which the structure function (second-order) measured by Dutton and Deaven behaves approximately like $r^{2/3}$. A similar conclusion may be drawn for their HICAT case, which exhibits self-similarity for $r = 165$ m and $r = 330$ m with $\alpha = 1/3$. They do not present any comparisons for smaller values of r, which in the present data define the largest fraction of the inertial subrange and are therefore crucial in determining the degree of self-similarity. The obvious severe failure of the self-similarity hypothesis using $\alpha = 1/3$ for the other larger lags (70 m and 700 m) considered by Dutton and Deaven, is, as they have indicated, probably connected with the absence of inertial subrange-like behavior of the structure functions and spectra for these values of r.

STRUCTURE FUNCTIONS

An attempt was made to compute higher-order structure functions, and these results are shown in Figs. 5 and 6. A natural normalization, dependent on n, is evident from Eq. (6), which may be written

$$\langle (u - u')^n \rangle / v_k^n (L/\eta)^{\mu n(n-3)/18} = \tilde{C}_n (r/\eta)^{n/3 - \mu n(n-3)/18}$$

A natural choice for L in the present case is the height above the mean water surface, although some other length might be found eventually to be more suitable for this purpose.

The measured structure functions up to ninth order are shown in Figs. 5 and 6. The lower-order functions (up to fourth) are almost identical with the earlier results. Up to seventh-order, each of the higher-order structure functions may be fairly well fitted by a power law, with the exception of the first two or three points at the smallest separations for the highest orders. Since the earlier results of Van Atta and Chen indicated that the data for the smallest separations may be influenced by viscosity, we have ignored these initial points in comparing with the theoretical inertial subrange results for the power law dependence of the individual functions. Up to, and including seventh-order, the structure functions for large r are closely fitted by power laws with the exponents predicted by the modified theory. This agreement may be at least partly fortuitous, however. It has been noted by H. Tennekes (private communication) that from his own experience it is very difficult to accurately compute moments higher than fourth order from data whose probability densities have tails decreasing as slowly relative to a Gaussian distribution as those for our smaller separations and for which the measured probability density extends roughly to only $\pm 10\,\sigma$. Taking $\Delta u/\sigma = x$, $\sigma p(\Delta u) = y$, the area under the $x^n y$-curve determines the moment $\langle x^n \rangle$. To test the reliability of the data, plots of the $x^n y$ curves should be examined to see how much of the area under the curves has been accounted for. Preliminary results for the present data indicate that for the largest values of r the scatter in these data is small, the maxima in the $x^n y$ curves are clearly defined, and the area is completely accounted for. But for the smallest lags the data for $n \geq 5$ are much more scattered and may not reach their maxima in the intervals considered, a behavior undoubtedly responsible for the large scatter previously noted for the smallest values of r, and not due simply to the effects of viscosity. As r is increased, the data becomes progressively smoother but the total area appears to be completely accounted for only for the last few points (the largest values of r), a behavior consistent with the observed smoother variation of the structure functions for the largest values of r. It is clearly necessary to use more data and possibly longer time series to obtain more satisfactory results at smaller r for the higher moments. In the present case the same total amount of data was used for each value of r. This restriction was, however, an artificial one. The amount of data available for the smallest values of r could be increased by a factor of four, since only every second point was used as the first point in each interval, and these points were chosen only from the first half of each data record. The latter limitation, imposed by the restriction that the largest value of r be equal to half the record length, can be progressively relaxed as smaller r values are considered. For the larger values of r, more of the available data could be used if longer computer records were used.

The data for the structure functions in Figs. 5 and 6 show large differences between the observed behavior and the original Kolmogorov theory. The dashed lines indicating the behavior predicted by the original Kolmogorov theory for fourth, sixth, and seventh orders also define the theoretical lower bounds found by Kuznetsov. For fourth order, the lower bound is roughly a factor of ten less than the measured values. For sixth and seventh orders, the theoretical lower bounds are less than the measured values over the entire range, but the two curves could cross at larger r because the rate of increase with r of the measured structure function is considerably slower than that predicted by the original theory. Such comparisons therefore become less valid for higher orders. Similar behavior is also evident regarding the prediction for the fifth-order structure function, which was calculated from the measured second-order structure function using the linear relation between C_5 and C_2 given by Kuznetsov.

CONCLUSIONS

The probability density of the increments of the turbulent velocity in the inertial subrange is not strictly self-similar but a remarkably close approximation to self-similarity does exist over a restricted range of velocity differences and separations. Under the assumptions of the original Kolmogorov theory, statistical self-similarity is consistent with the predicted behavior of the moments of the increments of the fluctuating velocity in the inertial subrange. Statistical self-similarity is, however, not consistent with the predictions of the modified theory. The measured behavior of the structure functions and the non-self similar behavior of the increments which generated them furnish suggestive evidence for the validity of the modified theory, but more extensive data are needed for $p(\Delta u; r)$, especially for large values of Δu and small values of r in order to accurately compute the higher-order structure functions over the entire range in r to produce a more conclusive test.

ACKNOWLEDGMENT

This research was supported by the Advanced Research Projects Agency of the Department of Defense and was monitored by the U.S. Army Research Office-Durham, Box CM, Duke Station, Durham, North Carolina 27706, under Contract DA-31-124-ARO-D-257. Partial support was also received from the Office of Sea Grant Programs - Sea Grant GH-112.

REFERENCES

Dutton, J. A., and Deaven, D. G., 1969: Radio Science 4, 1341-1349.

Gurvich, A. S., and Yaglom, A. M. 1967: Phys. Fluids Suppl., Boundary Layers and Turbulence, S59.

Kolmogorov, A. N., 1941a: C. R. Acad. Sci., USSR 30, 301.

Kolmogorov, A. N., 1941b: C. R. Acad. Sci., USSR 31, 538.

Kolmogorov, A. N., 1941c: C. R. Acad. Sci., USSR 32, 16.

Kolmogorov, A. N., 1962: J. Fluid Mech. 13, 82.

Kraichnan, R. H., 1964: Phys. Fluids 7, 1030.

Kuznetsov, V. R., 1967: PMM 31, 1069.

Landau, L. D., and Lifshitz, E. M., 1959: Fluid Mechanics. Reading, Mass.: Addison-Wesley.

Mandelbrot, B., and Van Ness, J. W., 1968: SIAM Rev. 10, 422-437.

Van Atta, C. W., and Chen, W. Y., 1970: Fluid Mech. 44, 145-159.

Yaglom, A. M., 1966: Dokl. Acad. Sci., USSR 166, 49.

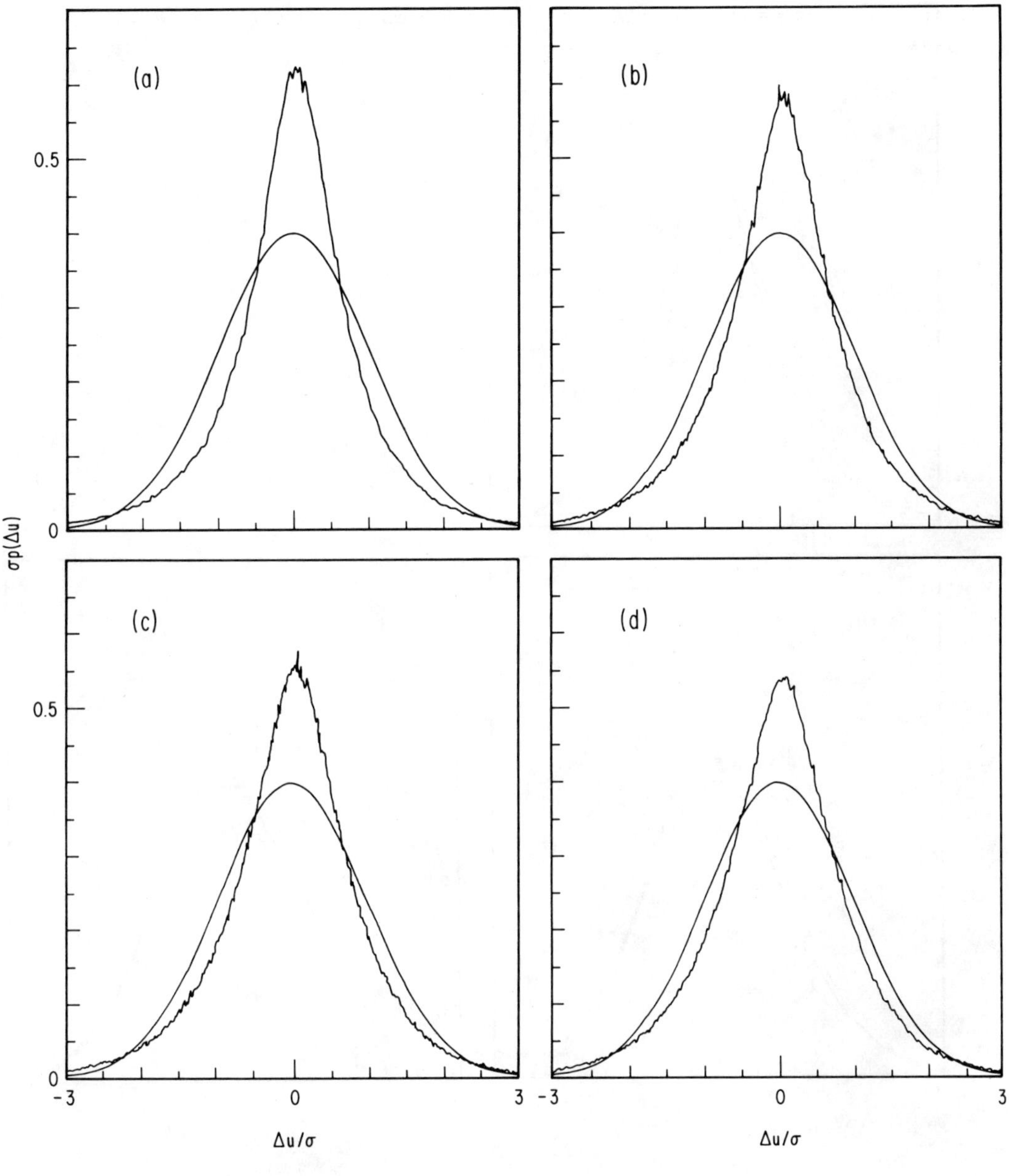

Fig. 1. Probability densities of the velocity difference for various values of r, with linear scale for $\sigma p(\Delta u)$ to emphasize behavior for smaller values of Δu. (a) r = 1.38 cm; (b) r = 4.14 cm; (c) r = 9.67 cm; (d) r = 20.7 cm.

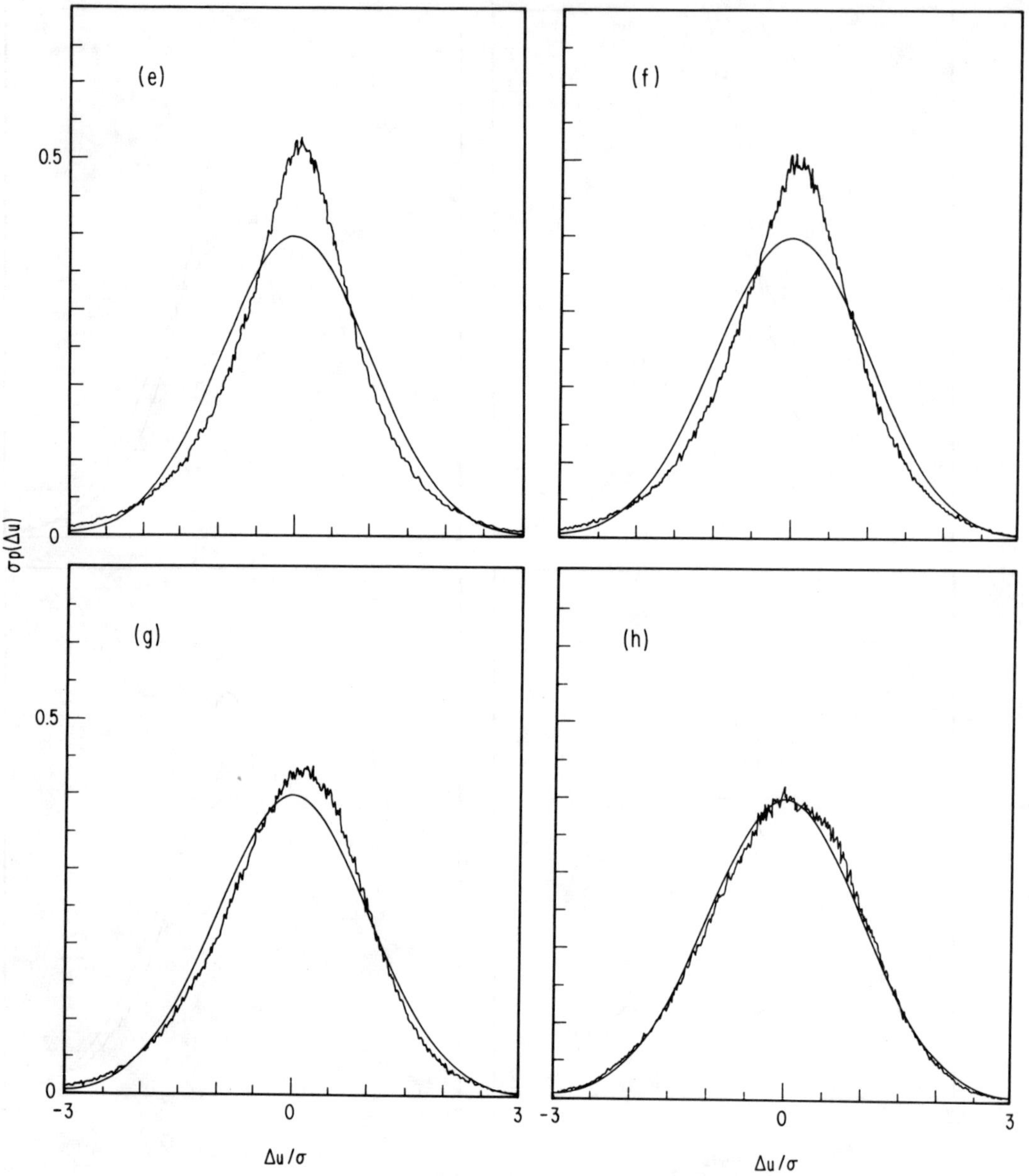

Fig. 2. Probability densities of the velocity difference as in Fig. 1, but for larger values of r. (e) r = 42.8 cm; (f) r = 87 cm; (g) r = 7.06 m; (h) r = 28.3 m.

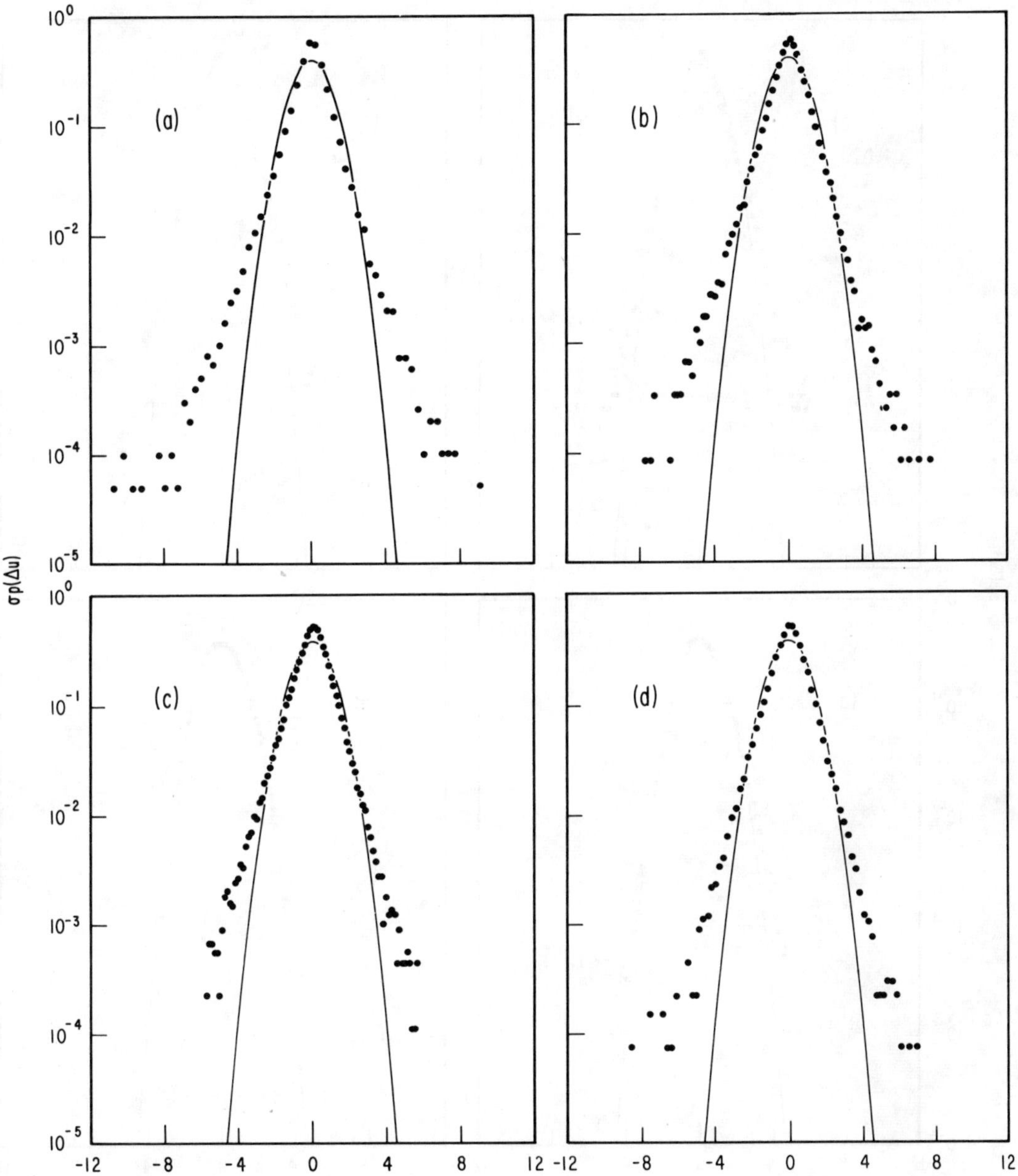

Fig. 3. Probability densities of the velocity difference for various values of r, with logarithmic scale for $\sigma p(\Delta u)$ to emphasize behavior for largest values of Δu. (a) r = 1.38 cm; (b) r = 4.14 cm; (c) r = 9.67 cm; (d) r = 20.7 cm.

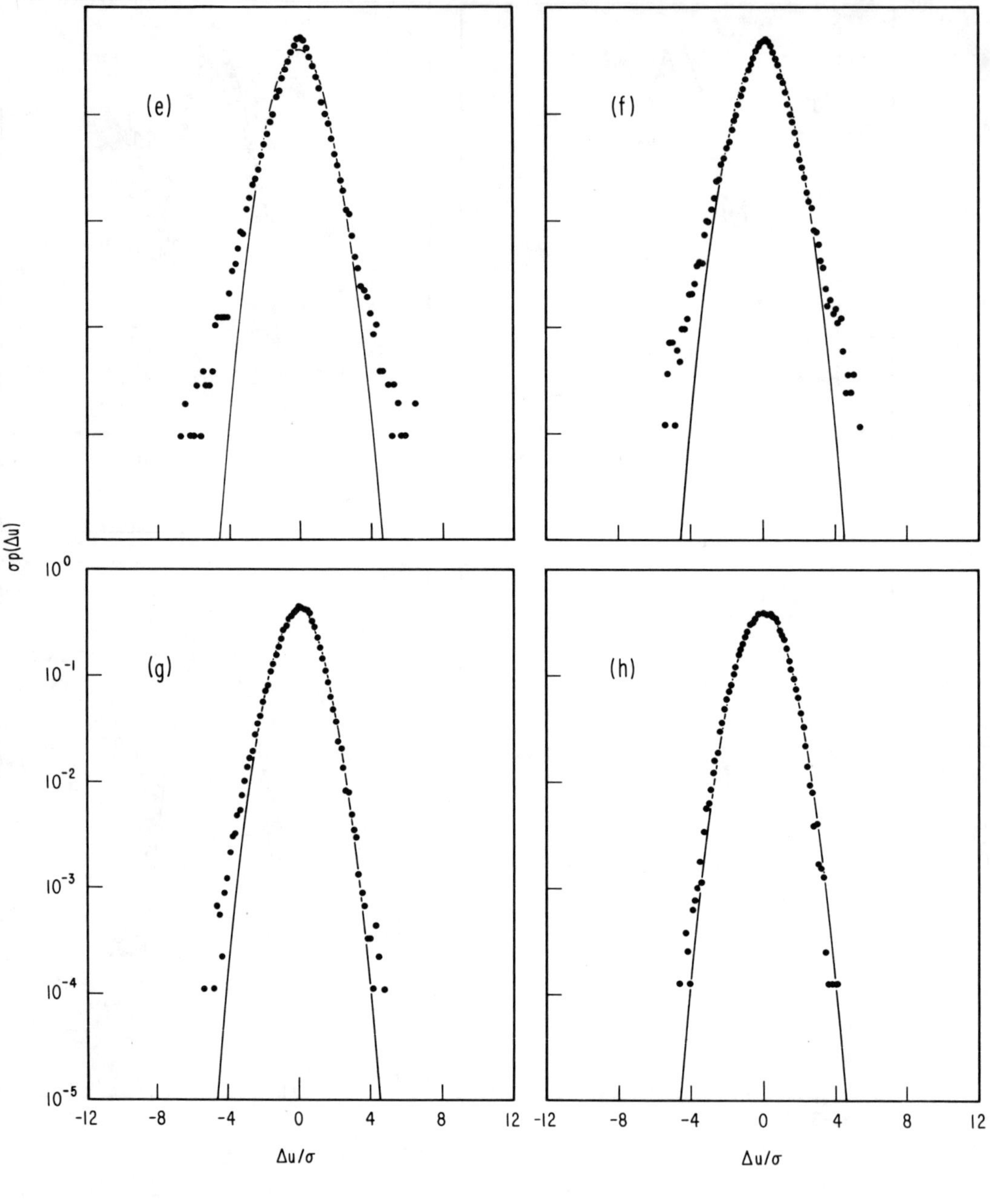

Fig. 4. Probability densities of the velocity difference as in Fig. 3, but for larger values of r. (e) r = 42.8 cm; (f) r = 87 cm; (g) r = 7.06 m; (h) r = 28.3 m.

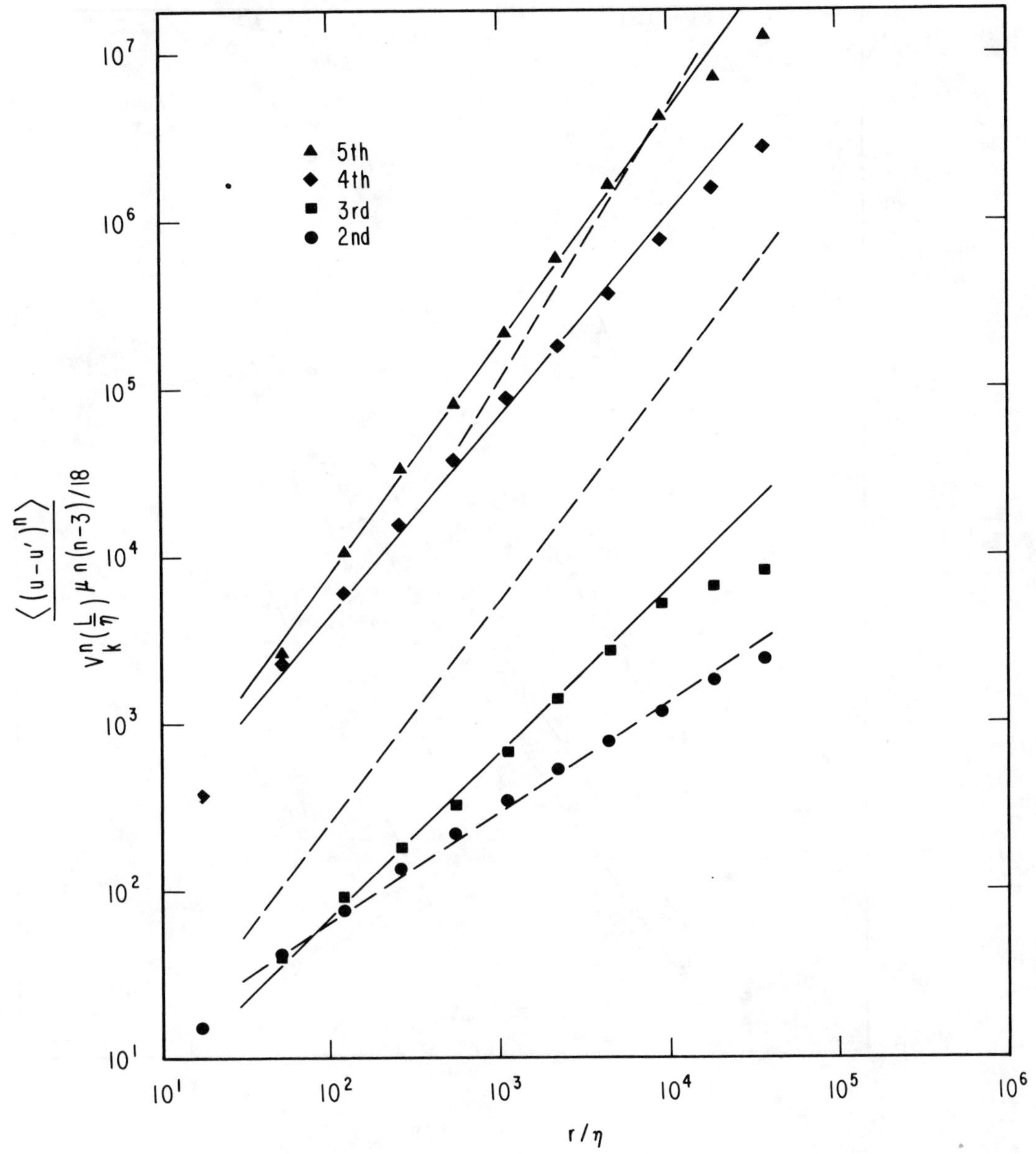

Fig. 5. Second-order through fifth-order structure functions. Order: ● 2, ■ 3; ♦ 4; ▲ 5. The solid lines through the data have slopes equal to those predicted by the modified theory. Dashed lines are corresponding slopes predicted by original theory, with magnitudes derived from Kuznetsov's lower bound (fourth-order) and from Kuznetsov's relation between C_5 and C_2 (fifth-order). All odd-order functions here and in Figure 6 are negative.

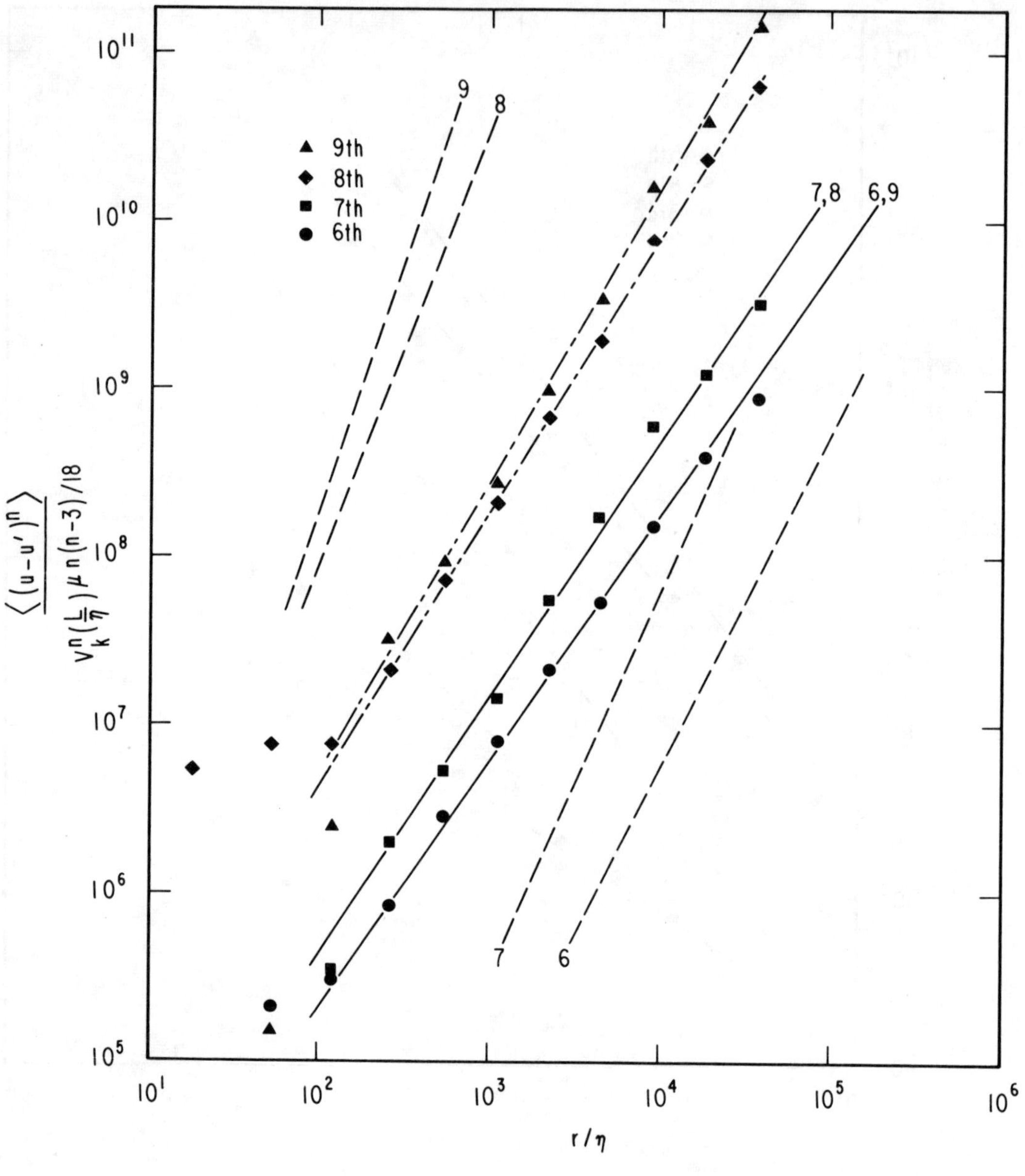

Fig. 6. Sixth-order through ninth-order structure functions. Order: ● 6; ■ 7; ◆ 8; ▲ 9. The solid lines through the data have slopes predicted by the modified Kolmogorov theory. Dashed lines are corresponding slopes predicted by original theory. Dashed-dot lines have slopes intermediate to those predicted by the two theories.

OBSERVATIONS OF THE VARIABILITY OF DISSIPATION RATES OF TURBULENT VELOCITY AND TEMPERATURE FIELDS

by
Carl H. Gibson, Associate Professor
and
Paul J. Masiello, Graduate Student
University of California, San Diego
La Jolla, California

ABSTRACT

Velocity and temperature derivatives are squared and averaged for comparison with lognormality theories of Kolmogoroff, Obukhoff, Yaglom and Gurvich for the variability of local dissipation rates. Averaged squared derivatives were found to depart from lognormality for small values, especially velocity at lower Reynolds numbers, contrary to the proposal of Gurvich and Yaglom (1967). The universal constant μ of Kolmogoroff's third hypothesis was estimated to be 0.47 ± 0.03 from the variation of $\sigma^2_{\ln \dot{u}_r^2}$ with $\ln(L_o/r)$, and 0.49 ± 0.2 from the variance of the ratio $(\dot{u}_r^2/\dot{u}_{2r}^2)$ for various scales r. The departure from lognormality for small values may be due to the fact that squared derivatives are not always proportional to the local dissipation rates.

1. INTRODUCTION

An important property of high Reynolds number turbulence is the concentration of viscous and diffusive dissipation rates into a small volume fraction of the fluid. This tendency toward dissipation "intermittency" introduces a Reynolds number dependence to the Kolmogoroff[1] - Obukhoff[2] universal similarity hypotheses, as noted by Landau,[3] which has been taken into account by a refinement of the first two similarity hypotheses by Kolmogoroff,[4] Obukhoff,[5] Yaglom[6] and others, and by Kolmogoroff's third hypothesis.[4] Kolmogoroff's third hypothesis assumes the locally averaged viscous dissipation of turbulent kinetic energy will be a log-normal random variable. More specifically, he proposed that

$$\sigma^2_{\ln \varepsilon_r} = A + \mu \ln(L_o/r), \quad L_o >> r >> L_K \qquad (1)$$

where $\sigma_{\ln \varepsilon_r}$ is the standard deviation of the logarithm of the viscous dissipation rate ε_r averaged over a volume of dimension r. L_o is the energy scale of the turbulence, L_K is the Kolmogoroff scale $(\nu^3/\bar{\varepsilon})^{1/4}$, ν is the kinematic viscosity, A is a constant depending on the flow geometry, μ is a universal constant, and $\bar{\varepsilon}$ is the dissipation averaged over L_o.

Experimental tests of Kolmogoroff's third hypothesis and attempts to evaluate the constant μ have previously involved squared temperature[8] or velocity[8, 9] differences for small separations, or squared velocity[10, 11, 12, 13, 17] or temperature[11] derivatives as parameters representing the local dissipation in high Reynolds number atmospheric boundary layer flows over the land,[8] bays[10] and ocean.[9, 11, 12, 13]

Yaglom has provided a physical basis for Eq. (1). His model involves ratios of averaged dissipation rates in a hierarchy of averaging volumes between length scales L_K and L_o, as well as hypotheses that the dissipation ratios for the same ratio of averaging scales should be independent and identically distributed. The purpose of the present paper is to extend the previous tests of Kolmogoroff's third hypothesis and also to test Yaglom's assumptions regarding its physical basis by examining these and other statistical properties of averaged dissipation parameters of the turbulence for various length scales. Most of the data examined is from hot wire anememeters about 30 meters over the open ocean, and covers four decades of length scale between L_o and L_K. Some of the measurements of velocity and temperature were made in a heated laboratory jet.

Averaged dissipation rates have also recently been studied by Chen,[14] who measured two distributions for 8 cm averages of streamwise velocity derivative; one at three meters above the ocean and one at 31 meters, from a tape supplied by C. Van Atta. Remarkably, both distributions were identical despite the order of magnitude increase in Reynolds number, and both were perfectly lognormal. The identity of the two distributions is remarkable because previous studies[11,13] have shown a systematic increase in streamwise derivative kurtosis with Reynolds number to values several times as large as reported by Chen, who obtained K values of 8.4 and 8.5 for the high and low positions respectively. Comparable K values for the present study are about 40 and 23. Distributions of averaged squared streamwise derivatives from the present study do not generally display the same degree of lognormality reported by Chen, for reasons which are not understood at the present time. Departures from lognormality occur for the smaller averaged $(du/dx)^2$ values and are of magnitude comparable to the departures of the $(du/dx)^2$ distribution. Experimental tests and discussion of the departures are given in a later section.

1.1 Lognormality of Scalar Fields

Equivalent arguments leading to the conclusion that the temperature dissipation rate should be a lognormal random variable can be made. Yaglom's analysis states that the viscous dissipation rate ε_r averaged over a region of scale r may be written as the product of averaged ratios of dissipation for length scales increasing to the energy scale L_o

$$\varepsilon_r = \left(\frac{\varepsilon_r}{\varepsilon_{n^{1/3}r}}\right)\left(\frac{\varepsilon_{n^{1/3}r}}{\varepsilon_{n^{2/3}r}}\right)\cdots\left(\frac{\varepsilon_{n^{\frac{j-1}{3}}r}}{\varepsilon_{n^{j/3}r}}\right)\varepsilon_{L_o}\ ;\ n^{j/3}r = L_o,\ n > 1\ . \qquad (2)$$

If j is large and the ratios of dissipation are independent identically distributed random variables, as proposed by Yaglom,[6] then the logarithm of ε_r should be Gaussian by virtue of the central limit theorem of probability theory. Yaglom also showed that μ should be $3\sigma^2/\ln n$ where σ^2 is the variance of the logarithm of each of the j dissipation ratios for volumes of ratio n. Thus by measuring σ for a given n one may measure the universal constant μ, and this is done in the present study. Clearly neither j nor n can be large for most turbulent flows. For example, if $n = 8$ and $L_o/r = 50$, corresponding to an 8 cm average at a 4 meter height in a boundary layer, the number of added random variables is only $j = 5$. Consequently we might conclude that evidence for perfect lognormality under these conditions must be somewhat fortuitous.

Similarly, we may write the diffusive dissipation rate χ_r averaged over a volume of size r, where r is larger than L_K as

$$\chi_r = \left(\frac{\chi_r}{\chi_{n^{1/3}r}}\right)\left(\frac{\chi_{n^{1/3}r}}{\chi_{n^{2/3}r}}\right)\cdots\left(\frac{\chi_{n^{(j-1)/3}r}}{\chi_{n^{j/3}r}}\right)\chi_{L_o} \tag{3}$$

where again n represents the ratio of averaging volumes for the hierarchy of dissipation ratios, and $n^{j/3}r = L_o$, $n > 1$. Yaglom refers to the volume of size r as a jth order cube and writes (2) as

$$\varepsilon_j = e_j \ldots e_1 \bar{\varepsilon} , \tag{4a}$$

so (3) may also be written

$$\chi_j = c_j \ldots c_1 \overline{\chi} . \tag{4b}$$

Note that if $n^{1/3}r$ is less than L_K, then $e_j = 1$, and if $n^{1/3}r$ is less than $L_B = L_K (D/\nu)^{1/2}$, where D is the thermal diffusivity, then $c_j = 1$. Hence $\ln \chi_r$ and $\ln \varepsilon_r$ are independent of r for r less than L_B and L_K, respectively.

Arguments identical to those given by Kolmogoroff,[4] Obukhoff[5] and Yaglom[6] apply to the dissipation of scalar variance χ with the conclusion that the random variable χ_r should also be lognormal, with variance

$$\sigma^2_{\ln \chi_r} = A_T + \mu_T \ln (L_o/r) \tag{5}$$

where A_T is a constant depending on the flow geometry and μ_T is a universal constant. If may also be shown (see reference 6 for the parallel velocity argument) that the correlation

$$\langle \chi_r(\underline{x}+\underline{r})\, \chi_r(\underline{x})\rangle \sim \langle \chi_r^2\rangle \sim \langle \chi\rangle^2 (L/r)^{\mu_T} \tag{6}$$

with corresponding spectral inertial subrange

$$\Phi_{\chi\chi} \sim k^{-1+\mu_T} , \quad \frac{1}{L_o} < k < \frac{1}{L_K} \tag{7}$$

where $\Phi_{\chi\chi}$ is the one dimensional spectrum of the scalar dissipation and k is the wavenumber. Simple dimensional analysis of $\Phi_{\chi\chi}$ assuming dependence only on χ, ϵ and k leads to $\Phi_{\chi\chi} \sim k^{-1}$, so we see a direct dependence of the spectrum of dissipation on the intermittency of the turbulence.

Equation (6) follows from the general expression[5] for moments of lognormal random variables which leads to $\langle \chi_r^p\rangle \sim \langle \chi\rangle^p (L_o/r)^{1/2\, \mu_T p(p-1)}$ using Eq. (5).

The universal intermittency constant for scalar fields μ_T may be evaluated using Eq. (7) from measurements of $\phi_{\chi\chi}$, or using Eq. (5) from measurements of the variation of $\sigma^2_{\ln \chi_r}$ with $\ln(L_o/r)$. Measurements[15] of $\phi_{\chi\chi}$ in the marine boundary layer and heated jet indicate $\mu_T = 0.50 \pm 0.05$, close to the value of μ evaluated from $\phi_{\epsilon\epsilon}$.[13]

1.2 Choice of Dissipation Parameters

One difficulty basic to the problem of experimental tests of lognormality theories is that of choosing statistical parameters representative of the averaged dissipation rates. The viscous dissipation rate at any point is given by

$$\epsilon = \frac{1}{2}\nu\left(\frac{\partial u_i}{\partial x_j} + \frac{\partial u_j}{\partial x_i}\right)^2 = \nu\left[\frac{\partial u_i}{\partial x_j}\frac{\partial u_j}{\partial x_i} + \left(\frac{\partial u_i}{\partial x_j}\right)^2\right] \tag{8}$$

which is a very complicated combination of squares and cross products of various components of the velocity gradient tensor. By assuming local isotropy it may be shown that

$$\langle \epsilon \rangle = 15\,\nu\,\langle (\partial u/\partial x)^2\rangle . \tag{9}$$

However, quantities with the same mean value need not be identically distributed, and it does not follow from Eq. (9) that the distribution of ϵ must be identical to the distribution of $(du/dx)^2$, even though it seems likely that they will be closely related. The distributions should be most similar for large values where the cross terms are dominated by $(du/dx)^2$, and most different for small values of $(du/dx)^2$ where the second order terms dominate ϵ but are neglected. Averaging $(du/dx)^2$ over a number of samples along a line of length r should also give a good representation of ϵ_r only when the $(du/dx)_r^2$ values are large. However, when $(du/dx)_r^2$ values are small they would seem to be subject to similar difficulties since more of the terms going into the average must be small, and therefore questionable, and it is not clear that the average of non-representative samples becomes representative. The problem of representing ϵ by $(du/dx)^2$ is not simply due to the fact, as suggested by some authors, that du/dx crosses zero with spatial frequency about L_K^{-1} whereby ϵ may never be precisely zero. It is quite possible to construct a continuous function with mean zero whose square is lognormal; for example, the temperature derivatives measured in this study, Figure 10b.

The expression for the diffusive dissipation of temperature variance is simpler than Eq. (8) for viscous dissipation:

$$\chi = 2D(\partial T/\partial x_i)^2 \quad . \tag{10}$$

All the three terms in Eq. (10) are of the same order of magnitude, and there are only three compared to nine for viscous dissipation, including the six smaller $\partial u_i/\partial x_j \; \partial u_j/\partial x_i$ cross terms. Consequently we might expect the quantity $(\partial T/\partial x)^2$ to be more representative of $\langle \chi \rangle$ using the isotropic expression

$$\langle \chi \rangle = 6D\langle (\partial T/\partial x)^2 \rangle \tag{11}$$

though similar caution must be exercised in interpreting the results. The sum of lognormal random variables need not be lognormal, and conversely, even though χ might be lognormal, it is not necessary that $(\partial T/\partial x)^2$ be.

Departures from lognormality of $(du/dx)^2$ have been observed by nearly everyone who has made the measurement. Generally the measured distribution of $\ln(du/dx)^2$ or $\ln(\Delta u)^2$ shows curvature for small values on a Gaussian plot (such as $\dot{u}^2$ given in Fig. 5) with a relatively straight (or Gaussian) asymptote for large values. Various explanations have been offered for the departures.[12, 14, 15, 17] The position taken in this paper is to ignore the departures from lognormality at small values of $(du/dx)^2$ and $(du/dx)_r^2$ since these may not represent the distribution of ϵ or ϵ_r. The extreme values are used to infer the variance of $\ln \epsilon_r$ as a function of $\ln(1/r)$ for comparison with Eq. (1), so that a value of the universal constant μ can be determined from the slope of the resulting curve.

2. EXPERIMENTAL ARRANGEMENT

Measurements of streamwise velocity derivative were made with constant temperature hot wire anemometer equipment on board the Scripps Floating Instrument Platform FLIP in the Atlantic Ocean in 1969 during the Barbados Oceanographic and Meteorological Experiment, Project BOMEX, and off the coast of Mexico in the summer of 1971. Probes were mounted at 30 meters and 4 meters over the open ocean. Frequency response of the circuits was flat to 10 Kc in a wind tunnel test compared to another circuit with short cables and response beyond this range. Vertical velocity fluctuations were detected at 4 meters with an X-wire anemometer calibrated in the FLIP laboratory. Further details of the FLIP measurements are given in Ref. 11.

Temperature measurements were accomplished using a 0.6 micron platinum cold wire in the same AC bridge used in Ref. 11. Data from BOMEX as well as from a heated jet[15, 16] are described. The jet Reynolds number was 10^5. Velocity and temperature data were obtained on the jet axis at 40 diameters.

Signals were recorded by FM analog tape recorders, and later played back and analyzed by an IBM 1130 computing system. To permit continuous data analysis during averaging operations the playback speed was reduced by a factor of four. The real time filter cutoff frequency was set at 480 hz, 36 db/octave, for the 30 meter data, and at 1000 hz, 36 db/octave, for the 4 meter data unless otherwise noted. Signals were monitored by strip chart, oscilloscope and by plotting the digitized signal stored on the computer disk to avoid clipping or distortion of extreme values while preserving the widest possible dynamic range. Squaring and averaging operations were accomplished on groups of samples before disk storage to conserve space. Non-overlapping groups were averaged to ensure sample independence in the evaluation of distribution functions for $(\dot{u})_r^2$.

Before squaring, a predetermined overall derivative average was subtracted from each derivative sample. This nonzero derivative average results from shifts in zero settings of the tape recorder and digitizer. Except for averages over lengths greater than 25 cm of air, 128, 000 mean values were computed from 10 to 24 samples per mean. For larger scale averages, 40-1000 samples were averaged, but sometimes fewer than 128, 000 mean values would be accumulated, because of the limitation of 1-1/2 hours of total continuous recorded data. The largest scale average investigated was 700 cm of air averaged from 1000 samples, for which 4, 800 values were computed, for a total of 4, 800, 000 samples.

3. RESULTS

In comparing distribution functions with the Gaussian curve, it is useful to use a Gaussian plot, or a plot on "probability paper" (for example, Keuffel and Esser Company 4680003 graph paper). This plot is based on the Gaussian density function

$$g(x) = \frac{1}{(2\pi)^{1/2}\sigma} \exp\left[-\frac{1}{2}\frac{(x-\overline{x})^2}{\sigma^2}\right] \tag{12}$$

where

$$\sigma^2 = \langle (x-\overline{x})^2 \rangle \quad . \tag{13}$$

The cumulative distribution function for the Gaussian function is

$$P[X \leq Z] = \int_{-\infty}^{Z} g(z')\, dz' , \tag{14}$$

which may be inverted to obtain $Z[P]$. Thus, when the cdf $P[Y \leq y]$ of a random variable Y is estimated by integrating its histogram, $Z[P]$ may be plotted versus y to determine if Y is Gaussian. If it is, the curve will be straight with slope $1/\sigma_Y$, and with $y = \bar{y}$ at $Z = 0$, corresponding to $P = 50\%$. Clearly $Z = (y - \bar{y})/\sigma_Y$ for Y Gaussian.

Figure 1 shows such a plot for the measured cumulative distribution function of the voltage of a random noise generator, normalized about the mean by the standard deviation. Data analysis and generation of the Gaussian plot were done by the computer for 256,000 samples, giving a kurtosis of 3.011 and skewness of -0.0007 compared to 3 and 0, respectively, for a Gaussian.

Two methods were used in calculating distribution functions of dissipation parameters for comparison with lognormality. In the first, the logarithms of the samples were calculated before computing their histogram. In the second, the histogram of the dissipation samples was found, and then converted to the histogram of the logarithm by the appropriate identity. The latter method is much faster, but loses resolution for the small values when extreme values are present, as they are when the kurtosis of the random variable is high.

Figure 2 shows typical derivative signals for velocity and temperature over the ocean measured by Gibson and Stegen during Project BOMEX. The extreme intermittency of both signals is apparent, especially the temperature derivative. Kurtosis values were 27 and 43 for velocity and temperature, respectively. The curves are computer plots of several thousand data points prior to processing.

Cumulative distribution functions for $(du/dt)_r^2$ averaged over several length scales are shown in the Gaussian plot of Fig. 3. The curves have been separated

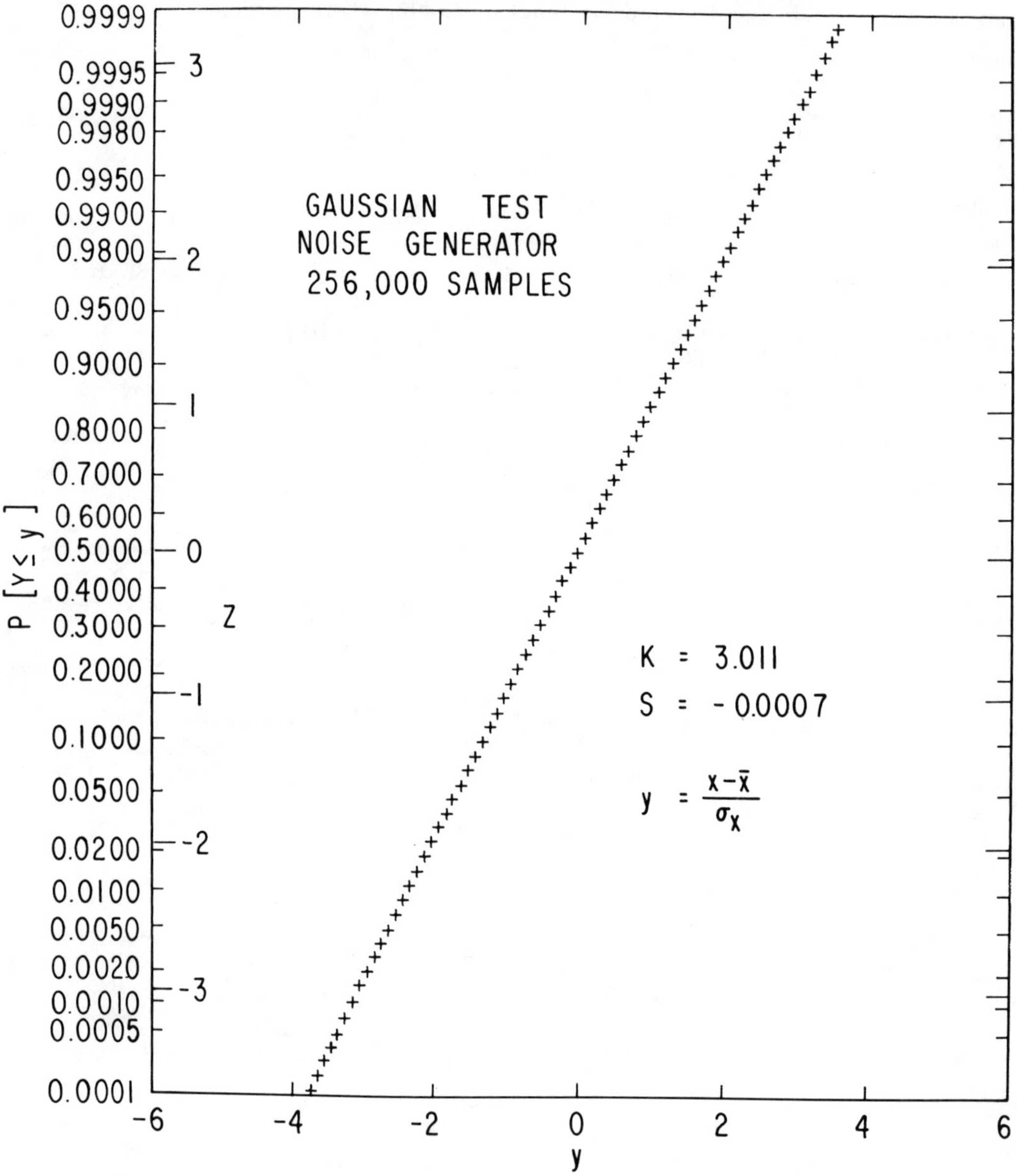

Figure 1. Gaussian test of noise generator signal, 256K samples.

by arbitrary shift constants C_r added to the $\ln(\dot{u})^2$ values. Calculations were made using the second method previously described in some cases, so the low values are missing. Tests were made on the same data using both methods, and only the first point differs signficantly. A number of observations can be made concerning the

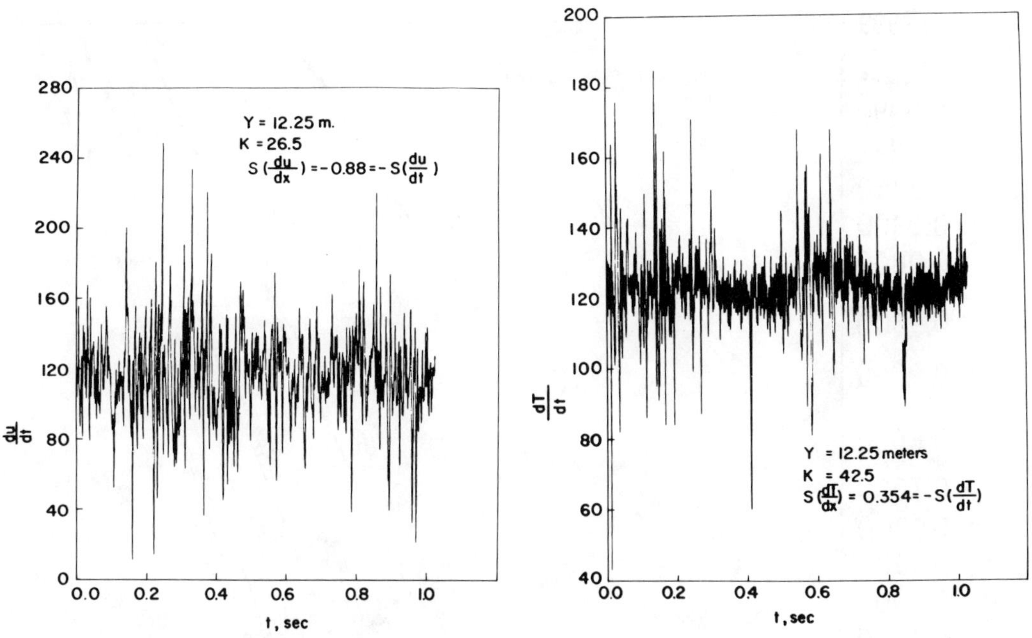

Figure 2. Derivative signals for velocity and temperature 12 meters over the ocean.

shapes of the curves in Fig. 3. Most exhibit a relatively straight portion for intermediate probability values with departures at the extremely low and high values. The departure at high $(du/dt)^2$ values can be explained in terms of inadequate sampling, which causes the scatter. Flat regions in the cdf correspond to a series of empty boxes in the measured histogram, indicating that more samples are needed to provide a meaningful estimate of the distribution in this region. Low values characteristically depart from the Gaussian straight line by first curving up to the left, as the unaveraged $(du/dt)^2$, but then sharply decreasing below the line for the smallest values.

As previously mentioned, the distribution function for low values of $(du/dt)_r^2$ may not be as representative as high values for the corresponding portions of the distribution function of ε_r. Consequently, tangent lines such as those shown in Fig. 3 were drawn for some twenty different length scales ranging from

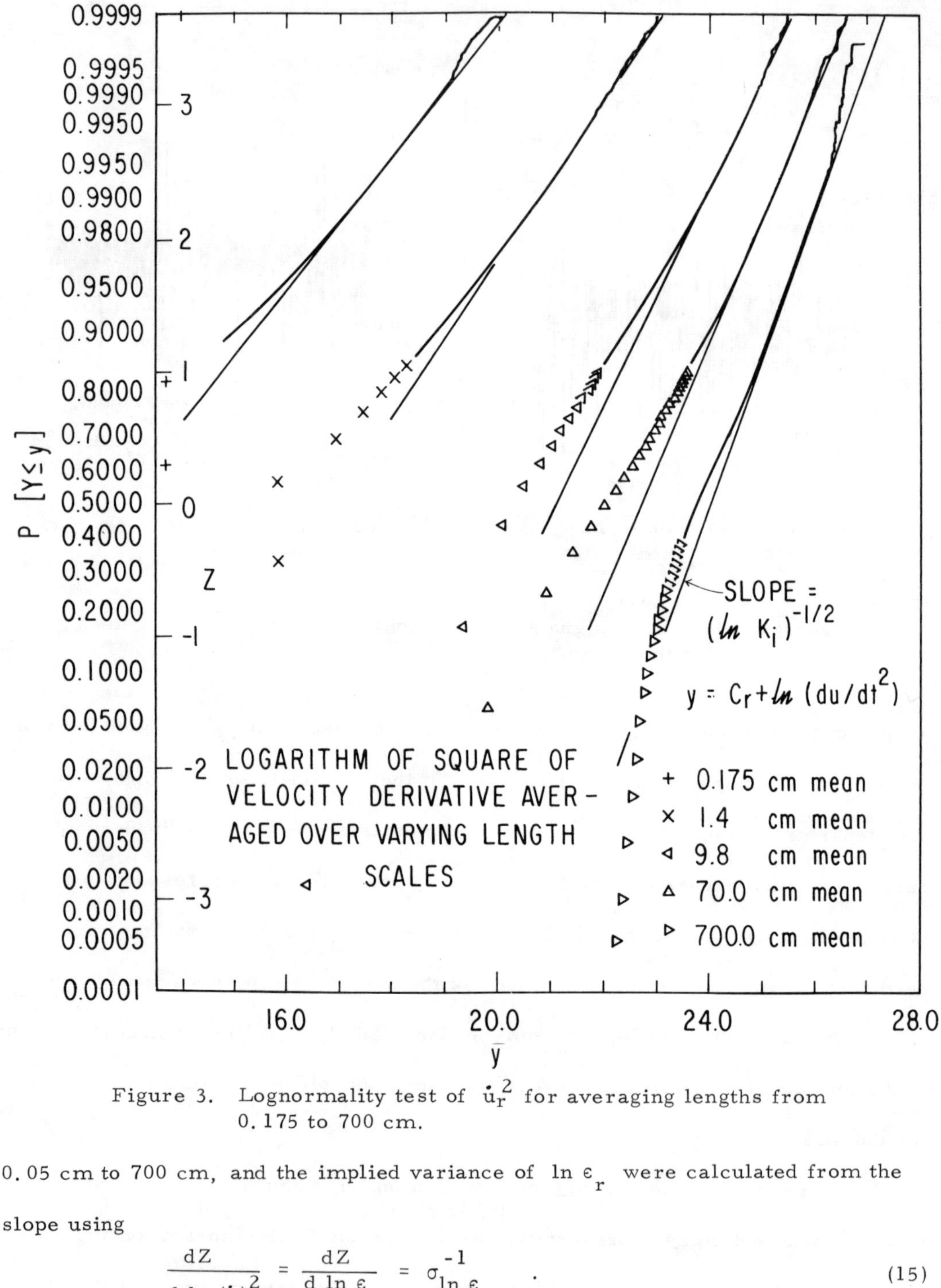

Figure 3. Lognormality test of $\dot{u}_r^2$ for averaging lengths from 0.175 to 700 cm.

0.05 cm to 700 cm, and the implied variance of $\ln \varepsilon_r$ were calculated from the slope using

$$\frac{dZ}{d \ln (\dot{u})_r^2} = \frac{dZ}{d \ln \varepsilon_r} = \sigma_{\ln \varepsilon_r}^{-1} \quad . \tag{15}$$

The resulting plot of $\sigma^2_{\ln \varepsilon_r}$ vs $\ln(L_o/r)$ is shown in Fig. 4, as well as the corresponding averaging scale r and implied kurtosis values $K_i = \exp(\sigma^2_{\ln \varepsilon_r})$. For r

values smaller than the Kolmogoroff scale, $\sigma^2_{\ln \epsilon_r}$ should approach $\ln K_{\dot{u}^2}$, which is constant, and there does seem to be a tendency to level off near $r = L_K$. For very large r values the variance of $\ln \epsilon_r$ should go to zero asymptotically. For $L_o \gg r \gg L_K$ a straight line is predicted by Kolmogoroff's third hypothesis, with slope equal to the universal constant μ according to Eq. (1).

The dashed line in Fig. 4 corresponds to Eq. (1) with a μ value of 1/2 and an A value of -1.2. Data points are consistent with the curve for averaging lengths ranging from about 0.1 cm to 20 cm. Implied kurtosis values for the averaging lengths consistent with Eq. (1) range from a minimum of about 3 to an upper bound of 40-60.

The kurtosis of $(\dot{u}_r^2)^{1/2}$ is closely related to the kurtosis implied by

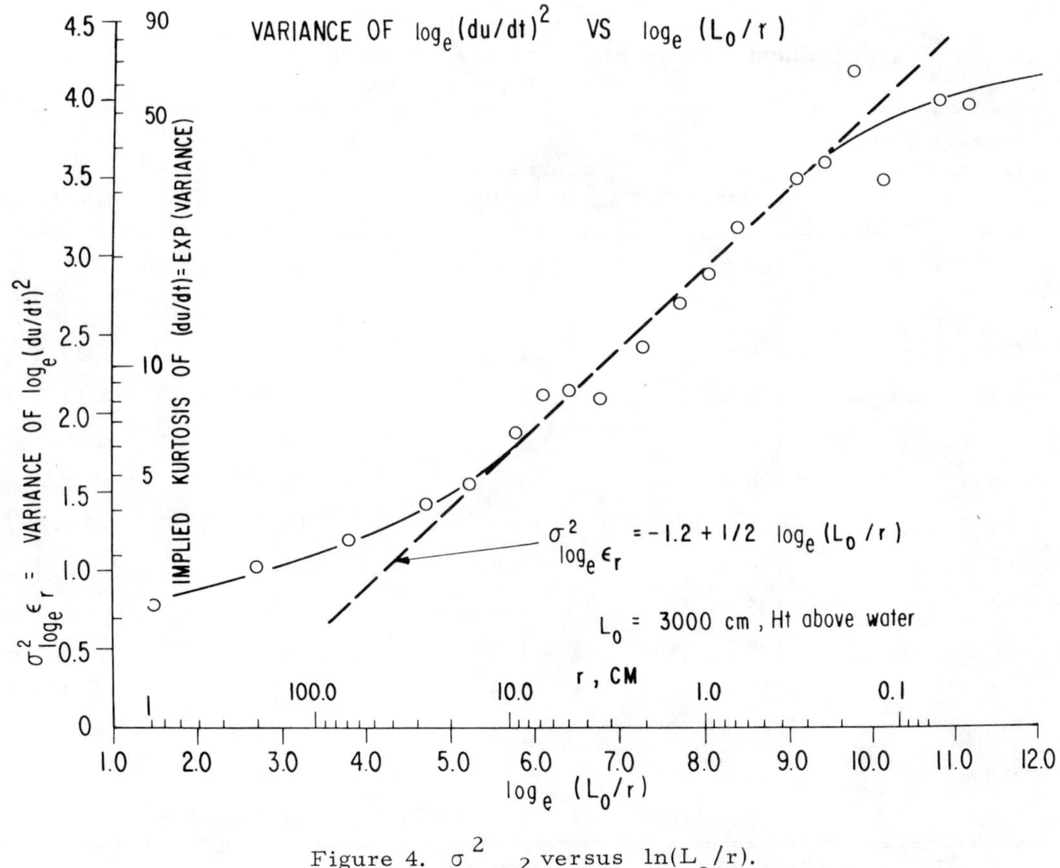

Figure 4. $\sigma^2_{\ln \dot{u}_r^2}$ versus $\ln(L_o/r)$.

lognormality, just as the kurtosis of $\dot{u}$ has been shown[11] to be closely approximated by the implied kurtosis from the unaveraged $\dot{u}^2$, where overdots denote time derivatives. We may write

$$K_{(\dot{u}_r^2)^{1/2}} = \frac{\langle (\dot{u}_r^2)^2 \rangle}{\langle (\dot{u}_r^2) \rangle^2} = \frac{\sigma^2_{\dot{u}_r^2} + \langle (\dot{u}_r^2) \rangle^2}{\langle (\dot{u}_r^2) \rangle^2} \tag{16}$$

Values of K_i and $K_{(\dot{u}_r^2)^{1/2}}$ are summarized in Table 1 for 14 averaging lengths from 0.044 to 700 cm. They are in agreement within experimental scatter over the full range of averaging lengths, with values from 2.2 to over 50. The large K value indicated as $r \to 0$ is consistent with $K \sim R_\lambda^{3/4}$ (Ref. 17) rather than $R_\lambda^{1/2}$, and extends the relation $\ln K = \text{const} + \ln(y_\varepsilon^{1/4})^\mu$ of Ref. 13.

TABLE 1
Fourth Moment Parameters for Averaged Energy Dissipation Function, 30 Meters Above the Ocean

Averaging Length r (cm)	Total Samples	Samples/cm in mean	Implied Kurtosis K_i	$K[(\dot{u}_r^2)^{1/2}]$
0.044	128 K	227	50	39
0.175	128 K	57	66	33
0.7	128 K	14	24.3	27.4
1.4	128 K	37	14.6	14.6
2.1	128 K	5	13.1	10.2
3.5	128 K	3	10.4	7.4
4.9	128 K	2	10.9	6.4
7.0	128 K	1.4	8.3	14.3
9.8	128 K	1.4	6.6	1.7
16.8	128 K	1.4	4.7	6.5
28.0	116 K	1.4	4.2	4.9
70.0	51 K	1.4	3.3	3.9
210.0	17 K	1.4	2.8	3.2
700.0	5 K	1.4	2.2	2.2

$K_i = \exp[(\text{slope})^{-2}]$; $K[(\dot{u}_r^2)^{1/2}] = \sigma^2_{\dot{u}_r^2} / \langle \dot{u}_r^2 \rangle^2 + 1$

Table 2 summarizes some published μ values, calculated by various methods. The value obtained from Fig. 4 is in good agreement with the best estimates using other techniques.

TABLE 2
Summary of μ Computations for High Reynolds Number Atmospheric Turbulence

Investigators	Date	Method	μ Value
Gurvich and Zubkovski[7]	1963	a. From slope of $(du_3/dx_1)^2$ spectrum	0.38
		b. Data within isotropic range pointed out by Pond and Stewart[10]	0.50
Pond, Stewart and Burling[20]	1963	From slope of $(du_1/dx_1)^2$ spectrum	0.38
Pond and Stewart[10]	1965	a. From slope of $(du_1/dx_1)^2$ spectrum, ocean data, 1.5 m	0.38 ± 0.05
		b. Damped by low pass filter	0.62 ± 0.024
Sheih et al.[19]	1969	Spectrum of $(du_1/dx_1)^2$, atmospheric data	0.70
Stewart, Wilson and Burling[12]	1970	Spectrum of $(du_1/dx_1)^2$ ocean data, 2 m	0.35
Wyngaard and Tennekes[17]	1970	Spectrum of $(du/dt)^2$ curved mixing layer, moderate R_e	0.85
Gibson, Stegen and McConnell[13]	1970	a. Slope of $(du_1/dx_1)^2$ ocean data, 2-12 m	0.51 ± 0.02
		b. From Kolmogoroff third hypothesis $r = L_K$	0.44 ± 0.25
Gibson and Masiello (present work)	1971	a. From slope of $\sigma^2_{\ln \epsilon_r}$ versus $\ln(L_o/r)$, 30 m ocean data	0.47 ± 0.03
		b. From $3\,\sigma_e^2/\ln n$	0.49 ± 0.20

μ = constant in Kolmogoroff's third hypothesis,

$\sigma^2_{\ln \epsilon_r} = A(\underline{x}, t) + \mu \ln(L_o/r)$, $E_{\epsilon\epsilon}(k) = k^{-1+\mu}$, $\epsilon \sim (du/dt)^2$

The fact that the straight portion of the curve ends at 30 cm may be used to get a rough idea of the limitations of lognormality, since it implies that $j \ln n = 3 \ln(L_o/r) = 14$, using a formula given by Yaglom,[6] where j is the number of subdivisions of volume ratio n used in Yaglom's model. Thus if $n = 8$ it would appear that a minimum of about 7 subdivisions are necessary before the random variable ε_r can be expected to approximate lognormality.

All of probability plots in Fig. 3 were lognormal only in the limit of very large values of $\dot{u}_r^2$, whereas Chen[14] found complete lognormality for averaged squared derivatives. Figure 5 shows the results of an experiment designed to reproduce the conditions of Chen's calculation using the present 30 meter data and calculational techniques.

The velocity for the present test at 30 m was 7.82 m/sec compared to 11 and 9.9 m/sec reported by Chen[14] for his measurements at 31 and 3 meters above sea level. Probes for both tests at about 30 m were mounted on the vertical mast fixed to FLIP's upper deck. The sampling rate was adjusted to give the same spatial separation of points 1.05 cm, and 8 points were averaged to compute the mean $\dot{u}_r^2$ values. The signal was filtered at nearly the same frequency as that used by Chen: 480 hz versus 500. The cut off wave length for the present test was at the Kolmogoroff scale, whereas Chen's was at twice the Kolmogoroff scale, which does represent some difference between the two. Figure 5 shows Gaussian plots for the logarithms of averaged and unaveraged $\dot{u}^2$ signals from Chen and from the present comparison test. The signal values are normalized by the variance after subtracting the mean. A Gaussian curve is a line with slope + 1 passing through $z = 0$, $y = 0$, when $y = (\ln \dot{u}^2 - \langle \ln \dot{u}^2 \rangle)/\sigma_{\ln \dot{u}^2}$.

The unaveraged plots are apparently in good agreement, although the calculated kurtosis of $\dot{u}$ for the present test was 14 compared to 8.5 for the two Chen runs, partly because of the difference in cutoff scales compared to the Kolmogoroff scale.

The averaged plots are in fair agreement for the larges values, but are in

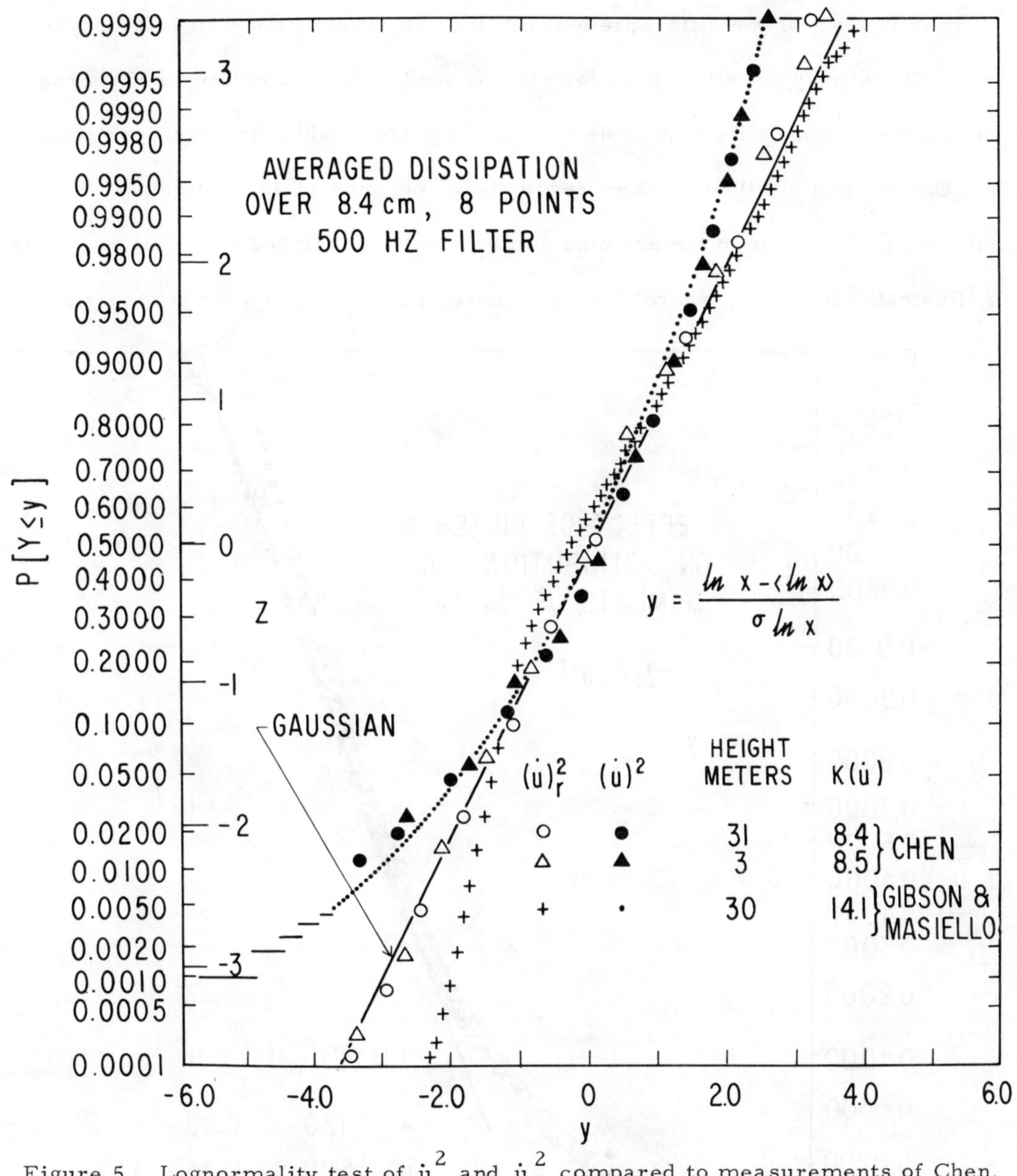

Figure 5. Lognormality test of $\dot{u}^2$ and $\dot{u}_r^2$ compared to measurements of Chen.

poor agreement as $\dot{u}_r^2$ approaches zero. Both of the averaged runs reported by Chen show nearly perfect lognormality within the resolution of the distribution. The plot for the test run shows the same characteristic shape as all the curves of Fig. 3; they curve up and then fall down sharply as the values approach zero. In order to achieve resolution for the small values, the logarithm was calculated for each sample prior to calculation of the histogram.

The reason for the difference between the two results is not clear at the present time. There is apparently a need for similar independent calculations of averaged dissipation lognormality to provide a basis for distinguishing the correct result.

One difficulty with the Chen result as a measure of local dissipation intermittency is that some of the extreme values have been damped by the filter. This is illustrated in Fig. 6, where $\dot{u}^2$ lognormality plots are given for several filter

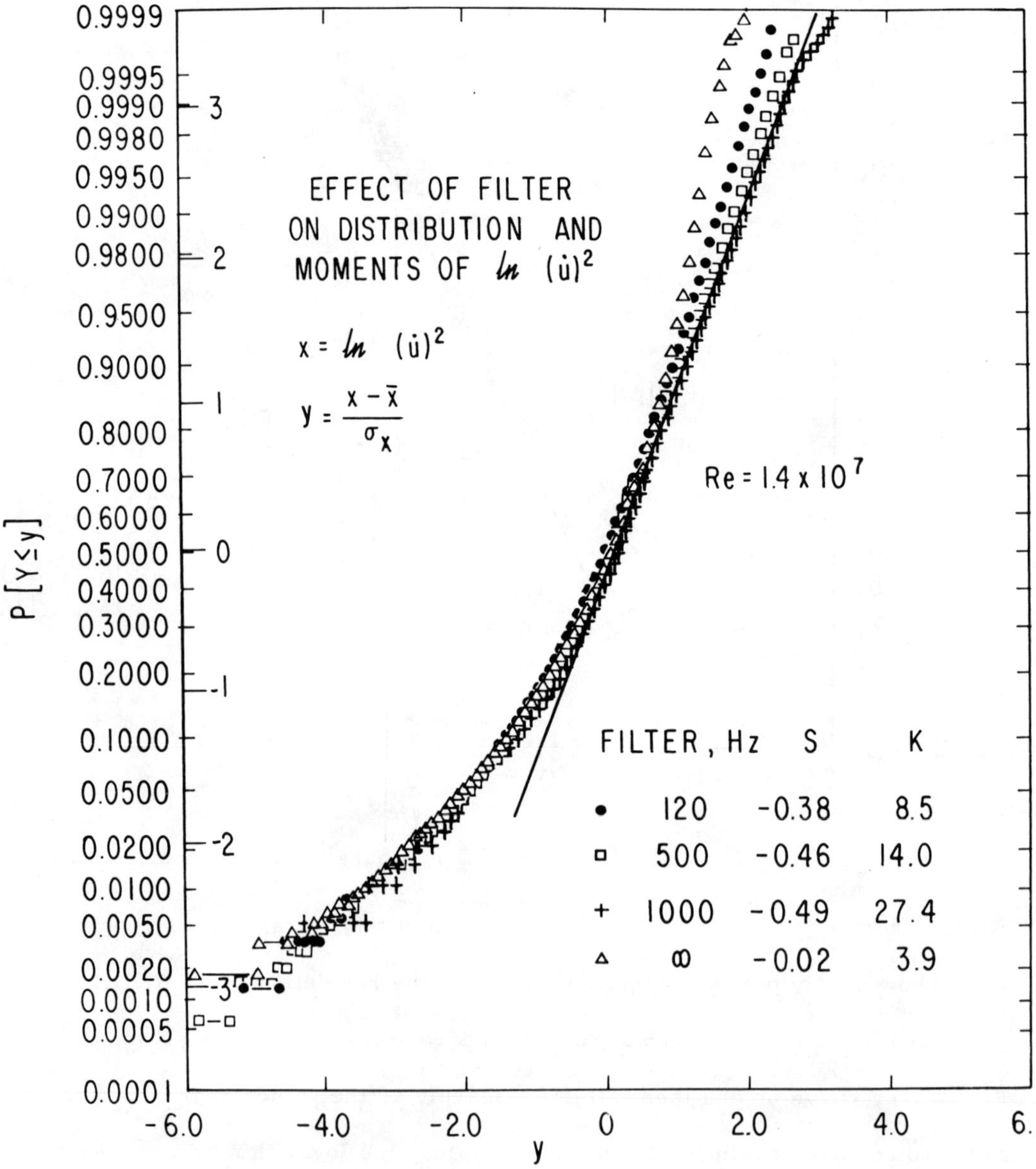

Figure 6. Influence of low pass filter on lognormality test of $\dot{u}^2$.

frequencies between 120 hz and the unfiltered case. The plots show a maximum range of lognormality, as well as a maximum kurtosis value of 27 for filter frequency about 1000 hz. The true value is probably nearly 40. The low cutoff frequency attenuates any large spikes, and the highest filter frequency permits contamination by noise. The low values for all filter frequencies are virtually identical, which suggests that the characteristic curvature in this region is not dependent on the noise, but is the true distribution function for $\ln \dot{u}^2$. This conclusion was reached by Stewart et al.[12] based on computer simulated noise tests.

One of Yaglom's hypotheses in formulating his model leading to lognormality of ϵ_r was that the random variables $\ln e_j$ in Eq. (3) should be independent and identically distributed. To test this hypothesis, the dissipation ratio parameter $e_r = u_r^2/u_{2r}^2$ was formed by averaging squared $\dot{u}$ samples corresponding to a length scale r, centrally located in a series of samples twice as long, from which the average $\dot{u}_{2r}^2$ was computed. Since the ratio of averaging lengths is 2/1 the ratio of volumes is 2^3 or 8/1, so the parameter n in Yaglom's theory is 8.

Figure 7 shows probability density functions for $\ln e_r$ calculated for $n = 8$ and for averaging lengths $r = 1, 2, 4$ and 8 centimeters. It is clear that the distributions are qualitatively quite similar, and are close to being identical. The variance values of the four $\ln e_r$ functions were calculated, and were used to compute the universal constant μ from Yaglom's formula $\mu = 3\sigma^2/\ln n$, where σ^2 is the variance of $\ln e_r$ which are shown in Fig. 7. They range from 0.29 to 0.65 with mean 0.49 ± 0.2. The mean is close to the values reported in Table 2 measured by other techniques, although the scatter is larger than most. The density functions are bounded by the fact $(\dot{u}_r^2/\dot{u}_{2r}^2)_{max} = 1/2$, corresponding to all of the nonzero values of $\dot{u}_{2r}^2$ being confined to the smaller volume. Similarly, $\left[\dot{u}_r^2/\dot{u}_{n^{1/3}r}^2\right]_{max} = 1/\sqrt[3]{n}$ for n values other than 8.

Figure 8 shows a Gaussian plot of the four $\ln e_r$ distributions. Clearly e_r is independent of r, but is not lognormal. One of Yaglom's assumptions was that

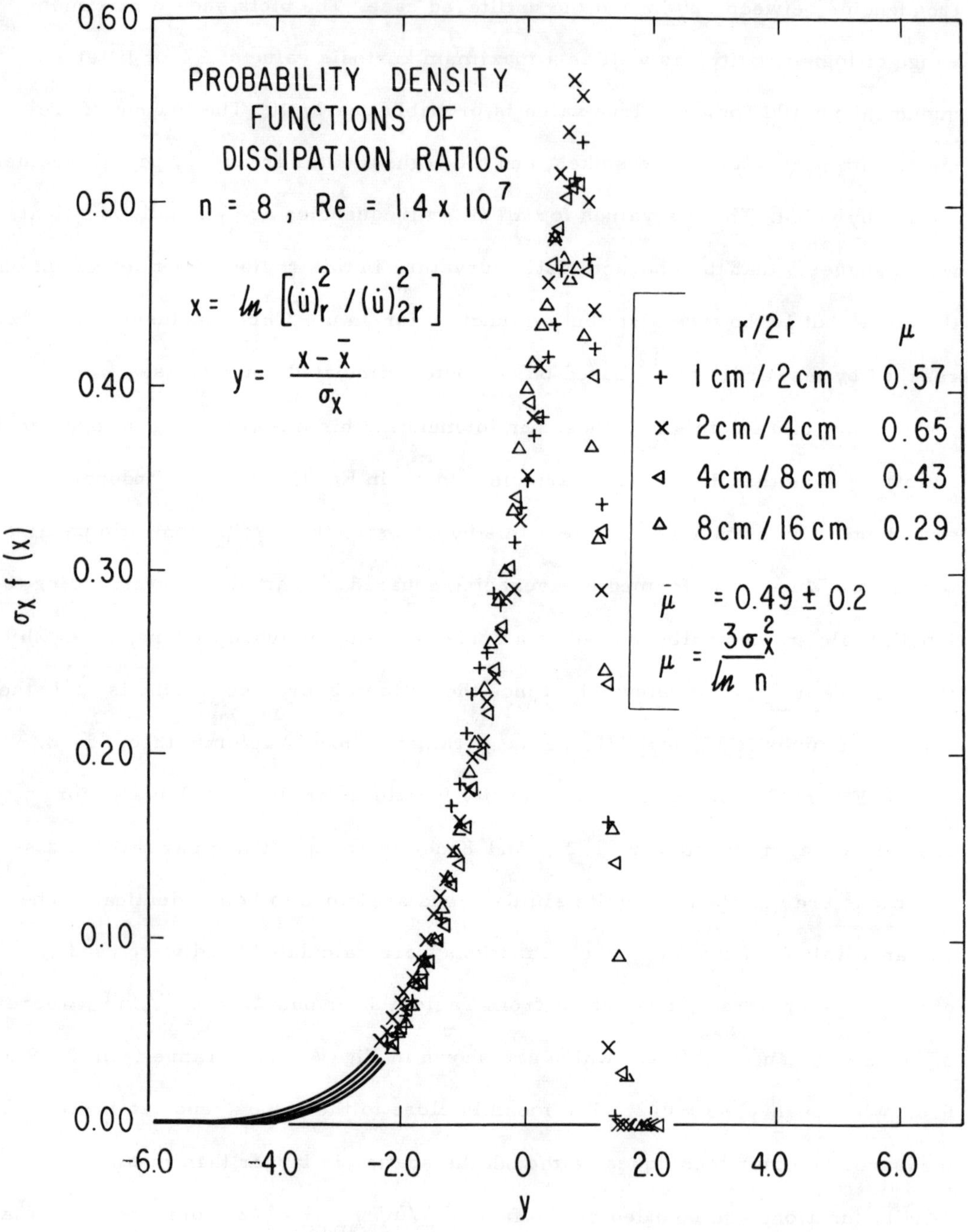

Figure 7. Probability density of $\ln(\dot{u}_r^2/\dot{u}_{2r}^2) \approx \ln e_r$.

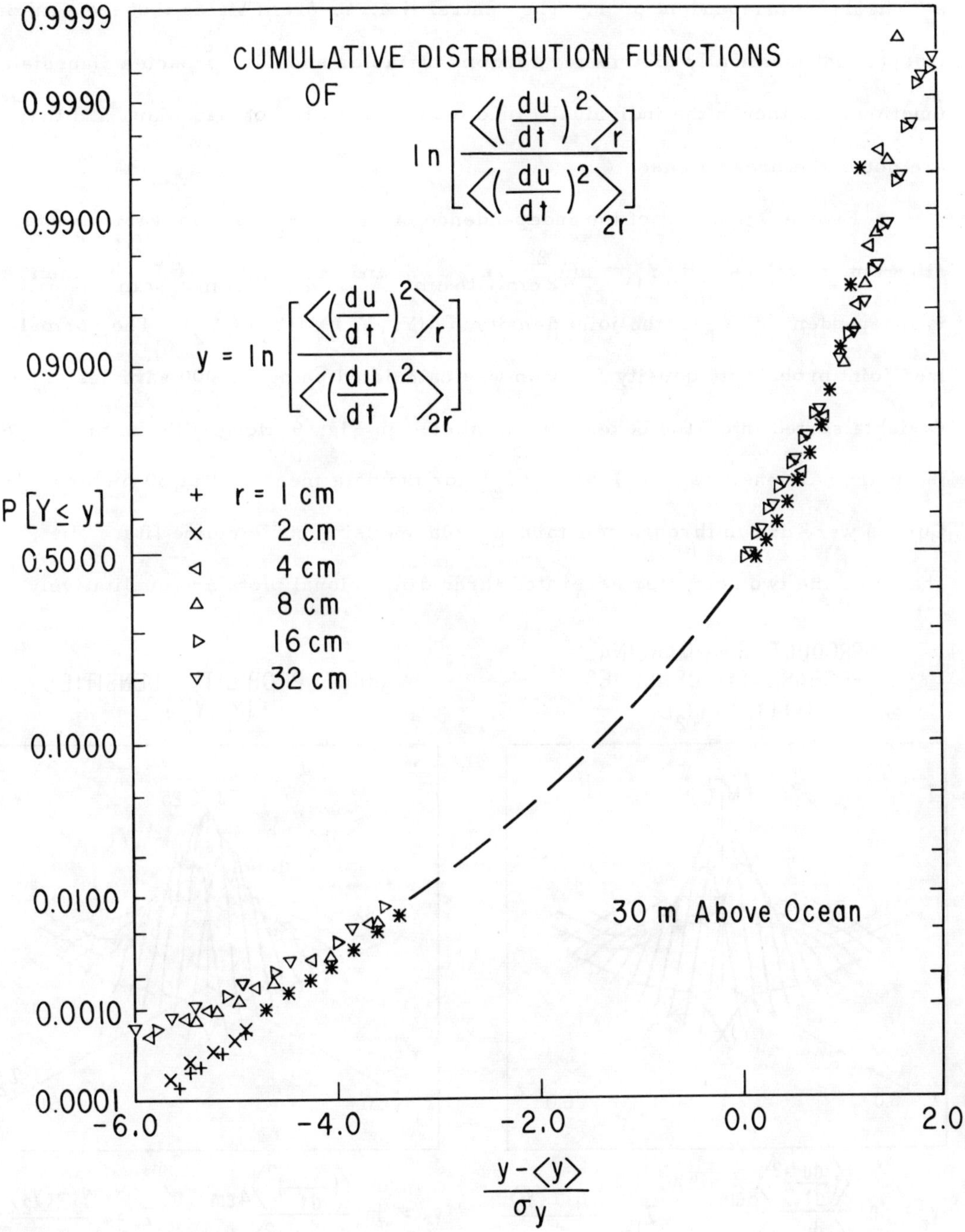

Figure 8. Lognormality test of $\ln e_r$.

e_r should be independent of r. The central limit theorem states that the sum of independent identically distributed random variables should approach a Gaussian function even though the individual random variables are not Gaussian, and they are not in the present case.

Figure 9 is a test of the independence of the $\ln e_r$ random variables for different r values. If $Y_1 = \ln(\dot{u}^2_{8\,cm}/\dot{u}^2_{16\,cm})$ and $Y_2 = (\dot{u}^2_{4\,cm}/\dot{u}^2_{8\,cm})$, then Y_2 is independent of Y_1 if the joint density $f(Y_1, Y_2) = f(Y_1)\,f(Y_2)$. The normalized joint probability density function was calculated using 64,000 samples of each variable sorted into 5184 boxes, and is plotted in Fig. 9 along with the product of the marginal densities $f(Y_1)$ and $f(Y_2)$ for the data measured at 30 meters. Curves were drawn through constant e_r planes using cubic spline fits. The shapes of the two computer generated three dimensional plots are qualitatively

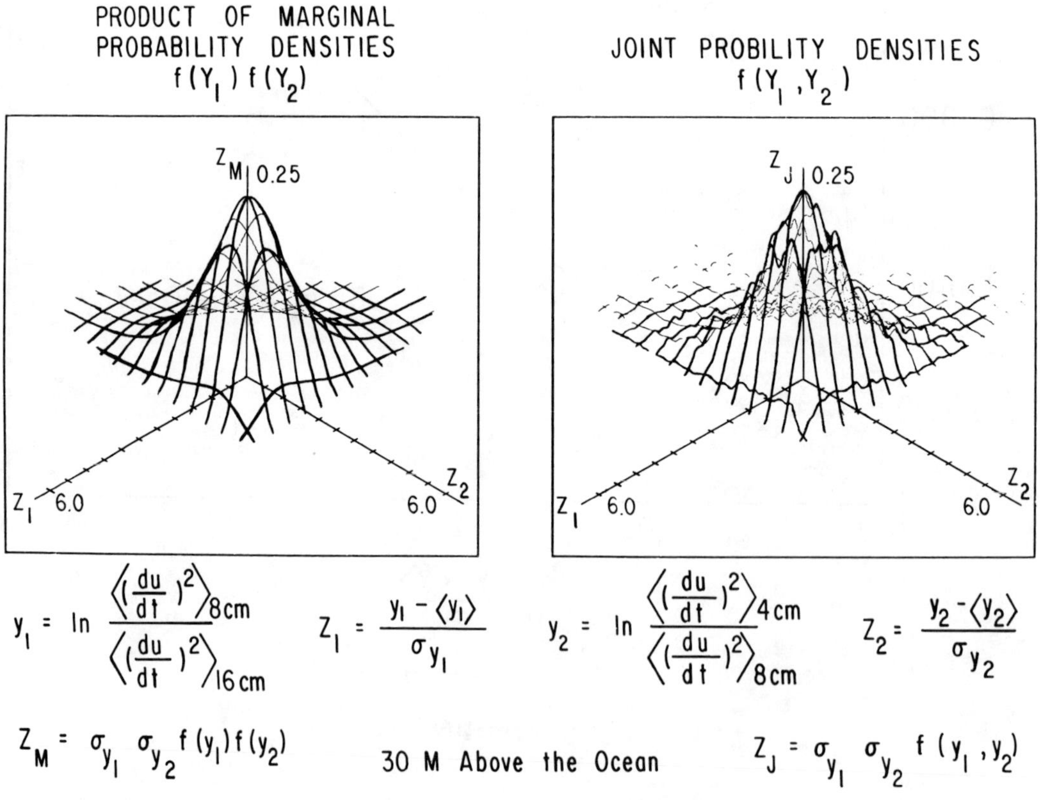

Figure 9. Independence test of e_r and e_{2r}.

similiar, and a similar plot of the difference shows rather uniform deviation with amplitude comparable to the fluctuations of $f(Y_1, Y_2)$ shown in Fig. 9.

Figure 10 shows 5 cm averaged and unaveraged lognormal tests of $\dot{w}^2$ and $\dot{u}^2$ measured at 4 meters during Expedition SOMA 2 in August 1971, where w is the vertical velocity fluctuation. Both sets of cdf curves are similar. $\dot{w}^2$ shows the same upward curvature to the left for small values that has been frequently observed for $\dot{u}^2$. The distribution of $\dot{w}^2_{5\,cm}$ seems to be more lognormal than the distribution for $\dot{u}^2_{5\,cm}$, for reasons that are not clear at present.

Figure 11 shows a lognormal comparison of the distributions for velocity and temperature derivatives measured in a heated laboratory jet. Probes were mounted 40 diameters along the axis of the jet. The jet Reynolds number was 120,000.

A striking difference exists between the two curves for the lower values. The characteristic curvature begins at about 90% for $\dot{u}^2$ at this low Reynolds number. However the distribution for $\dot{T}^2$ is lognormal all the way down to about 30%. Since both fields experience the same range of length scales between energy and viscous scales, one might expect the degree of lognormality for both ε and χ to be the same. The fact that the distributions of $\dot{u}^2$ and $\dot{T}^2$ are so different might be interpreted as evidence for the assumption that $\dot{u}^2$ may not be as representative of the local value of dissipation rate ε as $\dot{T}^2$ is representative of the local scalar dissipation rate χ, and that the true local dissipation rate ε may be more lognormal than $\dot{u}^2$. The strong departure of averaged and unaveraged $\dot{u}^2$ from lognormality is inconsistent with the extension of Yaglom's (1966) hypotheses for ε_r to include any non-negative small scale characteristic of the turbulence (such as $\dot{u}^2$) as proposed by Gurvich and Yaglom[18] (1967).

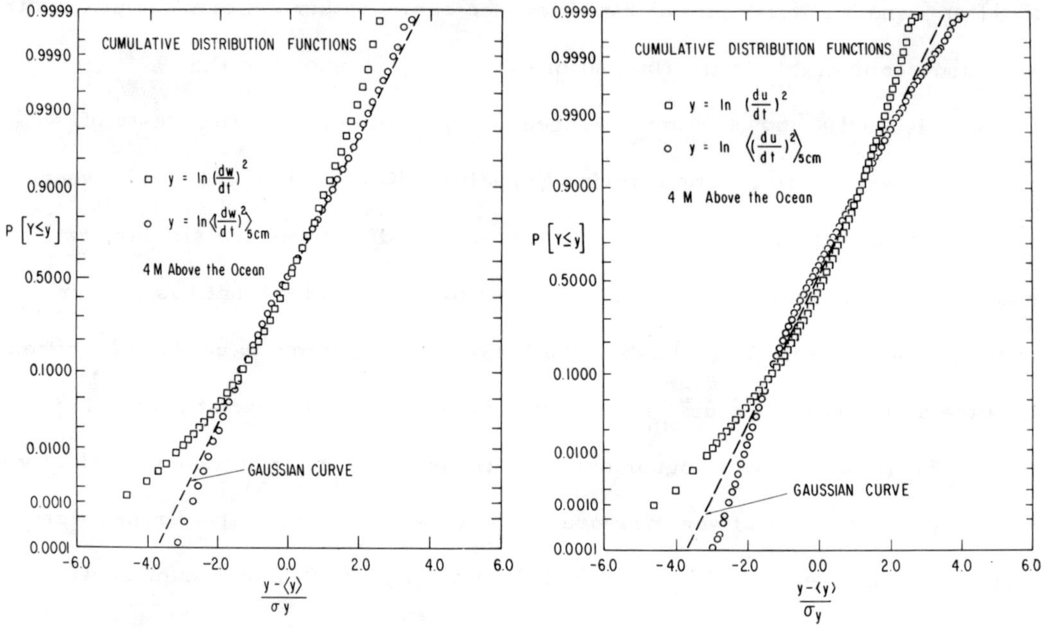

Figure 10. a. Lognormality test of $\dot{w}^2$ and $\dot{w}^2_{5\,cm}$, 4 meters.
b. Lognormality test of $\dot{u}^2$ and $\dot{u}^2_{5\,cm}$, 4 meters.

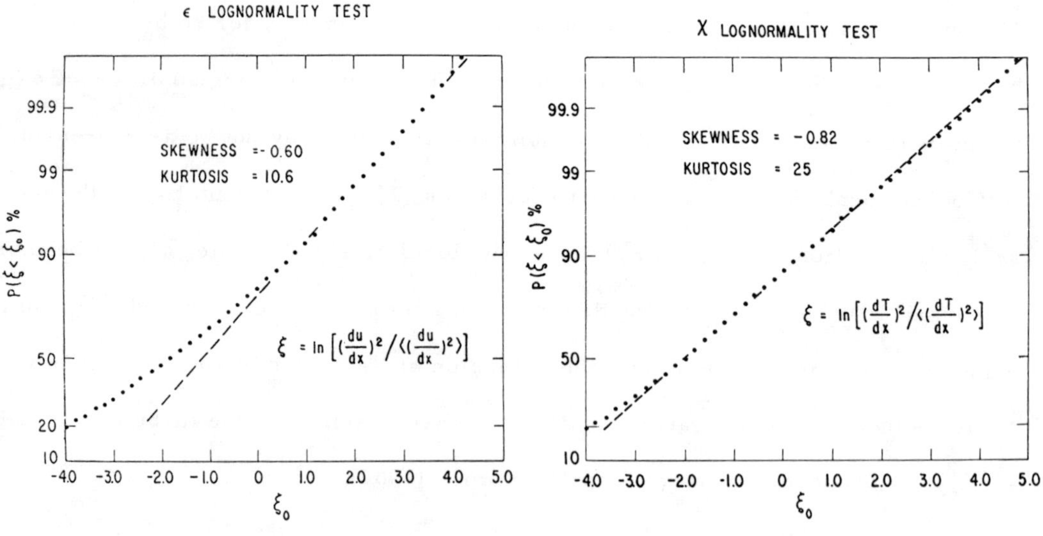

Figure 11. a. Lognormality test of $\dot{u}^2$ in jet, Re = 10^5, L/D = 40.
b. Lognormality test of $\dot{T}^2$ in a heated jet.

4. SUMMARY AND CONCLUSIONS

Measurements of averaged and unaveraged $\dot{u}^2$, $\dot{w}^2$ and $\dot{T}^2$ have been compared to lognormality under various conditions, averaging scales and Reynolds numbers. None of these quantities appear to be lognormal over the full range of values accurately measurable, especially the smaller values, contrary to the hypotheses of Gurvich and Yaglom[18] which include all non-negative small scale parameters of the turbulence in Kolmogoroff's[4] third hypothesis for the lognomality of the averaged dissipation rate ϵ_r.

An attempt to confirm Chen's[14] result that $\dot{u}_r^2$ is lognormal over the full range of values was unsuccessful, for reasons that are not clear at present. The low kurtosis values reported by Chen indicate that his $\dot{u}^2$ values were probably influenced by filtering, but a comparison test with the same filter frequency did not reproduce his result.

Averaged dissipation ratios were computed for various averaging lengths, and the distributions compared with Yaglom's hypotheses that they should be identical and independent. Although they are found to be neither perfectly identical or independent, the differences are small, and the indicated Kolmogoroff's third hypothesis constant μ is 0.49 ± 0.2, which is close to the value calculated by other methods. Again the departure from similarity occurs for small values, where $\dot{u}^2$ may fail to be representative of ϵ.

Based on the assumption that the extreme values of $\dot{u}^2$ are more representative of ϵ, $\sigma^2_{\ln \epsilon_r}$ values were inferred from the extreme values of $\dot{u}_r^2$, and the resulting plot versus $\ln(L_o/r)$ was used to test Kolmogoroff's hypothesis that $\sigma^2_{\ln \epsilon_r} = A + \mu \ln(L_o/r)$, and to estimate μ and A. In this measurement, $A = -0.84 \pm 0.2$ and $\mu = 0.47 \pm 0.03$ where the mean and standard deviations were computed from the points of constant indicated slope. Again, μ is quite close to the value of 0.51 ± 0.02 measured by Gibson, Stegen and McConnell.[13]

It appears that the present measurements are consistent with lognormality of ε_r only under the assumption that extreme values of $\dot{u}^2$ or $\dot{w}^2$ are representative of ε. The Gurvich-Yaglom extension to include all non-negative local parameters is apparently invalid. Averaging $\dot{u}^2$ and $\dot{w}^2$ has little effect on the range of lognormality. Measurements of quantities more closely related to ε and χ are very desirable in the future.

ACKNOWLEDGMENTS

The authors wish to acknowledge the valuable advice and assistance of several colleagues at UCSD during the course of this work. Steve McConnell made the measurements of temperature fluctuations in the heated jet, Gregory Dreyer and Carl Friehe made some of the ocean measurements analyzed in this study. Professor Murray Rosenblatt gave valuable advice on statistical analysis.

This work was supported by National Science Foundation Grant No. 3981, by the Advanced Research Projects Agency of the Department of Defense (monitored by the U.S. Army Research Office-Durham, Box CM, Duke Station, Durham, North Carolina 27706 under Contract DA-31-124-ARO-D-257), by ONR Contract N00014-69-A-0200-6006, and by Sea Grant Program/NSF/GH112.

REFERENCES

1. Kolmogoroff, A. N. 1941. Dokl. AN SSSR 30, 301.

2. Obukhoff, A. M. 1941. Dokl. AN SSSR 32, 19.

3. Landau, L. D. and Lifshitz, E. M. 1959. Fluid Mechanics, Addison-Wesley.

4. Kolmogoroff, A. N. 1962. J. Fluid Mech. 13, 82.

5. Obukhoff, A. M. 1962. J. Fluid Mech. 13, 77.

6. Yaglom, A. M. 1966. Dokl. AN SSSR 166, 49.

7. Gurvich, A. S. and Zubkovskii, S. L. 1963. Izv. AN SSR, Geofiz, 1856.

8. Gurvich, A. S. 1967. Dokl. AN SSSR 172, 554 [Sov. Phys. Dokl. 12, 17 (1967)].

9. Gibson, C. H. and Williams, R. B. 1969. Proceedings AGARD Specialists Symposium on the Aerodynamics of Atmoshperic Shear Flow, Munich, Sept. 15-17, Paper No. 5.

10. Pond, S. and Stewart, R. W. 1965. Izv. AN SSSR, Geofiz. 1,914.

11. Gibson, C. H., Stegen, G. R. and Williams, R. B. 1970. J. Fluid Mech. 41, 153.

12. Stewart, R. W., Wilson, J. R. and Burling, R. W. 1970. J. Fluid Mech. 41, 141.

13. Gibson, C. H., Stegen, G. R. and McConnell, S. 1970. Phys. Fluids 13, 2448.

14. Chen, W. Y. 1971. Phys. Fluids 14, No. 8.

15. Gibson, C. H., Friehe, C. A. and McConnell, S. 1971. "Measurements of Sheared Turbulent Scalar Fields" (to be published).

16. Friehe, C. A., Van Atta, C. and Gibson, C. H. 1971. Proceedings AGARD Specialists' Meeting on Turbulent Shear Flows, London, Sept. 13-15.

17. Wyngaard, J. C. and Tennekes, H. 1970. Phys. Fluids 13, 1962.

18. Gurvich, A. S. and A. M. Yaglom. 1967. Phys. Fluids Suppl. 10, 559.

19. Sheih, C. M., Tennekes, H. and Lumley, J. L. 1971. Phys. Fluids 14, 201.

20. Pond, S., R. W. Stewart and R. W. Burling. 1963. J. Atm. Sci. 20, 319.

PROBABILITY DISTRIBUTIONS IN TURBULENT FIELDS

François N. Frenkiel, Naval Ship Research and Development Center

and

Philip S. Klebanoff, National Bureau of Standards

Abstract

One-and two-dimensional density probability distributions for turbulent velocities and for gradients of turbulent velocities were measured using high-speed computing methods. Measurements made downstream of a grid and in the turbulent boundary layer are compared. The higher-order correlations

$$T_{m,n} = \frac{\overline{[f(t)]^m [f(t+h)]^n}}{\overline{[f^2(t)]}^{m/2}\,\overline{[f^2(t+h)]}^{n/2}}$$

where h *is a time interval and* f *is either the fluctuating turbulent velocity* $u(t)$ *or its gradient* $\partial u/\partial t$ *, are determined. The probability density distribution is represented as a Gram Charlier distribution of the type*[1]

$$\mathcal{P}(f(t), f(t+h)) = \mathcal{P}_0(f(t), f(t+h)) \sum_0^{j+k} A_{j,k} H_{j,k}(f(t), f(t+h))$$

where

$$\mathcal{P}_0 = \frac{1}{2\pi[1-(T_{1,1})^2]^{1/2}} \cdot$$

$$\cdot \exp\left\{-\frac{1}{2[1-(T_{1,1})^2]}[f^2(t) - 2f(t)f(t+h) + f^2(t+h)]\right\}$$

The coefficients $A_{j,k}$ *in the above expressions are determined as functions of the measured correlations* $T_{m,n}$ *and the resulting probability distribution derived from the Gram-Charlier distribution is compared to the measured distribution.*

In comparing the statistical characteristics of the turbulent field downstream of a grid to those in the turbulent boundary layer their interrelation with the theoretical ideas involving the intermittency of the small-scale structure is discussed.

This work was performed with partial support of the Division of Biology and Medicine of the Atomic Energy Commission.

---

[1] *Frenkiel, F. N. and Klebanoff, P. S.* *Phys. Fluids* *10, 507-520 (1967); Kampé de Fériet, J. David Taylor Model Basin Report 2013 (1966)*

TURBULENCE IN A STRATIFIED OCEAN

Walter H. Munk
Institute of Geophysics and Planetary Physics
and
Scripps Institution of Oceanography
University of California, La Jolla, California

Vertical distributions of potential temperature and salinity in the interior Pacific (excluding the top and bottom few hundred meters) can be reasonably fitted by exponential-like distributions with a scale height of 1 km. The observed distribution of C^{14} and other radio isotopes suggests an "overturn time" of somewhat less than 1000 years. A simple model balancing the effects of upward convection (velocity w) and downward turbulent diffusion (diffusivity κ) gives the exponential distribution and relates scale height to κ/w and scale time to κ/w^2. The result is $w = 0(1\ \text{cm/day})$ and $\kappa = 0(1\ \text{cm}^2/\text{sec})$. The diffusivity so inferred is 10^3 and 10^5 times the molecular coefficients for heat and salt, respectively.

The foregoing model relies exclusively on vertical transport processes. A rival hypothesis depends on intense mixing along coastal boundaries, islands, and seamounts, with the mixed waters readily communicated into the oceanic interior along surfaces of fixed potential density. It is a sad documentary to the state of our knowledge that the relative effectiveness of these two models is not known.

We shall be concerned with the vertical model; here the parameterization of diffusion in terms of κ tells us nothing of the processes that are involved. Visualization in terms of isotropic turbulence and related folklore is not very useful except perhaps near the top and bottom boundaries (which we ignore) and for very small scales of turbulence. Two phenomenologies are intriguing at this time:

(i) *Double diffusive processes* occur in a stably stratified two-phase system when one of the two components [1] is counter the overall stable configuration: warm [salty] over cold [fresh] water leads to "salt fingers"; [cold] fresh over [warm] salty water leads to micro-layers. A requirement for these processes is that the two components have different rates of molecular diffusion, as in fact they have. This field has been pioneered by Stommel, Stern and Turner. In the upper North Atlantic, large areas of salty over sweet water indeed suggest double-diffusive mixing as an important factor. But this cannot be the whole story, for in the Pacific and elsewhere the gross distribution is stable with regard to *both* temperature and salt, and so excludes double-diffusion processes, though small-scale inversions (on the scale of meters) have been observed and might suggest *local* double-diffusion.

(ii) *Dynamic instabilities* leading to local internal wave breaking and associated mixing have been studied by Phillips, Bretherton, Woods, Garrett and me.

The most obvious source of instability is shear. Dyson and Case, Miles and Howard have shown that, for a broad class of problems, if the shear $s = du/dz$ exceeds the $2n(z)$, where $n^2 = (g/\rho)\, d\rho/dz$ is the squared Brunt-Väisälä frequency, there will be a growth of infinitesimal waves leading presumably to wave breaking. This corresponds to a critical Richardson number $Ri = (n/s)^2 = \frac{1}{4}$. There have been many laboratory and field experiments which indeed suggest that instability arises if $s/n > S$ (a number of order 1).

The question now arises as to whether sufficient shear exists to produce instabilities and significant mixing. In the atmosphere the jet stream-connected shear can indeed produce unstable billows, as observed by RADAR scatterometry techniques, and is now believed responsible for clear air turbulence. In the oceans, shear associated with "steady" currents is not adequate, but oscillatory shear associated with internal wave activity may be. To put this question into quantitative terms, Garrett and I have contrived a model for the distribution of internal wave energy in wavenumber frequency-space that is consistent with the observations as of this instant. Our model (taking horizontal isotropy) is

$$E(\alpha,\omega) \propto \omega^{-1}(\omega^2-\omega_i^2)^{-1}, \qquad 0 < \alpha < 20\pi(\omega^2-\omega_i^2)^{\frac{1}{2}}$$

between the inertial frequency ω_i and the Väisälä frequency n, and zero otherwise. The α-bandwidth increases with ω and accounts for the poor coherence at high frequencies of horizontally-separated stations. The spectrum of vertical shear turns out to be proportional to $\alpha^2\omega^{-3}(\omega^2+\omega_i^2)(\omega^2-\omega_i^2)^{-2}$ and so is strongly peaked near the wavenumber cutoff at inertial frequencies. Integrating over α,ω-space we obtain an rms shear of $0.4\ n$ at the thermocline (less below) as compared to $2n$ for the Miles-Howard instability. For Gaussian statistics the probability of exceeding five times the rms value is too small to lead to important contributions. But we cannot exclude intermittent events that occasionally lead to large-scale internal wave breaking and associated eddy fluxes.

Another consideration leads to persistent instabilities on a much smaller scale. A fine-structure in the density distribution is associated with a very wiggly $n(z)$ so that $rms(n)$ exceeds $\overline{n}(z)$ by an order of magnitude. For dynamic reasons the shear is proportional to n^2. Accordingly the local instability number

$$\frac{s}{n} = \frac{n^2/\overline{n}^2}{n/\overline{n}}\ \frac{\overline{s}}{\overline{n}}$$

is amplified relative to the gross number in the ratio $n/\overline{n}$, and the stable "sheets" are sources of dynamic instability. Using Rice statistics for an envelope we find that the distribution of shear *maxima* is almost Gaussian and occur at intervals of three per hour. Garrett and I assume that with each peak those sheets that are unstable will be thickened so that at any instant the ratio $s/n = S$ (a constant). The model leads to an expression for the (so-called) eddy diffusivity in terms of the probabilities of shear (internal wave statistics) and of stratification (fine-structure statistics). Our final estimate is

$$\kappa = \text{order } 10 \; (\overline{s^2}/\overline{n^2})^2 \text{ cgs}$$

and for $\overline{s} = 0.4\ \overline{n}$ this value is far smaller than our early estimate of $\kappa = 1$ cgs.

Finally, I wish to report on some kinematic effects of fine-structure which are less important, but can be discussed with more precision. At a fixed depth, fluctuation of temperature with time due to internal waves will have a steppy appearance (due to fine-structure) that increases spectral energy at high frequencies. The temperature covariance can be written

$$\langle T(t)\ T(t+\tau)\rangle = \iint_{-\infty}^{\infty} T(z_o-\zeta_1)\ T(z_o-\zeta_2)\ p(\xi_1,\xi_2;\tau)\ d\xi_1\ d\xi_2$$

where $\xi(t)$ are the vertical displacements due to internal waves, and $p(\xi_1,\xi_2;\tau)$ is the joint probability density of a displacement ξ_1 at time t and ξ_2 at $t+\tau$. Assuming Gaussian wave statistics and band-limited spectra (in vertical wavenumber) of the fine-structure, Garrett and I obtain a fine-structure contribution to the spectrum $\sim \omega^{-2}$ (previously found by Phillips). Under some circumstances this contribution dominates at high frequencies. Further, measurements separated vertically by a distance Z become incoherent due to fine-structure at frequencies above W/Z, where W is the rms vertical velocity associated with internal waves. This result agrees in form and to order-of-magnitude with measurements by Webster. It is necessary to allow for such fine-structure contamination in the analysis of internal wave records.

References

1. Agar, J. N. and J. C. R. Turner (1960). Proc. Roy. Soc. A, 255, 307.
2. Cox, C., Y. Nagata and T. Osborn (1969) Oceanic fine structure and internal waves. Bull. Japanese Soc. Fisheries Oceanogr., "Papers in Dedication to Prof. Michitaka Uda," 67-71.
3. Davis, R. E. and A. Acrivos (1967) The stability of oscillatory internal waves. J. Fluid Mech., 30, 723-736.
4. Garrett, C. J. R. and W. H. Munk (1971) Internal wave spectra in the presence of fine-structure. J. Phys. Oceanog. (in press).
5. Garrett, C. J. R. and W. H. Munk (1971) Space-time scales of internal waves. To be submitted to Geophys. Fl. Dynamics.
6. Munk, W. H. (1966) Abyssal recipes. Deep-Sea Research 113, 707-730.
7. Munk, W. H. and N. Phillips (1968) Coherence and band-structure of inertial motion in the sea. Rev. Geophys. 6, 447-472.
8. Orlanski, I. and K. Bryan (1969) Formation of the thermocline step structure by large-amplitude internal gravity waves. J. Geophys. Res., 74, 6975-6983.
9. Phillips, O. M. (1971) On spectra measured in an undulating layered medium. J. Phys. Oceanogr., 1, 1-6.
10. Reid, R. O. (1971) A special case of Phillips' general theory of sampling statistics for a layered medium. J. Phys. Oceanogr., 1, 61-62.
11. Roden, G. I. (1971) Spectra of North Pacific temperature and salinity perturbations in the depth domain. J. Phys. Oceanogr., 1, 25-33.
12. Snowden, P. N. and J. C. R. Turner (1960). Trans. Faraday Soc., 59, 1409, 1812.
13. Stern, M. E. (1960). Tellus, 12, 172.
14. Stommel, H., A. B. Arons and D. Blanchard (1956). Deep-Sea Research, 3, 152.
15. Stommel, H. and K. N. Fedorov (1967) Small scale structure in temperature and salinity. Tellus, 19, 306.
16. Velarde, M. G. and R. S. Schechter (1971) Thermal diffusion and convective stability (II): an analysis of the convected fluxes. Phys. of Fl. (in press).

17. Webster, T. F. (1971) Estimates of the coherence of ocean currents over vertical distances (in press).

18. Woods, J. D. (1968) Wave-induced shear instability in the summer thermocline. J. Fl. Mech., 32, 791-800.

SPECTRA CHANGING OVER NARROW BANDS

Edward James Hannan
The Australian National University

1. Introduction

The purpose of this paper is to discuss some general theories concerning the way inferences are to be made in situations where the phenomenon of interest is one which manifests itself through a change in spectra and cross spectra over a narrow band of frequencies. A central example, with which we shall largely be concerned, is that where p; p = 1,2,...; antennae or recorders are arranged, usually in a symmetrical fashion, so as to receive a signal with a pattern of lags, from which characteristics of the signal are to be determined. Of course each recorder also receives noise or there would be no statistical problem. Thus we are basically concerned, in this example, with an interferometer in which noise effects are not negligible. An example with some relevance to the present symposium might be a pattern of anemometers. The theory which we present includes more general situations also, such as those which arise, for example, with an echoed signal received by a number of recorders, the difference between this and the previous example being due to the fact that now each recorder receives a mixture of lagged (and presumably attenuated) forms of the signal.

We consider a situation where the spectrum of the signal (and of the noise) is not uniform and is not known, for that would usually be the case. For the same reason we also allow the pattern of lags to depend upon frequency as would be so if the signal is propagated through a dispersive medium. To describe our situation more precisely let us introduce the matrix function, $f_s(\lambda)$, which is the matrix of spectra and cross spectra of the signal, and the corresponding matrix $f_n(\lambda)$ for the noise. Here λ is angular frequency so that, $s_j(t)$ being the jth signal component, $j = 1,\ldots,q$, we have, for example,

$$E\{s_j(t)s_k(t+\tau)\} - E\{s_j(t)\}\{Es_k(t)\} = \int_{-\infty}^{\infty} e^{i\tau\lambda} f_{s,jk}(\lambda)d\lambda$$

where $f_{s,jk}$ is the $(j,k)^{th}$ component of the Hermitian matrix f_s. We shall often also speak of $f_s(\lambda)$ as the spectrum of $s(t)$ and similarly for $f_n(\lambda)$. The received signal, $x(t)$, (with p components $x_j(t)$) is got from the sum of a filtered form of the transmitted signal $s(t)$ plus noise $n(t)$. Thus its spectrum is

$$f_x(\lambda) = k(\lambda)\, f_s(\lambda)k(\lambda)^* + f_n(\lambda) \tag{1}$$

where $k(\lambda)$ is composed of p rows and q columns and is the response of the system. The star indicates conjugation combined with transposition. In our first example q might be unity and $k(\lambda)$ might have $\beta_j(\lambda) \exp i\lambda\tau_j(\lambda)$ in the jth place so that $\beta_j(\lambda)$ describes the attenuation of the signal at frequency λ which is received after a lag $\tau_j(\lambda)$ relative to the first recorder, for which $\tau_1(\lambda) = 0$. The $\tau_j(\lambda)$ need not be positive. We allow the $\beta_j(\lambda)$ to be complex since apart from the phase effect due to the lag in the receipt of the signal there might be other phase distortion due, for example, to the nature of the interaction of signal and recorder. It is clear that without further assumptions the components f_s, f_n and k cannot be separated. The basic assumption which we make is that $f_s(\lambda)$ and $f_n(\lambda)$ are relatively stable and can be taken as effectively constant over a narrow band around λ but that $k(\lambda)$ is changing even over such a narrow band. In the example which we discussed we have in mind that $\beta_j(\lambda)$, $\tau_j(\lambda)$ are also stable but that $\tau_j(\lambda)$ is sufficiently large for there to be perceptible variation in $\lambda\tau_j(\lambda)$ across the band. This type of assumption appears to accord with the facts in a range of applications if only because $\tau_j(\lambda)$ is a function of the spacing of the recorders which can be chosen so as to make it fairly large.

We take our observations to be made at unit time intervals (i.e. we take the interval of observation as the time unit) and assume that aliasing effects are negligible, so that $f(\lambda)$ is negligible outside of $(-\pi,\pi]$. We shall base our calculations on what we call the finite Fourier transform, namely

$$(2) \qquad w_j(u) = \frac{1}{\sqrt{(2\pi N)}} \sum_{n=1}^{N} x_j(n) e^{in(\lambda+\omega_u)}, \quad \omega_u = 2\pi u/N, \ -m/2 < u \leq [m/2].$$

We have used N for the number of time points and n for a time point at which an observation is made. By $[x]$ we mean the largest integer not greater than x. The factor $\sqrt{(2\pi N)}$ is of course purely conventional. It is well known that these quantities may be calculated by fast Fourier transform techniques so that even very large values of N may be handled. We have centred our grid of frequencies at λ, the mid band frequency of interest, and have chosen the band to be of width $2\pi m/N$. The transformation (2) may be viewed as a procedure to facilitate the study of a narrow frequency band. Thus we might now invert the transformation so as to form

$$\zeta_j(n) = \frac{1}{\sqrt{(2\pi N)}} \sum_u w_j(u) e^{-in(\lambda+\omega_u)}$$

where here and below, unless we say otherwise we sum over $-m/2 < u \leq [m/2]$. Clearly $\zeta_j(n)$ (or rather its real part) represents the part of $x_j(n)$ due to the narrow band of frequencies around $\pm\lambda$. Equally clearly there is no real point in computing $\zeta_j(n)$ at more than m values of n. If $s = N/m$ is an integer these may as well be chosen to be the equidistant points sj. We thus introduce, whether or not N/m is an integer, the quantities

$$(3) \qquad \xi_j(n) = \frac{1}{\sqrt{m}} \sum_u w_j(u) \exp(-in2\pi u/m), \quad -m/2 < n \leq [m/2],$$

which shall play a part in our later calculations. As we shall later emphasise, the estimates which we obtain using the $w_j(u)$ will sometimes have, when expressed approximately by means of the $\xi_j(n)$ or the $\zeta_j(n)$, a simple interpretation in terms of the maximisation of a correlation (or, rather, a coherence). However there seems to be no very good reason for forming the $\xi_j(n)$ and one can as well work directly with the $w_j(u)$.

Our purpose here is to construct a general theory on the basis of which to make inferences, in a more or less optimal manner, in the kind of circumstance which we have described. To this end let us first explain what might by now be called the classical theory of narrow band spectral inference. We call $w(u)$ the vector with $w_j(u)$ in the jth place. Under very general conditions concerning the vectors $x(n)$ (including stationarity, ergodicity and continuity of f_x at λ, but also a little more) it may be shown that the vectors $w(u)$ are asymptotically independently distributed with the same normal distribution having density

$$(4) \qquad \frac{1}{\pi^p \det\{f_x(\lambda)\}} e^{-w(u)^* f_x(\lambda)^{-1} w(u)}, \quad -m/2 < u \leq [m/2], \quad \lambda \neq 0, \pi.$$

For a discussion see Rosenblatt(1956), Hannan (1970), Chapters IV, V. Commencing from this distribution all of the standard estimation procedures for spectra and cross spectra may be obtained by the method of maximum likelihood i.e. by maximising with respect to the unknown constants the likelihood function, which is the product of the m factors described by (4). The relevant (asymptotic) distributions then follow. However it is apparent that this approach cannot form a basis for the solution of our present problem since (4) is a function only of $f_x(\lambda)$ so that variation of that function across the band is not reflected in the likelihood.

Together with Mr P. Thomson I have sought to solve this problem in a series of papers (Hannan and Thomson 1971 a, b, c) and it is this work and some elaborations of it which I wish to discuss here. Our idea is to introduce a generalisation of (4) which covers the cases we wish to discuss. From this we obtain the estimates we require by maximum likelihood. Finally we establish the asymptotic distribution of these estimates, as m is increased. Some comments are called for here. In the first place the new likelihood is rather complicated and the estimates are highly non linear functions of the $w_j(u)$. As a result an exact distribution theory is nearly impossible. Allowing m to increase with N is not a deficiency since it appears to make the results more relevant. However it has associated with it a difficulty. It is apparent that there will be <u>some</u> rate of increase of N and m, with N dependent on m, which makes the asymptotic results valid. In fact for confidence intervals for the lags $\tau_j(\lambda)$ to be asymptotically useful it will be necessary that $m^3/N^2 \to \infty$. However so far we have not shown that this may be so. It must be emphasised that the results almost certainly do hold under this condition. Moreover any additional information about the permissible relative rates of increase of m and N would be of little use in practice and always a judgment would have to be made as to whether m was large enough and N large enough relative to m to validate the asymptotic results as approximations. There is little doubt also that the methods we present are near to optimal under a wide range of circumstances.

In the next section we fairly quickly survey the basic inferential procedures. In section 3 we illustrate them by relating them to examples of the type discussed in this section.

2. The Basic Estimation and Distribution Theory

We construct our asymptotic theory as follows. We imagine our observations $x(n)$; $n = 1,\ldots,N$; to be imbedded in a sequence of such sets of observations indexed by N. The typical set is thus $x^{(N)}(n)$; $n = 1,\ldots,N$; and this is taken to be part of a realization of a stationary process obtained by a filter with response $k^{(N)}(\lambda)$ from a fixed process with spectrum $f_s(\lambda)$. To the result is added noise with spectrum $f_n(\lambda)$ so that the observed process has spectrum $k^{(N)}(\lambda)f_s(\lambda)k^{(N)}(\lambda)^*$ $+ f_n(\lambda)$. Of course only one such set is observed, corresponding to the N observations to hand, and the sequence of sets is considered only as a device for the construction of a relevant theory. We are concerned with a band of width of order N^{-1} about λ (though later we allow it to increase so that the band is of order m/N in width). In the neighbourhood of λ we allow $k^{(N)}$ to vary with N but in such a way as to approach a limiting form as N increases. Of course in applications we must hope that our N is sufficiently large for us to be near enough to this limiting form for our results to be applicable. The limiting function we call h and require it to have typical element $h_{k\ell}$ of the form

$$h_{k\ell}(\theta) = \int_0^1 e^{iv\theta}\, dG_{k\ell}(v). \tag{5}$$

The sense in which $k^{(N)}$ is to approach h is that

$$\lim_{N\to\infty} \int_{-\infty}^{\infty} \| k^{(N)}(\lambda+\theta/N) - h(\theta)\|^2 \sin^2\theta/\theta^2 \, d\theta = 0.$$

Here by $\|A\|$ we mean the square root of the largest eigenvalue of A^*A. We have suppressed reference to λ in $h(\theta)$ for convenience, since λ is being held fixed. The case we principally have in mind is that where

$$h_{k\ell}(\theta) = \sum_{j=1}^{r} \alpha_{k\ell}(j)\, e^{iv_j\theta}, \quad 0 \le v_1 < \ldots < v_r < 1, \tag{6}$$

for this is the most relevant to the applications which suggested this work. However the theory is general. The range of integration in (5) may be taken to be $(-\infty, \infty)$. However it is not hard to see that a form of aliasing obtains so that all values of $\alpha_{k\ell}(v)$ differing by an integer from each other are not distinguishable. We have chosen our "principal aliases" in the interval $[0,1)$ but other intervals of unit length (e,g. $(-1/2, 1/2]$) might be more relevant to some applications.

Under rather mild conditions on the process generating the signal $s(t)$ and the noise $n(t)$ we obtained (see Hannan and Thomson, 1971a) the limiting distribution of the $w^{(N)}(u)$, constructed via (2) with $x_j^{(N)}(n)$ replacing $x_j(n)$. Of course we shall observe only one set which we shall call $x_j(n)$ as before. To describe this distribution we introduce the matrix $Q(\theta)$ having

$$\frac{4 \sin^2 \tfrac{1}{2}\theta}{(\theta-2\pi k)(\theta-2\pi\ell)}$$

in row k, column ℓ, $-m/2 < k, \ell \leq [m/2]$. By the Kronecker product, $A \otimes B$, of a $(p \times q)$ matrix A and an $(r \times s)$ matrix B we mean the $pr \times qs$ matrix with $a_{ij}\, b_{k\ell}$ in row $(i-1)r+k$, column $(j-1)s+\ell$. We call $w^{(N)}$ the vector with $w_j^{(N)}(u)$ in the $((j-1)m+u)$th place, $-m/2 < u \leq [m/2]$. Thus the m values $w_1^{(N)}(n)$ are in the first m places, then the $w_2^{(N)}(n)$ follow and so on. Then as N increases the distribution of the vector $w^{(N)}$ approaches a form with density

$$\frac{1}{\pi^{pm}\det(\Gamma)} \exp(-w^*\Gamma^{-1}w) \tag{7}$$

where

$$\Gamma = \frac{1}{2\pi} \int_{-\infty}^{\infty} \{h(\theta) f_s h(\theta)^* + f_n\} \otimes Q(\theta)\, d\theta. \tag{8}$$

We have suppressed reference to λ, for convenience. It is this function (7) which replaces the likelihood obtained as the product of the m factors (4). Of course if $h(\theta)$ is a constant then (7) reduces to that product since

$$\frac{1}{2\pi} \int_{-\infty}^{\infty} Q(\theta)\, d\theta = I_m.$$

In case $h(\theta)$ is of the form (6) then, calling $A(j)$ the matrix with entries $\alpha_{k\ell}(j)$, we easily find that

$$\Gamma = \sum_{j,k=1}^{r}\sum \{A(j) f_s A(k)^* \otimes \Psi(v_k - v_j)\} + f_n \otimes I_m \tag{8)'$$

where $\Psi(v) = \Phi(v)\Lambda(v)$ while $\Lambda(v)$ is diagonal with $\exp(-i2\pi kv)$ in the kth place, $-m/2 < k \leq [m/2]$, and $\Phi(v)$ has elements, for $-m/2 < j, k \leq [m/2]$,

$$\phi_{jk}(v) \begin{cases} = \int_0^{1-v} e^{i2\pi(j-k)\theta}\, d\theta, & v \geq 0 \\ = \int_{-v}^{1} e^{i2\pi(j-k)\theta}\, d\theta, & v \leq 0. \end{cases} \tag{9}$$

Not all parameters in Γ are individually distinguishable (i.e. identifiable) but we refer the reader to Hannan and Thomson (1971c) for further details.

It is (7) with Γ given by (8), or more specially (8)', which is to be maximised (for fixed w and the unknown parameters in Γ varying) to provide estimates of these parameters. In this way we have constructed a general theory for the estimation of models of the type introduced in section 1.

To illustrate we consider the special case of the p recorders discussed in section 1. Now $k(\lambda)$ is a column vector with $\beta_j(\lambda) \exp i\lambda\tau_j(\lambda)$ to the jth place. Since we are regarding $\beta_j(\lambda)$ and $\tau_j(\lambda)$ as fixed over the narrow band we have

$$k_j(\lambda+\theta/N) = \beta_j(\lambda) \exp(i\lambda\tau_j(\lambda)) \exp\{i\theta\tau_j(\lambda)/N\}.$$

If we put $v_j = \tau_j(\lambda)/N$ then $h_j(\theta) = [\beta_j(\lambda) \exp i\{\lambda\tau_j(\lambda)\}] \exp iv_j\theta$. In this case the parameters we estimate are (again suppressing reference to λ)

$$f_j = |\beta_j|^2 f_s + f_n,\ j = 1,\ldots,p;\quad \sigma_{jk} = |\beta_j(\lambda)\beta_k(\lambda)| f_s/\{f_j f_k\}^{\frac{1}{2}},\quad j\neq k, j,k=1,\ldots,p;$$

$$v_j, j = 2,\ldots,p,$$

together with the $p-1$ real numbers, θ_j, which are the arguments of the complex numbers $\beta_j(\lambda) \exp i\lambda\tau_j(\lambda)$, $j = 2,\ldots,p$. These are the identifiable parameters. We have put $v_1 = 0$ and the argument of $\beta_1(\lambda) \exp i\lambda\tau_1(\lambda)$ also equal to zero. We assume $\sigma_{jk} > 0$ for all j,k as otherwise the model is not identified. We call ρ the vector of all identifiable parameters. The estimate got by maximising (7) we call $\hat\rho$. We have in Hannan and Thomson (1971c) proved the following result, which describes the limiting distribution of $\hat\rho$ as m and N increase. We have confined ourselves to the case (6) so that Γ is given by (8)' because this is most relevant. It is also the most difficult case. We divide ρ (and $\hat\rho$) into subvectors β, v ($\hat\beta$, $\hat v$) so that $\rho' = (\beta' \vdots v')$. Here v comprises the p parameters v_j while β comprises all other parameters. Then we form $\hat x' = (m^{\frac{1}{2}}(\hat\beta-\beta)' \vdots m^{3/2}(\hat v - v)')$. As m and N increase (in a suitably related fashion, see the end of section 1) the distribution of $\hat x$ approaches a multivariate normal distribution with zero mean vector and covariance matrix estimated as having the typical element

$$m^{-s}\ \mathrm{tr}\{\hat\Gamma^{-1}\frac{\partial\hat\Gamma}{\partial\rho_i}\hat\Gamma^{-1}\frac{\partial\hat\Gamma}{\partial\rho_k}\}\ .$$

Here s is unity if ρ_i, ρ_j both belong to β, is 2 if one but not the other belongs to β and is 3 if both ρ_i, ρ_j belong to v. By $\partial\hat\Gamma/\partial\rho_i$ we mean that Γ is to be differentiated element by element and that its parameters are to be estimated by $\hat\rho$. The true limiting covariance matrix is got by taking the limit in the above formula.

This completes the discussion of the general theory. In the next section we go on to study particular cases at more detail.

3. Approximate Forms of the Likelihood Equations

We now wish to make our results more perspicuous by (approximately) reinterpreting the likelihood equations (namely the equations got by equating the gradient of (7), with respect to ρ, to zero)*. The approximation is got first by replacing $\Phi(v)$ (see (9)), in the case where $v \geq 0$ for example, by an approximating sum i.e.

$$\sum_{0}^{[m(1-v)]} \chi_k \chi_k^*$$

where χ_k has as j^{th} element $m^{-\frac{1}{2}} \exp(i2\pi jk/m)$. The error we introduce by replacing $\Phi(v)$ by this matrix may be shown to be negligible for large m. Alternatively we

* We do not devote any space to a precise justification of these approximations because one may as well proceed from (7) and (8), or (8)', and the approximations are basically introduced for perspicuity and to relate our methods to others in use.

may write $\Phi(v) \simeq PJ(v)P^*$ where P is unitary with χ_k as the kth column, $-m/2 < k \leq [m/2]$, and $J(v)$ is diagonal with units or zeros in the main diagonal, the zeros being in the $[mv]$ places $k = -1,-2,\ldots-[mv]$, for $v \geq 0$, for example. (We shall be considering only $|v|$ small in this section). For $v \leq 0$ they will be in the places $k = 0,1,\ldots[mv]-1$. Then we obtain

$$\Psi(v_k - v_j) = \Lambda(v_j)^* P J(v_j, v_k) P^* \Lambda(v_k),$$

where, for example, $J(v_j,v_k) = [\int_{-v_j}^{1-v_k} e^{i(r-s)\theta} d\theta]$, $v_k \geq v_j \geq 0$. Now Γ becomes, using this approximation

$$\text{(10)} \qquad \sum_{j,k}\sum \{I_p \otimes \Lambda^*(v_j)P\}\{A(j)\, f_s A(k)^* \otimes J(v_j,v_k)\}\{I_p \otimes \Lambda^*(v_k)P\}^* + f_n \otimes I_m.$$

We now restrict ourselves to the case where $A(j)\, f_s A(k)^*$ has non null element only in the $(j,k)^{th}$ place (which includes the case of the p recorders discussed in section 1). Then putting $B = \sum_{j,k}\sum\{A(j)f_s A(k)^*\} + f_n$ we define the typical element to be

$$b_{jk} = \surd(f_j f_k)\sigma_{jk} \exp i\theta_{jk}.$$

Then (7) reduces to a matrix of p^2 blocks of which a typical one is $b_{jk}\, \Psi(v_k - v_j)$. Thus the approximation (10) now has a typical block

$$b_{jk}\, \Lambda(v_j)^* P J(v_j,v_k) P^* \Lambda(v_k).$$

If $P^*\Lambda(v_k)$ is applied to the vector having the m quantities $w_k(u)$ down it we obtain the vector with the m quantities $\xi_k(n+mv_k)$ down it, $-m/2 < n \leq [m/2]$. If we call ξ the vector, comprised of these $\xi_k(n+mv_k)$ as w was of the $w_j(u)$, then to the same order of approximation as is involved in (10) we have

$$w^* \Gamma^{-1} w \simeq \xi^* \Omega^{-1} \xi$$

where Ω is comprised of p^2 blocks of which a typical one is $b_{jk}\, J(v_j,v_k)$. Now if the v_j are small we may approximately invert Ω and obtain the matrix with typical block $b^{jk}\, J(v_j,v_k)$ where b^{jk} is the typical element B^{-1}. Thus we have, approximately,

$$w^* \Gamma^{-1} w \simeq m\, \mathrm{tr}(B^{-1}\, C(v))$$

where $C(v)$ has typical element

$$c_{jk}(v) = \frac{1}{m} \sum \xi_j(n+mv_j)\, \overline{\xi_k(n+mv_k)}$$

and the sum is over n such that

$$-m/2 < n+mv_j,\ n+mv_k \leq [m/2].$$

To the same order of approximation we have $\det(\Gamma) = \{\det(B)\}^m$ so that the maximisation of the likelihood function is much the same as the minimisation of

$$\text{(11)} \qquad \mathrm{tr}(B^{-1}C(v)) + \log\{\det(B)\}.$$

It is worth pointing out that

$$c_{jk}(v) \simeq \frac{1}{N} \Sigma\ \zeta_j(n+Nv_j)\ \overline{\zeta_k(n+Nv_k)}$$

where the sum is over $-N/2 \le n+Nv_j,\ n+Nv_k \le [N/2]$, the approximation being due to end effects. (There would be no difference if both sums ran over a full range of, respectively, m and N values.) We have expressed the $c_{jk}(v)$ in terms of the ξ_j since clearly it will be computationally easier to use them. Nevertheless the expression in terms of the ζ_j has an even more direct meaning. The estimates of f_j are now the natural ones

$$\hat{f}_j = \frac{1}{m} \sum_u |\xi_j(u)|^2 = \frac{1}{m} \sum_u |w_j(u)|^2.$$

(If $|v|$ is not small these can be improved). Moreover putting

$$\hat{\sigma}_{jk}(v) = \frac{|c_{jk}(v)|}{\surd(\hat{f}_j\ \hat{f}_K)}$$

and $\hat{\theta}_{jk}(v)$ as the argument of $c_{jk}(v)$ (11) is minimised when the determinant of the matrix $\hat{\Sigma}(v)$, having $\hat{\sigma}_{jk}(v)$ in the typical place, is minimised. If $\hat{v}$ is the vector achieving this then $\hat{\sigma}_{jk}(\hat{v})$, $\hat{\theta}_{jk}(\hat{v})$ are the (approximate) maximum likelihood estimators of the other parameters. In this case the asymptotic distribution theory for the estimates of f_j, σ_{jk}, ϕ_{jk} is the same as that which obtains when (4) provides the true likelihood function. Of course $\hat{f}_j$ is a spectral estimate, $\hat{\theta}_{jk}$ is an estimate of phase and $\hat{\sigma}_{jk}$ is an estimate of coherence. Since this "classical" theory is fairly fully surveyed in Hannan (1970) I shall not discuss it further here. The $\hat{v}_j$ are, asymptotically, independent of the other parameters and the statistics $m^{3/2}(\hat{v}_j - v_j)$, $j = 2,\ldots,p$, are asymptotically jointly normal with covariance matrix having as typical element

(12) $$3/\pi^2 [b^{jk}\ b_{kj} + b_{jk}\ b^{kj} - 2\delta_{jk}]^{-1}, \quad j,k = 2,\ldots,p.$$

This is, of course, estimated by inserting the estimates of σ_{jk}, f_j, θ_{jk} in B. The interpretation of the estimation procedure in this case is fairly direct. To illustrate further we consider the important case where $p = 2$. This case came first to our attention because when $\tau(\lambda)$ is large it had been observed that the standard estimate of σ_{12}, which we now call σ, is badly biased downwards. (By the standard estimate we mean $\hat{\sigma}_{jk}(0)$.) For references to these observations see Hannan and Thomson 1971b. This problem is overcome by using an estimate of v in forming $\hat{\sigma}_{jk}(\hat{v})$. The formula (12) now becomes $3(1-\sigma^2)/\{2\pi^2\sigma^2\}$. If v is not negligibly small the more exact treatment of section 2 adjoins to this a factor $(1-|v|)^{-1}$. The case studied in this section is basically that where v is small but mv is not small and this case is also certainly one in which the estimate of coherence will be badly biased down if the formula $\hat{\sigma}(\hat{v})$ is not used. There seems to be some argument for using this statistic in any case. As we have already said its asymptotic distribution is the same as that obtained by $\hat{\sigma}(0)$ on the assumption $Nv = 0$.

In the case $p > 2$ a more relevant case might be that where all f_j are equal, all σ_{jk} are equal and $\theta_{jk} = \theta_j - \theta_k$. This would hold if the noise spectrum (at frequency λ) was the same at each recorder so that the signal to noise ratio also was constant. The assumption $\theta_{jk} = \theta_j - \theta_k$ seems more reasonable than the introduction of general θ_{jk}. Now modifying the definition of B accordingly we find, on minimising (11), that we must form

(13) $$\hat{\sigma}(v,\theta) = \frac{u^*C(v)u - \mathrm{tr}\{C(v)\}}{(p-1)\,\mathrm{tr}\{C(v)\}},$$

where u has $\exp i\theta_j$ in the jth place, $j = 1,\dots,p$, $(\theta_0 = 0)$. This has to be maximised by variation of the v_j and θ_j. The maximising values, $\hat{v}_j, \hat{\theta}_j$ estimate the v_j and θ_j and $\hat{\sigma}(\hat{v}, \hat{\theta})$ estimates σ. Finally

$$\hat{f} = \frac{1}{p} \cdot \frac{1}{1-\hat{\sigma}} \cdot [\mathrm{tr}\{C(\hat{v})\} - \frac{\hat{\sigma}}{1+(p-1)\hat{\sigma}} \hat{u}^*C(\hat{v})\hat{u}].$$

here, of course, $\hat{u}$ has $\exp i\hat{\theta}_j$ in the jth place. The distribution of the $\hat{v}_j$ is as before, of course, but the distribution of the θ_{jk} and f_j have to be modified. The estimator $\hat{\sigma}(\hat{v}, \hat{\theta})$ will be distributed as a coherence. We shall not go into details. In this case the interpretation is very direct. The parameters θ, v are chosen so as to make $u^*\hat{C}(v)u/\mathrm{tr}(\hat{C}(v))$ as large as possible. This corresponds to adjusting the v_j and θ_j so as to maximise the "variance" of

(14) $$\sum_{j=1}^{p} \exp -i\theta_j\, \xi_j(n-mv_j)$$

relative to the average of the "variances" of the $\xi_j(n)$.

In some cases it might be more relevant to put $\theta_j = v_j\alpha$ (where α is known, for example as $N\lambda$). Again the v_j might be prescribed to be of the form jv. Other cases could be considered in the same way. For example one might have two linearly arranged sets of recorders in two dimensions arranged in the form of a cross with the distance between adjacent recorders the same (or perhaps only the same for the recorders in each individual arm of the cross). Now the v_j for the first arm will be of the form ju_1 where u_1 is a simply determined function of the distance apart of the recorders and the two parameters prescribing the velocity of the signal. The same would be true for the other arm. It might be true that the signal and noise power were constant along each arm individually. The optimal procedure would now consist in forming the expression analogous to (14) individually for each arm and then adjusting the θ_j, and the parameters prescribing the velocity of the signal so as to minimise the determinant of a 2×2 matrix. There are many variations upon the possibilities and the existence of these variations makes necessary the general theory. If the v_j are small one can equivalently begin from (11), with B parametrised according to the prescriptions that are made, and minimise that expression.

We have not dealt, in this section with the more elaborate case where the various recorders may receive echoes of the signal (or of a number of signals for that matter) so that signals with different lags contribute to each recorder. We refer the reader to Hannan and Thomson (1971c) for a discussion. We have also restricted ourselves, almost entirely, to the case where the matrices $G(v)$ (see (5)) increase only by jumps at the points v_j. (We have called these jumps $A(j)$). A considerably simpler case is that $G(v) = g(v)dv$ and $g(v)$ is a well behaved function of certain parameters (e.g. sufficiently often differentiable). The functions $\{\sin \theta/2\}/\{(\theta-2\pi j)/2\}$ now play a premier role since they are dense, in the L^2 norm, in the space of square integrable functions of exponential type $\leq 1/2$ and $h(\theta)$ will be of this form.

One final point may be made. With simple symmetric patterns of recorders such as those briefly discussed in this section and with simple (but presumably reasonable) assumptions concerning noise and signal spectra at the different recorders, it is quite easy to determine the asymptotic form of the covariance structure of the best estimates of the v_j. (Indeed (12) gives such a formula for general v_j and special f_s, f_n.) It will then be possible to compare different patterns of recorders to determine their efficiency as estimators of (for example) signal velocity. In principle it would be possible to investigate an optimum arrangement of antennae under suitable restrictions (on the number of antennae, the maximum distance apart and so on).

References

Hannan, E.J. Multiple Time Series. John Wiley. (1970).

Hannan, E.J. and Thomson, P. Spectral inference over narrow bands. J. Appl. Probability, 8, 157-169 (1971a).

Hannan, E.J. and Thomson, P. The estimation of coherence and group delay. Biometrika, 58 (1971b).

Hannan, E.J. and Thomson, P. The estimation of echo times. (To be published. Preprints available from Department of Statistics, Institute of Advanced Studies, Australian National University, Canberra, Australia).

Rosenblatt, M. A central limit theorem and a strong mixing condition. Proc. Nat. Acad. Sci. U.S.A., 42, 43-47 (1956).

SOME RECENT ADVANCES IN TIME SERIES ANALYSIS[1]

by Emanuel Parzen

Department of Statistics
State University of New York at Buffalo

0. Introduction and Summary

Research in time series analysis is conducted from many different viewpoints by researchers concerned with a wide range of concrete problems. Because researchers are spread so thinly over the field one may not perceive any feeling of progress in the theory of structure, analysis, and synthesis of time series models. It is my view that the last few years have seen many highly applicable developments.

This account of some recent advances necessarily reflects my personal knowledge (or ignorance) for which I apologize both to the reader and to researchers whose work I may have unwittingly slighted.

The section headings are: 1. Time Series, Covariance, and Spectra; 2. Reproducing Kernel Hilbert Spaces; 3. Spectral Estimation; 4. Autoregressive Spectral Estimation; 5. Estimation of Parameters of Moving Average and Mixed Schemes; 6. Equivalences Among Time Series, Approximation, and Control; 7. Probability Density Estimation; 8. Time Series Modeling and Identification.

1. Time Series, Covariances, and Spectra

In the statistical theory of turbulence, it is customary to assume that the velocity field of a fluid in turbulent flow may be represented by a random vector field $U = (u_1, u_2, u_3)$ with components u_j, $j = 1, 2, 3$. Each component u_j is a function of (1) time t, (2) spatial coordinates $x = (x_1, x_2, x_3)$, and (3) a variable or parameter ω representing the field randomly chosen from the space of all possible fields.

---

[1] Research supported by the Office of Naval Research under Contract NONR-225 80.

The meaning of random choice of ω is defined by postulating an abstract probability space $(\Omega, \mathcal{A}, P)$ such that, for $A \in \mathcal{A}$, $P[A]$ is the probability that $\omega \in A$.

If we let T denote the index set of time, R^3 denote the index set of 3-dimensional space, then U can be regarded as a 3-vector valued function of (t, x, ω) ; in symbols,

(1) $$U: T \times R^3 \times \Omega \to R^3$$

If U were a deterministic function of (t, x) we would write $U: T \times R^3 \to R^3$.

Note that the space of all possible fields is the family of all functions with domain $T \times R^3$ and range R^3.

The notion that U is a random function of (t, x) is therefore best expressed by writing

(2) $$U: T \times R^3 \to L^o(\Omega, \mathcal{A}, P; R^3)$$

where $L^o(\Omega, \mathcal{A}, P; R^3)$ is the space of all <u>random</u> (3-dimensional row) <u>vectors</u> on the probability space $(\Omega, \mathcal{A}, P)$.

We call U a <u>time series</u> (rather than a random function or stochastic process) when instead of (2) we assume

(3) $$U: T \times R^3 \to L^2(\Omega, \mathcal{A}, P; R^3)$$

where $L^2(\Omega, \mathcal{A}, P; R^3)$ is the subset of random vectors in $L^o(\Omega, \mathcal{A}, P; R^3)$ whose components have finite second moments; it is a Hilbert space under various inner products, such as $(U, V)_P := E[UV'] := \int_\Omega \sum_{i=1}^{3} U_i(\omega) V_i(\omega) P(d\omega)$. By the symbol $:=$ we mean equal by definition; the expression next to the colon is being defined as equal to the other side of the equality.

A time series has two main statistical characteristics: its mean function

(4) $$m_U: T \times R^3 \to R^3,$$

$$m_U(t, x) := \{E[u_j(t, x)]\} = (m_1(t, x), m_2(t, x), m_3(t, x))$$

and its covariance kernel

$$(5)\qquad \begin{aligned} &R_U: (T \times R^3) \times (T \times R^3) \to \text{space of } (3,3) \text{ positive definite matrices} \\ &R_U((t^{(1)}, x^{(1)}), (t^{(2)}, x^{(2)})) := \{\mathrm{Cov}[u_j(t^{(1)}, x^{(1)}), u_k(t^{(2)}, x^{(2)})]\\ &\qquad =: \{R_{jk}(t^{(1)}, x^{(1)}; t^{(2)}, x^{(2)})\} \end{aligned}$$

Modeling a time series is equivalent to modeling its mean function and covariance kernel. If one assumes a representation for the mean and covariance as a linear combinations of known functions with unknown coefficients, then modeling a time series is equivalent to modeling, or estimating, these coefficients.

One important mode of representation of covariances is of the form

$$(6)\qquad R_{jk}(t^{(1)}, x^{(1)}; t^{(2)}, x^{(2)}) = \int_\Lambda g(t^{(1)}, x^{(1)}; \lambda) g^*(t^{(2)}, x^{(2)}; \lambda) \mu_{jk}(d\lambda)$$

where the functions g are assumed known, and the measures μ_{ij} are assumed unknown; the asterisk on g indicates its complex conjugate.

In discussing the properties of a turbulent field it is often convenient to consider it at a single instant of time $t^{(0)}$ and write

$$(7)\qquad R_{jk}(x^{(1)}; x^{(2)}) := \mathrm{Cov}[u_j(t^{(0)}, x^{(1)}), u_k(t^{(0)}, x^{(2)})] .$$

The random field is called space-homogeneous if all R_{jk} are functions only of $x^{(1)} - x^{(2)}$; then there are functions $\rho_{jk}(h)$, called covariance functions, such that

$$(8)\qquad R_{jk}(x^{(1)}; x^{(2)}) = \rho_{jk}(x^{(1)} - x^{(2)}) .$$

The basic property of ρ_{jk} are their spectral representations

$$(9)\qquad \rho_{jk}(h) = \int_\Lambda e^{i\lambda h} dS_{jk}(\lambda)$$

where $e^{i\lambda h} = e^{i(\lambda_1 x_1 + \lambda_2 x_2 + \lambda_3 x_3)}$, and S_{jk} are functions of bounded variation; if they are absolutely continuous then

$$(10)\qquad \rho_{jk}(h) = \int_\Lambda e^{i\lambda h} \varphi_{jk}(\lambda) d\lambda .$$

One calls $\{\varphi_{jk}(\lambda)\}$ the spectral density tensor. The diagonal functions $\varphi_{jj}(\lambda)$ are non-negative and the off-diagonal functions have the Hermitian property

$$\varphi_{jk}(\lambda) = \varphi^*_{kj}(\lambda) .$$

For proofs and discussion of the foregoing assertions, see Kampé de Feriet (1953) or Lumley (1970).

Modeling, or estimation, of the spectral density tensor is equivalent to modeling of the time series.

2. Reproducing Kernel Hilbert Spaces

The assumption that a turbulent fluid is incompressible implies that the random vector field satisfies the continuity equation

$$\nabla u := \sum_j \frac{\partial}{\partial x_j} u_j(x) = 0 . \tag{1}$$

Thus one of the problems of the foundations of time series is to define the meaning of derivatives of time series such as in (1). Now all random variables which are obtained by addition, limits, integrals, or derivatives of a time series U are members of a certain Hilbert space, denoted H_U and called the Hilbert space <u>spanned</u> by the time series U .

H_U is defined to be the smallest Hilbert subspace of $L^2(\Omega, \mathcal{A}, P; R^3)$ containing $u_j(t,x)$ for all j, t, x .

As long as one confines oneself to linear operations on a time series one encounters only random variables in H_U . When one wants to consider non-linear operations one introduces a Hilbert space $H_{\mathcal{A}(U)}$, called the Hilbert space <u>generated</u> by U .

$H_{\mathcal{A}(U)}$ is defined to be $L^2(\Omega, \mathcal{A}(U), P; R^3)$, where $\mathcal{A}(U)$ is the smallest sub sigma-field of $\mathcal{A}$ with respect to which $u_j(t,x)$ is measurable, for all j, t, x .

A derivative $\frac{\partial}{\partial x_k} u_k(x)$ can be specified by its indexing function in another Hilbert space H which is used to index H_U (that is, there is a one-one correspondence between H_U and H). The spectral representation of the covariance tensor of a spatially homogeneous turbulent field provides an indexing in terms of a Hilbert space of functions of wave number λ .

Corresponding to a covariance R there is a unique Hilbert space denoted H(R) called the reproducing kernel Hilbert space with reproducing kernel R [see Parzen (1967a)]. Its members are R^3-valued functions with domain the index set of the time series U .

Reproducing kernel Hilbert spaces play many roles in time series analysis [see Parzen (1970)].

It seems natural to require that the mean m_U of a time series belong to H(R) . Then one can set up a one-one correspondence between H_U and H(R) which associates $u_j(t, x)$ and $\{R_{jk}(t, x; ., .), k = 1, 2, 3\}$. This correspondence has the intuitive meaning that functional equations for the time series are equivalent to functional equations for the covariances.

As an example, (1) is equivalent to the following relations for the covariance tensor ρ of a space-homogeneous random field:

(2) $$\sum_j \frac{\partial}{\partial h_j} \rho_{j,k} = 0 , \quad k = 1, 2, 3 .$$

The Navier Stokes equation for an incompressible viscous field are

(3) $$\frac{\partial}{\partial t} u_j + \sum_k \frac{\partial}{\partial x_k} u_j u_k = - \frac{1}{\rho} \frac{\partial p}{\partial x_j} + \nu \Delta u_j , \quad j = 1, 2, 3$$

where ρ and ν are characteristic constants of the fluid (namely, its specific mass and kinematic viscosity), $p(t, x)$ is the pressure at x at time t, and Δ is the Laplacian.

From the point of view of linear operations on time series the Navier Stokes equation involves 7 time series: p, u_1 , u_2, u_3 , $u_1 u_2$, $u_1 u_3$, $u_2 u_3$. Therefore the equations of the Navier Stokes equation are equivalent to 21 equations involving covariances of these 7 time series.

An alternative approach to developing relations among probability law parameters equivalent to the Navier Stokes equations, which gives clearer recognition to their non-linear nature, is to employ the one-one correspondence between the

Hilbert space $H_{\mathfrak{A}(U)}$ generated by a time series U and the reproducing kernel Hilbert space with reproducing kernel the characteristic functional Φ of the time series U, defining

(4) $$\Phi_U(V) := E(e^{iV}) , \quad V \in H_U \quad \text{or} \quad \Phi_U(f) := E(e^{i f\sim}) , \quad f \in H(R)$$

where $f\sim$ is the random variable in H_U corresponding to $f \in H(R)$.

Recent applications of reproducing kernel Hilbert spaces to non-linear representations of time series are given by Duttweiler (1970) and Kallianpur (1970).

For recent advances in the theory of time series from the point of view of detection and estimation see Kailath (1970).

For recent advances in the abstract theory of time series see Dudley (1971) and Feldman (1971).

3. Spectral Estimation

For purposes of estimation of probability law parameters it seems to me natural to adopt as the basic problem for investigation the analysis of a multiple stationary time series $X(t) = (X_1(t), \ldots, X_k(t))$ since one might begin the study of a random field by observing it at a finite set of points in space $x^{(1)}, x^{(2)}, \ldots, x^{(k/3)}$. Further one may assume $X(t)$ is discrete parameter (with index $t = 0, \pm 1, \pm 2, \ldots$) since for analysis on a digital computer one would often sample regularly in time.

This paper is concerned with some basic issues in estimation of probability law parameters of stationary time series; for ease of exposition, we discuss only the analysis of a single time series $X(\cdot)$ of which we have a finite sample $\{X(t), t = 0, 1, \ldots T-1\}$. This section focuses on estimation of the spectral density function $f(\omega)$ of a stationary zero mean time series [compare Rosenblatt (1971)].

We omit discussion of the transformations (such as second differencing, substracting a suitable straight line, or "tapering") which may be necessary to "transform" an observed sample so that it approximates a stationary zero mean time series or satisfies other ideal conditions. For discussions of these points, see for example Tukey (1967), Parzen (1967e), and Jenkins and Watts (1968).

We assume the time series $X(t)$ is zero mean stationary with summable covariance function $R(v)$ and continuous spectral density function $f(\omega)$; one has the relations

(1) $$E[X(t)\,X(t+v)] = R(v) = \int_0^{2\pi} e^{iv\omega} f(\omega)\,d\omega = \int_{-\pi}^{\pi} e^{iv\omega} f(\omega)\,d\omega \;;$$

(2) $$f(\omega) = \frac{1}{2\pi} \sum_{v=-\infty}^{\infty} e^{-iv\omega} R(v)\,, \quad -\infty < \omega < \infty\,.$$

Before discussing the estimation of $R(v)$ or $f(\omega)$, let us note that it seems natural to analyze the sample by representing it as a sum of harmonics (spectral representation). One can always write

(3) $$X(t) = \sum_{n=0}^{T-1} \hat{X}(n) \exp\left(-itn \frac{2\pi}{T}\right), \qquad t = 0, 1, \ldots, T-1\,,$$

defining

(4) $$\hat{X}(n) := \frac{1}{T} \sum_{t=0}^{T-1} X(t) \exp\left(i\, n \frac{2\pi}{T} t\right), \qquad n = 0, 1, \ldots, T-1\,.$$

As indicators of the dominant frequencies in the sample one would plot (as a function of n) the normalized intensities $I_n := |\hat{X}(n)|^2 \div \overline{X^2}$,

where

(5) $$\overline{X^2} := \frac{1}{T} \sum_{t=0}^{T-1} X^2(t) = \sum_{n=0}^{T-1} |\hat{X}(n)|^2\,.$$

No account of recent advances in time series analysis could be complete without mentioning the Fast Fourier Transform (FFT) which is an algorithm [see Tukey (1967)] for computing $\hat{X}$ from X in the order of $T \log T$ operations rather than T^2 operations. Along with the FFT comes the advice (to be justified below) that one should add T zeroes to the array X, and compute

(6) $$\hat{\hat{X}}(n) := \frac{1}{2T} \sum_{t=0}^{T-1} X(t) \exp\left(i\, n \frac{2\pi}{2T} t\right), \qquad n = 0, 1, \ldots, 2T-1\,.$$

The spectral representation of $X(t)$ then becomes

(7) $$X(t) = \sum_{n=0}^{2T-1} \hat{\hat{X}}(n) \exp\left(-itn \frac{2\pi}{2T}\right), \qquad t = 0, 1, \ldots, T-1\,.$$

Our definitions of $\hat{X}(n)$ and $\hat{\hat{X}}(n)$ point out that n is a meaningless dummy index. A physically meaningful variable is ω, the <u>frequency</u> of a harmonic in the spectral representation of $X(\bullet)$. Therefore we define the functions

(8) $$f_T(\omega) := \frac{1}{2\pi T}\left| \sum_{t=0}^{T-1} X(t) \exp(it\omega)\right|^2 , \quad \bar{f}_T(\omega) = f_T(\omega) \div \overline{X^2}$$

called the sample <u>spectral</u> density function and <u>spectrum</u> density function of the sample $\{X(t),\ t = 0, \ldots, T-1\}$. We point out that we use "spectrum" as a synonym for "normalized spectral."

Computing the FFT in the sense of $\hat{\hat{X}}(\bullet)$ enables us to compute $\bar{f}_T(\omega)$ at a grid of $2T$ frequencies in the interval $0 \le \omega \le 2\pi$ by

(9) $$\frac{2\pi}{2T}\bar{f}_T\left(n\frac{2\pi}{2T}\right) = \left|\hat{\hat{X}}(n)\right|^2 \div \sum_{v=0}^{2T-1} \left|\hat{\hat{X}}(v)\right|^2 , \qquad n = 0, 1, \ldots, 2T-1 ;$$

note that

(10) $$\frac{2\pi}{T}\bar{f}_T\left(n\frac{2\pi}{T}\right) = \left|\hat{X}(n)\right|^2 \div \sum_{v=0}^{T-1} \left|\hat{X}(v)\right|^2 , \qquad n = 0, 1, \ldots, T-1 .$$

One might feel content to compute $\bar{f}\left(n\frac{2\pi}{T}\right)$, $n = 0, 1, \ldots, T-1$. To illustrate the difference between the two approaches, consider the case of $\{X(t) = 1,\ t = 0, 1, \ldots, 7\}$:

$\{2\pi\ \bar{f}\left(n\frac{2\pi}{8}\right),\ n = 0, 1, \ldots, 7\} = \{8, 0, 0, 0, 0, 0, 0, 0\}$

$\{2\pi\ \bar{f}\left(n\frac{2\pi}{16}\right),\ n = 0, 1, \ldots, 15\} = \{8,\ 3.284,\ 0,\ .405,\ 0,\ .181,\ 0,\ 130,\ 0,\ .130,$
$0,\ .181,\ 0,\ .405,\ 0,\ 3.284\}$

The normalized spectral distribution function is defined by

(11) $$\bar{F}(\omega) := \int_0^{\omega} f(\omega')\,d\omega' \div R(0) , \quad 0 \le \omega \le 2\pi .$$

As its estimator I believe

(12) $$\hat{\hat{\bar{F}}}\left(n\frac{2\pi}{2T}\right) := \sum_{k=0}^{n} \frac{2\pi}{2T}\bar{f}_T\left(k\frac{2\pi}{2T}\right) , \qquad n = 0, 1, \ldots, 2T-1$$

is preferable to

(13) $$\hat{\bar{F}}\left(n\frac{2\pi}{T}\right) := \sum_{k=0}^{n} \frac{2\pi}{T}\bar{f}_T\left(k\frac{2\pi}{T}\right) , \qquad n = 0, 1, \ldots, T-1 .$$

For $\{X(t) = 1 , t = 0, 1, \ldots, 7\}$, (12) yields the sequence .5, .71, .71, .73, .73, .74, .74, .75, .75, .76, .76, .77, .77, .79, .79, 1.0 while (13) yields the sequence 1, 1, 1, 1, 1, 1, 1, 1, . Actually it is most meaningful to consider $\{\overline{F}(\omega) + 1 - \overline{F}(2\pi - \omega) , \ 0 \le \omega \le \pi\}$ as the normalized spectral distribution with estimators the $T + 1$ vector

(14)
$$\frac{\pi}{T}\overline{f}(0), \frac{\pi}{T}\overline{f}(0) + 2\frac{2\pi}{2T}\overline{f}(\frac{2\pi}{2T}), \ldots, \overline{f}(0) + 2\sum_{k=1}^{T-1}\frac{2\pi}{2T}\overline{f}(k\frac{2\pi}{2T}), \frac{\pi}{T}\overline{f}(0) + 2\sum_{k=1}^{T-1}\frac{2\pi}{2T}\overline{f}(k\frac{2\pi}{2T}) + \frac{\pi}{T}\overline{f}(\pi)$$

which in our example is .5, .91, .91, .96, .96, .98, .98, 1, 1 .

To discuss methods of estimation of $f(\omega)$ one must first introduce the notion of the sample covariance function $R_T(v)$, $v = 0, \pm 1, \ldots,$ defined as follows. For $v = 0, 1, \ldots, T - 1$

(15)
$$R_T(v) := \frac{1}{T}\sum_{t=0}^{T-v-1} X(t)\, X(t + v) .$$

For $v \ge T$, $R_T(v) := 0$. For $v < 0$, $R_T(v) := R_T(-v)$.

Many writers on time series analysis persist in a divisor of $T - v$ rather than T in (15), a practice motivated by an astonishingly uncritical devotion to the criterion of unbiased estimation. The mean of $R_T(v)$ is $(1 - \frac{|v|}{T})R(v)$ which makes it a biased estimator of $R(v)$.

What makes $R_T(v)$ desirable is that it is positive definite, which seems a highly desirable property.

The sample correlation function is defined by

(16)
$$\rho_T(v) := R_T(v) \div R_T(0) .$$

In the 1950's, the sample covariance function was computed directly as a convolution of the data with itself in order to avoid having to compute the Fourier transform of the original data. With the advent of the FFT one can compute $\rho_T(v)$ [or, if desired, $R_T(v)$] most quickly by Fourier transforming the sample spectrum density function at $2T$ points:

(17) $$\rho_T(v) = \frac{2\pi}{2T} \sum_{k=0}^{2T-1} \exp\left(ivk \frac{2\pi}{2T}\right) \bar{f}_T\left(k \frac{2\pi}{2T}\right), \quad v = 0, \pm 1, \ldots, \pm (T-1) .$$

This formula deserves to be widely known, if only because of the intimate involvement of such formulas with the recently increased awareness of the FFT. The FFT was buried in the literature for many years but not widely appreciated. It seems it was the claim of Sande (1965) that he could compute the covariance function up to lag 500 of a series of length 4096 on an IBM 7094 computer in one-twentieth the computer time previously needed that brought home to researchers the incredible advance in speed of computation the FFT could make available.

Because of the importance of (17) it merits a proof; we offer a short one. Note first that

(18) $$\bar{f}_T(\omega) = \frac{1}{2\pi} \sum_{|j| < T} \exp(ij\omega)\, \rho_T(j) , \qquad 0 \le \omega \le 2\pi .$$

Therefore the right hand side of (17) equals

$$\frac{2\pi}{2T} \sum_{k=0}^{2T-1} \exp\left(ivk \frac{2\pi}{2T}\right) \frac{1}{2\pi} \sum_{|j|<T} \exp\left(-ijk \frac{2\pi}{2T}\right) \rho_T(j)$$

(19) $$= \sum_{|j|<T} \rho_T(j) \frac{1}{2T} \sum_{k=0}^{2T-1} \exp\left(i(v-j)\, k \frac{2\pi}{2T}\right)$$

$$= \rho_T(v) \quad \text{for } |v| < T .$$

There appears to be a paradox here concerning when one needs $2T$ samples and when one needs T samples. The sample spectral density needs to be computed at $2T$ equi-spaced points in the interval $0 \le \omega \le 2\pi$ because it is basically a function on $0 \le \omega \le \pi$ with T non-vanishing Fourier coefficients; by sampling theory it is determined by its values at T multiples of π/T. On the other hand, the Fourier transform of the original data,

(20) $$\hat{X}(\omega) = \frac{1}{T} \sum_{t=0}^{T-1} X(t) \exp(i\omega t) ,$$

regarded as a function of all frequencies ω in $0 \le \omega \le 2\pi$, is determined by its values at T multiples of $2\pi/T$.

If one desires $\rho_T(v)$ for only a few values of v one can use a generalization of formula (17): for any $S > T$, and $|v| < S - T$

$$\rho_T(v) = \frac{2\pi}{S} \sum_{k=0}^{S-1} \exp\left(ivk\frac{2\pi}{S}\right)\bar{f}_T\left(k\frac{2\pi}{S}\right) . \tag{21}$$

An estimator of $f(\omega)$ (which I call a covariance average) is defined by

$$f_{T,M}(\omega) := \frac{1}{2\pi} \sum_{|v|<T} \exp(-iv\omega)\, k\left(\frac{v}{M}\right) R_T(v) \tag{22}$$

where M is a constant called the truncation point and $k(v)$ is a weighing function called a lag window generator. A covariance average can be equivalently expressed as a spectral average by

$$f_{T,M}(\omega) = \int_{-\pi}^{\pi} f_T(\lambda) K_{T,M}(\omega - \lambda)\, d\lambda \tag{23}$$

defining

$$K_{T,M}(\omega) := \frac{1}{2\pi} \sum_{|v|<T} e^{-iv\omega} k\left(\frac{v}{M}\right) . \tag{24}$$

One will have to choose a grid of frequencies at which to compute the estimator. If one computes it at the frequencies $n(2\pi/S)$, $n = 0, 1, \ldots, 2S-1$ (where $S > T$) then one can use the FFT to compute $f_{T,M}(\omega)$ using (22). Equivalently one could compute it using a discrete spectral average

$$f_{T,M}\left(n\frac{2\pi}{S}\right) = \frac{2\pi}{S} \sum_{k=0}^{S-1} f_T\left(k\frac{2\pi}{S}\right) K_{T,M}\left((k-n)\frac{2\pi}{S}\right) \tag{25}$$

assuming that $k(\frac{v}{M}) = 0$ for $S - T \leq |v| < T$; one proves (25) using (21).

One could use (25) directly to estimate $f(\omega)$ without ever computing the sample covariance function $R_T(v)$. In this approach (which I call the method of direct spectral averages) one forms $f(n\frac{2\pi}{S})$ at S equi-spaced frequencies in 0 to 2π. Note however that the usual advice [see Tukey (1967), p 41] is only that $S \geq T$. Further one is advised to choose $K_{T,M}(\omega)$ to be effectively a square wave; to paraphrase Tukey "for the case of a sample size 389 enlarged to size 512 by adding zeroes, average adjacent values of $f(n\frac{2\pi}{S})$ in blocks of from 5 to 50 values to provide the desired spectrum estimates."

I do not believe that in the era of the FFT spectral estimation via covariance averages (22) has been superseded by direct spectral averages (25). Computationally one can argue in favor of (22). However their common theoretical properties may lead one to prefer a newly appreciated method (to be described in the next section) which I call autoregressive spectral estimation.

The asymptotic variability theory of covariance averages can be summarized: in the limit as $M \to \infty$ and $T \to \infty$ in such a way that $M/T \to 0$

$$\text{Var}[f_{T,M}(\omega)] = \frac{M}{T}\overline{k^2} f^2(\omega), \quad 0 < \omega < \pi, \tag{26}$$

where $\overline{k^2} = \int_{-\infty}^{\infty} k^2(u)\,du$. For $\omega = 0$ or π, the above formula has a factor of 2 on the right hand side.

Since variances often decrease to 0 at a rate of $1 \div T$, my intuitive interpretation of the factor M is that it represents the effective number of parameters we are estimating. In other words, the probability law of a stationary time series is parametrized by an infinite number of parameters. Given a finite sample of size T one seems to be approximating the probability law by one characterized by M parameters, where $M < T$; this becomes clear when one considers autoregressive spectral estimation or when one seeks a principle to unify variability formulas in the theory of probability density estimation.

The bias theory of covariance averages studies the behavior of the bias

$$b[f_{T,M}(\omega)] = E[f_{T,M}(\omega)] - f(\omega) \tag{27}$$

or its integral $\int_{-\pi}^{\pi} b^2[f_{T,M}(\omega)]\,d\omega$, as a function of M and $k(v)$. Now

$$2\pi b[f_{T,M}(\omega)] = \sum_{|v|<T} e^{-iv\omega}\left(1 - \frac{|v|}{T}\right) k\left(\frac{v}{M}\right) R(v) - \sum_{v=-\infty}^{\infty} e^{-iv\omega} R(v). \tag{28}$$

The bias theory of covariance averages in the form in which it was developed in the 1950's (see Parzen (1967a)) turns out to be completely analogous to the theory of approximation and smoothing (as developed say in the recent book by Shapiro (1970)).

Covariance averages suffer a defect which in approximation theory is called a saturation effect.

Suppose that the expansion of $k(v)$ about $v = 0$ is of the form

(29) $$k(v) = 1 - k_r |v|^r + \dots \; .$$

The exponent r is called the characteristic exponent of the kernel; some examples are (all these kernels vanish for $|v| > 1$) :

Bartlett	$k_B(v) := 1 - \lvert v \rvert$	$r = 1$
Tukey	$k_T(v) := \frac{1}{2}(1 + \cos \pi v)$	$r = 2$
Parzen	$k_p(v) := \begin{cases} 1 - 6v^2 + 6\lvert v\rvert^3, & \lvert v\rvert \le 0.5 \\ 2(1 - \lvert v\rvert)^3 & 0.5 \le \lvert v\rvert \le 1 \end{cases}$	$r = 2$
Parzen bias reducing	$k_{pp}(v) := (1 + 6v^2)k_p(v)$	$r = 4$

As M tends to ∞ an approximate expression for bias is

(30) $$b[f_{T,M}(\omega)] \doteq \frac{1}{M^r} k_r f^{[r]}(\omega) ,$$

where

(31) $$f^{[r]}(\omega) = \frac{-1}{2\pi} \sum_{v=-\infty}^{\infty} e^{-iv\omega} |v|^r R(v)$$

is called the generalized r-th derivative of $f(\omega)$; to write (30) we assume that the time series $X(\cdot)$ satisfies

(32) $$\sum_{v=-\infty}^{\infty} |v|^r R(v) < \infty$$

One sees that the rate of falloff of bias is basically determined by the characteristic exponent r of the kernel $k(v)$ rather than how "nice" or "smooth" $f(\omega)$ is, as long as it is sufficiently differentiable to satisfy (32). This situation is called a saturation effect.

The flat kernel $k(v) = 1$ or 0 as $|v| \le 1$ or $|v| > 1$, does not possses the saturation effect. But it suffers from other defects, among which are lack of positive-definiteness.

The kernel $k(v)$ is positive definite if and only if $f_{T,M}(\omega)$ is non-negative for all possible data. The characteristic exponent of a kernel must satisfy $r \leq 2$ if $k(v)$ is positive definite. The Parzen kernel is positive definite while the Tukey kernel is not, although for both $r=2$. The Parzen bias reducing kernel has $r = 4$ and may be preferable to the Tukey kernel; for the considerations leading to this conclusion, see Akaike (1968).

Window carpentry may actually be of no importance if we confine ourselves to autoregressive spectral estimators to be described in the next section. However one more warning seems to be in order. Tukey (1967), p. 28 mentions the Bartlett window as a widely used reasonable alternative to other windows and it continues to be so regarded (particularly in the econometric literature). The Bartlett kernel seems to me to have nothing to recommend its use; as I pointed out in Parzen (1967b) it can lead to spectral estimators with many ripples and misleading conclusions.

4. Autoregressive Spectral Estimation

The method which we call autoregressive spectral estimation seems to offer a method of approximating a non-negative function given its Fourier coefficients which is (1) saturation free, and (2) non-negative.

First let us state the method in a way that makes clear its contribution to approximation theory. The approximation problem can be formulated: given the first M Fourier coefficients,

$$R(v) = \int_{-\pi}^{\pi} e^{iv\omega} f(\omega)\,d\omega\,, \qquad v = 0,\ 1, \ldots, M \tag{1}$$

of a non-negative even function $f(\cdot)$ of period 2π, find a function $f_M(\omega)$ which approximates $f(\omega)$ whose rate of convergence (as $M \to \infty$) is as quick as possible. Usually one considers approximants of the form

$$f_M(\omega) = \frac{1}{2\pi} \sum_{|v| \leq M} e^{iv\omega} k_M(v) R(v) \tag{2}$$

for suitable weights $k_M(v)$. Functions of the form (2) can be called smoothed Fourier series (they are the analogues of covariance averages).

A function of the following form we call an autoregressive approximant:

(3) $$f_M(\omega) := \frac{\sigma_M^2}{2\pi} \left|1 - \sum_{j=1}^{M} a_{j,M} e^{-i\omega j}\right|^{-2}, \quad -\pi \le \omega \le \pi ,$$

where $\{a_{j,M}, j = 1, \ldots, M\}$ are the solutions of the (so-called Yule Walker) equations.

(4) $$\sum_{j=1}^{M} a_{j,M} R(j-k) = R(k), \quad k = 1, 2, \ldots, M$$

and

(5) $$\sigma_M^2 := R(0) - \sum_{j=1}^{M} a_{j,M} R(j) .$$

The $a_{j,M}$ are the coefficients of the minimum mean square error finite memory M linear predictor $X_M(t)$ of $X(t)$

(6) $$X_M(t) := a_{1,M} X(t-1) + \ldots + a_{M,M} X(t-M) ,$$

and σ_M^2 is the memory M one-step prediction error,

(7) $$\sigma_M^2 = E[|X_M(t) - X(t)|^2] .$$

As M tends to ∞, σ_M^2 decreases monotonically to a limit σ_∞^2, the infinite memory one-step prediction error.

The rate of convergence of σ_M^2 to σ_∞^2 has been shown by Ibragimov (1964) to depend on the differentiability of $f(\omega)$; in particular

(8) $$\sigma_M^2 - \sigma_\infty^2 \le \frac{C_r}{M^{2(r+\eta)}}$$

for a suitable constant C_r if

(9) $$\sum_{v=-\infty}^{\infty} |v|^{r+\eta} |R(v)| < \infty$$

where $0.5 < \eta \le 1$.

Let us return to the problem of the rate of convergence of autoregressive approximants. Assume in addition to (9) that $f(\omega)$ is continuous and non-vanishing. Then there exists a sequence $\{a_j\}$ such that

(10) $$f(\omega) = \frac{\sigma_\infty^2}{2\pi} \left| 1 - \sum_{j=1}^{\infty} a_j e^{-i\omega j} \right|^{-2} , \qquad -\pi \leq \omega \leq \pi$$

and

(11) $$\sigma_M^2 - \sigma^2 = \int_{-\pi}^{\pi} \left| \sum_{j=1}^{\infty} (a_{j,M} - a_j)\, e^{i\omega j} \right|^2 f(\omega) d\omega ,$$

defining $a_{j,M} = 0$ for $j > M$. Starting with these facts Kromer (1969) has shown that for all $\xi < \eta - \frac{1}{2}$

(12) $$|f_M(\omega) - f(\omega)| = o(M^{-(r+\xi)}) .$$

One should compare the conclusion (12) with (3.30) and the condition (9) with (3.32).

The rate of convergence of σ_M^2 to σ^2, and therefore of $f_M(\omega)$ to $f(\omega)$, can be discussed under a variety of hypotheses. These results lead to the heuristic conclusion: autoregressive approximants almost achieve best possible rates of convergence (such as exponential decrease for rational spectral densities).

Spectral estimation via covariance averages is tantamount to approximating the true probability law of a time series by a moving average scheme(of moderate order). However a reasonable low order approximation can often be achieved by an autoregressive scheme.

A stationary time series $X(\bullet)$ is an autoregressive scheme of order M if it obeys a stochastic difference equation of order M (with constant coefficients $a_1, \ldots, a_M$).

(13) $$X(t) - a_1 X(t-1) - \ldots - a_M X(t-M) = \varepsilon(t)$$

where $\{\varepsilon(t)\}$ is a white noise sequence (independent with zero mean and common variance σ^2) and the difference equation's characteristic polynomial

(14) $$g(z) = 1 - a_1 z - \ldots - a_M z^M$$

has all its roots outside the unit circle in the complex plane.

The probability law of an autoregressive scheme can be parameterized by $\sigma^2, a_1, \ldots, a_M$. Given these parameters its spectral density $f_M(\omega)$ can be obtained by

(15) $$f_M(\omega) = \frac{\sigma^2}{2\pi} \left|1 - a_1 e^{-i\omega} - \dots - a_M e^{-iM\omega}\right|^{-2} .$$

The probability law can also be parameterized by $R(0), \dots, R(M)$. Given these parameters, $\sigma^2, a_1, \dots, a_M$ can be obtained by solving

(16) $$\sum_{j=1}^{M} a_j R(j-k) = R(k) , \quad j = 1, \dots, M$$

$$\sigma^2 = R(0) - \sum_{j=1}^{M} a_j R(j) .$$

It is accepted that the solution $\{a_j\}$ of (16) has a characteristic polynomial g(z) with all its roots outside the unit circle; for a proof of this assertion and its time series history, see Pagano (1971).

To estimate the probability law parameters of an autoregressive scheme from a sample $\{X(t), t = 1, 2, \dots, T\}$ a variety of algorithms are available (described by adjectives such as least squares, maximum likelihood, Bayesian, recursive). One approach is to first form the sample covariances $R_T(0), \dots, R_T(M)$ and obtain estimators $\hat{a}_1, \dots, \hat{a}_M, \hat{\sigma}^2$ by solving

(17) $$\sum_{j=1}^{M} \hat{a}_j R_T(j-k) = R_T(k), \quad j = 1, \dots, M$$

$$\hat{\sigma}^2 = R_T(0) - \sum_{j=1}^{M} \hat{a}_j R_T(j) .$$

These estimators can be shown to be asymptotically efficient.

Whether or not $X(\cdot)$ is truly an autoregressive scheme one could compute $\hat{a}_j$ and $\hat{\sigma}^2$. Suppose one then computes

(18) $$\hat{f}_{T,M}(\omega) = \frac{\hat{\sigma}^2}{2\pi} \left|1 - \sum_{j=1}^{M} \hat{a}_j e^{-ij\omega}\right|^{-2} .$$

Problem: what are the properties of $\hat{f}_{T,M}(\omega)$ as an estimator of $f(\omega)$? This problem has been extensively investigated by Kromer (1969) who shows that it is asymptotically normal with asymptotic variance

$$2 \frac{M}{T} f^2(\omega) , \quad 0 < \omega < \pi$$

in the sense that $\{\hat{f}_{T,M}(\omega) - f_M(\omega)\} \div \{2 \frac{M}{T} f^2(\omega)\}^{\frac{1}{2}}$ is asymptotically normal (with zero mean, unit variance) as first $T \to \infty$ and then $M \to \infty$, where $f_M(\omega)$ is defined by (15) and (16). Using relations such as (12) to study the "bias" term $f_M(\omega) - f(\omega)$, one concludes that autoregressive spectral estimators are always non-negative, do not have a saturation effect, and are asymptotically efficient.

Autoregressive spectral estimation for multiple time series is described in Parzen (1969). I believe its value will be especially great for estimation of coherence because it seems to avoid many of the pitfalls associated with coherence estimation using covariance averages or direct spectral averages; for example see Gersch (1971).

Finally, it should be noted that choosing the truncation point in the autoregressive spectral estimator is the basic problem in practice; much research needs to be done on this point. Hopefully guidance can be obtained from goodness of fit tests for autoregressive schemes and especially from procedures for stepwise regression. Akaike (1970) has suggested a possible criterion for choosing M.

5. Estimation of Parameters of Moving Average and Mixed Schemes

Great progress seems to me to have been made recently in the theory of efficient estimation of parameters of moving average schemes and mixed autoregressive-moving average schemes. Methods have been proposed by Box-Jenkins (1970), Clevenson (1970), Hannan (1970), and Parzen (1971a). See Anderson (1971) for a summary of methods suggested in the early 1960's by Durbin and Walker.

The work of Box and Jenkins is especially noteworthy for the attention it has focused on mixed schemes as appropriate models when the aim is forecasting or control.

6. Equivalences Among Time Series, Approximation, and Control

Time series analysis originated out of the need to fit models to observed data. I believe it will have in addition important applications to numerical analysis, to provide algorithms for interpolation, function approximation, calculation of integrals and derivatives, and solution of integral equations. For some of the basic theoretical ideas underlying this development see Kimeldorf and Wahba (1969) and Parzen (1971b).

It seems appropriate at a conference in LaJolla to note that links exist between the work of Backus (1970) on inference from inadequate and inaccurate data and the theory of equivalence of approximation and filtering of time series.

7. Probability Density Estimation

Estimation of the probability law generating a process is perhaps the basic statistical problem. There is an extensive literature on the theory and applications of non-parametric probability density estimation. Almost all of the techniques of time series modeling and estimation provide useful approaches to the problem of non-parametric probability density estimation. The kernel method (analogous to covariance averages) has been widely applied. Awaiting application are the methods of autoregressive spectral estimations, and methods based on the equivalence of approximation theory and time series filtering theory. That the latter method will be of use seems to be indicated by the recent heuristic work of Boneva, Kendall, and Stefanov (1971).

8. Time Series Modeling and Identification

In my opinion some of the most important applications of the theory of time series will be to modeling of phenomena for which well accepted theories do not exist, as well as checking of proposed theories. I have called this field empirical time series analysis. Progress in this field will be greatly helped by developments in interactive computing and programming languages such as APL.

I am currently concerned with writing an account of the main theorems of time series analysis in the form of a collection of APL functions for transforming between the various modes of parametrization of a stationary zero mean time series.

An account of recent advances in time series analysis would not be complete without taking into account advances on the modeling problem or the identification problem. Noteworthy papers with many references are Nieman, Fisher, Seborg (1970), Astrom, Wittermark (1971), Wilson (1971). There ought to be a conference bringing together the entire broad spectrum of researchers using time series methods to review and coordinate progress in this area.

Among possible recent advances in identification techniques are: (1) the direct methods of estimation of parameters of mixed schemes [Parzen (1971a)] combined with stepwise regression techniques [illustrated in Parzen (1967c)]; (2) new methods [Pagano and Parzen (1971)] for efficient estimation of parameters of models of signal plus noise [such as those described in Parzen (1967d)].

Finally it should be noted that one of the main applications of time series analysis is still the study of relations between time series; review papers on this problem are being given at the 1971 Washington meeting of The International Statistical Institute [see especially Priestley (1971)].

REFERENCES

Akaike, H. (1968). Low pass filter design. Ann. Inst. Stat. Math., 20, 271-297.

Akaike, H. (1970). Statistical predictor identification, Ann. Inst. Stat. Math., 22, 203-217.

Anderson, T. W. (1971). The Statistical Analysis of Time Series, Wiley: New York.

Astrom, K. J. and B. Wittenmark. (1971). Problems of identification and control. Journal Math. Analysis and Applications, 34, 90-113.

Backus, George. (1970). Inference from inadequate and inaccurate data, I and II, Proc. Nat. Acad. Sci., 65, 1-7, 281-287.

Boneva, L., D. Kendall, and L. Stefanov. (1971). Spline transformations: Three new diagnostic aids for the statistical data analyst, Journal of the Royal Statistical Society, 33, No. 1. Discussion by Parzen outlines equivalences between probability density estimation and time series filtering theory.

Box, G. P. and G. M. Jenkins. (1970). Time Series Analysis Forecasting and Control, Holden Day: San Francisco.

Clevenson, M. L. (1970). Asymptotically Efficient Estimates of the Parameters of a Moving Average Time Series. Technical Report No. 15, July 13, 1970 (ONR Research Project in Time Series Analysis, directed by E. Parzen).

Dudley, R. M. (1970). Random Linear Functionals: Some Recent Results. pp. 60-70 in Lectures in Modern Analysis and Applications III, C. Taam, editor. Springer: Berlin.

Duttweiler, D. (1970). Reproducing kernel Hilbert space techniques for detection and estimation problems. Technical Report No. 7050-18 (Stanford University Information Systems Laboratory). December 1970.

Feldman, J. (1970). Absolute Continuity of Stochastic Processes, pp. 71-86 in Lectures in Modern Analysis and Applications, III, C. Taam, editor. Springer: Berlin.

Gersch, W. (1971). Spectral Analysis of EEG's by Autoregressive Decomposition of Time Series. To be published.

Hannan, E. J. (1970). Multiple Time Series, Wiley: New York.

Ibragimov, I. A. (1964). On the asymptotic behavior of the prediction error. Theory of Probability and Its Applications, 9, 627-633.

Jenkins, G. M. and D. G. Watts. (1968). Spectral Analysis and Its Applications. Holden Day: San Francisco.

Kailath, T. (1970). An RKHS approach to detection and estimations. (In several parts), IEEE Trans. on Information Theory.

Kailath, T. (1970). Likelihood ratios for Gaussian processes. IEEE Trans. on Information Theory. IT-16, 276-288.

Kallianpur, G. (1970). The Role of Reproducing Kernel Hilbert Spaces in the Study of Gaussian Processes. pp. 49-83 in Advances in Probability and Related Topics, P. Ney, editor. Marcel Dekker: New York.

Kampé de Fériet, J. (1953). Random Functions and the Statistical Theory of Turbulence, pp. 568-623 in A. Blanc Pierre and R. Fortet, Théorie des Functions Aléatoires, Masson: Paris (English translation by J. Gani, Gordon and Breach: New York, 1968, Vol. II, pp. 249-308).

Kimeldorf, G. S. and Wahba, Grace. (1970). A correspondence between Bayesian estimation on stochastic processes and smoothing by splines, Annals Math. Stat., 41, 495-502.

Kromer, R. E. (1969). Asymptotic properties of the autoregressive spectral estimator. Technical Report No. 13, December 15, 1969. (ONR Research Project in Time Series Analysis, directed by E. Parzen).

Lumley, J. L. (1970). Stochastic Tools in Turbulence. Academic Press: New York.

Nieman, R. E., D. G. Fisher, and D. E. Seborg. (1971). A review of process identification and parameter estimation techniques. Inst. J. Control, 13, 209-264.

Pagano, M. (1971). When is an autoregressive scheme stationary? In preparation.

Pagano, M. and E. Parzen. (1971). Estimation of models of autoregressive signal plus white noise. To be presented at Institute of Mathematical Statistics Annual Meeting, August 1971.

Parzen, E. (1967a). Time Series Analysis Papers. Holden Day: San Francisco.

Parzen, E. (1967b). On Empirical Multiple Time Series Analysis, pp. 305-340 in Proc. Fifth Berkeley Symp., Vol. I, edited by L. LeCam, University of California Press: Berkeley.

Parzen, E. (1967c). The role of spectral analysis in time series analysis, Review of the International Statistical Institute, 35, 125-141.

Parzen, E. (1967d). Time series analysis for models of signal plus white noise, pp. 233-257 in Spectral Analysis of Time Series, edited by Bernard Harris. Wiley: New York.

Parzen, E. (1967e). Comments on spectral analysis, Transactions of IEEE Audio and Electro-acoustics Group, AU-15, 75-76.

Parzen, E. (1968). Statistical Spectral Analysis (Single Channel Case) in 1968. Technical Report 11, June 10, 1968 (ONR Research Project in Time Series Analysis directed by E. Parzen).

Parzen, E. (1969). Multiple Time Series Modeling, pp. 389-409 in Multivariate Analysis - II, edited by P. R. Krishnaiah, Academic Press: New York.

Parzen, E. (1970). Statistical Inference on Time Series by RKHS Methods, pp. 1-39 in Proceedings of the Twelfth Biennial Seminar of the Canadian Mathematical Congress on Time Series and Stochastic Processes; Convexity and Combinatories, edited by R. Pyke, Canadian Mathematical Congress: Montreal.

Parzen, E. (1971a). Efficient Estimation of Stationary Time Series Mixed Schemes. Technical Report No. 16 (ONR Research Project in Time Series Analysis, directed by E. Parzen).

Parzen, E. (1971b). On the Equivalence Among Time Series Parameter Estimation, Approximation Theory, and Control Theory. Technical Report No. 17 (ONR Research Project in Time Series Analysis, directed by E. Parzen).

Priestley, M. B. (1971). Fitting linear relationships between time series. International Statistical Institute 1971, Washington Meeting Proceedings.

Rosenblatt, M. (1971). Curve estimates. Ann. Math. Stat., 42, No. 6.

Sande, G. (1965). On An Alternative Method of Calculating Covariance Functions. Unpublished.

Shapiro, H. S. (1970). Smoothing and Approximation of Functions, Van Nostrand Reinhold: New York.

Tukey, J. W. (1967). An Introduction to the Calculations of Numerical Spectrum Analysis, pp. 25-46 in Spectral Analysis of Time Series, edited by B. Harris. Wiley: New York.

Wilson, G. T. (1971). Recent developments in statistical process control. International Statistical Institute, Washington Meeting Proceedings.

Lecture Notes in Physics

Bisher erschienen / Already published

Vol. 1: J. C. Erdmann, Wärmeleitung in Kristallen, theoretische Grundlagen und fortgeschrittene experimentelle Methoden. 1969. DM 20,–

Vol. 2: K. Hepp, Théorie de la renormalisation. 1969. DM 18,–

Vol. 3: A. Martin, Scattering Theory: Unitarity, Analyticity and Crossing. 1969. DM 14,–

Vol. 4: G. Ludwig, Deutung des Begriffs physikalische Theorie und axiomatische Grundlegung der Hilbertraumstruktur der Quantenmechanik durch Hauptsätze des Messens. 1970. DM 28,–

Vol. 5: M. Schaaf, The Reduction of the Product of Two Irreducible Unitary Representations of the Proper Orthochronous Quantummechanical Poincaré Group. 1970. DM 14,–

Vol. 6: Group Representations in Mathematics and Physics. Edited by V. Bargmann. 1970. DM 24,–

Vol. 7: R. Balescu, J. L. Lebowitz, I. Prigogine, P. Résibois, Z. W. Salsburg, Lectures in Statistical Physics. 1971. DM 18,–

Vol. 8: Proceedings of the Second International Conference on Numerical Methods in Fluid Dynamics. Edited by M. Holt. 1971. DM 28,–

Vol. 9: D. W. Robinson, The Thermodynamic Pressure in Quantum Statistical Mechanics. 1971. DM 14,–

Vol. 10: J. M. Stewart, Non-Equilibrium Relativistic Kinetic Theory. 1971. DM 14,–

Vol. 11: O. Steinmann, Perturbation Expansions in Axiomatic Field Theory. 1971. DM 14,–

Vol. 12: Statistical Models and Turbulence. Edited by M. Rosenblatt and C. Van Atta. 1972. DM 28,–